STATA GRAPHICS REFERENCE MANUAL

RELEASE 12

A Stata Press Publication
StataCorp LP
College Station, Texas

Published by Stata Press, 4905 Lakeway Drive, College Station, Texas 77845
Typeset in TEX
Printed in the United States of America
10 9 8 7 6 5 4 3 2 1

ISBN-10: 1-59718-081-5
ISBN-13: 978-1-59718-081-8

The suggested citation for this software is

StataCorp. 2011. *Stata: Release 12*. Statistical Software. College Station, TX: StataCorp LP.

Table of contents

[G-1] Introduction and overview

[G-2] Commands

[G-3] Options

[G-4] Styles, concepts, and schemes

Cross-referencing the documentation

When reading this manual, you will find references to other Stata manuals. For example,

[U] **26 Overview of Stata estimation commands**
[R] **regress**
[D] **reshape**

The first example is a reference to chapter 26, *Overview of Stata estimation commands*, in the *User's Guide*; the second is a reference to the `regress` entry in the *Base Reference Manual*; and the third is a reference to the `reshape` entry in the *Data-Management Reference Manual*.

All the manuals in the Stata Documentation have a shorthand notation:

[GSM] *Getting Started with Stata for Mac*
[GSU] *Getting Started with Stata for Unix*
[GSW] *Getting Started with Stata for Windows*
[U] *Stata User's Guide*
[R] *Stata Base Reference Manual*
[D] *Stata Data-Management Reference Manual*
[G] *Stata Graphics Reference Manual*
[XT] *Stata Longitudinal-Data/Panel-Data Reference Manual*
[MI] *Stata Multiple-Imputation Reference Manual*
[MV] *Stata Multivariate Statistics Reference Manual*
[P] *Stata Programming Reference Manual*
[SEM] *Stata Structural Equation Modeling Reference Manual*
[SVY] *Stata Survey Data Reference Manual*
[ST] *Stata Survival Analysis and Epidemiological Tables Reference Manual*
[TS] *Stata Time-Series Reference Manual*
[I] *Stata Quick Reference and Index*

[M] *Mata Reference Manual*

Detailed information about each of these manuals may be found online at

http://www.stata-press.com/manuals/

[G-1] Introduction and overview

Title

[G-1] intro — Introduction to graphics manual

Description

This entry describes this manual and what has changed since Stata 11. See the next entry, [G-1] **graph intro**, for an introduction to Stata's graphics capabilities.

Remarks

This manual documents Stata's `graph` commands and is referred to as [G] in references.

Following this entry, [G-1] **graph intro** provides an overview of Stata's `graph` command, and [G-1] **graph editor** describes the Stata Graph Editor. The remaining manual is divided into three sections:

Commands	This section is arranged alphabetically by `graph` subcommand and documents all the families of graphs (e.g., twoway, bar, or box) and the graph management commands (e.g., `graph drop` or `graph use`). All references to this section appear in the text as bolded command names, e.g., [G-2] **graph twoway**.
Options	This section is arranged alphabetically by option type (e.g., *marker_options* or *legend_options*) and documents the options available to `graph`. All references to this section appear in the text as bolded, italicized option names with *_options* appended, e.g., [G-3] ***axis_label_options***.
Styles, concepts, and schemes	This section is arranged alphabetically by style name and documents the valid arguments for graph options; e.g., *colorstyle* shows all the valid arguments for options that take a color. Almost all references to this section appear in the text as bolded, italicized style names with *style* appended, e.g., [G-4] ***linestyle***. Concept entries are the exception; these references appear in the text as bold text, such as [G-4] **concept: lines** or [G-4] **schemes intro**.

Only the `graph` command is documented in this manual, though the statistical graph commands documented in [MV], [R], [ST], [TS], and [XT] often refer to the *Options* and *Styles and concepts* sections of this manual.

When using this manual as documentation for the `graph` command and its families, you will typically begin in the *Commands* section and be referred to the *Options* and *Styles and concepts* sections as needed. If you are an experienced user, you might sometimes refer directly to the *Options* section for entries such as *legend_options*, where the 35 options for controlling where a legend appears and how it looks are documented. Similarly, you may jump directly to entries such as *colorstyle* in *Styles and concepts* to determine the valid arguments to an option specifying the color of a graph object. If you are new to Stata's graphics, see [G-1] **graph intro** for a suggested reading order.

Stata is continually being updated, and Stata users are continually writing new commands. To ensure that you have the latest features, you should install the most recent official update; see [R] **update**.

What's new

This section is intended for previous Stata users. If you are new to Stata, you may as well skip it.

1. **Graphs of margins, marginal effects, contrasts, and pairwise comparisons**. Margins and effects can be obtained from linear or nonlinear (for example, probability) responses. See [R] **marginsplot**.
2. **Contour plots**. Filled and outlined plots are available. See [G-2] **graph twoway contour** and [G-2] **graph twoway contourline**.
3. **PDF export for graphs and logs** lets you directly create PDFs from your Stata graphs. See [G-2] **graph export**.
4. **Time-series operators now supported** by `twoway lfit`, `twoway lfitci`, `twoway qfit`, and `twoway qfitci`. See [G-2] **graph twoway lfit**, [G-2] **graph twoway lfitci**, [G-2] **graph twoway qfit**, and [G-2] **graph twoway qfitci**.
5. **Graphs of marginal and covariate-specific ROC curves.** New command `rocregplot` plots the fitted ROC curve after `rocreg`. See [R] **rocregplot**.
6. **Option addplot() now places added graphs above or below.** Graph commands that allow option `addplot()` can now place the added plots above or below the command's plots.

For a complete list of all the new features in Stata 12, see [U] **1.3 What's new**.

Also see

[U] **1.3 What's new**

[R] **intro** — Introduction to base reference manual

Title

[G-1] graph intro — Introduction to graphics

Remarks

Remarks are presented under the following headings:

Suggested reading order
A quick tour
Using the menus

Suggested reading order

We recommend that you read the entries in this manual in the following order:

Read *A quick tour* below, then read *Quick start* in [G-1] **graph editor**, and then ...

Entry	Description
[G-2] **graph**	Overview of the `graph` command
[G-2] **graph twoway**	Overview of the `graph twoway` command
[G-2] **graph twoway scatter**	Overview of the `graph twoway scatter` command

When reading those sections, follow references to other entries that interest you. They will take you to such useful topics as

Entry	Description
[G-3] ***marker_label_options***	Options for specifying marker labels
[G-3] ***by_option***	Option for repeating graph command
[G-3] ***title_options***	Options for specifying titles
[G-3] ***legend_options***	Option for specifying legend

We could list many, many more, but you will find them on your own. Follow the references that interest you, and ignore the rest. Afterward, you will have a working knowledge of twoway graphs. Now glance at each of

Entry	Description
[G-2] **graph twoway line**	Overview of the `graph twoway line` command
[G-2] **graph twoway connected**	Overview of the `graph twoway connected` command
etc.	

Turn to [G-2] **graph twoway**, which lists all the different `graph twoway` plottypes, and browse the manual entry for each.

Now is the time to understand schemes, which have a great effect on how graphs look. You may want to specify a different scheme before printing your graphs.

Entry	Description
[G-4] **schemes intro**	Schemes and what they do
[G-2] **set printcolor**	Set how colors are treated when graphs are printed
[G-2] **graph print**	Printing graphs the easy way
[G-2] **graph export**	Exporting graphs to other file formats

Now you are an expert on the `graph twoway` command, and you can even print the graphs it produces.

To learn about the other types of graphs, see

Entry	Description
[G-2] **graph matrix**	Scatterplot matrices
[G-2] **graph bar**	Bar and dot charts
[G-2] **graph box**	Box plots
[G-2] **graph dot**	Dot charts (summary statistics)
[G-2] **graph pie**	Pie charts

To learn tricks of the trade, see

Entry	Description
[G-2] **graph save**	Saving graphs to disk
[G-2] **graph use**	Redisplaying graphs from disk
[G-2] **graph describe**	Finding out what is in a .gph file
[G-3] ***name_option***	How to name a graph in memory
[G-2] **graph display**	Display graph stored in memory
[G-2] **graph dir**	Obtaining directory of named graphs
[G-2] **graph rename**	Renaming a named graph
[G-2] **graph copy**	Copying a named graph
[G-2] **graph drop**	Eliminating graphs in memory
[P] **discard**	Clearing memory

For a completely different and highly visual approach to learning Stata graphics, see Mitchell (2008). Hamilton (2009) offers a concise 50-page overview within the larger context of statistical analysis with Stata. Excellent suggestions for presenting information clearly in graphs can be found in Cleveland (1993 and 1994), in Wallgren et al. (1996), and even in chapters of books treating larger subjects, such as Good and Hardin (2009).

A quick tour

graph is easy to use:

```
. use http://www.stata-press.com/data/r12/auto
(1978 Automobile Data)

. graph twoway scatter mpg weight
```

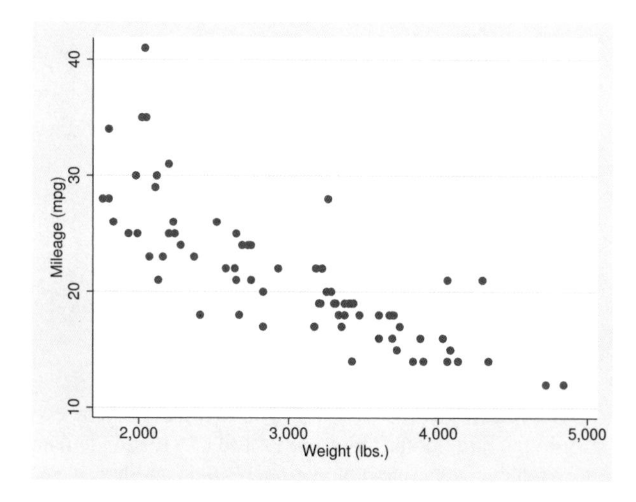

All the commands documented in this manual begin with the word graph, but often the graph is optional. You could get the same graph by typing

```
. twoway scatter mpg weight
```

and, for scatter, you could omit the twoway, too:

```
. scatter mpg weight
```

We, however, will continue to type twoway to emphasize when the graphs we are demonstrating are in the twoway family.

Twoway graphs can be combined with by():

```
. twoway scatter mpg weight, by(foreign)
```

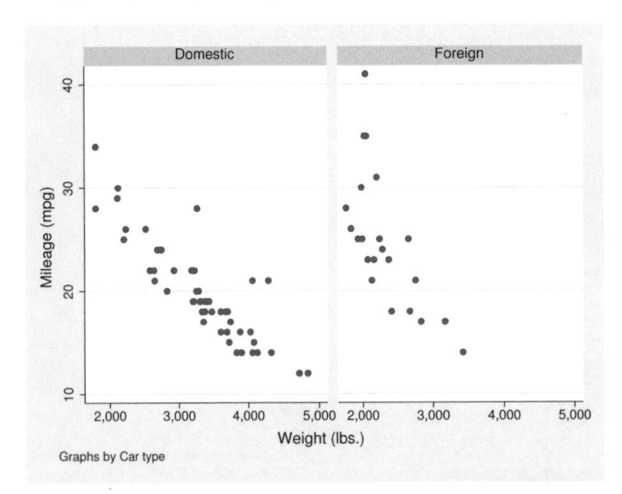

Graphs in the twoway family can also be overlaid. The members of the twoway family are called *plottypes*; scatter is a plottype, and another plottype is lfit, which calculates the linear prediction and plots it as a line chart. When we want one plottype overlaid on another, we combine the commands, putting || in between:

```
. twoway scatter mpg weight || lfit mpg weight
```

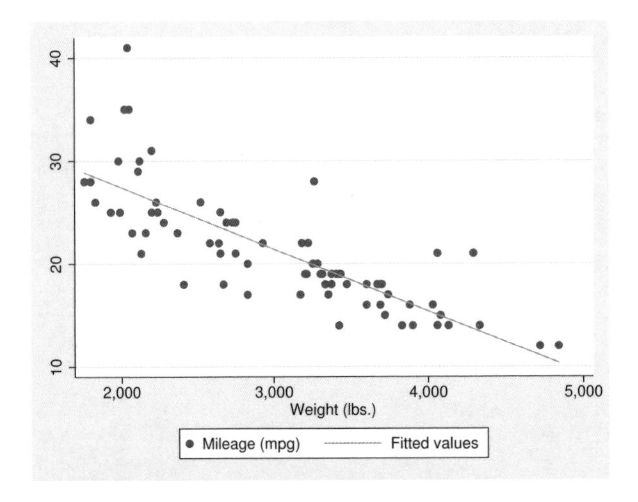

Another notation for this is called the ()-binding notation:

```
. twoway (scatter mpg weight) (lfit mpg weight)
```

It does not matter which notation you use.

Overlaying can be combined with `by()`. This time, substitute `qfitci` for `lfit`. `qfitci` plots the prediction from a quadratic regression, and it adds a confidence interval. Then add the confidence interval on the basis of the standard error of the forecast:

```
. twoway (qfitci mpg weight, stdf) (scatter mpg weight), by(foreign)
```

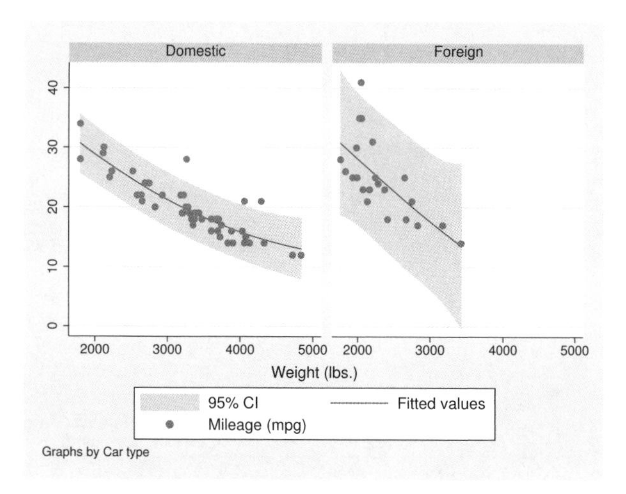

We used the `()`-binding notation just because it makes it easier to see what modifies what:

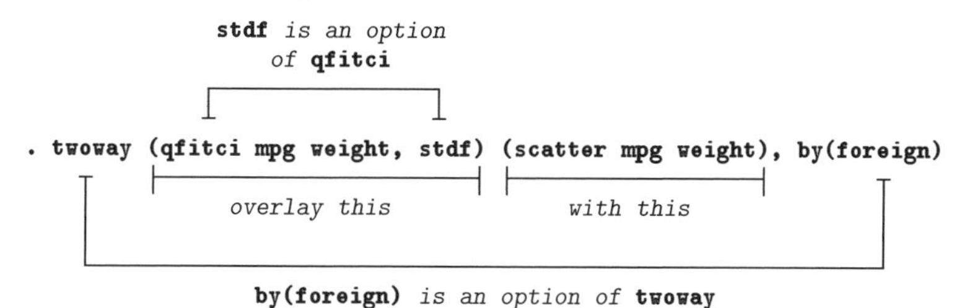

We could just as well have typed this command with the `||`-separator notation,

```
. twoway qfitci mpg weight, stdf || scatter mpg weight ||, by(foreign)
```

and, as a matter of fact, we do not have to separate the `twoway` option `by(foreign)` (or any other `twoway` option) from the `qfitci` and `scatter` options, so we can type

```
. twoway qfitci mpg weight, stdf || scatter mpg weight, by(foreign)
```

or even

```
. twoway qfitci mpg weight, stdf by(foreign) || scatter mpg weight
```

All these syntax issues are discussed in [G-2] **graph twoway**. In our opinion, the `()`-binding notation is easier to read, but the `||`-separator notation is easier to type. You will see us using both.

It was not an accident that we put `qfitci` first and `scatter` second. `qfitci` shades an area, and had we done it the other way around, that shading would have been put right on top of our scattered points and erased (or at least hidden) them.

Plots of different types or the same type may be overlaid:

```
. use http://www.stata-press.com/data/r12/uslifeexp
(U.S. life expectancy, 1900-1999)
. twoway line le_wm year || line le_bm year
```

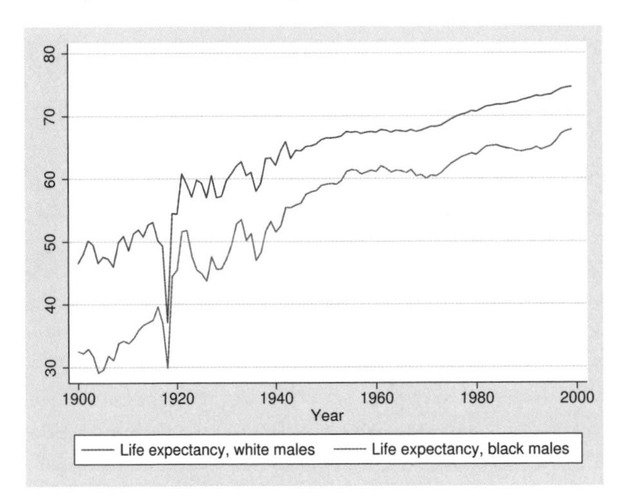

Here is a rather fancy version of the same graph:

```
. generate diff = le_wm - le_bm
. label var diff "Difference"
. twoway line le_wm year, yaxis(1 2) xaxis(1 2)
      || line le_bm year
      || line diff  year
      || lfit diff  year
      ||,
         ytitle( "",          axis(2) )
         xtitle( "",          axis(2) )
         xlabel( 1918,        axis(2) )
         ylabel( 0(5)20,      axis(2) grid gmin angle(horizontal) )
         ylabel( 0 20(10)80,          gmax angle(horizontal) )
         ytitle( "Life expectancy at birth (years)" )
         ylabel(, axis(2) grid)
         title( "White and black life expectancy" )
         subtitle( "USA, 1900-1999" )
         note( "Source: National Vital Statistics, Vol 50, No. 6"
               "(1918 dip caused by 1918 Influenza Pandemic)" )
         legend( label(1 "White males") label(2 "Black males") )
```

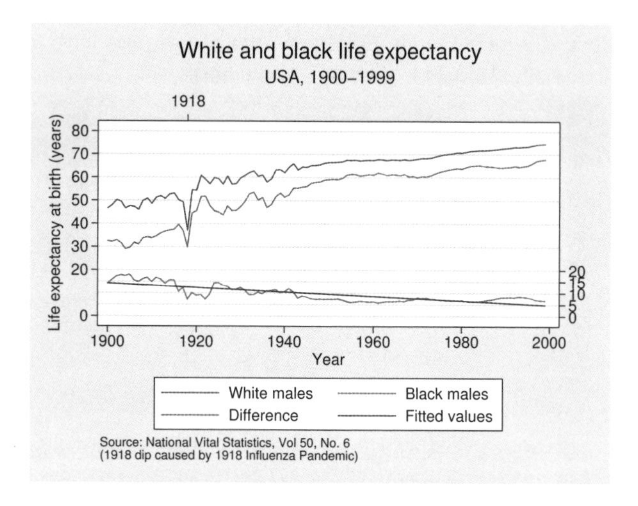

There are many options on this command. (All except the first two options could have been accomplished in the Graph Editor; see [G-1] **graph editor** for an overview of the Editor.) Strip away the obvious options, such as `title()`, `subtitle()`, and `note()`, and you are left with

```
. twoway line le_wm year, yaxis(1 2) xaxis(1 2)
      || line le_bm year
      || line diff  year
      || lfit diff  year
      ||,
         ytitle( "",          axis(2) )
         xtitle( "",          axis(2) )
         xlabel( 1918,        axis(2) )
         ylabel( 0(5)20,      axis(2) grid gmin angle(horizontal) )
         ylabel( 0 20(10)80,              gmax angle(horizontal) )
         legend( label(1 "White males") label(2 "Black males") )
```

Let's take the longest option first:

```
ylabel( 0(5)20,      axis(2) grid gmin angle(horizontal) )
```

The first thing to note is that options have options:

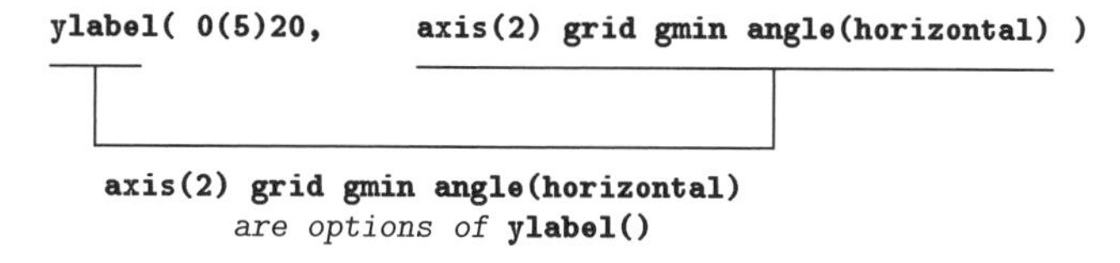

Now look back at our graph. It has two y axes, one on the right and a second on the left. Typing

```
ylabel( 0(5)20,      axis(2) grid gmin angle(horizontal) )
```

caused the right axis—`axis(2)`—to have labels at 0, 5, 10, 15, and 20—`0(5)20`. `grid` requested grid lines for each labeled tick on this right axis, and `gmin` forced the grid line at 0 because, by default, `graph` does not like to draw grid lines too close to the axis. `angle(horizontal)` made the 0, 5, 10, 15, and 20 horizontal rather than, as usual, vertical.

You can now guess what

```
ylabel( 0 20(10)80,          gmax angle(horizontal) )
```

did. It labeled the left y axis—axis(1) in the jargon—but we did not have to specify an axis(1) suboption because that is what ylabel() assumes. The purpose of

```
xlabel( 1918,       axis(2) )
```

is now obvious, too. That labeled a value on the second x axis.

So now we are left with

```
. twoway line le_wm year, yaxis(1 2) xaxis(1 2)
      || line le_bm year
      || line diff  year
      || lfit diff  year
      ||,
         ytitle( "",          axis(2) )
         xtitle( "",          axis(2) )
         legend( label(1 "White males") label(2 "Black males") )
```

Options ytitle() and xtitle() specify the axis titles. We did not want titles on the second axes, so we got rid of them. The legend() option,

```
legend( label(1 "White males") label(2 "Black males") )
```

merely respecified the text to be used for the first two keys. By default, legend() uses the variable label, which in this case would be the labels of variables le_wm and le_bm. In our dataset, those labels are “Life expectancy, white males” and “Life expectancy, black males”. It was not necessary—and undesirable—to repeat “Life expectancy”, so we specified an option to change the label. It was either that or change the variable label.

So now we are left with

```
. twoway line le_wm year, yaxis(1 2) xaxis(1 2)
      || line le_bm year
      || line diff  year
      || lfit diff  year
```

and that is almost perfectly understandable. The yaxis() and xaxis() options caused the creation of two y and two x axes rather than, as usual, one.

Understand how we arrived at

```
. twoway line le_wm year, yaxis(1 2) xaxis(1 2)
      || line le_bm year
      || line diff  year
      || lfit diff  year
      ||,
         ytitle( "",          axis(2) )
         xtitle( "",          axis(2) )
         xlabel( 1918,        axis(2) )
         ylabel( 0(5)20,      axis(2) grid gmin angle(horizontal) )
         ylabel( 0 20(10)80,               gmax angle(horizontal) )
         ytitle( "Life expectancy at birth (years)" )
         title( "White and black life expectancy" )
         subtitle( "USA, 1900-1999" )
         note( "Source: National Vital Statistics, Vol 50, No. 6"
               "(1918 dip caused by 1918 Influenza Pandemic)" )
         legend( label(1 "White males") label(2 "Black males") )
```

We started with the first graph we showed you,

```
. twoway line le_wm year || line le_bm year
```

and then, to emphasize the comparison of life expectancy for whites and blacks, we added the difference,

```
. generate diff = le_wm - le_bm
. twoway line le_wm year,
      || line le_bm year
      || line diff  year
```

and then, to emphasize the linear trend in the difference, we added "`lfit diff year`",

```
. twoway line le_wm year,
      || line le_bm year
      || line diff  year,
      || lfit diff  year
```

and then we added options to make the graph look more like what we wanted. We introduced the options one at a time. It was rather fun, really. As our command grew, we switched to using the Do-file Editor, where we could add an option and hit the **Do** button to see where we were. Because the command was so long, when we opened the Do-file Editor, we typed on the first line

```
#delimit ;
```

and we typed on the last line

```
;
```

and then we typed our ever-growing command between.

Many of the options we used above are common to most of the graph families, including `twoway`, `bar`, `box`, `dot`, and `pie`. If you understand how the `title()` or `legend()` option is used with one family, you can apply that knowledge to all graphs, because these options work the same across families.

While we are on the subject of life expectancy, using another dataset, we drew

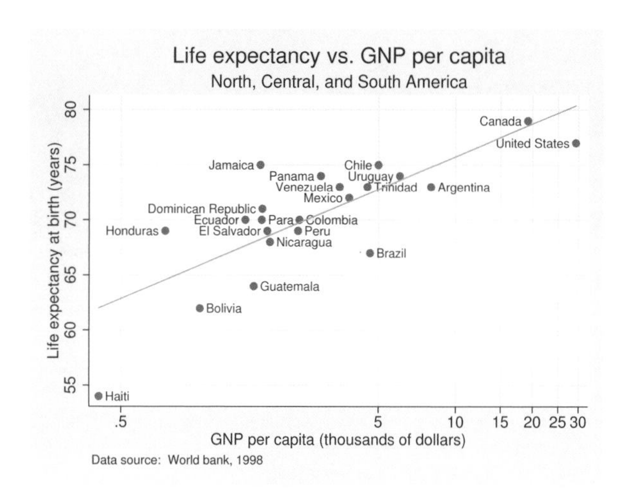

See [G-3] ***marker_label_options*** for an explanation of how we did this. Staying with life expectancy, we produced

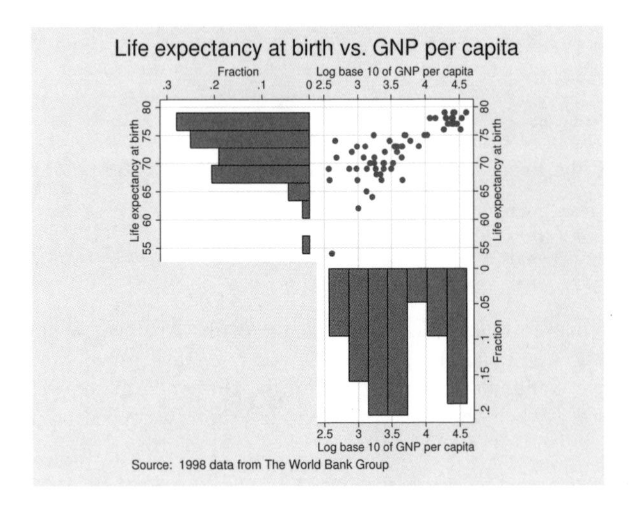

which we drew by separately drawing three rather easy graphs

```
. twoway scatter lexp loggnp,
        yscale(alt) xscale(alt)
        xlabel(, grid gmax)              saving(yx)
. twoway histogram lexp, fraction
        xscale(alt reverse) horiz        saving(hy)
. twoway histogram loggnp, fraction
        yscale(alt reverse)
        ylabel(,nogrid)
        xlabel(,grid gmax)               saving(hx)
```

and then combining them:

```
. graph combine hy.gph yx.gph hx.gph,
        hole(3)
        imargin(0 0 0 0) grapharea(margin(l 22 r 22))
        title("Life expectancy at birth vs. GNP per capita")
        note("Source:  1998 data from The World Bank Group")
```

See [G-2] **graph combine** for more information.

Back to our tour, twoway, by() can produce graphs that look like this

```
. use http://www.stata-press.com/data/r12/auto, clear
(1978 Automobile Data)
. scatter mpg weight, by(foreign, total row(1))
```

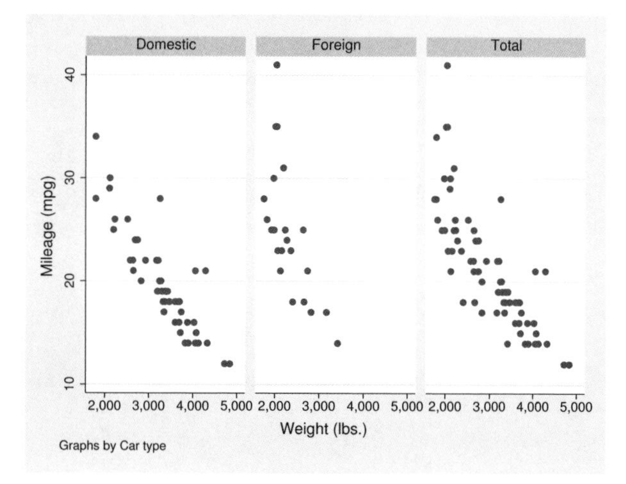

or this

```
. scatter mpg weight, by(foreign, total col(1))
```

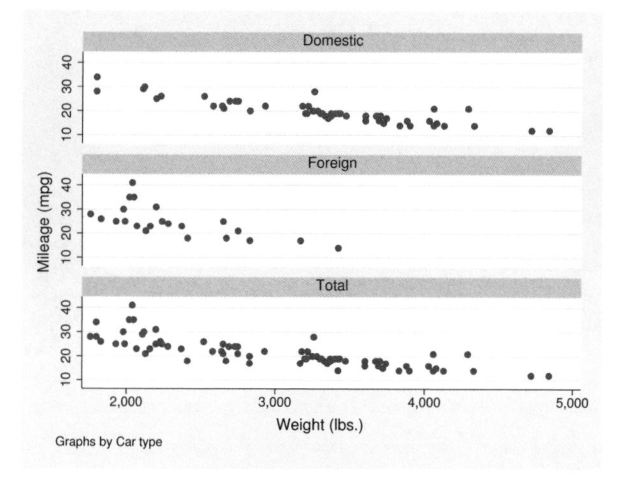

or this

```
. scatter mpg weight, by(foreign, total)
```

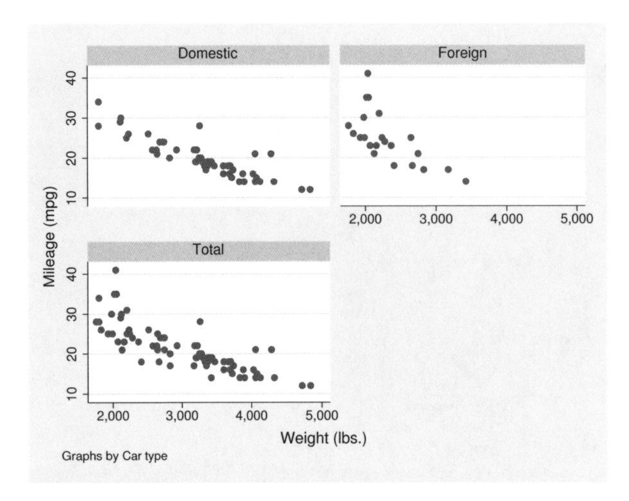

See [G-3] ***by_option***.

by() is another of those options that is common across all graph families. If you know how to use it on one type of graph, then you know how to use it on any type of graph.

There are many plottypes within the twoway family, including areas, bars, spikes, dropped lines, and dots. Just to illustrate a few:

```
. use http://www.stata-press.com/data/r12/sp500
(S&P 500)
. replace volume = volume/1000
. twoway
        rspike hi low date ||
        line   close  date ||
        bar    volume date, barw(.25) yaxis(2) ||
  in 1/57
  , yscale(axis(1) r(900 1400))
    yscale(axis(2) r(  9   45))
    ytitle("                          Price -- High, Low, Close")
    ytitle(" Volume (millions)", axis(2) bexpand just(left))
    legend(off)
    subtitle("S&P 500", margin(b+2.5))
    note("Source:  Yahoo!Finance and Commodity Systems, Inc.")
```

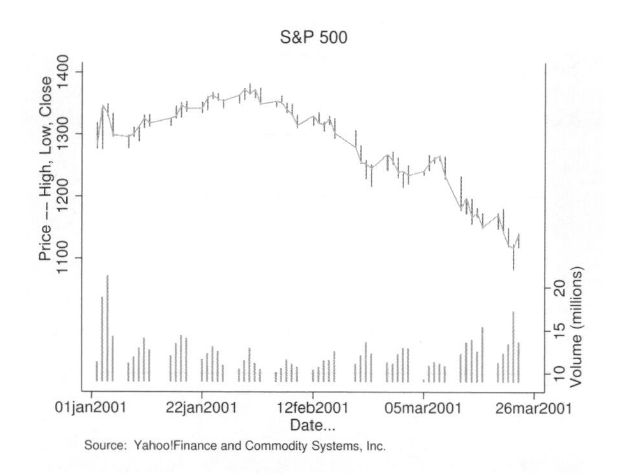

The above graph is explained in [G-2] **graph twoway rspike**. See [G-2] **graph twoway** for a listing of all available `twoway` plottypes.

Moving outside the `twoway` family, `graph` can draw scatterplot matrices, box plots, pie charts, and bar and dot plots. Here are examples of each.

A scatterplot matrix of the variables `popgr`, `lexp`, `lgnppc`, and `safe`:

```
. use http://www.stata-press.com/data/r12/lifeexp, clear
(Life expectancy, 1998)
. generate lgnppc = ln(gnppc)
. graph matrix popgr lgnppc safe lexp
```

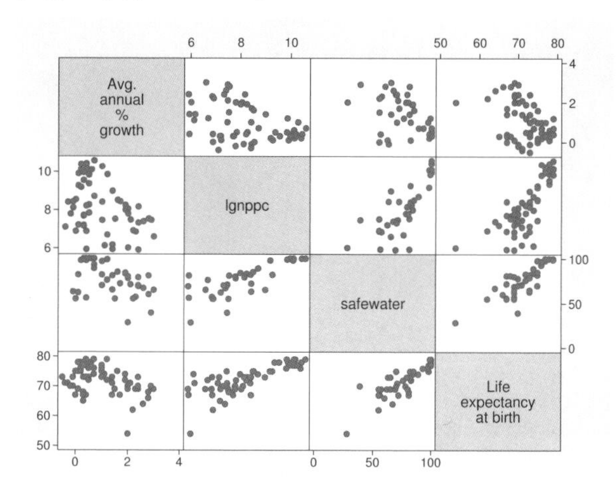

Or, with grid lines and more axis labels:

```
. graph matrix popgr lgnppc safe lexp, maxes(ylab(#4, grid) xlab(#4, grid))
```

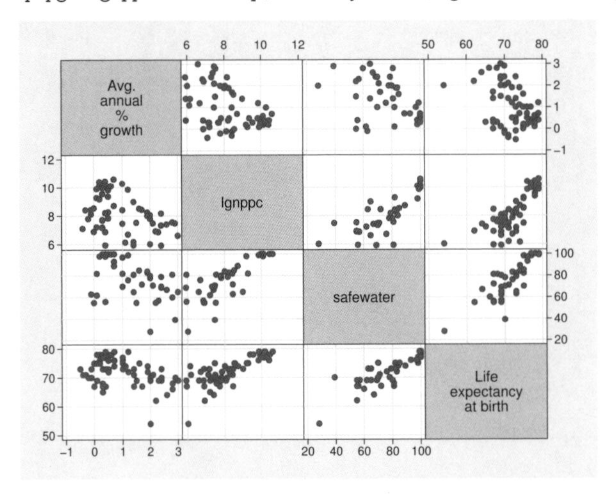

See [G-2] **graph matrix**.

A box plot of blood pressure, variable bp, over each group in the variable when and each group in the variable sex:

```
. use http://www.stata-press.com/data/r12/bplong, clear
(fictional blood pressure data)
. graph box bp, over(when) over(sex)
```

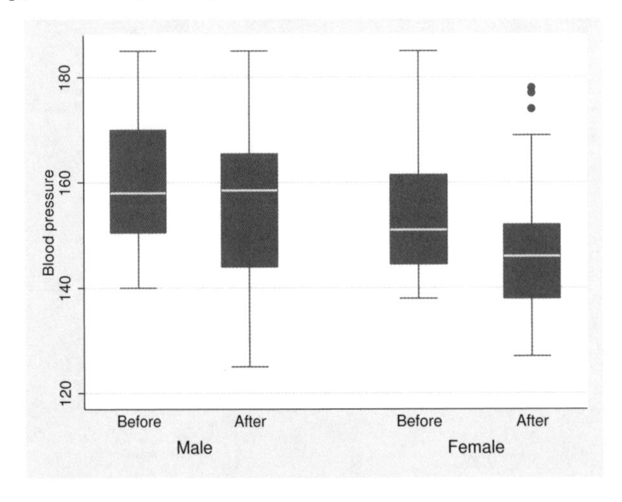

Or, for a graph with complete titles:

```
. graph box bp, over(when) over(sex)
        ytitle("Systolic blood pressure")
        title("Response to Treatment, by Sex")
        subtitle("(120 Preoperative Patients)" " ")
        note("Source:  Fictional Drug Trial, StataCorp, 2003")
```

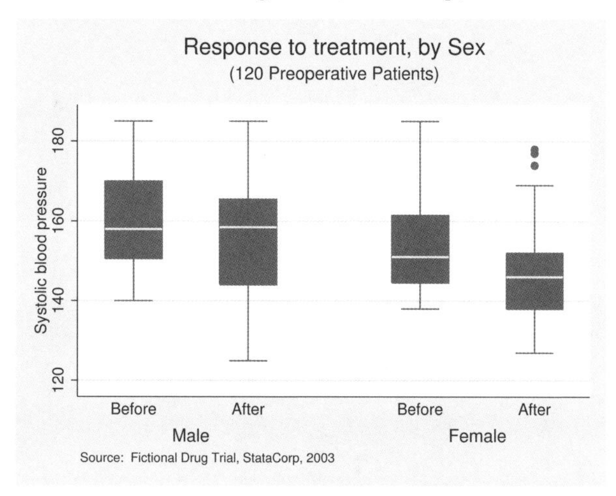

See [G-2] **graph box**.

A pie chart showing the proportions of the variables `sales`, `marketing`, `research`, and `development`:

```
. graph pie sales marketing research development
```

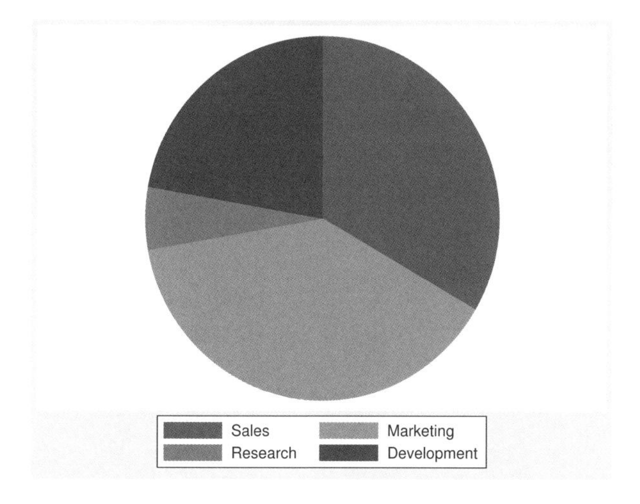

Or, for a graph with nice titles and better labeling of the pie slices:

```
. graph pie sales marketing research development,
        plabel(_all name, size(*1.5) color(white))
        legend(off)
        plotregion(lstyle(none))
        title("Expenditures, XYZ Corp.")
        subtitle("2002")
        note("Source:  2002 Financial Report (fictional data)")
```

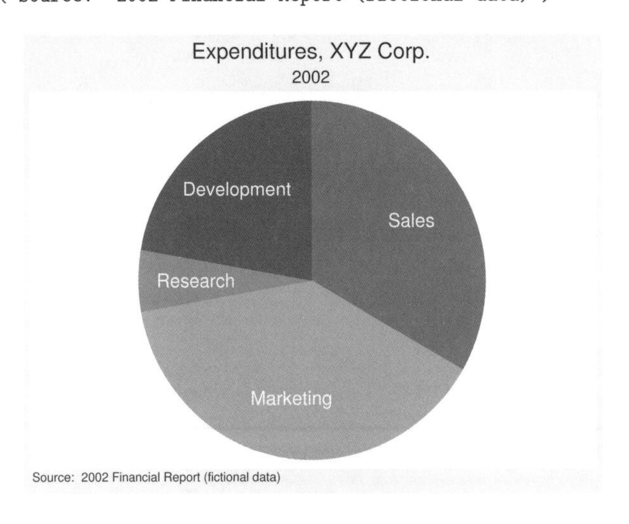

See [G-2] **graph pie**.

A vertical bar chart of average wages over each group in the variables smsa, married, and collgrad:

```
. use http://www.stata-press.com/data/r12/nlsw88
(NLSW, 1988 extract)
. graph bar wage, over(smsa) over(married) over(collgrad)
```

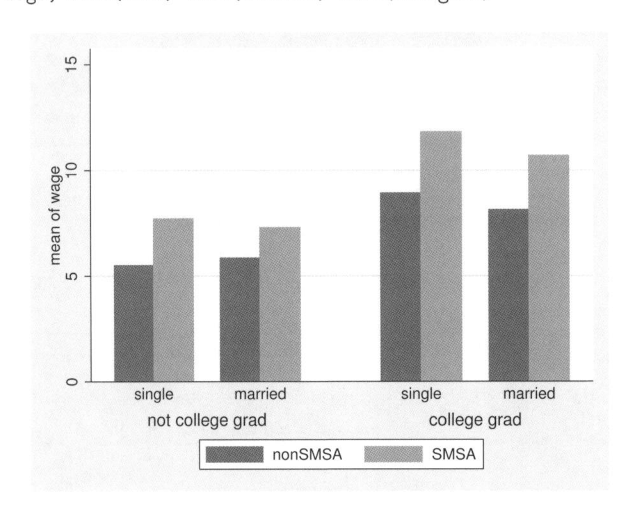

Or, for a prettier graph with overlapping bars, titles, and better labels:

```
. graph bar wage,
                over( smsa, descend gap(-30) )
                over( married )
                over( collgrad, relabel(1 "Not college graduate"
                                        2 "College graduate"    ) )
                ytitle("")
                title("Average Hourly Wage, 1988, Women Aged 34-46")
                subtitle("by College Graduation, Marital Status,
                          and SMSA residence")
                note("Source:  1988 data from NLS, U.S. Dept of Labor,
                      Bureau of Labor Statistics")
```

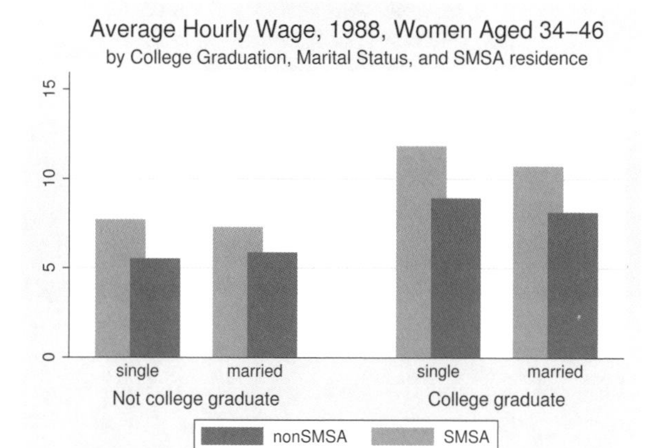

See [G-2] **graph bar**.

A horizontal bar chart of private versus public spending over countries:

```
. use http://www.stata-press.com/data/r12/educ99gdp
(Education and GDP)
. generate total = private + public
. graph hbar (asis) public private, over(country)
```

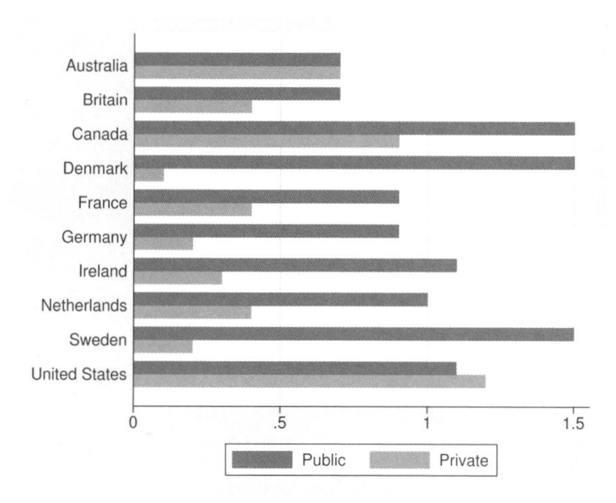

Or, the same information with stacked bars, an informative sorting of total spending, and nice titles:

```
. graph hbar (asis) public private,
                over(country, sort(total) descending)
                stack
                title("Spending on tertiary education as % of GDP,
                        1999", span position(11) )
                subtitle(" ")
                note("Source:  OECD, Education at a Glance 2002", span)
```

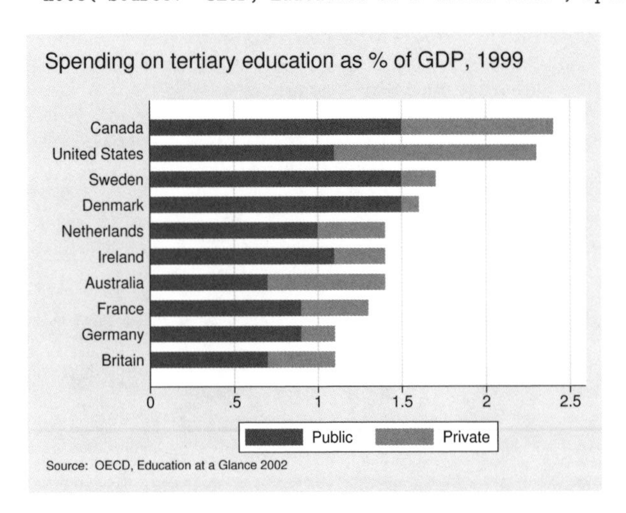

See [G-2] **graph bar**.

A dot chart of average hourly wage over occupation, variable `occ`, with separate subgraphs for college graduates and not college graduates, variable `collgrad`:

```
. use http://www.stata-press.com/data/r12/nlsw88, clear
(NLSW, 1988 extract)
. graph dot wage, over(occ) by(collgrad)
```

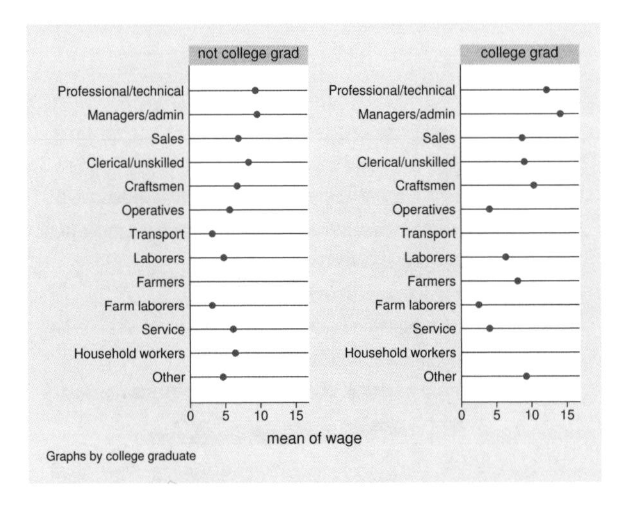

Or, for a plot that orders the occupations by wage and has nice titles:

```
. graph dot wage,
        over(occ, sort(1))
        by(collgrad,
             title("Average hourly wage, 1988, women aged 34-46", span)
             subtitle(" ")
             note("Source:  1988 data from NLS, U.S. Dept. of Labor,
                    Bureau of Labor Statistics", span)
        )
```

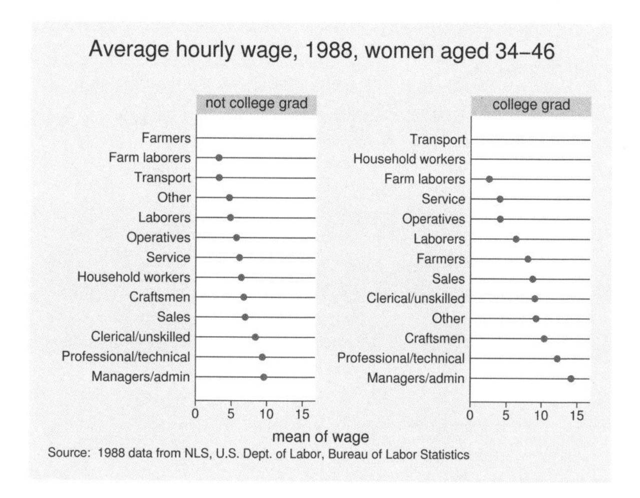

See [G-2] **graph dot**.

Have fun. Follow our advice in the *Suggested reading order* above: turn to [G-2] **graph**, [G-2] **graph twoway**, and [G-2] **graph twoway scatter**.

Using the menus

In addition to using the command-line interface, you can access most of `graph`'s features by Stata's pulldown menus. To start, load a dataset, select **Graphics**, and select what interests you.

When you have finished filling in the dialog box (do not forget to click on the tabs—lots of useful features are hidden there), rather than click on **OK**, click on **Submit**. This way, once the graph appears, you can easily modify it and click on **Submit** again.

Feel free to experiment. Clicking on **Submit** (or **OK**) never hurts; if you have left a required field blank, you will be told. The dialog boxes make it easy to spot what you can change.

References

Cleveland, W. S. 1993. *Visualizing Data*. Summit, NJ: Hobart.

——. 1994. *The Elements of Graphing Data*. Rev. ed. Summit, NJ: Hobart.

Cox, N. J. 2004a. Speaking Stata: Graphing distributions. *Stata Journal* 4: 66–88.

——. 2004b. Speaking Stata: Graphing categorical and compositional data. *Stata Journal* 4: 190–215.

——. 2004c. Speaking Stata: Graphing agreement and disagreement. *Stata Journal* 4: 329–349.

——. 2004d. Speaking Stata: Graphing model diagnostics. *Stata Journal* 4: 449–475.

Good, P. I., and J. W. Hardin. 2009. *Common Errors in Statistics (and How to Avoid Them)*. 3rd ed. New York: Wiley.

Hamilton, L. C. 2009. *Statistics with Stata (Updated for Version 10)*. Belmont, CA: Brooks/Cole.

Mitchell, M. N. 2008. *A Visual Guide to Stata Graphics*. 2nd ed. College Station, TX: Stata Press.

Wallgren, A., B. Wallgren, R. Persson, U. Jorner, and J.-A. Haaland. 1996. *Graphing Statistics and Data: Creating Better Charts*. Newbury Park, CA: Sage.

Also see

[G-2] **graph** — The graph command

[G-2] **graph other** — Other graphics commands

[G-1] **graph editor** — Graph Editor

Title

[G-1] graph editor — Graph Editor

Remarks

Remarks are presented under the following headings:

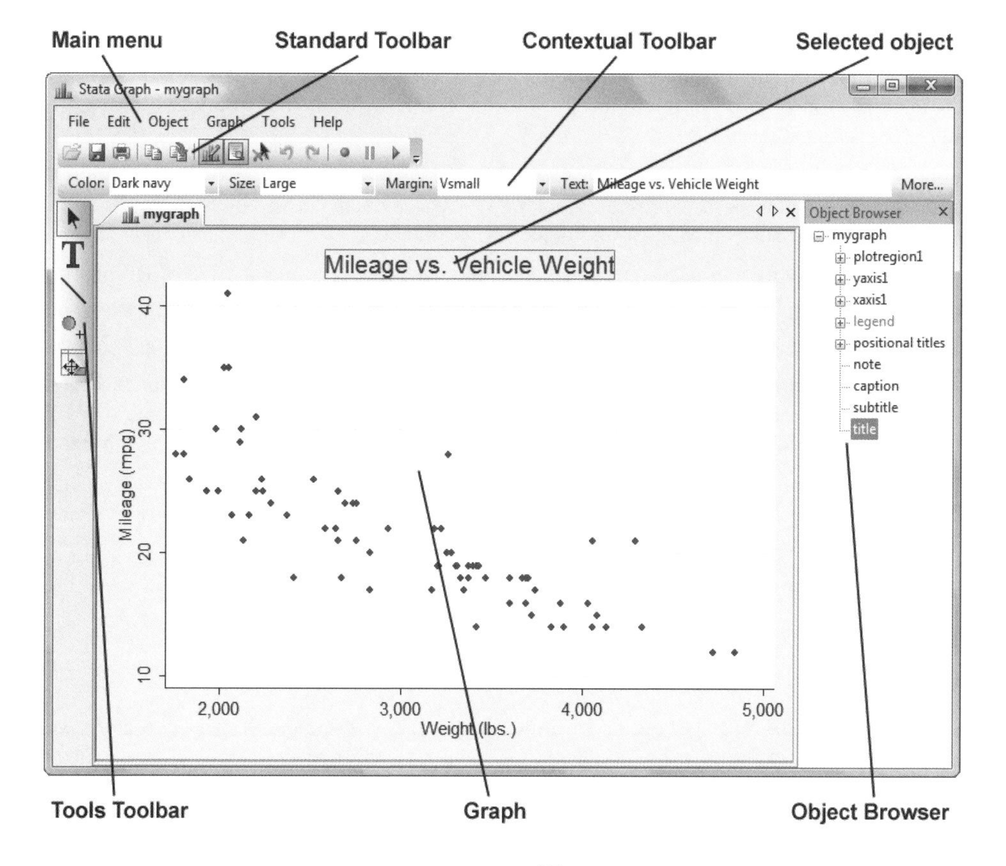

Quick start

Start the Editor by right-clicking on any graph and selecting **Start Graph Editor**. Select any of the tools along the left of the Editor to edit the graph. The Pointer Tool, , is selected by default.

Change the properties of objects or drag them to new locations by using the Pointer. As you select objects with the Pointer, a Contextual Toolbar will appear just above the graph. Use any of the controls on the Contextual Toolbar to immediately change the most important properties of the selected object. Right-click on an object to access more properties and operations. Hold the *Shift* key when dragging objects to constrain the movement to horizontal, vertical, or 90-degree angles.

Do not be afraid to try things. If you do not like a result, change it back with the same tool, or click on the **Undo** button, , in the Standard Toolbar (below the main menu). **Edit > Undo** in the main menu does the same thing.

Add text, lines, or markers (with optional labels) to your graph by using the three *Add...* tools— T, , and . Lines can be changed to arrows with the Contextual Toolbar. If you do not like the default properties of the added objects, simply change their settings in the Contextual Toolbar before adding the text, line, or marker. The new setting will then be applied to all subsequently added objects, even in future Stata sessions.

Remember to reselect the Pointer Tool when you want to drag objects or change their properties.

Move objects on the graph and have the rest of the objects adjust their position to accommodate the move with the Grid Edit Tool, . With this tool, you are repositioning objects in the underlying grid that holds the objects in the graph. Some graphs, for example, by-graphs, are composed of nested grids. You can reposition objects only within the grid that contains them; they cannot be moved to other grids.

You can also select objects in the Object Browser to the right of the graph. This window shows a hierarchical listing of the objects in the graph. Clicking or right-clicking on an object in the Browser is the same as clicking or right-clicking on the object in the graph.

You can record your edits and play them back on other graphs. Click on the **Start Recording** button, , in the Standard Toolbar to begin recording; all ensuing edits are recorded. Click the same button, , to end the recording. You will be prompted to name the recording. While editing another graph, click the **Play Recording** button, , and select your recording from the list. Your recorded edits will be applied to the graph. You can also play recorded edits from the command line when a graph is created or used from disk. See the `play(`*recordingname*`)` option in [G-3] ***std_options*** and [G-2] **graph use**.

Stop the Editor by selecting **File > Stop Graph Editor** from the main menu. You must stop the Graph Editor to enter Stata commands.

Start editing graphs now, or read on for a gentler introduction that discusses some nuances of the Editor.

Introduction

With Stata's **Graph Editor** you can change almost anything on your graph. You can add text, lines, arrows, and markers wherever you would like. As you read through this documentation (or at least on your second reading), we recommend that you open Stata, draw a graph, and try what is described. If you are surprised by a result, or do not like how something looks, you can always undo the operation by pressing the **Undo** button, , on the Standard Toolbar (more on that later) or by selecting **Edit > Undo** from the main menu.

Starting and stopping the Graph Editor

To start the Editor, 1) right-click within the Graph window and select **Start Graph Editor**, 2) select **File > Start Graph Editor** from the main menu, or 3) click on the **Start Graph Editor** button, , in the toolbar.

To close the Editor, 1) right-click within the Graph window and select **Stop Graph Editor**, 2) select **File > Stop Graph Editor** from the main menu, or 3) click on the **Stop Graph Editor** button, , in the toolbar.

When in the Editor, you cannot execute Stata commands. In fact, the Command window is grayed out and will not accept input.

The tools

When the Graph Editor starts, you will notice several changes to the Graph window. The most important is the addition of a Tools Toolbar to the left of the graph. (You can move this toolbar under Microsoft Windows, and if you have previously moved it, it will appear wherever you last placed it.) This toolbar holds the tools you use to edit graphs. There are other changes to the window, but ignore these for now.

To use any tool, simply click on that tool. The selected tool will remain in effect until you select another tool.

You are always using one of the tools. When you first start the Editor, the Pointer Tool is active.

The Pointer Tool

With the Pointer Tool you can select objects, drag objects, or modify the properties of objects. For example, you can select a title and by holding the left mouse button drag that title to another position on the graph. Hold the *Shift* key while dragging to constrain the direction to horizontal, vertical, or a 90-degree angle from the original position.

A few graph objects cannot be moved with the Pointer, in particular, axes, plot regions, and plots. Moving these objects would almost certainly distort the information in the graph. You can reposition these objects by using the Grid Editor Tool with a better chance of not distorting the information; more on that later.

Some objects cannot by default be repositioned, but you can right-click on many of these objects and select **Unlock Position** from the resulting menu. The object can then be repositioned by dragging. If you want to relock the object's position on the graph, just right-click on the object and select **Lock Position**. In the same way, you can lock the position of objects that can normally be dragged.

When you select an object—whether a title, axis, legend, scatterplot, line plot, etc.—you will notice that a toolbar appears (or changes) immediately above the graph. This is the Contextual Toolbar, with which you can immediately change the most important properties of the selected object: color, text size, or even text for titles and other text objects; marker color, marker size, or marker symbol for scatterplots; etc. Try it. Select something in the graph and change one of the properties in the Contextual Toolbar: the change is immediately reflected on the graph.

Only the most important properties are shown on the Contextual Toolbar. Most objects have many more settable properties. You can see and change all of an object's properties by selecting the **More...** button, More..., on the right of the Contextual Toolbar or by double-clicking on the object. You will be presented with a dialog box with all the object's properties. Change one or more of those properties. Then click on **Apply** if you want to see the changes on the graph and continue changing properties, or click on **OK** to apply the properties and dismiss the dialog box. Click on **Cancel** to dismiss the dialog without applying any of the edits since you last clicked on **Apply**.

Almost anything that you can do to change the look of an object with the `graph` command, you can also do with the object's dialog box.

As with dragging, any changes made from the object toolbar or the dialog boxes can be reversed by clicking on the **Undo** button, , or by selecting **Edit > Undo** from the main menu.

Add Text Tool T

You add text by using the Add Text Tool in the Tools Toolbar. Select the Add Text Tool and then click anywhere in your graph that you would like to add some text. You will be presented with the text dialog box. Type your text in the **Text** control. You can change how the text looks on the graph by changing the properties on the dialog, or select the text later with the Pointer and make changes then.

If the text is not exactly where you want it, switch to the Pointer and drag the text to the desired position.

As with text options in the `graph` command, you can create multiline text by placing each line in quotation marks. For example, `"The little red hen" "baked bread"` will appear as two lines of text. If you have text with embedded quotes, use compound quotes to bind the line, for example, `` `"She said to "use compound quotes" in such cases"' ``.

When you select the Add Text Tool, the Contextual Toolbar shows the properties for the tool. Any changes you make to the properties on the toolbar will be recorded as new default settings. These new settings are then used on all added text. In fact, these settings are stored and will be used on added text whenever you reopen the Graph Editor, either in your current Stata session or in future Stata sessions. When setting new default properties, you are not limited to the settings available on the Contextual Toolbar; you can also select the **More...** button to bring up a dialog box with the complete set of text properties. Any changes made and saved there will also become new defaults for adding text. If you want to change back to the default settings from when Stata was installed, select the **Advanced** tab on the dialog and click on **Reset Defaults**.

Add Line Tool 

You add lines and arrows by using the Add Line Tool, which is located below the Add Text Tool in the Tools Toolbar. To add a line, click within the graph to establish a starting point, and hold the left mouse button while dragging to the ending point. The line's path is shown as you drag, and the line is added when you release the left button. If you want an arrow rather than a line, click on the

Pointer Tool and then select whether you want the arrowhead at the beginning or at the end of the line from the **Arrowhead** control in the Contextual Toolbar.

After adding a line, you can use the Pointer to drag not only the entire line but also either endpoint separately.

As with the Add Text Tool, you can change the default properties for added lines by changing the settings in the Contextual Toolbar or associated dialog box while the Add Line Tool is active. As with the text settings, these settings are retained until you change them again.

If you draw more arrows than lines, this may be the time to change your default setting for the Add Line Tool. Select the tool and then select **Head** in the **Arrowhead** control of the Contextual Toolbar. Now, whenever you draw a line, an arrowhead will be drawn on the endpoint where you release the mouse.

Add Marker Tool 

You add markers by using the Add Marker Tool, which is located below the Add Line Tool. With the Add Marker Tool active, simply click anywhere you wish to add a marker. As with text and lines, you can change the marker's properties immediately or later by using the Pointer Tool and the Contextual Toolbar or the associated dialog box.

As with markers on plots, added markers can be labeled. Double-click on an added marker with the Pointer Tool (or select **More...** from its Contextual Toolbar) and use the controls on the **Label** tab of the dialog box.

As with the other *Add...* tools, you can change any of the properties of the default marker by changing settings in the Contextual Toolbar or the associated dialog when the tool is in use.

Grid Edit Tool 

The final tool on the Tools Toolbar is the Grid Edit Tool. This is an advanced tool that moves objects within their containing grid. See *Grid editing* for details; we mention it here only because it is part of the toolbar.

The Object Browser

To the right of the Graph window (unless you have moved it elsewhere or turned it off) is the Object Browser, or just Browser. The Browser is for advanced use, but we mention it here because it comes up when discussing some other tools and because there is not much to say. The Browser shows a hierarchical listing of all the objects in your graph. At the top of the hierarchy is the name of your graph, and within that is typically a plot region (**plotregion1**), the axes (**yaxis1** and **xaxis1**), the **legend**, a **note**, a **caption**, a **subtitle**, a **title**, and the **positional titles**. Some of these objects contain other objects. Most importantly, the plot region contains all of the plots, for example, scatterplots, line plots, and area plots. These plots are simply numbered 1 through N, where N is the number of plots on your graph. In addition to containing its own titles, the **legend** contains a **key region** that holds the legend's components: keys and labels.

Some graphs, such as bar charts, box plots, dot charts, and pie charts, have slightly different sets of objects. Combined graphs, in addition to their own set of titles, have a plot region that contains other graphs, which themselves nest all the objects listed earlier. By-graphs are particularly messy in the Browser because they are constructed with many of their objects hidden. Showing these objects rarely leads to anything interesting.

Although you may be able to largely ignore the Browser, it has several features that are helpful.

First, if two or more objects occupy the same space on the graph, you will be able to select only the topmost object. You would have to move the upper objects to reach a lower object. With the Browser, you can directly select any object, even one that is hidden by another object. Just select the object's name in the Browser. That object will stay selected for dragging or property changes through the Contextual Toolbar or associated dialog.

Second, the Browser is the quickest way to add titles, notes, or captions to a graph. Just select one of them in the Browser and then type your title, note, or caption in the **Text** control of the Contextual Toolbar.

As you select objects with the Pointer, those objects are also selected and highlighted in the Browser. The reverse is also true: as you select objects in the Browser, they will also be selected on the graph and their Contextual Toolbar will be displayed. There is no difference between selecting objects by name in the Browser and selecting them directly on the graph with the Pointer. In fact, you can right-click on an object in the Browser to access its properties.

If you find the Browser more of a distraction than a help, select **Tools > Hide Object Browser** from the main menu. You can always reshow the Browser from the same place.

Right-click menus, or Contextual menus

You can right-click on any object to see a list of operations specific to the object and tool you are working with. This feature is most useful with the Pointer Tool. For almost all objects, you will be offered the following operations:

Hide Hide an object that is currently shown. This will also gray the object in the Browser.

Show Show an object that is currently hidden. Available only when selecting grayed objects in the Browser.

Lock Lock the object, making it unselectable and unchangeable by the Pointer. When you lock an object, a lock icon will appear beside the object in the Browser. This is another way to select an object that is underneath another object. Lock the upper object and you will be able to select the lower object.

Unlock Unlock the object, making it selectable and its properties changeable. Available only when selecting locked objects in the Browser.

xyz **Properties** Open the properties dialog box for object *xyz*. The same dialog is opened by double-clicking on an object or clicking on the **More...** button from its Contextual Toolbar.

When you have selected an object that can be repositioned, you will also see the following:

Lock Position Lock the position of an object so that it cannot be dragged to a different position. This type of lock is not reflected in the Browser.

Unlock Position Unlock the position of an object so that it may be dragged to a different position by using the Pointer. Some objects are created with their position locked by default to avoid accidental dragging, but many may be manually unlocked with this menu item.

When you select a plot where individual observations are visible—for example, scatterplots, connected plots, spike plots, range bar plots, arrow plots—you will also see

Observation Properties Change the properties of the currently selected observation without affecting the rendition of the remaining plot. You can further customize the observation later by reselecting it with the Pointer. Once changed, the observation's custom properties become available in the Contextual Toolbar and properties dialog box.

When you select an axis, you will also see the following:

Add Tick/Label Add a tick, label, or tick and label to the selected axis.

Tick/Label Properties Change the properties of the tick or label closest to your current Pointer position. This is a quicker way to customize a tick or label than navigating to it through the **Edit or add individual ticks and labels** button in the axis properties dialog box.

Many objects with shared properties—such as plots and labels on a scatterplot matrix, bars and labels on a bar chart, and pie slices and labels on a pie chart—will also add

Object-specific Properties Change the properties of only the selected object, not all the objects that by default share its properties.

With **Object-specific Properties**, you can customize one bar, label, or other object that you would normally want to look the same as related objects.

Many of the operations come in pairs, such as **Hide/Show**. You are offered only the appropriate operations, for example, to **Hide** a shown object or to **Show** a hidden object.

The Standard Toolbar

The Standard Toolbar normally resides at the top of the Graph window (just below the main menu on Unix and Windows systems). In addition to standard operations—such as **Open Graph**, ; **Save Graph**, ; and **Print Graph**, —there are several graph and Graph Editor–specific operations available. You can **Rename** graphs, ; **Start/Stop Graph Editor**, ; **Show/Hide Object Browser**, ; **Deselect**, , the selected object; **Undo**, , or **Redo**, , edits; **Record**, , edits; and **Play**, , previously recorded edits.

You can undo and redo up to 300 consecutive edits.

The main Graph Editor menu

On Unix and Windows systems, the Graph Editor menus reside on the menubar at the top of the Graph window. Menu locations on the Mac are a little different than on other operating systems. On the Mac, all the menus referenced throughout this documentation except **File**, **Edit**, and **Help** are located under the **Graph Editor** menu. In addition, items found under the **Tools** menu on Windows and Unix systems are found under the **Graph Editor** menu on the Mac.

File In addition to opening, closing, saving, and printing graphs, you can start and stop the Graph Editor from this menu. The **Save As...** menu not only saves graphs in Stata's standard "live" format, which allows future editing in the Graph Editor, but also exports graphs in formats commonly used by other applications: PostScript, Encapsulated PostScript (EPS), TIFF, and Portable Network Graphics (PNG) on all computers; Windows Metafile (WMF) and Windows Enhanced Metafile (EMF) on Microsoft Windows computers; and Portable Document Format (PDF) on Mac computers.

Object Mirrors the operations available in the right-click menu for an object, with two additions: 1) you can unlock all objects by using **Object > Unlock All Objects** and 2) you can deselect a selected object by using **Object > Deselect**. On the Mac, this menu is located under the **Graph Editor** menu.

Graph Launches the dialog boxes for changing the properties of the objects that are common to most graphs (titles, axes, legends, etc.). You can also launch these dialogs by double-clicking on an object in the graph, by double-clicking on the object's name in the Object Browser, by selecting **Properties** from the object's right-click menu, or by clicking on **More...** in the object's Contextual Toolbar. On the Mac, this menu is located under the **Graph Editor** menu.

Tools Selects the tool for editing: **Pointer**, **Add Text**, **Add Line**, **Add Marker**, **Grid Edit**. These can also be selected from the Tools Toolbar. Under **Tools**, you can also control the Graph Recorder. From here you can also hide and show the Object Browser. On the Mac, this menu is named **Graph Editor** and also contains the **Object** and **Graph** menus.

Help Provides access to this documentation, **Help > Graph Editor**; advice on using Stata, **Help > Advice**; a topical overview of Stata's commands, **Help > Contents**; searching, **Help > Search...**; and help on specific commands, **Help > Stata Command...**.

Grid editing

Click on the Grid Edit Tool, , to begin grid editing. When you drag objects with this tool, you are rearranging them on the underlying grid where `graph` placed them.

When you select an object, it will be highlighted in red and the grid that contains the object will be shown. You can drag the object to other cells in that grid or to new cells that will be created between the existing cells. As you drag an object to other cells, those cells will appear darker red. If you drop the object on a darker red cell, you are placing it in that cell along with any objects already in the cell. As you drag over cell boundaries, the boundary will appear darker red. If you drop the object on a cell boundary, a new row or column is inserted into the grid and the object is dropped into a cell in this new row or column.

Regardless of whether you drag the object to an existing cell or to a new cell, the other objects in the graph expand or contract to make room for the object in its new position.

This concept sounds more difficult than it is in practice. Draw a graph and try it.

Some graphs, such as by-graphs and combined graphs, are composed of nested grids. You can drag objects only within the grid that contains them; you cannot drag them to other grids.

One of the more useful things you can do when grid editing is to drag a title or legend to a new position on the graph. See *Tips, tricks, and quick edits* for more examples.

You can also expand or contract the number of cells that the selected object occupies by using the Contextual Toolbar. Most objects occupy only one cell by default, but there are exceptions. If you specify the `span` option on a title, the title will occupy all the columns in its row; see [G-3] ***title_options***. To make an object occupy more or fewer cells, click on **Expand Cell** or **Contract Cell** in the Contextual Toolbar and then select the desired direction to expand or contract.

You can use the Object Browser to select objects when grid editing. With the Browser, you can individually select among objects that occupy the same cell. Selecting in the Browser is often easier for objects like legends, which are themselves a grid. In the graph, you must click on the edge of the legend to select the whole legend and not just one of its cells. If you have difficulty selecting such objects in the graph, pick their name in the Object Browser instead.

Graph Recorder

You can record your edits and play them back on other graphs by using the Graph Recorder. To start recording your edits, click on the **Start Recording** button, , in the Standard Toolbar. All ensuing edits are saved as a recording. To end a recording, click the same button, ; you will be prompted to name your recording. The recorded edits can be replayed on other graphs.

To play the edits from a recording, click on the **Play Recording** button, . You will be presented with a list of your recordings. Select the recording you want to play and the edits will be applied to your current graph.

You can also play recordings from the command line. Play a recording on the current graph using the `graph play` command; see [G-2] **graph play**. Play a recording as a graph is being used from disk; see [G-2] **graph use**. Or, play a recording by using the `play()` option at the time a graph is created; see [G-3] ***std_options***.

Some edits from a recording may not make sense when applied to another graph, for example, changes to a plotted line's color when played on a scatterplot. Such edits are ignored when a recording is played, though a note is written to the Results window for any edits that cannot be applied to the current graph.

If you want to make some edits that are not saved in the recording, select the **Pause Recording** button, . Make any edits you do not want recorded. When you are ready to record more edits, click again on the **Pause Recording** button.

You cannot **Undo** or **Redo** edits while recording. If you set a property and do not like the result, simply reset the property. If you add an object (such as a line) incorrectly, delete the added object.

❑ Technical note

Where are recordings stored?

By default, all recordings are stored in the `grec` subdirectory of your `PERSONAL` directory. (See [P] **sysdir** for information about your `PERSONAL` directory.) The files are stored with a `.grec` extension and are text files that can be opened in any standard editor, including Stata's Do-file Editor. They are not, however, meant to be edited. To remove a recording from the list of recordings shown when the **Play Recording** button, ▶, is clicked, remove it from this directory.

Most recordings are meant to be used across many graph files and so belong in the standard place. You may, however, make some recordings that are specific to one project, so you do not want them shown in the list presented by **Play Recording** button, ▶. If you want to save a recording with a project, just browse to that location when you are prompted to save the recording. Recordings stored this way will not be listed when you select **Play Recording**. To play these recordings, select Browse from the list, change to the directory where you stored the recording, and open the recording. Your recording will be played and its edits applied.

❑

Tips, tricks, and quick edits

Because you can change anything on the graph by using the Editor and because many of these changes can be done from the Contextual Toolbar, there is no end to the tips, tricks, and especially quick edits we might discuss. Here are a few to get you started.

Save your graph to disk
Make your Graph Editor bigger
Use the Apply button on dialogs
Change a scatterplot to a line plot
Add vertical grid lines
Left-justify a centered title
Reset rather than Undo
Think relative
Add a reference line
Move the y axis to the right of the graph
Move the legend into the plot region
Change the aspect ratio of a graph
Use the Graph Recorder to create a custom look for graphs
Rotate a bar graph

When you try these tips, remember that while the Graph Editor is open you cannot execute Stata commands. Exit the Editor to enter and run commands.

Save your graph to disk. It is a simple and obvious suggestion, but people with years of experience using only Stata's command-line graphics might lose precious work in the Graph Editor if they do not save the edited graph. However, Stata will prompt you when you leave the Graph Editor to save any graph that has been changed.

You can draw a graph, edit it, save it to disk, restore it in a later Stata session, and continue editing it.

Make your Graph Editor bigger. Stata recalls the size of Graph windows and the size of the Graph Editor window separately, so you can have a larger window for editing graphs. It is easier to edit graphs if you have more room to maneuver, and they will return to their normal size when you exit the Editor.

Use the Apply button on dialogs. If you are unsure of a change you are considering, you want to continue making changes using a dialog, or you just want to see what something does, click on the **Apply** button rather than the **OK** button on a dialog. The **Apply** button does not dismiss the dialog, so it is easy to change a setting back or make other changes.

Change a scatterplot to a line plot. This one is truly easy, but we want you to explore the Contextual Toolbar, and this might be an enticement.

If you do not have a scatterplot handy, use one of U.S. life expectancy versus year,

```
. use http://www.stata-press.com/data/r12/uslifeexp
(U.S. life expectancy, 1900-1999)
. scatter le year
```

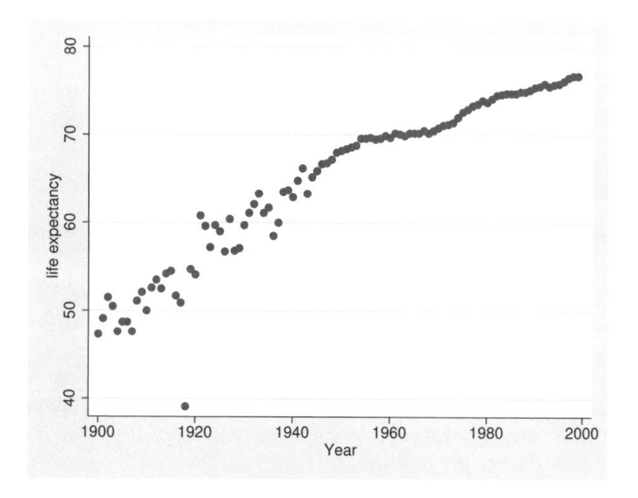

and start the Graph Editor.

1. Select the scatterplot by clicking on any of its markers.
2. Select **Line** from the **Plottype** control in the Contextual Toolbar.

That's it!

This method works for all plottypes that use the same number of variables. Scatters, lines, connecteds, areas, bars, spikes, and droplines can all be interchanged. So can the range plots: rareas, rbars, rspikes, rcapsyms, rscatters, rlines, and rconnecteds. So, too, can the paired-coordinate plots: pcspikes, pccapsyms, pcarrows, pcbarrows, and pcscatters. See [G-2] **graph twoway** for a description of all the plottypes.

Add vertical grid lines. This one is easy too, but we really do want you to explore the Contextual Toolbar. Most graph schemes show horizontal grid lines aligned with each tick on the y axis, but they do not show vertical grid lines. To add vertical grid lines,

1. Select the Pointer Tool, [pointer icon], and then click on the x axis.
2. Click in the **Show Grid** button (or checkbox under Mac and Windows) in the Contextual Toolbar.

That's it!

Left-justify a centered title. If your graph does not have a title, click on **title** in the Object Browser and add a title by typing in the **Text** field of the Contextual Toolbar (enter *Enter* to see the title).

1. Select the Pointer Tool, , and then click on the title.
2. Look for a control that justifies the title on the Contextual Toolbar. There is not one. We need more control than the toolbar offers.
3. Click on **More...** in the Contextual Toolbar to launch the dialog that controls all title properties.
4. Click on the **Format** tab in the dialog, and then select **West** from the **Position** control and click on the **Apply** button.

That's it!

This might be a good time to explore the other tabs and controls on the **Textbox Properties** dialog. This is the dialog available for almost all the text appearing on a graph, including any that you add with the Add Text Tool, **T**.

Reset rather than Undo. If you are using the Contextual Toolbar or a dialog to change the properties of an object and you want to reverse a change you have just made, simply change the setting back rather than clicking on the **Undo** button. **Undo** must completely re-create the graph, which takes longer than resetting a property.

Think relative. On dialogs, you can often enter anything in a control that you could enter in the option for the associated style or property. For example, in a size or thickness control, in addition to selecting a named size, you could enter an absolute number in percentage of graph height, or you could enter a multiple like `*.5` to make the object half its current size or `*2` to make it twice its current size.

Add a reference line. Reference lines are often added to emphasize a particular value on one of the axes, for example, the beginning of a recession or the onset of a disease. With the Add Line Tool, you could simply draw a vertical or horizontal line at the desired position, but this method is imprecise. Instead,

1. Using the Pointer Tool, double-click on the x axis.
2. Click on the **Reference line** button.
3. Enter the x value where the reference line is to be drawn and click on **OK**.

That's it!

Move your y axis to the right of the graph.

1. Click on the Grid Edit Tool, .
2. Drag the axis to the right until the right boundary of the plot region glows red, and then release the mouse button. The plot region is in the right spot, but the ticks and labels are still on the wrong side.
3. Right-click on the axis and select **Axis Properties**.
4. Click on the **Advanced** button, and then select **Right** from the **Position** control in the resulting dialog.

That's it!

Move the legend into the plot region. If you do not have a graph with a legend handy, consider this line plot of female and male life expectancies in the United States.

```
. use http://www.stata-press.com/data/r12/uslifeexp
(U.S. life expectancy, 1900-1999)
. scatter le_female le_male year
```

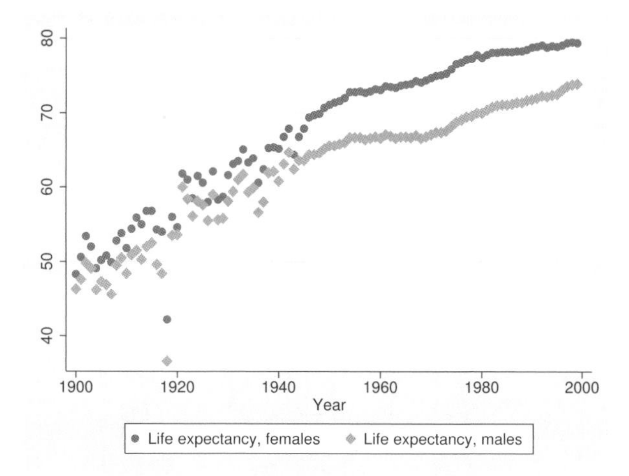

You could just use the Pointer to drag the legend into the plot region, but doing so would leave unwanted space at the bottom of the graph where the legend formerly appeared. Instead, use the Grid Edit Tool, , to place the legend atop the plot region, and then use the Pointer to fine-tune the position of the legend.

1. Click on the Grid Edit Tool, .
2. Drag the legend over the plot region. (The plot region should appear highlighted before you release the mouse button.) If you have trouble selecting the whole legend, click on its name in the Object Browser, and then drag it over the plot region.
3. Position the legend exactly where you want it by selecting the Pointer, , and dragging the legend.

That's it!

If you are using the line plot of life expectancies, you will find that there is no good place in the plot region for the wide and short default legend. To remedy that, just change the number of columns in the legend from 2 to 1 by using the **Columns** control in the legend's Contextual Toolbar. With its new shape, the legend now fits nicely into several locations in the plot region.

Change the aspect ratio of a graph. Some graphs are easier to interpret when the y and x axes are the same length, that is, the graph has an aspect ratio of 1. We might check the normality of a variable, say, trade volume stock shares in the S&P 500, by using `qnorm`; see [R] **diagnostic plots**.

```
. use http://www.stata-press.com/data/r12/sp500
(S&P 500)
. qnorm volume
```

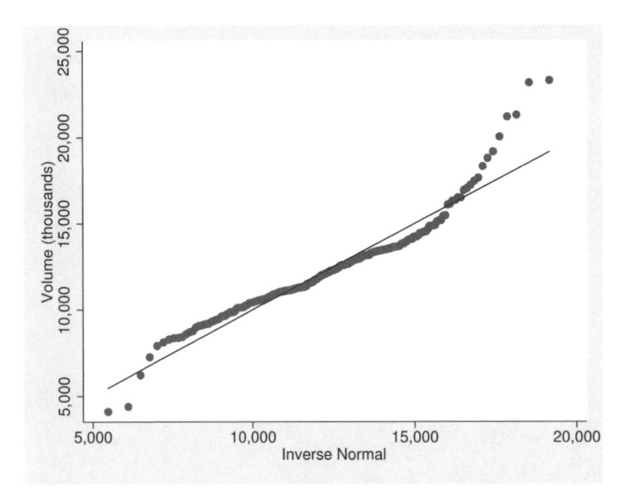

The qnorm command does not by default restrict the plot region to an aspect ratio of 1, though arguably it should. We can fix that. Start the Editor and

1. Click on **Graph** in the Object Browser. We could click directly on the graph, but doing so requires missing all the objects on the graph, so using the Browser is easier.
2. Type **1** in the **Aspect ratio** field of the Contextual Toolbar and press *Enter*.

That's it!

Use the Graph Recorder to create a custom look for graphs. If you want your graphs to have a particular appearance, such as specific colors for each plotted line or the legend being to the right of the plot region, you can automate this process by using the Graph Recorder.

The specific steps for creating the recording depend on the look you want to achieve. Here is a general outline.

1. Create the type of graph you want to customize—scatterplot, line plot, pie graph, etc. Be sure you draw as many plots as you will ever want to draw on a graph of this type, and also be sure to include all the other plot elements you wish to customize—titles, notes, etc. For a line plot, you might type

   ```
   . sysuse uslifeexp, clear
   . line le* year, title(my title) subtitle(my subtitle)
   > note(my note) caption(my caption)
   ```

   Because there are nine variables beginning with le, this will create a line plot with nine lines—probably more than you need. The graph will also have all the basic plot elements.
2. Start the Graph Editor. Then start the Recorder by clicking on the **Start Recording** button, [icon], in the Standard Toolbar.

3. Use the Editor to make the graph look the way you want line graphs to look.

   - Change the color, thickness, or pattern of the first line by selecting the line and using the Contextual Toolbar or any of the options available from the Contextual menus.

     Repeat this for every line you want to change.

     With so many lines, you may find it easier to select lines in the legend, rather than in the plot region.

   - Change the size, color, etc., of titles and captions.
   - Change the orientation of axis tick labels, or even change the suggested number of ticks.
   - Change the background color of the graph or plot region.
   - Move titles, legends, etc., to other locations—for example, move the legend to the right of the plot region. This is usually best done with the Grid Edit Tool, which allows the other graph elements to adjust to the repositioning.
   - Make any other changes you wish using any of the tools in the Graph Editor.

4. End the Recorder by clicking on the **Recording** button, , again, and give the recording a name—say, `mylineplot`.
5. Apply the recorded edits to any other line graph either by using the **Play Recording** button, , on the Graph Editor or by using one of the methods for playing a recording from the command line: `graph play` or `play()`.

If you wish to change the look of plots—markers, lines, bars, pie slices, etc.—you must create a separate recording for each graph family or plottype. You need separate recordings because changes to one plottype do not affect other plottypes. That is, changing markers does not affect lines. If you wish to change only overall graph features—background colors, titles, legend position, etc.—you can make one recording and play it back on any type of graph.

For a more general way to control how graphs look, you can create your own scheme (see [G-4] **schemes intro**). Creating schemes, however, requires some comfort with editing control files and a tolerance for reading through the hundreds of settings available from schemes. See `help scheme files` for details on how to create your own scheme.

Note: We said in step 1 that you should include titles, notes, and other graph elements when creating the graph to edit. Creating these elements makes things easier but is usually not required. Common graph elements always appear in the Object Browser, even if they have no text or other contents to show on the graph; you can select them in the Browser and change their properties, even though they do not appear on the graph. Such invisible elements will still be difficult to manipulate with the Grid Edit Tool. If you need an invisible object to relocate, click on the **Pause Recording** button, , add the object, unpause the recording, and then continue with your edits.

Rotate a bar graph. You can rotate the over-groups of a bar, dot, or box chart. This is easier to see than to explain. Let's create a bar graph of wages over three different sets of categories.

```
. use http://www.stata-press.com/data/r12/nlsw88
(NLSW, 1988 extract)

. graph bar wage, over(race) over(collgrad) over(union)
```

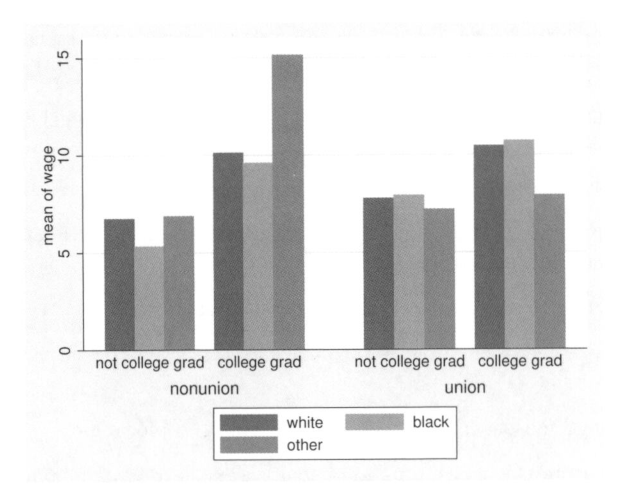

Start the Graph Editor.

1. Using the Pointer, click within the plot region, but not on any of the bars.
2. Click on the **Rotate** button in the Contextual Toolbar.
3. Click on **Rotate** a few more times, watching what happens on the graph.

To see some other interesting things, click on the **More...** button. In the resulting dialog, check **Stack bars** and click on **Apply**. Then check **Graph percentages** and click on **Apply**.

During rotation, sometimes the labels on the x axis did not fit. Select the **Horizontal** radio button in the dialog and click on **Apply** to flip the bar graph to horizontal, and then repeat the rotation. Bar graphs requiring long labels typically work better when drawn horizontally.

❑ Technical note

When the Add Text Tool, Add Line Tool, and Add Marker Tool add things to a graph, the new object can be added to a plot region, a subgraph, or the graph as a whole. They will be added to the plot region if the starting point for the added object is within a plot region. The same is true of subgraphs. Otherwise, the new objects will be added to the overall graph.

Why do you care? When a line, for example, is added to a graph, its endpoints are recorded on the generic metric of the graph, whereas when a line is added to a plot region, the endpoints are recorded in the metric of the x and y axes. That is, in the plot region of a graph of `mpg` versus `weight`, the endpoints are recorded in "miles per gallon" and "curb weight". If you later change the range of the graph's axes, your line's endpoints will still be at the same values of `mpg` and `weight`. This is almost always what you want.

If you prefer your added object to not scale with changes in the axes, add it outside the plot region. If you still want it on the plot region, drag it into the plot region after adding it outside the region.

If your x or y axis is on a log scale, you may be surprised at how lines added to the plot region react when drawn. When you are dragging the endpoints, all will be fine. When you drag the line as a whole, however, the line will change its length and angle. This happens because dimensions in a log metric are not linear and dragging the line affects each endpoint differently. The Graph Editor is not smart enough to track this nonlinearity, and the actual position of the line appears only after you drop it. We recommend that you drag only the endpoints of lines added to plot regions whose dimensions are on a log scale.

❑

Also see

[G-2] **graph twoway**

[G-2] Commands

Title

[G-2] graph — The graph command

Syntax

```
graph ...
```

The commands that draw graphs are

Command	Description
`graph twoway`	scatterplots, line plots, etc.
`graph matrix`	scatterplot matrices
`graph bar`	bar charts
`graph dot`	dot charts
`graph box`	box-and-whisker plots
`graph pie`	pie charts
other	more commands to draw statistical graphs

See [G-2] **graph twoway**, [G-2] **graph matrix**, [G-2] **graph bar**, [G-2] **graph dot**, [G-2] **graph box**, [G-2] **graph pie**, and [G-2] **graph other**.

The commands that save a previously drawn graph, redisplay previously saved graphs, and combine graphs are

Command	Description
`graph save`	save graph to disk
`graph use`	redisplay graph stored on disk
`graph display`	redisplay graph stored in memory
`graph combine`	combine multiple graphs

See [G-2] **graph save**, [G-2] **graph use**, [G-2] **graph display**, and [G-2] **graph combine**.

The commands for printing a graph are

Command	Description
`graph print`	print currently displayed graph
`set printcolor`	set how colors are printed
`graph export`	export `.gph` file to PostScript, etc.

See [G-2] **graph print**, [G-2] **set printcolor**, and [G-2] **graph export**.

The commands that deal with the graphs currently stored in memory are

Command	Description
`graph display`	display graph
`graph dir`	list names
`graph describe`	describe contents
`graph rename`	rename memory graph
`graph copy`	copy memory graph to new name
`graph drop`	discard graphs in memory

See [G-2] **graph manipulation**; [G-2] **graph display**, [G-2] **graph dir**, [G-2] **graph describe**, [G-2] **graph rename**, [G-2] **graph copy**, and [G-2] **graph drop**.

The commands that describe available schemes and allow you to identify and set the default scheme are

Command	Description
`graph query, schemes`	list available schemes
`query graphics`	identify default scheme
`set scheme`	set default scheme

See [G-4] **schemes intro**; [G-2] **graph query**, and [G-2] **set scheme**.

The command that lists available styles is

Command	Description
`graph query`	list available styles

See [G-2] **graph query**.

The command for setting options for printing and exporting graphs is

Command	Description
`graph set`	set graphics options

See [G-2] **graph set**.

The command that allows you to draw graphs without displaying them is

Command	Description
`set graphics`	set whether graphs are displayed

See [G-2] **set graphics**.

Description

`graph` draws graphs.

Remarks

See [G-1] **graph intro**.

Also see

[G-1] **graph intro** — Introduction to graphics

[G-2] **graph other** — Other graphics commands

[G-2] **graph export** — Export current graph

[G-2] **graph print** — Print a graph

Title

[G-2] graph bar — Bar charts

Syntax

graph bar *yvars* [*if*] [*in*] [*weight*] [, *options*]

graph hbar *yvars* [*if*] [*in*] [*weight*] [, *options*]

where *yvars* is

(asis) *varlist*

or is

[(*stat*)] *varname* [[(*stat*)] ...]

[(*stat*)] *varlist* [[(*stat*)] ...]

[(*stat*)] [*name*=]*varname*[...] [[(*stat*)] ...]

where *stat* may be any of

mean median p1 p2 ...p99 sum count min max

or

any of the other *stats* defined in [D] **collapse**

mean is the default. p1 means 1st percentile, p2 second, and so on; p50 means the same as median. count means the number of nonmissing values of the specified variable.

options	Description
group_options	groups over which bars are drawn
yvar_options	variables that are the bars
lookofbar_options	how the bars look
legending_options	how *yvars* are labeled
axis_options	how the numerical y axis is labeled
title_and_other_options	titles, added text, aspect ratio, etc.

Each is defined below.

group_options	Description
over(*varname*[, *over_subopts*])	categories; option may be repeated
nofill	omit empty categories
missing	keep missing value as category
allcategories	include all categories in the dataset

yvar_options	Description
`ascategory`	treat *yvars* as first `over()` group
`asyvars`	treat first `over()` group as *yvars*
`percentages`	show percentages within *yvars*
`stack`	stack the *yvar* bars
`cw`	calculate *yvar* statistics omitting missing values of any *yvar*

lookofbar_options	Description
`outergap(`[*]#`)`	gap between edge and first bar and between last bar and edge
`bargap(`#`)`	gap between *yvar* bars; default is 0
`intensity(`[*]#`)`	intensity of fill
`lintensity(`[*]#`)`	intensity of outline
`pcycle(`#`)`	bar styles before pstyles recycle
`bar(`#, *barlook_options*`)`	look of #th *yvar* bar

See [G-3] ***barlook_options***.

legending_options	Description
legend_options	control of *yvar* legend
`nolabel`	use *yvar* names, not labels, in legend
`yvaroptions(`*over_subopts*`)`	*over_subopts* for *yvars*; seldom specified
`showyvars`	label *yvars* on x axis; seldom specified
`blabel(...)`	add labels to bars

See [G-3] ***legend_options*** and [G-3] ***blabel_option***.

axis_options	Description
`yalternate`	put numerical y axis on right (top)
`xalternate`	put categorical x axis on top (right)
`exclude0`	do not force y axis to include 0
`yreverse`	reverse y axis
axis_scale_options	y-axis scaling and look
axis_label_options	y-axis labeling
`ytitle(...)`	y-axis titling

See [G-3] ***axis_scale_options***, [G-3] ***axis_label_options***, and [G-3] ***axis_title_options***.

title_and_other_options	Description
`text(...)`	add text on graph; x range [0, 100]
`yline(...)`	add y lines to graph
aspect_option	constrain aspect ratio of plot region
std_options	titles, graph size, saving to disk
`by(`*varlist*`, ...)`	repeat for subgroups

See [G-3] ***added_text_options***, [G-3] ***added_line_options***, [G-3] ***aspect_option***, [G-3] ***std_options***, and [G-3] ***by_option***.

The *over_subopts*—used in over(*varname*, *over_subopts*) and, on rare occasion, in yvaroptions(*over_subopts*)—are

over_subopts	Description
relabel(# "*text*" ...)	change axis labels
label(*cat_axis_label_options*)	rendition of labels
axis(*cat_axis_line_options*)	rendition of axis line
gap([*]#)	gap between bars within over() category
sort(*varname*)	put bars in prespecified order
sort(#)	put bars in height order
sort((*stat*) *varname*)	put bars in derived order
descending	reverse default or specified bar order
reverse	reverse scale to run from maximum to minimum

See [G-3] ***cat_axis_label_options*** and [G-3] ***cat_axis_line_options***.

aweights, fweights, and pweights are allowed; see [U] **11.1.6 weight** and see note concerning weights in [D] **collapse**.

Menu

Graphics > Bar chart

Description

graph bar draws vertical bar charts. In a vertical bar chart, the y axis is numerical, and the x axis is categorical.

```
. graph bar (mean) numeric_var, over(cat_var)
```

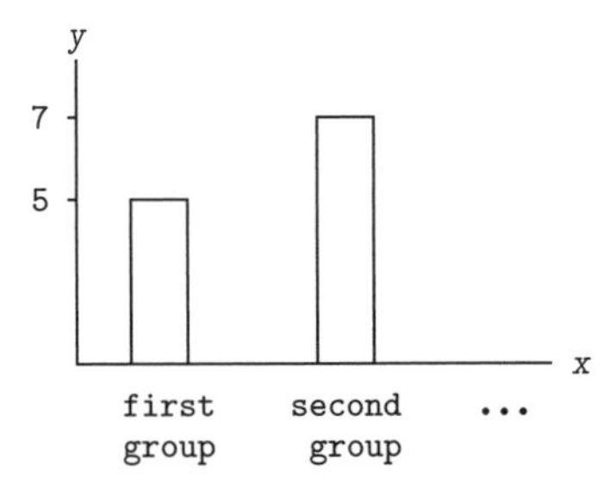

numeric_var must be numeric; statistics of it are shown on the *y* axis.

cat_var may be numeric or string; it is shown on the categorical *x* axis.

graph hbar draws horizontal bar charts. In a horizontal bar chart, the numerical axis is still called the y axis, and the categorical axis is still called the x axis, but y is presented horizontally, and x vertically.

```
. graph hbar (mean) numeric_var, over(cat_var)
```

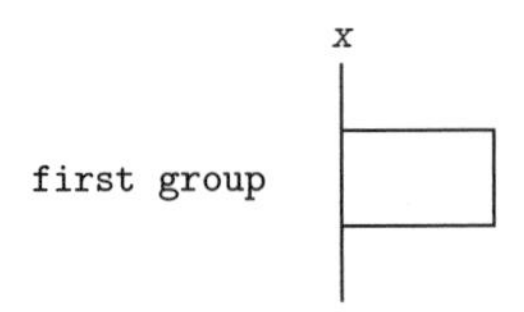

same conceptual layout: *numeric_var* still appears

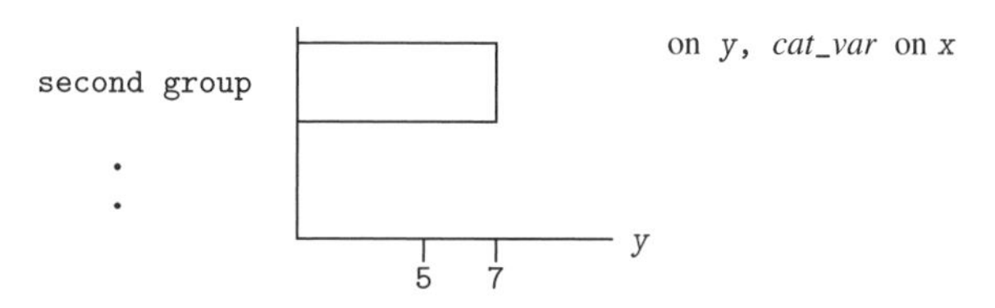

The syntax for vertical and horizontal bar charts is the same; all that is required is changing `bar` to `hbar` or `hbar` to `bar`.

group_options

`over(`*varname* [`,` *over_subopts*]`)` specifies a categorical variable over which the *yvars* are to be repeated. *varname* may be string or numeric. Up to two `over()` options may be specified when multiple *yvars* are specified, and up to three `over()`s may be specified when one *yvar* is specified; options may be specified; see *Examples of syntax* and *Multiple over( )s (repeating the bars)* under *Remarks* below.

`nofill` specifies that missing subcategories be omitted. For instance, consider

```
. graph bar (mean) y, over(division) over(region)
```

Say that one of the divisions has no data for one of the regions, either because there are no such observations or because `y==.` for such observations. In the resulting chart, the bar will be missing:

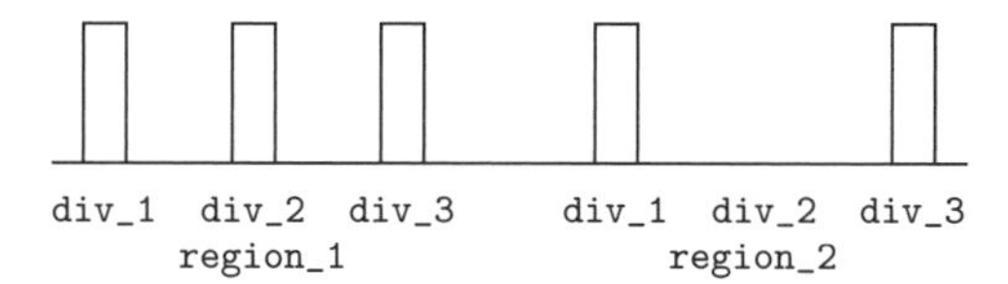

If you specify `nofill`, the missing category will be removed from the chart:

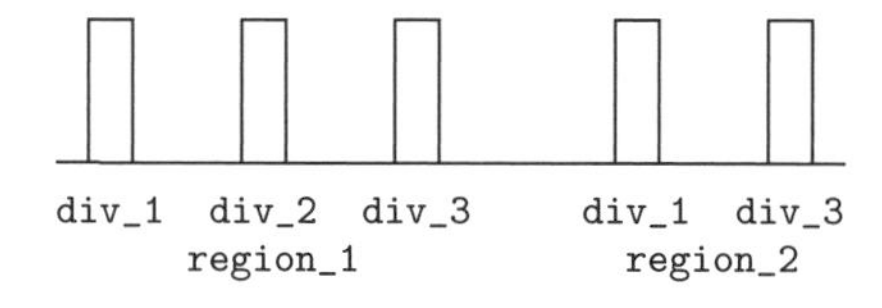

`missing` specifies that missing values of the `over()` variables be kept as their own categories, one for `.`, another for `.a`, etc. The default is to act as if such observations simply did not appear in the dataset; the observations are ignored. An `over()` variable is considered to be missing if it is numeric and contains a missing value or if it is string and contains "".

`allcategories` specifies that all categories in the entire dataset be retained for the `over()` variables. When `if` or `in` is specified without `allcategories`, the graph is drawn, completely excluding any categories for the `over()` variables that do not occur in the specified subsample. With the `allcategories` option, categories that do not occur in the subsample still appear in the legend, and zero-height bars are drawn where these categories would appear. Such behavior can be convenient when comparing graphs of subsamples that do not include completely common categories for all `over()` variables. This option has an effect only when `if` or `in` is specified or if there are missing values in the variables. `allcategories` may not be combined with `by()`.

yvar_options

`ascategory` specifies that the *yvars* be treated as the first `over()` group; see *Treatment of bars* under *Remarks* below. `ascategory` is a useful option.

When you specify `ascategory`, results are the same as if you specified one *yvar* and introduced a new first `over()` variable. Anyplace you read in the documentation that something is done over the first `over()` category, or using the first `over()` category, it will be done over or using *yvars*.

Suppose that you specified

```
. graph bar y1 y2 y3, ascategory whatever_other_options
```

The results will be the same as if you typed

```
. graph bar y, over(newcategoryvariable) whatever_other_options
```

with a long rather than wide dataset in memory.

`asyvars` specifies that the first `over()` group be treated as *yvars*. See *Treatment of bars* under *Remarks* below.

When you specify `asyvars`, results are the same as if you removed the first `over()` group and introduced multiple *yvars*. If you previously had k *yvars* and, in your first `over()` category, G groups, results will be the same as if you specified $k \times G$ *yvars* and removed the `over()`. Anyplace you read in the documentation that something is done over the *yvars* or using the *yvars*, it will be done over or using the first `over()` group.

Suppose that you specified

```
. graph bar y, over(group) asyvars whatever_other_options
```

Results will be the same as if you typed

```
. graph bar y1 y2 y3 ..., whatever_other_options
```

with a wide rather than a long dataset in memory. Variables *y1*, *y2*, ..., are sometimes called the virtual *yvars*.

`percentages` specifies that bar heights be based on percentages that *yvar_i* represents of all the *yvars*. That is,

```
. graph bar (mean) inc_male inc_female
```

would produce a chart with bar height reflecting average income.

```
. graph bar (mean) inc_male inc_female, percentage
```

would produce a chart with the bar heights being $100 \times$ `inc_male`/(`inc_male` + `inc_female`) and $100 \times$ `inc_female`/(`inc_male` + `inc_female`).

If you have one *yvar* and want percentages calculated over the first `over()` group, specify the `asyvars` option. For instance,

```
. graph bar (mean) wage, over(i) over(j)
```

would produce a chart where bar heights reflect mean wages.

```
. graph bar (mean) wage, over(i) over(j) asyvars percentages
```

would produce a chart where bar heights are

$$100 \times \left(\frac{\text{mean}_{ij}}{\sum_i \text{mean}_{ij}} \right)$$

Option `stack` is often combined with option `percentage`.

`stack` specifies that the *yvar* bars be stacked.

```
. graph bar (mean) inc_male inc_female, over(region) percentage stack
```

would produce a chart with all bars being the same height, 100%. Each bar would be two bars stacked (percentage of `inc_male` and percentage of `inc_female`), so the division would show the relative shares of `inc_male` and `inc_female` of total income.

To stack bars over the first `over()` group, specify the `asyvars` option:

```
. graph bar (mean) wage, over(sex) over(region) asyvars percentage stack
```

`cw` specifies casewise deletion. If `cw` is specified, observations for which any of the *yvars* are missing are ignored. The default is to calculate the requested statistics by using all the data possible.

lookofbar_options

`outergap(*#)` and `outergap(#)` specify the gap between the edge of the graph to the beginning of the first bar and the end of the last bar to the edge of the graph.

`outergap(*#)` specifies that the default be modified. Specifying `outergap(*1.2)` increases the gap by 20%, and specifying `outergap(*.8)` reduces the gap by 20%.

`outergap(#)` specifies the gap as a percentage-of-bar-width units. `outergap(50)` specifies that the gap be half the bar width.

`bargap(#)` specifies the gap to be left between *yvar* bars as a percentage-of-bar-width units. The default is `bargap(0)`, meaning that bars touch.

`bargap()` may be specified as positive or negative numbers. `bargap(10)` puts a small gap between the bars (the precise amount being 10% of the width of the bars). `bargap(-30)` overlaps the bars by 30%.

`bargap()` affects only the *yvar* bars. If you want to change the gap for the first, second, or third `over()` groups, specify the *over_subopt* `gap()` inside the `over()` itself; see *Suboptions for use with over() and yvaroptions()* below.

`intensity(#)` and `intensity(*#)` specify the intensity of the color used to fill the inside of the bar. `intensity(#)` specifies the intensity, and `intensity(*#)` specifies the intensity relative to the default.

By default, the bar is filled with the color of its border, attenuated. Specify `intensity(*#)`, $\#< 1$, to attenuate it more and specify `intensity(*#)`, $\#> 1$, to amplify it.

Specify `intensity(0)` if you do not want the bar filled at all. Specify `intensity(100)` if you want the bar to have the same intensity as the bar's outline.

`lintensity(#)` and `lintensity(*#)` specify the intensity of the line used to outline the bar. `lintensity(#)` specifies the intensity, and `lintensity(*#)` specifies the intensity relative to the default.

By default, the bar is outlined at the same intensity at which it is filled or at an amplification of that, which depending on your chosen scheme; see [G-4] **schemes intro**. If you want the bar outlined in the darkest possible way, specify `intensity(255)`. If you wish simply to amplify the outline, specify `intensity(*#)`, $\# > 1$, and if you wish to attenuate the outline, specify `intensity(*#)`, $\# < 1$.

pcycle(*#*) specifies how many variables are to be plotted before the pstyle (see [G-4] ***pstyle***) of the bars for the next variable begins again at the pstyle of the first variable—p1bar (with the bars for the variable following that using p2bar and so). Put another way: # specifies how quickly the look of bars is recycled when more than # variables are specified. The default for most schemes is pcycle(15).

bar(*#*, *barlook_options*) specifies the look of the *yvar* bars. bar(1, ...) refers to the bar associated with the first *yvar*, bar(2, ...) refers to the bar associated with the second, and so on. The most useful *barlook_option* is color(*colorstyle*), which sets the color of the bar. For instance, you might specify bar(1, color(green)) to make the bar associated with the first *yvar* green. See [G-4] ***colorstyle*** for a list of color choices, and see [G-3] ***barlook_options*** for information on the other *barlook_options*.

legending_options

legend_options controls the legend. If more than one *yvar* is specified, a legend is produced. Otherwise, no legend is needed because the over() groups are labeled on the categorical x axis. See [G-3] ***legend_options***, and see *Treatment of bars* under *Remarks* below.

nolabel specifies that, in automatically constructing the legend, the variable names of the *yvars* be used in preference to "mean of *varname*" or "sum of *varname*", etc.

yvaroptions(*over_subopts*) allows you to specify *over_subopts* for the *yvars*. This is seldom done.

showyvars specifies that, in addition to building a legend, the identities of the *yvars* be shown on the categorical x axis. If showyvars is specified, it is typical also to specify legend(off).

blabel() allows you to add labels on top of the bars; see [G-3] ***blabel_option***.

axis_options

yalternate and xalternate switch the side on which the axes appear.

Used with graph bar, yalternate moves the numerical y axis from the left to the right; xalternate moves the categorical x axis from the bottom to the top.

Used with graph hbar, yalternate moves the numerical y axis from the bottom to the top; xalternate moves the categorical x axis from the left to the right.

If your scheme by default puts the axes on the opposite sides, then yalternate and xalternate reverse their actions.

exclude0 specifies that the numerical y axis need not be scaled to include 0.

yreverse specifies that the numerical y axis have its scale reversed so that it runs from maximum to minimum. This option causes bars to extend down rather than up (graph bar) or from right to left rather than from left to right (graph hbar).

axis_scale_options specify how the numerical y axis is scaled and how it looks; see [G-3] ***axis_scale_options***. There you will also see option xscale() in addition to yscale(). Ignore xscale(), which is irrelevant for bar charts.

axis_label_options specify how the numerical y axis is to be labeled. The *axis_label_options* also allow you to add and suppress grid lines; see [G-3] ***axis_label_options***. There you will see that, in addition to options ylabel(), ytick(), ..., ymtick(), options xlabel(), ..., xmtick() are allowed. Ignore the x*() options, which are irrelevant for bar charts.

`ytitle()` overrides the default title for the numerical y axis; see [G-3] ***axis_title_options***. There you will also find option `xtitle()` documented, which is irrelevant for bar charts.

title_and_other_options

`text()` adds text to a specified location on the graph; see [G-3] ***added_text_options***. The basic syntax of `text()` is

`text(`*#_y #_x* `"`*text*`")`

`text()` is documented in terms of twoway graphs. When used with bar charts, the "numeric" x axis is scaled to run from 0 to 100.

`yline()` adds horizontal (`bar`) or vertical (`hbar`) lines at specified y values; see [G-3] ***added_line_options***. The `xline()` option, also documented there, is irrelevant for bar charts. If your interest is in adding grid lines, see [G-3] ***axis_label_options***.

aspect_option allows you to control the relationship between the height and width of a graph's plot region; see [G-3] ***aspect_option***.

std_options allow you to add titles, control the graph size, save the graph on disk, and much more; see [G-3] ***std_options***.

`by(`*varlist*`, ...)` draws separate plots within one graph; see [G-3] ***by_option*** and see *Use with by()* under *Remarks* below.

Suboptions for use with over() and yvaroptions()

`relabel(`*#* `"`*text*`" ...)` specifies text to override the default category labeling. Pretend that variable `sex` took on two values and you typed

```
. graph bar ..., ... over(sex, relabel(1 "Male" 2 "Female"))
```

The result would be to relabel the first value of `sex` to be "Male" and the second value, "Female"; "Male" and "Female" would appear on the categorical x axis to label the bars. This would be the result, regardless of whether variable `sex` were string or numeric and regardless of the codes actually stored in the variable to record `sex`.

That is, # refers to category number, which is determined by sorting the unique values of the variable (here `sex`) and assigning 1 to the first value, 2 to the second, and so on. If you are unsure as to what that ordering would be, the easy way to find out is to type

```
. tabulate sex
```

If you also plan on specifying `graph bar`'s or `graph hbar`'s `missing` option,

```
. graph bar ..., ... missing over(sex, relabel(...))
```

then type

```
. tabulate sex, missing
```

to determine the coding. See [R] **tabulate oneway**.

Relabeling the values does not change the order in which the bars are displayed.

You may create multiple-line labels by using quoted strings within quoted strings:

`over(`*varname*`, relabel(1 ‘" "Male" "patients" "’ 2 ‘" "Female" "patients" "’))`

When specifying quoted strings within quoted strings, remember to use compound double quotes `" and "' on the outer level.

relabel() may also be specified inside yvaroptions(). By default, the identity of the *yvars* is revealed in the legend, so specifying yvaroptions(relabel()) changes the legend. Because it is the legend that is changed, using legend(label()) is preferred; see *legending_options* above. In any case, specifying

```
yvaroptions(relabel(1 "Males" 2 "Females"))
```

changes the text that appears in the legend for the first *yvar* and the second *yvar*. # in relabel(# ...) refers to *yvar* number. Here you may not use the nested quotes to create multiline labels; use the legend(label()) option because it provides multiline capabilities.

label(*cat_axis_label_options*) determines other aspects of the look of the category labels on the x axis. Except for label(labcolor()) and label(labsize()), these options are seldom specified; see [G-3] ***cat_axis_label_options***.

axis(*cat_axis_line_options*) specifies how the axis line is rendered. This is a seldom specified option. See [G-3] ***cat_axis_line_options***.

gap(*#*) and gap(**#*) specify the gap between the bars in this over() group. gap(*#*) is specified in percentage-of-bar-width units, so gap(67) means two-thirds the width of a bar. gap(**#*) allows modifying the default gap. gap(*1.2) would increase the gap by 20%, and gap(*.8) would decrease the gap by 20%.

To understand the distinction between over(..., gap()) and option bargap(), consider

```
. graph bar revenue profit, bargap(...) over(division, gap(...))
```

bargap() sets the distance between the revenue and profit bars. over(,gap()) sets the distance between the bars for the first division and the second division, the second division and the third, and so on. Similarly, in

```
. graph bar revenue profit, bargap(...)
                            over(division, gap(...))
                            over(year,     gap(...))
```

over(division, gap()) sets the gap between divisions and over(year, gap()) sets the gap between years.

sort(*varname*), sort(*#*), and sort((*stat*) *varname*) control how bars are ordered. See *How bars are ordered* and *Reordering the bars* under *Remarks* below.

sort(*varname*) puts the bars in the order of *varname*; see *Putting the bars in a prespecified order* under *Remarks* below.

sort(*#*) puts the bars in height order. # refers to the *yvar* number on which the ordering should be performed; see *Putting the bars in height order* under *Remarks* under *Remarks* below.

sort((*stat*) *varname*) puts the bars in an order based on a calculated statistic; see *Putting the bars in a derived order* under *Remarks* below.

descending specifies that the order of the bars—default or as specified by sort()—be reversed.

reverse specifies that the categorical scale run from maximum to minimum rather than the default minimum to maximum. Among other things, when combined with bargap(-*#*), reverse causes the sequence of overlapping to be reversed.

Remarks

Remarks are presented under the following headings:

Introduction

Let us show you some bar charts:

```
. use http://www.stata-press.com/data/r12/citytemp
(City Temperature Data)

. graph bar (mean) tempjuly tempjan, over(region)
        bargap(-30)
        legend( label(1 "July") label(2 "January") )
        ytitle("Degrees Fahrenheit")
        title("Average July and January temperatures")
        subtitle("by regions of the United States")
        note("Source:  U.S. Census Bureau, U.S. Dept. of Commerce")
```

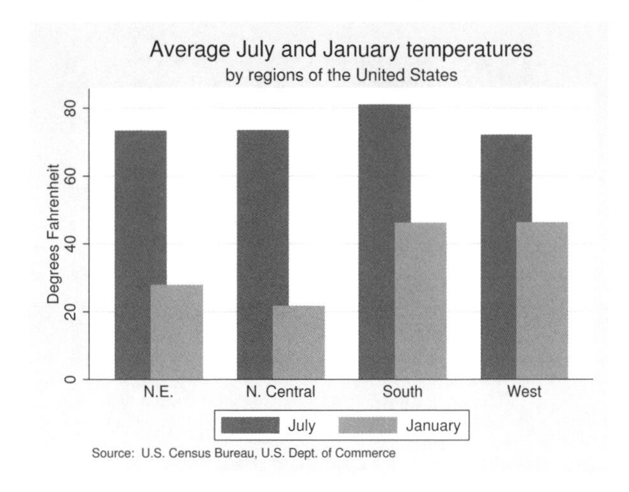

```
. use http://www.stata-press.com/data/r12/citytemp, clear
(City Temperature Data)
. graph hbar (mean) tempjan, over(division) over(region) nofill
        ytitle("Degrees Fahrenheit")
        title("Average January temperature")
        subtitle("by region and division of the United States")
        note("Source:  U.S. Census Bureau, U.S. Dept. of Commerce")
```

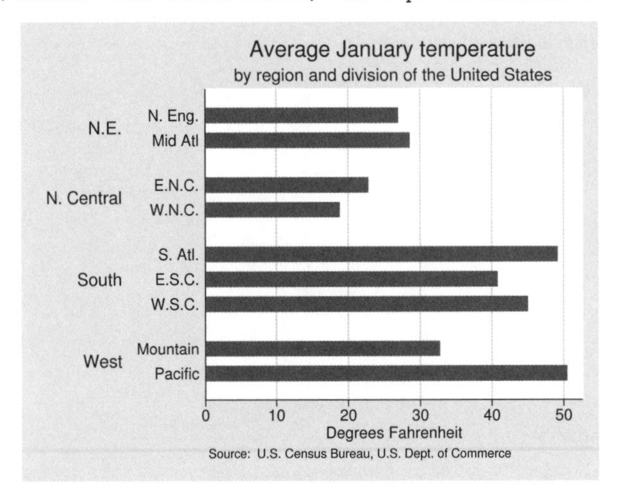


```
. use http://www.stata-press.com/data/r12/nlsw88, clear
(NLSW, 1988 extract)
. graph bar (mean) wage, over(smsa) over(married) over(collgrad)
        title("Average Hourly Wage, 1988, Women Aged 34-46")
        subtitle("by College Graduation, Marital Status,
                  and SMSA residence")
        note("Source:  1988 data from NLS, U.S. Dept. of Labor,
              Bureau of Labor Statistics")
```

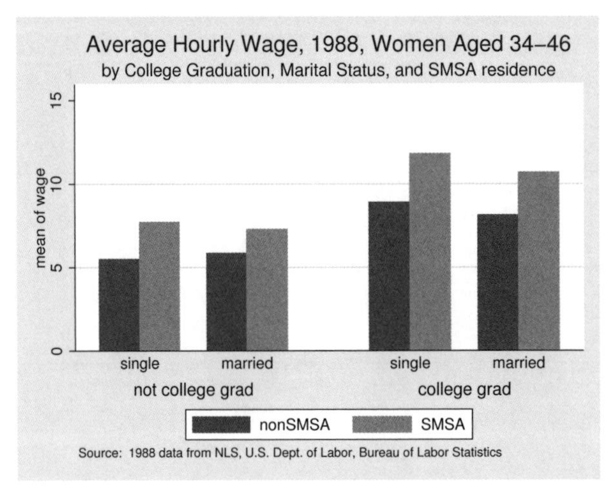

```
. use http://www.stata-press.com/data/r12/educ99gdp, clear
(Education and GDP)
. generate total = private + public
. graph hbar (asis) public private,
        over(country, sort(total) descending) stack
        title( "Spending on tertiary education as % of GDP,
                1999", span pos(11) )
        subtitle(" ")
        note("Source:  OECD, Education at a Glance 2002", span)
```

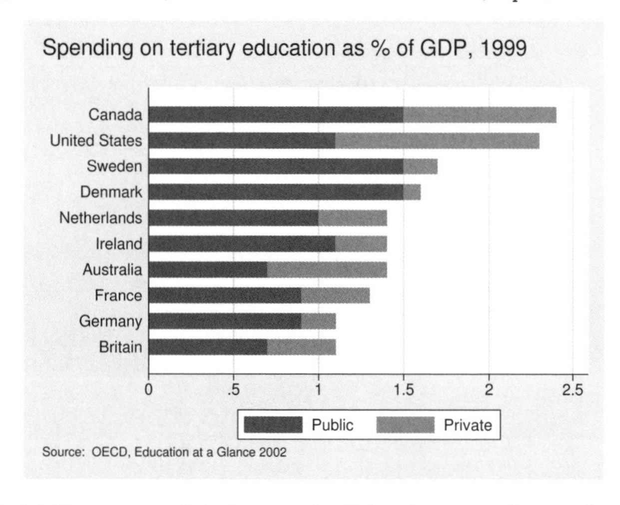

In the sections that follow, we explain how each of the above graphs—and others—are produced.

Examples of syntax

Below we show you some `graph bar` commands and tell you what each would do:

`graph bar revenue`
One big bar showing average revenue.

`graph bar revenue profit`
Two bars, one showing average revenue and the other showing average profit.

`graph bar revenue, over(division)`
#_of_divisions bars showing average revenue for each division.

`graph bar revenue profit, over(division)`
2 × *#_of_divisions* bars showing average revenue and average profit for each division. The grouping would look like this (assuming three divisions):

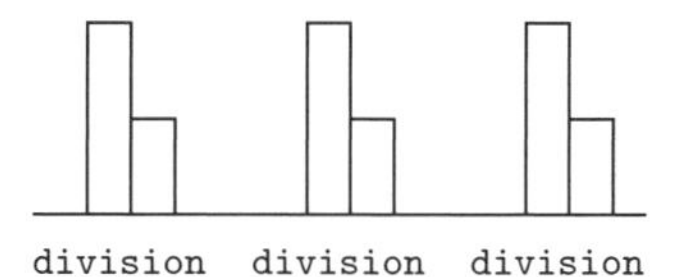

`graph bar revenue, over(division) over(year)`
#_of_divisions × *#_of_years* bars showing average revenue for each division, repeated for each of the years. The grouping would look like this (assuming three divisions and 2 years):

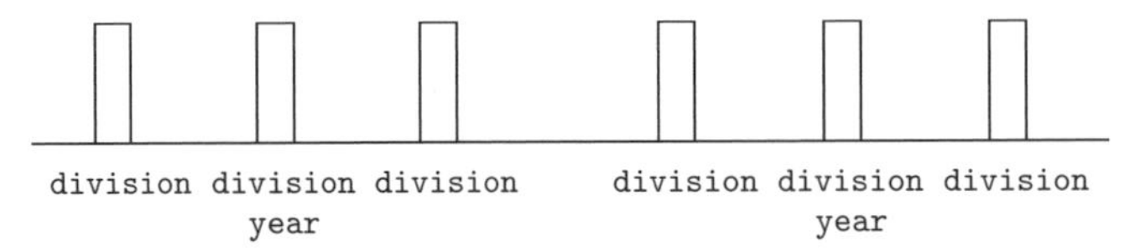

`graph bar revenue, over(year) over(division)`
same as above but ordered differently. In the previous example, we typed `over(division) over(year)`. This time, we reverse it:

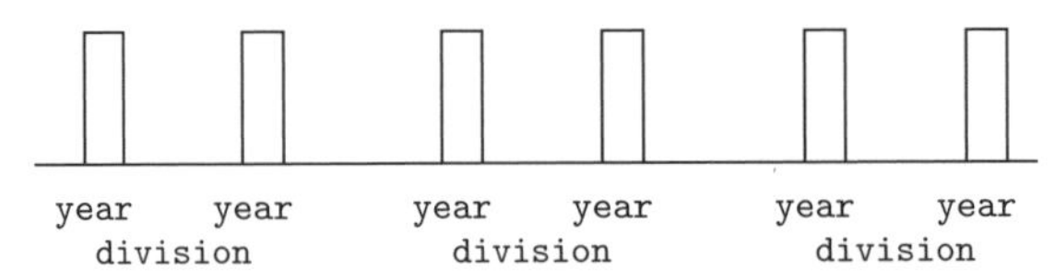

`graph bar revenue profit, over(division) over(year)`
2 × *#_of_divisions* × *#_of_years* bars showing average revenue and average profit for each division, repeated for each of the years. The grouping would look like this (assuming three divisions and 2 years):

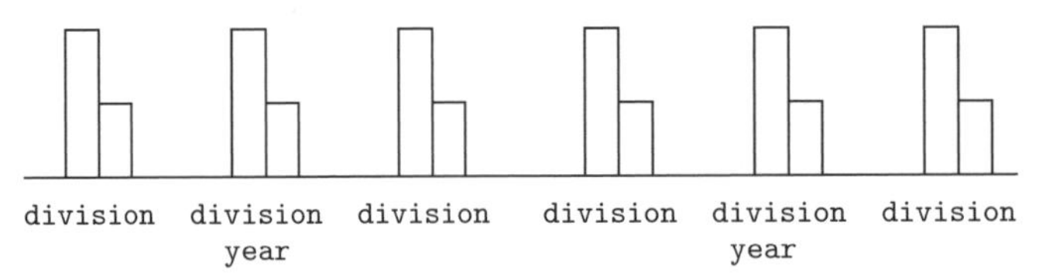

`graph bar (sum) revenue profit, over(division) over(year)`
2 × *#_of_divisions* × *#_of_years* bars showing the sum of revenue and sum of profit for each division, repeated for each of the years.

`graph bar (median) revenue profit, over(division) over(year)`
2 × *#_of_divisions* × *#_of_years* bars showing the median of revenue and median of profit for each division, repeated for each of the years.

`graph bar (median) revenue (mean) profit, over(division) over(year)`
2 × *#_of_divisions* × *#_of_years* bars showing the median of revenue and mean of profit for each division, repeated for each of the years.

Treatment of bars

Assume that someone tells you that the average January temperature in the Northeast of the United States is 27.9 degrees Fahrenheit, 27.1 degrees in the North Central, 46.1 in the South, and 46.2 in the West. You could enter these statistics and draw a bar chart:

```
. input ne nc south west
           ne          nc       south        west
  1. 27.9 21.7 46.1 46.2
  2. end
```

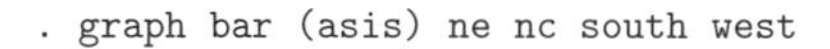

```
. graph bar (asis) ne nc south west
```

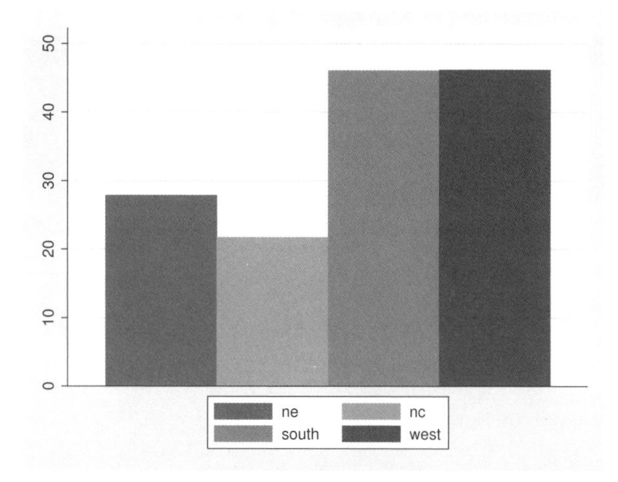

The above is admittedly not a great-looking chart, but specifying a few options could fix that. The important thing to see right now is that, when we specify multiple *yvars*, (1) the bars touch, (2) the bars are different colors (or at least different shades of gray), and (3) the meaning of the bars is revealed in the legend.

We could enter these data another way:

```
. clear
. input  str10 region  float tempjan
          region   tempjan
  1. N.E. 27.9
  2. "N. Central" 21.7
  3. South 46.1
  4. West 46.2
  5. end
. graph bar (asis) tempjan, over(region)
```

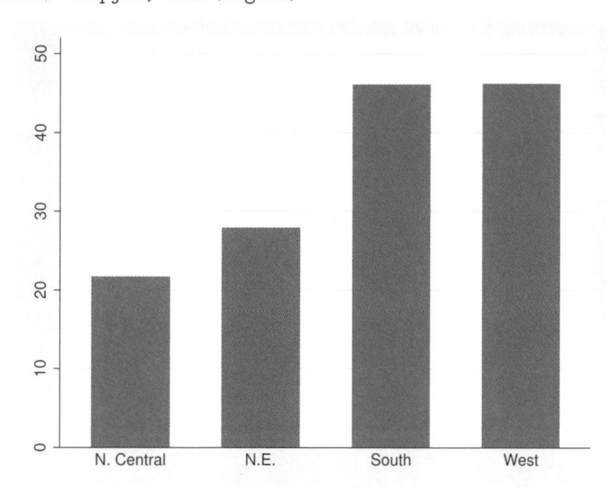

Observe that, when we generate multiple bars via an `over()` option, (1) the bars do not touch, (2) the bars are all the same color, and (3) the meaning of the bars is revealed by how the categorical x axis is labeled.

These differences in the treatment of the bars in the multiple *yvars* case and the `over()` case are general properties of `graph bar` and `graph hbar`:

	multiple *yvars*	`over()` groups
bars touch	yes	no
bars different colors	yes	no
bars identified via ...	legend	axis label

Option `ascategory` causes multiple *yvars* to be presented as if they were `over()` groups, and option `asyvars` causes `over()` groups to be presented as if they were *yvars*. Thus

```
. graph bar (asis) tempjan, over(region)
```

would produce the first chart and

```
. graph bar (asis) ne nc south west, ascategory
```

would produce the second.

Treatment of data

In the previous two examples, we already had the statistics we wanted to plot: 27.9 (Northeast), 21.7 (North Central), 46.1 (South), and 46.2 (West). We entered the data, and we typed

```
. graph bar (asis) ne nc south west
```

or

```
. graph bar (asis) tempjan, over(region)
```

We do not have to know the statistics ahead of time: `graph bar` and `graph hbar` can calculate statistics for us. If we had datasets with lots of observations (say, cities of the United States), we could type

```
. graph bar (mean) ne nc south west
```

or

```
. graph bar (mean) tempjan, over(region)
```

and obtain the same graphs. All we need do is change `(asis)` to `(mean)`. In the first example, the data would be organized the wide way:

cityname	ne	nc	south	west
name of city	42	.	.	.
another city	.	28	.	.
...				

and in the second example, the data would be organized the long way:

cityname	region	tempjan
name of city	ne	42
another city	nc	28
...		

We have such a dataset, organized the long way. In `citytemp.dta`, we have information on 956 U.S. cities, including the region in which each is located and its average January temperature:

```
. use http://www.stata-press.com/data/r12/citytemp, clear
(City Temperature Data)
. list region tempjan if _n < 3 | _n > 954
```

	region	tempjan
1.	NE	16.6
2.	NE	18.2
955.	West	72.6
956.	West	72.6

With these data, we can type

```
. graph bar (mean) tempjan, over(region)
```

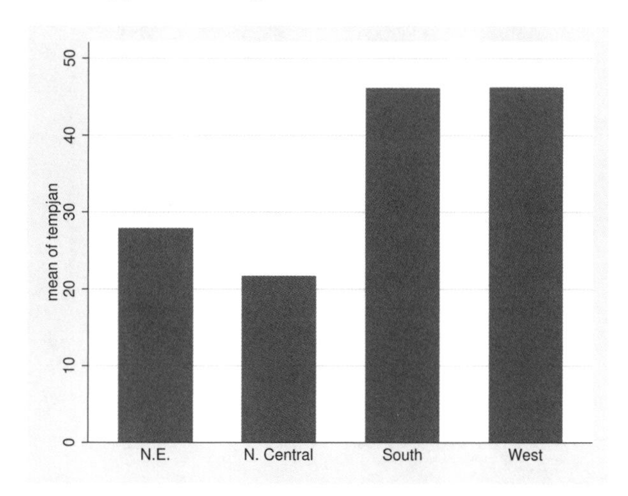

We just produced the same bar chart we previously produced when we entered the statistics 27.9 (Northeast), 21.7 (North Central), 46.1 (South), and 46.2 (West) and typed

```
. graph bar (asis) tempjan, over(region)
```

When we do not specify `(asis)` or `(mean)` (or `(median)` or `(sum)` or `(p1)` or any of the other *stats* allowed), `(mean)` is assumed. Thus `(...)` is often omitted when `(mean)` is desired, and we could have drawn the previous graph by typing

```
. graph bar tempjan, over(region)
```

Some users even omit typing `(...)` in the `(asis)` case because calculating the mean of one observation results in the number itself. Thus in the previous section, rather than typing

```
. graph bar (asis) ne nc south west
```

and

```
. graph bar (asis) tempjan, over(region)
```

We could have typed

```
. graph bar ne nc south west
```

and

```
. graph bar tempjan, over(region)
```

Multiple bars (overlapping the bars)

In `citytemp.dta`, in addition to variable `tempjan`, there is variable `tempjuly`, which is the average July temperature. We can include both averages in one chart, by `region`:

```
. use http://www.stata-press.com/data/r12/citytemp, clear
(City Temperature Data)
. graph bar (mean) tempjuly tempjan, over(region)
```

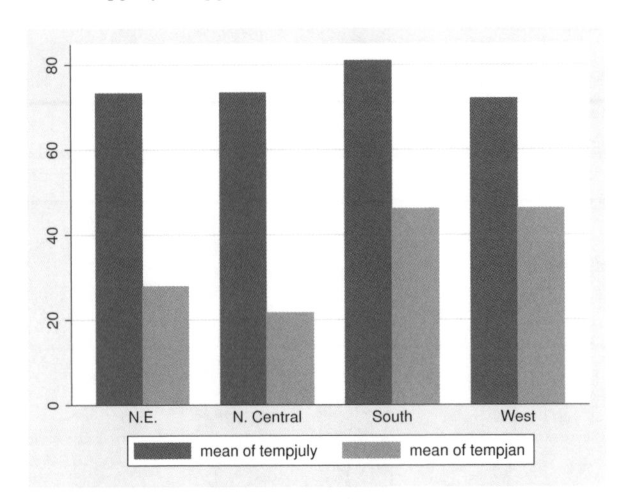

We can improve the look of the chart by

1. including the *legend_options* `legend(label())` to change the text of the legend; see [G-3] ***legend_options***;
2. including the *axis_title_option* `ytitle()` to add a title saying "Degrees Fahrenheit"; see [G-3] ***axis_title_options***;
3. including the *title_options* `title()`, `subtitle()`, and `note()` to say what the graph is about and from where the data came; see [G-3] ***title_options***.

Doing all that produces

```
. graph bar (mean) tempjuly tempjan, over(region)
        legend( label(1 "July") label(2 "January") )
        ytitle("Degrees Fahrenheit")
        title("Average July and January temperatures")
        subtitle("by regions of the United States")
        note("Source:  U.S. Census Bureau, U.S. Dept. of Commerce")
```

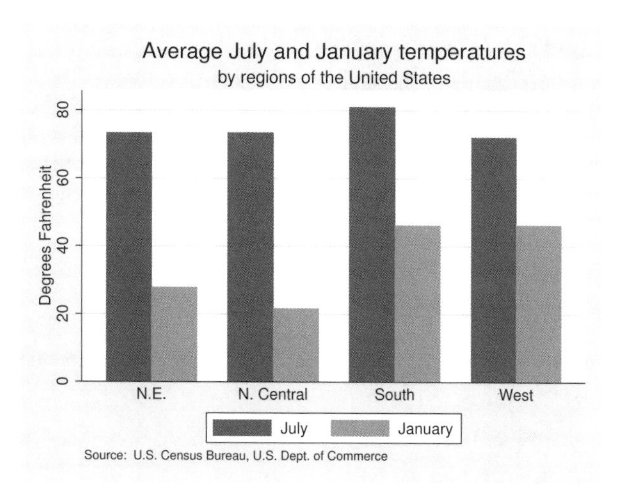

We can make one more improvement to this chart by overlapping the bars. Below we add the option `bargap(-30)`:

```
. graph bar (mean) tempjuly tempjan, over(region)
        bargap(-30)                                             ← new
        legend( label(1 "July") label(2 "January") )
        ytitle("Degrees Fahrenheit")
        title("Average July and January temperatures")
        subtitle("by regions of the United States")
        note("Source:  U.S. Census Bureau, U.S. Dept. of Commerce")
```

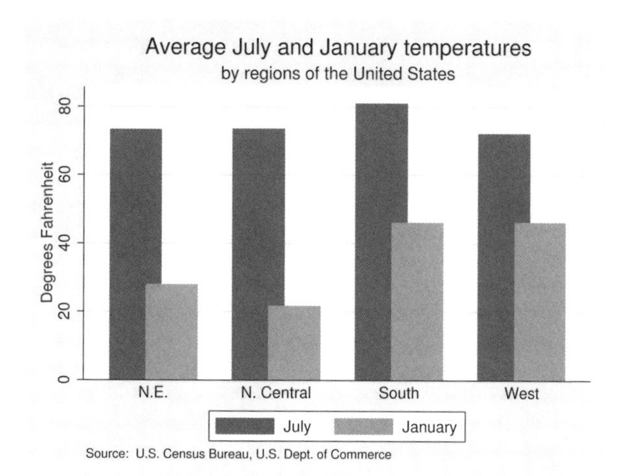

`bargap(#)` specifies the distance between the *yvar* bars (that is, between the bars for `tempjuly` and `tempjan`); # is in percentage-of-bar-width units, so `barwidth(-30)` means that the bars overlap by 30%. `bargap()` may be positive or negative; its default is 0.

Controlling the text of the legend

In the above example, we changed the text of the legend by specifying the legend option:

```
legend( label(1 "July") label(2 "January") )
```

We could just as well have changed the text of the legend by typing

```
yvaroptions( relabel(1 "July" 2 "January") )
```

Which you use makes no difference, but we prefer `legend(label())` to `yvaroptions(relabel())` because `legend(label())` is the way to modify the contents of a legend in a twoway graph; so why do bar charts differently?

Multiple over( )s (repeating the bars)

Option `over(`*varname*`)` repeats the *yvar* bars for each unique value of *varname*. Using `citytemp.dta`, if we typed

```
. graph bar (mean) tempjuly tempjan
```

we would obtain two (fat) bars. When we type

```
. graph bar (mean) tempjuly tempjan, over(region)
```

we obtain two (thinner) bars for each of the four regions. (We typed exactly this command in *Multiple bars* above.)

You may repeat the `over()` option. You may specify `over()` twice when you specify two or more *yvars* and up to three times when you specify just one *yvar*.

In dataset nlsw88.dta, we have information on 2,246 women:

```
. use http://www.stata-press.com/data/r12/nlsw88, clear
(NLSW, 1988 extract)
. graph bar (mean) wage, over(smsa) over(married) over(collgrad)
        title("Average Hourly Wage, 1988, Women Aged 34-46")
        subtitle("by College Graduation, Marital Status,
                  and SMSA residence")
        note("Source:  1988 data from NLS, U.S. Dept. of Labor,
              Bureau of Labor Statistics")
```

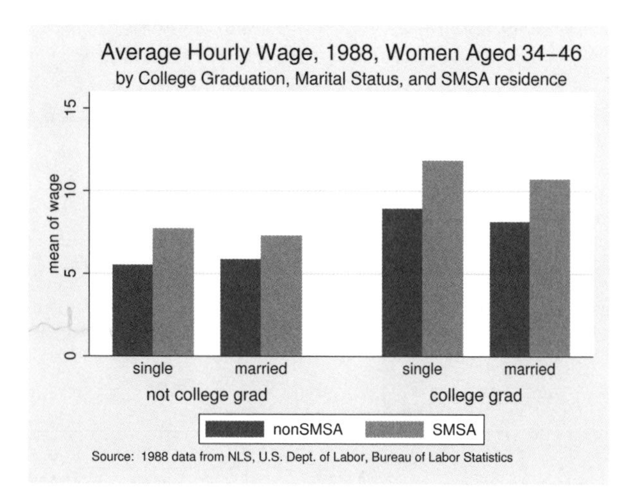

If you strip away the *title_options*, the above command reads

```
. graph bar (mean) wage, over(smsa) over(married) over(collgrad)
```

In this three-over() case, the first over() is treated as multiple *yvars*: the bars touch, the bars are assigned different colors, and the meaning of the bars is revealed in the legend. When you specify three over() groups, the first is treated the same way as multiple *yvars*. This means that if we wanted to separate the bars, we could specify option bargap(*#*), *#*>0, and if we wanted them to overlap, we could specify bargap(*#*), *#*<0.

Nested over()s

Sometimes you have multiple over() groups with one group explicitly nested within the other. In the citytemp.dta dataset, we have variables region and division, and division is nested within region. The Census Bureau divides the United States into four regions and into nine divisions, which work like this

Region	*Division*
1. Northeast	1. New England 2. Mid Atlantic
2. North Central	3. East North Central 4. West North Central
3. South	5. South Atlantic 6. East South Central 7. West South Central
4. West	8. Mountain 9. Pacific

Were we to type

```
. graph bar (mean) tempjuly tempjan, over(division) over(region)
```

we would obtain a chart with space allocated for 9*4 = 36 groups, of which only nine would be used:

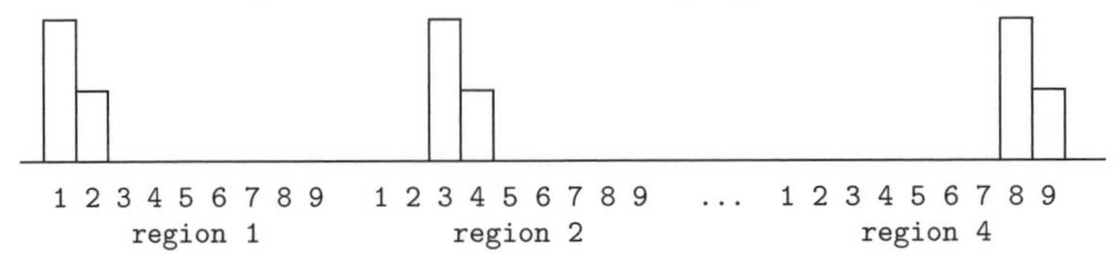

The `nofill` option prevents the chart from including the unused categories:

```
. use http://www.stata-press.com/data/r12/citytemp, clear
(City Temperature Data)
. graph bar tempjuly tempjan, over(division) over(region) nofill
        bargap(-30)
        ytitle("Degrees Fahrenheit")
        legend( label(1 "July") label(2 "January") )
        title("Average July and January temperatures")
        subtitle("by region and division of the United States")
        note("Source:  U.S. Census Bureau, U.S. Dept. of Commerce")
```

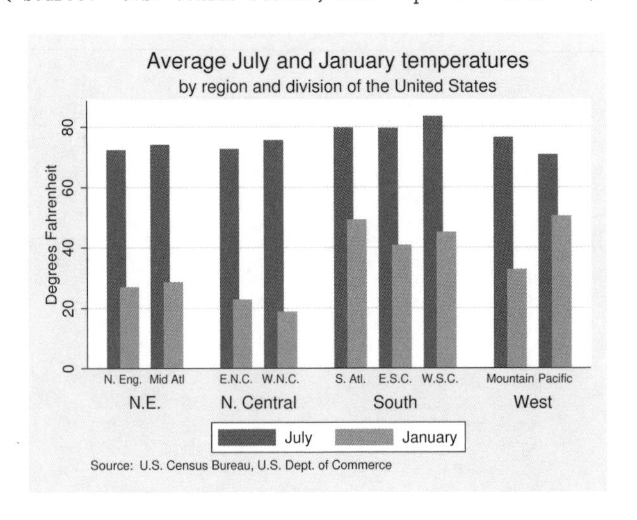

The above chart, if we omit one of the temperatures, also looks good horizontally:

```
. graph hbar (mean) tempjan, over(division) over(region) nofill
        ytitle("Degrees Fahrenheit")
        title("Average January temperature")
        subtitle("by region and division of the United States")
        note("Source:  U.S. Census Bureau, U.S. Dept. of Commerce")
```

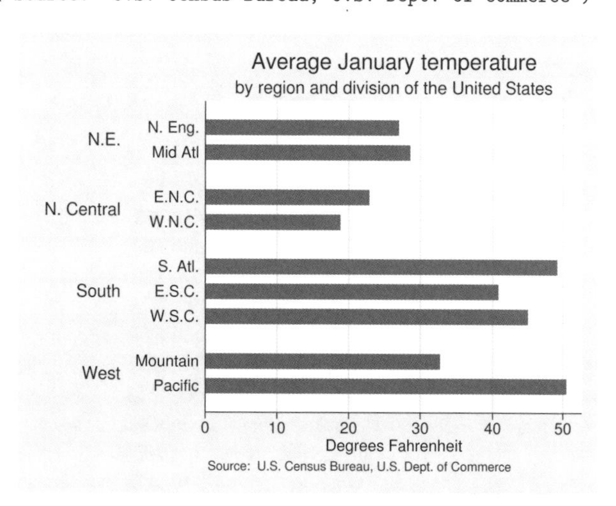

Charts with many categories

Using `nlsw88.dta`, we want to draw the chart

```
. use http://www.stata-press.com/data/r12/nlsw88
(NLSW, 1988 extract)
. graph bar wage, over(industry) over(collgrad)
```

Variable `industry` records industry of employment in 12 categories, and variable `collgrad` records whether the woman is a college graduate. Thus we will have 24 bars. We draw the above and quickly discover that the long labels associated with industry result in much overprinting along the horizontal x axis.

Horizontal bar charts work better than vertical bar charts when labels are long. We change our command to read

```
. graph hbar wage, over(ind) over(collgrad)
```

That works better, but now we have overprinting problems of a different sort: the letters of one line are touching the letters of the next.

Graphs are by default 4×5: 4 inches tall by 5 inches wide. Here we need to make the chart taller, and that is the job of the `region_option ysize()`. Below we make a chart that is 7 inches tall:

```
. use http://www.stata-press.com/data/r12/nlsw88, clear
(NLSW, 1988 extract

. graph hbar wage, over(ind, sort(1)) over(collgrad)
        title("Average hourly wage, 1988, women aged 34-46", span)
        subtitle(" ")
        note("Source:  1988 data from NLS, U.S. Dept. of Labor,
              Bureau of Labor Statistics", span)
        ysize(7)
```

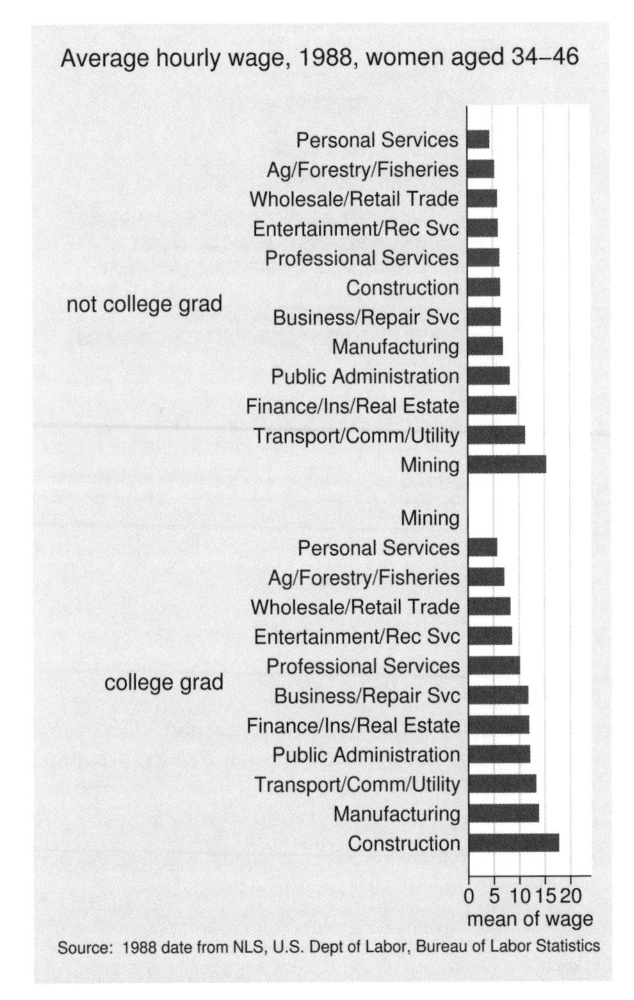

The important option in the above is `ysize(7)`, which made the graph taller than usual; see [G-3] ***region_options***. Concerning the other options:

`over(ind, sort(1)) over(collgrad)`

`sort(1)` is specified so that the bars would be sorted on mean wage. The `1` says to sort on the first *yvar*; see *Reordering the bars* below.

`title("Average hourly wage, 1988, women aged 34-46", span)`

`span` is specified so that the title, rather than being centered over the plot region, would be centered over the entire graph. Here the plot region (the part of the graph where the real chart appears, ignoring the labels) is narrow, and centering over that was not going to work. See [G-3] ***region_options*** for a description of the graph region and plot region, and see [G-3] ***title_options*** and [G-3] ***textbox_options*** for a description of `span`.

`subtitle(" ")`
We specified this because the title looked too close to the graph without it. We could have done things properly and specified a `margin()` suboption within the `title()`, but we often find it easier to include a blank subtitle. We typed `subtitle(" ")` and not `subtitle("")`. We had to include the blank, or the subtitle would not have appeared.

`note("Source: 1988 data from NLS, ...", span)`
`span` is specified so that the note would be left-justified in the graph rather than just in the plot region.

How bars are ordered

The default is to place the bars in the order of the *yvars* and to order each set of `over(`*varname*`)` groups according to the values of *varname*. Let us consider some examples:

`graph bar (sum) revenue profit`
Bars appear in the order specified: `revenue` and `profit`.

`graph bar (sum) revenue, over(division)`
Bars are ordered according to the values of variable `division`.

If `division` is a numeric variable, the lowest division number comes first, followed by the next lowest, and so on. This is true even if variable `division` has a value label. Say that division 1 has been labeled "Sales" and division 2 is labeled "Development". The bars will be in the order Sales followed by Development.

If `division` is a string variable, the bars will be ordered by the sort order of the values of `division` (meaning alphabetically, but with capital letters placed before lowercase letters). If variable `division` contains the values "Sales" and "Development", the bars will be in the order Development followed by Sales.

`graph bar (sum) revenue profit, over(division)`
Bars appear in the order specified, `revenue` and `profit`, and are repeated for each division, which will be ordered as explained above.

`graph bar (sum) revenue, over(division) over(year)`
Bars appear ordered by the values of division, as previously explained, and then that is repeated for each of the years. The years are ordered according to the values of the variable `year`, following the same rules as applied to the variable `division`.

`graph bar (sum) revenue profit, over(division) over(year)`
Bars appear in the order specified, profit and revenue, repeated for division ordered on the values of variable `division`, repeated for year ordered on the values of variable `year`.

Reordering the bars

There are three ways to reorder the bars:

1. You want to control the order in which the elements of each `over()` group appear. Your divisions might be named Development, Marketing, Research, and Sales, alphabetically speaking, but you want them to appear in the more logical order Research, Development, Marketing, and Sales.

2. You wish to order the bars according to their heights. You wish to draw the graph

```
. graph bar (sum) empcost, over(division)
```

   and you want the divisions ordered by total employee cost.

3. You wish to order on some other derived value.

We will consider each of these desires separately.

Putting the bars in a prespecified order

We have drawn the graph

```
. graph (sum) bar empcost, over(division)
```

Variable `division` is a string containing "Development", "Marketing", "Research", and "Sales". We want to draw the chart, placing the divisions in the order Research, Development, Marketing, and Sales.

To do that, we create a new numeric variable that orders `division` as we would like:

```
. generate order = 1 if division=="Research"
. replace  order = 2 if division=="Development"
. replace  order = 3 if division=="Marketing"
. replace  order = 4 if division=="Sales"
```

We can name the variable and create it however we wish, but we must be sure that there is a one-to-one correspondence between the new variable and the `over()` group's values. We then specify the `over()`'s `sort(`*varname*`)` option:

```
. graph bar (sum) empcost, over( division, sort(order) )
```

If you want to reverse the order, you may specify the `descending` suboption:

```
. graph bar (sum) empcost, over(division, sort(order) descending)
```

Putting the bars in height order

We have drawn the graph

```
. graph bar (sum) empcost, over(division)
```

and now wish to put the bars in height order, shortest first. We type

```
. graph bar (sum) empcost, over( division, sort(1) )
```

If we wanted the tallest first, type

```
. graph bar empcost, over(division, sort(1) descending)
```

The `1` in `sort(1)` refers to the first (and here only) *yvar*. If we had multiple *yvars*, we might type

```
. graph bar (sum) empcost othcost, over( division, sort(1) )
```

and we would have a chart showing employee cost and other cost, sorted on employee cost. If we typed

```
. graph bar (sum) empcost othcost, over( division, sort(2) )
```

the graph would be sorted on other cost.

We can use `sort(#)` on the second `over()` group as well:

```
. graph bar (sum) empcost, over( division, sort(1) )
                           over( country,  sort(1) )
```

Country will be ordered on the sum of the heights of the bars.

Putting the bars in a derived order

We have employee cost broken into two categories: `empcost_direct` and `empcost_indirect`. Variable `emp_cost` is the sum of the two. We wish to make a chart showing the two costs, stacked, over `division`, and we want the bars ordered on the total height of the stacked bars. We type

```
. graph bar (sum) empcost_direct empcost_indirect,
                  stack
                  over(division, sort((sum) empcost) descending)
```

Reordering the bars, example

We have a dataset showing the spending on tertiary education as a percentage of GDP from the 2002 edition of *Education at a Glance: OECD Indicators 2002*:

```
. use http://www.stata-press.com/data/r12/educ99gdp, clear
(Education and GDP)

. list
```

	country	public	private
1.	Australia	.7	.7
2.	Britain	.7	.4
3.	Canada	1.5	.9
4.	Denmark	1.5	.1
5.	France	.9	.4
6.	Germany	.9	.2
7.	Ireland	1.1	.3
8.	Netherlands	1	.4
9.	Sweden	1.5	.2
10.	United States	1.1	1.2

We wish to graph total spending on education and simultaneously show the distribution of that total between public and private expenditures. We want the bar sorted on total expenditures:

```
. generate total = private + public
. graph hbar (asis) public private,
        over(country, sort(total) descending) stack
        title( "Spending on tertiary education as % of GDP, 1999",
               span pos(11) )
        subtitle(" ")
        note("Source:  OECD, Education at a Glance 2002", span)
```

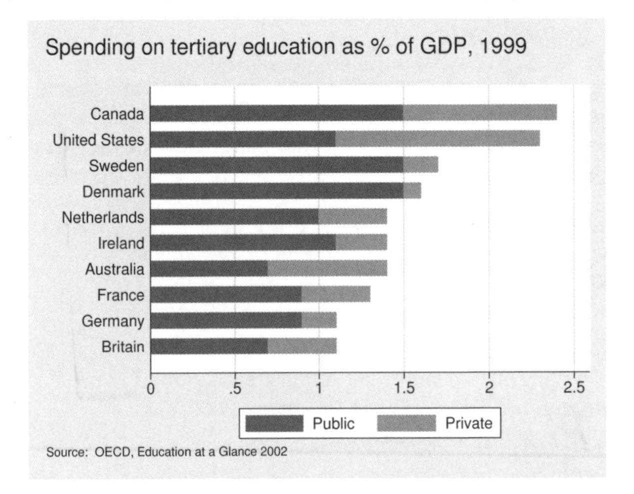

Or perhaps we wish to disguise the total expenditures and focus the graph exclusively on the share of spending that is public and private:

```
. generate frac = private/(private + public)
. graph hbar (asis) public private,
        over(country, sort(frac) descending) stack percent
        title("Public and private spending on tertiary education, 1999",
              span pos(11) )
        subtitle(" ")
        note("Source:  OECD, Education at a Glance 2002", span)
```

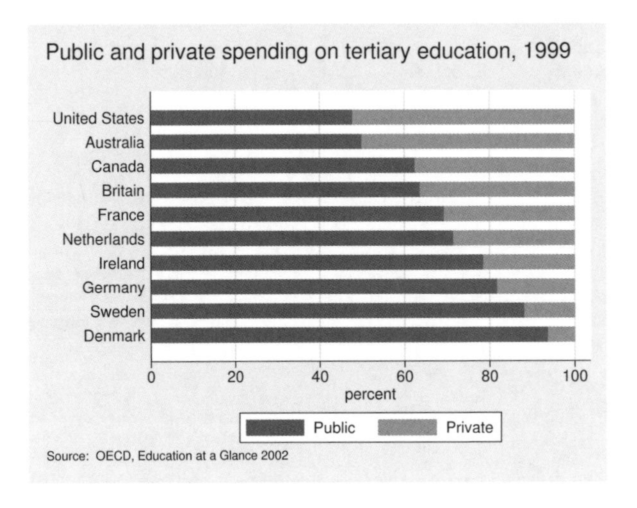

The only differences between the two `graph hbar` commands are as follows:

1. The `percentage` option was added to change the *yvars* `public` and `private` from spending amounts to percentages each is of the total.
2. The order of the bars was changed.
3. The title was changed.

Use with by()

`graph bar` and `graph hbar` may be used with `by()`, but in general, you want to use `over()` in preference to `by()`. Bar charts are explicitly categorical and do an excellent job of presenting summary statistics for multiple groups in one chart.

A good use of `by()`, however, is when you are ordering the bars and you wish to emphasize that the ordering is different for different groups. For instance,

```
. use http://www.stata-press.com/data/r12/nlsw88, clear
(NLSW, 1988 extract)
. graph hbar wage, over(occ, sort(1)) by(union)
```

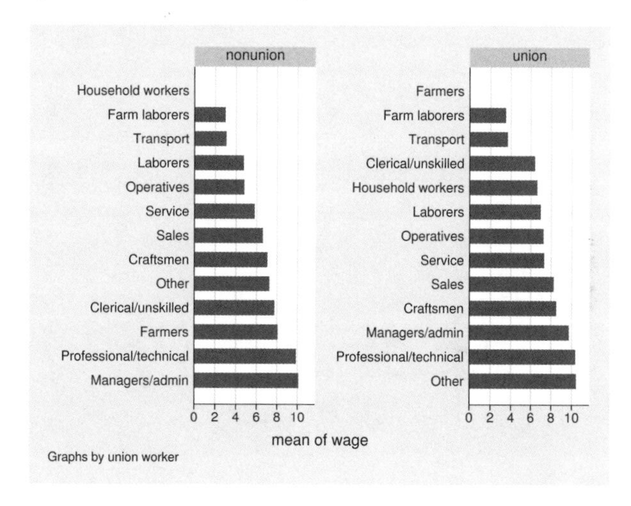

The above graph orders the bars by height (hourly wage); the orderings are different for union and nonunion workers.

History

The first published bar chart appeared in William Playfair's *Commercial and Political Atlas* (1786). See Tufte (2001, 32–33) or Beniger and Robyn (1978) for more historical information.

References

Beniger, J. R., and D. L. Robyn. 1978. Quantitative graphics in statistics: A brief history. *American Statistician* 32: 1–11.

Cox, N. J. 2004. Speaking Stata: Graphing categorical and compositional data. *Stata Journal* 4: 190–215.

——. 2005. Stata tip 24: Axis labels on two or more levels. *Stata Journal* 5: 469.

——. 2008. Speaking Stata: Spineplots and their kin. *Stata Journal* 8: 105–121.

Playfair, W. 1786. *Commercial and Political Atlas: Representing, by means of stained Copper-Plate Charts, the Progress of the Commerce, Revenues, Expenditure, and Debts of England, during the Whole of the Eighteenth Century.* London: Corry.

Tufte, E. R. 2001. *The Visual Display of Quantitative Information.* 2nd ed. Cheshire, CT: Graphics Press.

Also see

[G-2] **graph dot** — Dot charts (summary statistics)

[D] **collapse** — Make dataset of summary statistics

[R] **table** — Tables of summary statistics

Title

[G-2] graph box — Box plots

Syntax

graph box *yvars* [*if*] [*in*] [*weight*] [, *options*]

graph hbox *yvars* [*if*] [*in*] [*weight*] [, *options*]

where *yvars* is a *varlist*

options	Description
group_options	groups over which boxes are drawn
yvar_options	variables that are the boxes
boxlook_options	how the boxes look
legending_options	how variables are labeled
axis_options	how numerical y axis is labeled
title_and_other_options	titles, added text, aspect ratio, etc.

Each is defined below.

group_options	Description
over(*varname* [, *over_subopts*])	categories; option may be repeated
nofill	omit empty categories
missing	keep missing value as category
allcategories	include all categories in the dataset

yvar_options	Description
ascategory	treat *yvars* as first over() group
asyvars	treat first over() group as *yvars*
cw	calculate variable statistics omitting missing values of any variable

boxlook_options	Description
`nooutsides`	do not plot outside values
`box(`*#*`,` *barlook_options*`)`	look of #th box
`pcycle(`*#*`)`	box styles before pstyles recycle
`intensity(`[`*`]*#*`)`	intensity of fill
`lintensity(`[`*`]*#*`)`	intensity of outline
`medtype(line`|`cline`|`marker)`	how median is indicated in box
`medline(`*line_options*`)`	look of line if `medtype(cline)`
`medmarker(`*marker_options*`)`	look of marker if `medtype(marker)`
`cwhiskers`	use custom whiskers
`lines(`*line_options*`)`	look of custom whiskers
`alsize(`*#*`)`	width of adjacent line; default is 67
`capsize(`*#*`)`	height of cap on adjacent line; default is 0
`marker(`*#*, *marker_options* *marker_label_options*`)`	look of #th marker and label for outside values
`outergap(`[`*`]*#*`)`	gap between edge and first box and between last box and edge
`boxgap(`*#*`)`	gap between boxes; default is 33

See [G-3] ***barlook_options***, [G-3] ***line_options***, [G-3] ***marker_options***, and [G-3] ***marker_label_options***.

legending_options	Description
legend_options	control of *yvar* legend
`nolabel`	use *yvar* names, not labels, in legend
`yvaroptions(`*over_subopts*`)`	*over_subopts* for *yvars*; seldom specified
`showyvars`	label *yvars* on x axis; seldom specified

See [G-3] ***legend_options***.

axis_options	Description
`yalternate`	put numerical y axis on right (top)
`xalternate`	put categorical x axis on top (right)
`yreverse`	reverse y axis
axis_scale_options	y-axis scaling and look
axis_label_options	y-axis labeling
`ytitle(...)`	y-axis titling

See [G-3] ***axis_scale_options***, [G-3] ***axis_label_options***, and [G-3] ***axis_title_options***.

title_and_other_options	Description
text(...)	add text on graph; x range [0, 100]
yline(...)	add y lines to graph
aspect_option	constrain aspect ratio of plot region
std_options	titles, graph size, saving to disk
by(*varlist*, ...)	repeat for subgroups

See [G-3] ***added_text_options***, [G-3] ***added_line_options***, [G-3] ***aspect_option***, [G-3] ***std_options***, and [G-3] ***by_option***.

The *over_subopts*—used in over(*varname*, *over_subopts*) and, on rare occasion, in yvaroptions(*over_subopts*)—are

over_subopts	Description
total	add total group
relabel(# "*text*" ...)	change axis labels
label(*cat_axis_label_options*)	rendition of labels
axis(*cat_axis_line_options*)	rendition of axis line
gap([*]#)	gap between boxes within over() category
sort(*varname*)	put boxes in prespecified order
sort(#)	put boxes in median order
descending	reverse default or specified box order

See [G-3] ***cat_axis_label_options*** and [G-3] ***cat_axis_line_options***.

aweights, fweights, and pweights are allowed; see [U] **11.1.6 weight** and see note concerning weights in [D] **collapse**.

Menu

Graphics > Box plot

Description

graph box draws vertical box plots. In a vertical box plot, the y axis is numerical, and the x axis is categorical.

```
. graph box y1 y2, over(cat_var)
```

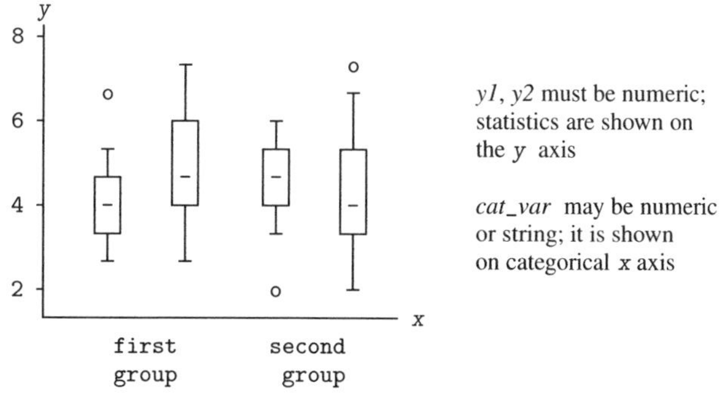

y
8
6
4
2
x
first group
second group

y1, *y2* must be numeric; statistics are shown on the *y* axis

cat_var may be numeric or string; it is shown on categorical *x* axis

The encoding and the words used to describe the encoding are

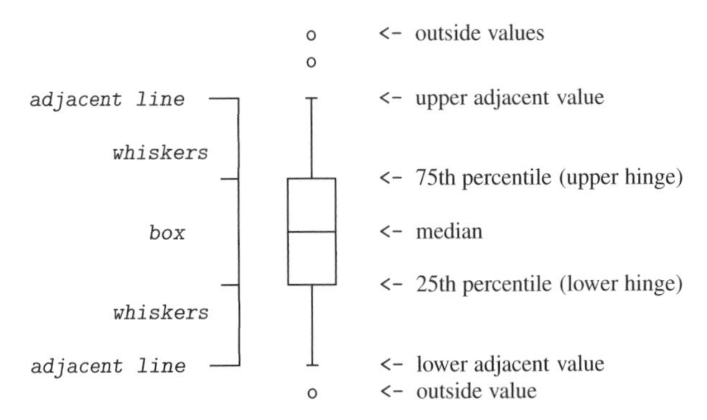

`graph hbox` draws horizontal box plots. In a horizontal box plot, the numerical axis is still called the y axis, and the categorical axis is still called the x axis, but y is presented horizontally, and x vertically.

. `graph hbox` *y1 y2*`, over(`*cat_var*`)`

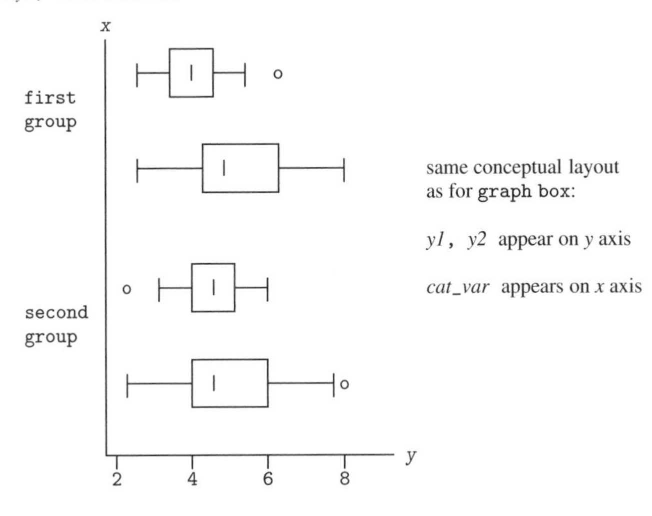

group_options

`over(`*varname*[`,` *over_subopts*]`)` specifies a categorical variable over which the *yvars* are to be repeated. *varname* may be string or numeric. Up to two `over()` options may be specified when multiple *yvars* are specified, and up to three `over()`s may be specified when one *yvar* is specified; see *Examples of syntax* under *Remarks* below.

`nofill` specifies that missing subcategories be omitted. See the description of the `nofill` option in [G-2] **graph bar**.

`missing` specifies that missing values of the `over()` variables be kept as their own categories, one for `.`, another for `.a`, etc. The default is to ignore such observations. An `over()` variable is considered to be missing if it is numeric and contains a missing value or if it is string and contains "".

`allcategories` specifies that all categories in the entire dataset be retained for the `over()` variables. When `if` or `in` is specified without `allcategories`, the graph is drawn, completely excluding any categories for the `over()` variables that do not occur in the specified subsample. With the `allcategories` option, categories that do not occur in the subsample still appear in the legend, and zero-height bars are drawn where these categories would appear. Such behavior can be convenient when comparing graphs of subsamples that do not include completely common categories for all `over()` variables. This option has an effect only when `if` or `in` is specified or if there are missing values in the variables. `allcategories` may not be combined with `by()`.

yvar_options

`ascategory` specifies that the *yvars* be treated as the first `over()` group. The important effect of this is to move the captioning of the variables from the legend to the categorical x axis. See the description of `ascategory` in [G-2] **graph bar**.

`asyvars` specifies that the first `over()` group be treated as *yvars*. The important effect of this is to move the captioning of the first over group from the categorical x axis to the legend. See the description of `asyvars` in [G-2] **graph bar**.

`cw` specifies casewise deletion. If `cw` is specified, observations for which any of the *yvars* are missing are ignored. The default is to calculate statistics for each box by using all the data possible.

boxlook_options

`nooutsides` specifies that the outside values not be plotted or used in setting the scale of the y axis.

`box(`*#*`,` *barlook_options*`)` specifies the look of the *yvar* boxes. `box(1, ...)` refers to the box associated with the first *yvar*, `box(2, ...)` refers to the box associated with the second, and so on.

You specify *barlook_options*. Those options are borrowed from `graph bar` for boxes. The most useful *barlook_option* is `color(`*colorstyle*`)`, which sets the color of the box. For instance, you might specify `box(1, color(green))` to make the box associated with the first *yvar* green. See [G-4] ***colorstyle*** for a list of color choices and see [G-3] ***barlook_options*** for information on the other *barlook_options*.

`pcycle(`*#*`)` specifies how many variables are to be plotted before the pstyle (see [G-4] ***pstyle***) of the boxes for the next variable begins again at the pstyle of the first variable—`p1box` (with the boxes for the variable following that using `p2box` and so on). Put another way: # specifies how quickly the look of boxes is recycled when more than # variables are specified. The default for most schemes is `pcycle(15)`.

`intensity(`*#*`)` and `intensity(*`*#*`)` specify the intensity of the color used to fill the inside of the box. `intensity(`*#*`)` specifies the intensity, and `intensity(*`*#*`)` specifies the intensity relative to the default.

By default, the box is filled with the color of its border, attenuated. Specify `intensity(*`*#*`)`, $\# < 1$, to attenuate it more and specify `intensity(*`*#*`)`, $\# > 1$, to amplify it.

Specify `intensity(0)` if you do not want the box filled at all. If you are using a scheme that draws the median line in the background color such as `s2mono`, also specify option `medtype(line)` to change the median line to be in the color of the outline of the box.

`lintensity(`*#*`)` and `lintensity(*`*#*`)` specify the intensity of the line used to outline the box. `lintensity(`*#*`)` specifies the intensity, and `lintensity(*`*#*`)` specifies the intensity relative to the default.

By default, the box is outlined at the same intensity at which it is filled or at an amplification of that, which depending on your chosen scheme; see [G-4] **schemes intro**. If you want the box outlined in the darkest possible way, specify `intensity(255)`. If you wish simply to amplify the outline, specify `intensity(*#)`, $\# > 1$, and if you wish to attenuate the outline, specify `intensity(*#)`, $\# < 1$.

`medtype()`, `medline()`, and `medmarker()` specify how the median is to be indicated in the box.

`medtype(line)` is the default. A line is drawn across the box at the median. Here options `medline()` and `medmarker()` are irrelevant.

`medtype(cline)` specifies a custom line be drawn across the box at the median. The default custom line is usually a different color. You can, however, specify option `medline(`*line_options*`)` to control exactly how the line is to look; see [G-3] ***line_options***.

`medtype(marker)` specifies a marker be placed in the box at the median. Here you may also specify option `medmarker(`*marker_options*`)` to specify the look of the marker; see [G-3] ***marker_options***.

`cwhiskers`, `lines(`*line_options*`)`, `alsize(`*#*`)`, and `capsize(`*#*`)` specify the look of the whiskers.

`cwhiskers` specifies that custom whiskers are desired. The default custom whiskers are usually dimmer, but you may specify option `lines(`*line_options*`)` to specify how the custom whiskers are to look; see [G-3] ***line_options***.

`alsize(`*#*`)` and `capsize(`*#*`)` specify the width of the adjacent line and the height of the cap on the adjacent line. You may specify these options whether or not you specify `cwhiskers`. `alsize()` and `capsize()` are specified in percentage-of-box-width units; the defaults are `alsize(67)` and `capsize(0)`. Thus the adjacent lines extend two-thirds the width of a box and, by default, have no caps. Caps refer to whether the whiskers look like

If you want caps, try `capsize(5)`.

`marker(`*#*`,` *marker_options marker_label_options*`)` specifies the marker and label to be used to display the outside values. See [G-3] ***marker_options*** and [G-3] ***marker_label_options***.

`outergap(*`*#*`)` and `outergap(`*#*`)` specify the gap between the edge of the graph to the beginning of the first box and the end of the last box to the edge of the graph.

`outergap(*`*#*`)` specifies that the default be modified. Specifying `outergap(*1.2)` increases the gap by 20%, and specifying `outergap(*.8)` reduces the gap by 20%.

`outergap(`*#*`)` specifies the gap as a percentage-of-box-width units. `outergap(50)` specifies that the gap be half the box width.

`boxgap(`*#*`)` specifies the gap to be left between *yvar* boxes as a percentage-of-box-width units. The default is `boxgap(33)`.

`boxgap()` affects only the *yvar* boxes. If you want to change the gap for the first, second, or third `over()` group, specify the *over_subopt* `gap()` inside the `over()` itself; see *Suboptions for use with over() and yvaroptions()* below.

legending_options

legend_options allows you to control the legend. If more than one *yvar* is specified, a legend is produced. Otherwise, no legend is needed because the `over()` groups are labeled on the categorical x axis. See [G-3] ***legend_options***, and see *Treatment of multiple yvars versus treatment of over() groups* under *Remarks* below.

`nolabel` specifies that, in automatically constructing the legend, the variable names of the *yvars* be used in preference to their labels.

`yvaroptions(`*over_subopts*`)` allows you to specify *over_subopts* for the *yvars*. This is seldom done.

`showyvars` specifies that, in addition to building a legend, the identities of the *yvars* be shown on the categorical x axis. If `showyvars` is specified, it is typical to also specify `legend(off)`.

axis_options

`yalternate` and `xalternate` switch the side on which the axes appear.

Used with `graph box`, `yalternate` moves the numerical y axis from the left to the right; `xalternate` moves the categorical x axis from the bottom to the top.

Used with `graph hbox`, `yalternate` moves the numerical y axis from the bottom to the top; `xalternate` moves the categorical x axis from the left to the right.

If your scheme by default puts the axes on the opposite sides, then `yalternate` and `xalternate` reverse their actions.

`yreverse` specifies that the numerical y axis have its scale reversed so that it runs from maximum to minimum.

axis_scale_options specify how the numerical y axis is scaled and how it looks; see [G-3] ***axis_scale_options***. There you will also see option `xscale()` in addition to `yscale()`. Ignore `xscale()`, which is irrelevant for box plots.

axis_label_options specify how the numerical y axis is to be labeled. The *axis_label_options* also allow you to add and suppress grid lines; see [G-3] ***axis_label_options***. There you will see that, in addition to options `ylabel()`, `ytick()`, ..., `ymtick()`, options `xlabel()`, ..., `xmtick()` are allowed. Ignore the `x*()` options, which are irrelevant for box plots.

`ytitle()` overrides the default title for the numerical y axis; see [G-3] ***axis_title_options***. There you will also find option `xtitle()` documented, which is irrelevant for box plots.

title_and_other_options

`text()` adds text to a specified location on the graph; see [G-3] ***added_text_options***. The basic syntax of `text()` is

`text(`*#_y #_x* `"`*text*`")`

`text()` is documented in terms of twoway graphs. When used with box plots, the "numeric" x axis is scaled to run from 0 to 100.

`yline()` adds horizontal (`box`) or vertical (`hbox`) lines at specified y values; see [G-3] ***added_line_options***. The `xline()` option, also documented there, is irrelevant for box plots. If your interest is in adding grid lines, see [G-3] ***axis_label_options***.

aspect_option allows you to control the relationship between the height and width of a graph's plot region; see [G-3] ***aspect_option***.

std_options allow you to add titles, control the graph size, save the graph on disk, and much more; see [G-3] ***std_options***.

`by(`*varlist*`, ...)` draws separate plots within one graph; see [G-3] ***by_option*** and see *Use with by()* under *Remarks* below.

Suboptions for use with over() and yvaroptions()

`total` specifies that, in addition to the unique values of `over(`*varname*`)`, a group be added reflecting all the observations. When multiple `over()`s are specified, `total` may be specified in only one of them.

`relabel(`*#* `"`*text*`" ...)` specifies text to override the default category labeling. See the description of the `relabel()` option in [G-2] **graph bar** for more information about this useful option.

`label(`*cat_axis_label_options*`)` determines other aspects of the look of the category labels on the x axis. Except for `label(labcolor())` and `label(labsize())`, these options are seldom specified; see [G-3] ***cat_axis_label_options***.

`axis(`*cat_axis_line_options*`)` specifies how the axis line is rendered. This is a seldom specified option. See [G-3] ***cat_axis_line_options***.

`gap(`*#*`)` and `gap(*`*#*`)` specify the gap between the boxes in this `over()` group. `gap(`*#*`)` is specified in percentage-of-box-width units, so `gap(67)` means two-thirds the width of a box. `gap(*`*#*`)` allows modifying the default gap. `gap(*1.2)` would increase the gap by 20% and `gap(*.8)` would decrease the gap by 20%.

To understand the distinction between `over(..., gap())` and option `boxgap()`, consider

```
. graph box before after, boxgap(...) over(sex, gap(...))
```

`boxgap()` sets the distance between the before and after boxes. `over(,gap())` sets the distance between the boxes for males and females. Similarly, in

```
. graph box before after, boxgap(...)
                          over(sex,     gap(...))
                          over(agegrp, gap(...))
```

`over(sex, gap())` sets the gap between males and females, and `over(agegrp, gap())` sets the gap between age groups.

`sort(`*varname*`)` and `sort(`*#*`)` control how the boxes are ordered. See *How boxes are ordered* and *Reordering the boxes* under *Remarks* below.

`sort(`*varname*`)` puts the boxes in the order of *varname*; see *Putting the boxes in a prespecified order* under *Remarks* below.

`sort(`*#*`)` puts the boxes in order of their medians. *#* refers to the *yvar* number on which the ordering should be performed; see *Putting the boxes in median order* under *Remarks* below.

`descending` specifies that the order of the boxes—default or as specified by `sort()`—be reversed.

Remarks

Remarks are presented under the following headings:

Also see [G-2] **graph bar**. Most of what is said there applies equally well to box plots.

Introduction

`graph box` draws vertical box plots:

```
. use http://www.stata-press.com/data/r12/bplong
(fictional blood pressure data)

. graph box bp, over(when) over(sex)
        ytitle("Systolic blood pressure")
        title("Response to Treatment, by Sex")
        subtitle("(120 Preoperative Patients)" " ")
        note("Source:  Fictional Drug Trial, StataCorp, 2003")
```

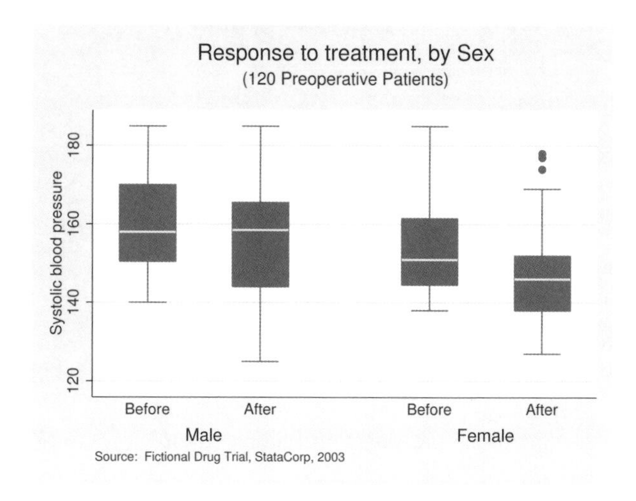

graph hbox draws horizontal box plots:

```
. use http://www.stata-press.com/data/r12/nlsw88, clear
(NLSW, 1988 extract)

. graph hbox wage, over(ind, sort(1)) nooutside
        ytitle("")
        title("Hourly wage, 1988, woman aged 34-46", span)
        subtitle(" ")
        note("Source:  1988 data from NLS, U.S. Dept of Labor,
                       Bureau of Labor Statistics", span)
```

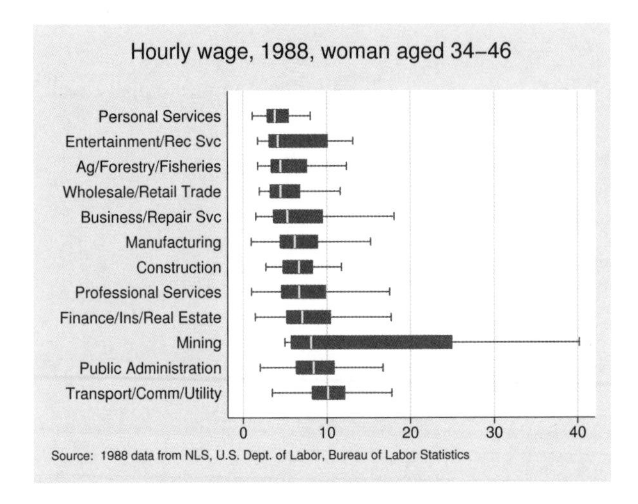

Examples of syntax

Below we show you some graph box commands and tell you what each would do:

graph box bp
: One big box showing statistics on blood pressure.

graph box bp_before bp_after
: Two boxes, one showing average blood pressure before, and the other, after.

graph box bp, over(agegrp)
: *#_of_agegrp* boxes showing blood pressure for each age group.

graph box bp_before bp_after, over(agegrp)
: 2×*#_of_agegrp* boxes showing blood pressure, before and after, for each age group. The grouping would look like this (assuming three age groups):

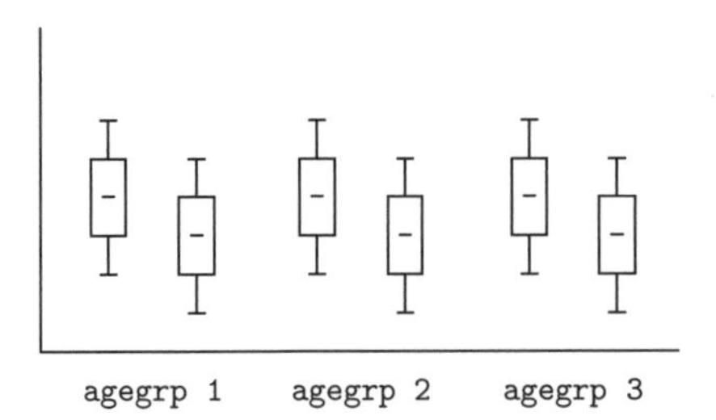

`graph box bp, over(agegrp) over(sex)`
#_of_agegrps × *#_of_sexes* boxes showing blood pressure for each age group, repeated for each sex. The grouping would look like this:

`graph box bp, over(sex) over(agegrp)`
Same as above, but ordered differently. In the previous example we typed `over(agegrp) over(sex)`. This time, we reverse it:

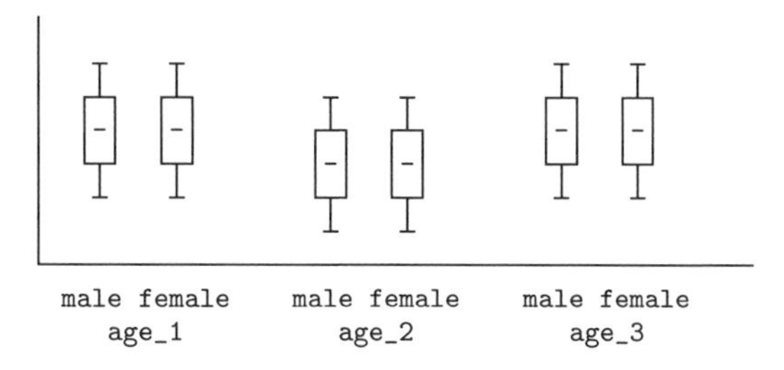

`graph box bp_before bp_after, over(agegrp) over(sex)`
2 × *#_of_agegrps* × *#_of_sexes* boxes showing blood pressure, before and after, for each age group, repeated for each sex. The grouping would look like this:

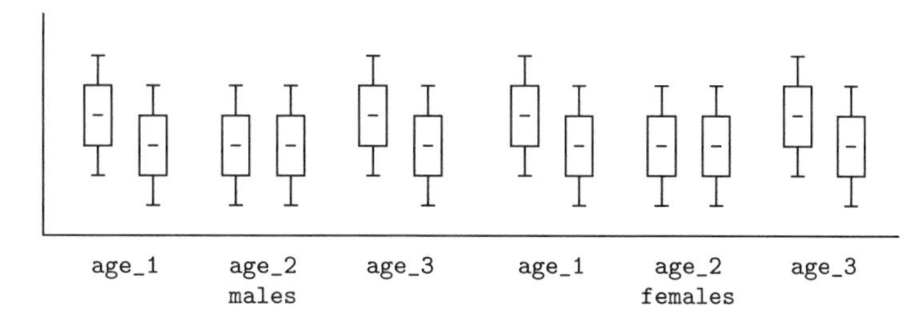

Treatment of multiple yvars versus treatment of over( ) groups

Consider two datasets containing the same data but organized differently. The datasets contain blood pressure before and after an intervention. In the first dataset, the data are organized the wide way; each patient is an observation. A few of the data are

patient	sex	agegrp	bp_before	bp_after
1	Male	30-45	143	153
2	Male	30-45	163	170
3	Male	30-45	153	168

In the second dataset, the data are organized the long way; each patient is a pair of observations. The corresponding observations in the second dataset are

patient	sex	agegrp	when	bp
1	Male	30-45	Before	143
1	Male	30-45	After	153
2	Male	30-45	Before	163
2	Male	30-45	After	170
3	Male	30-45	Before	153
3	Male	30-45	After	168

Using the first dataset, we might type

```
. use http://www.stata-press.com/data/r12/bpwide, clear
(fictional blood press data)
. graph box bp_before bp_after, over(sex)
```

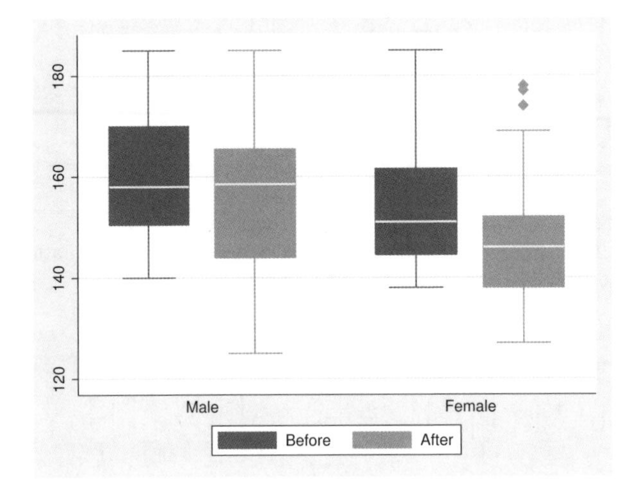

Using the second dataset, we could type

```
. use http://www.stata-press.com/data/r12/bplong, clear
. graph box bp, over(when) over(sex)
```

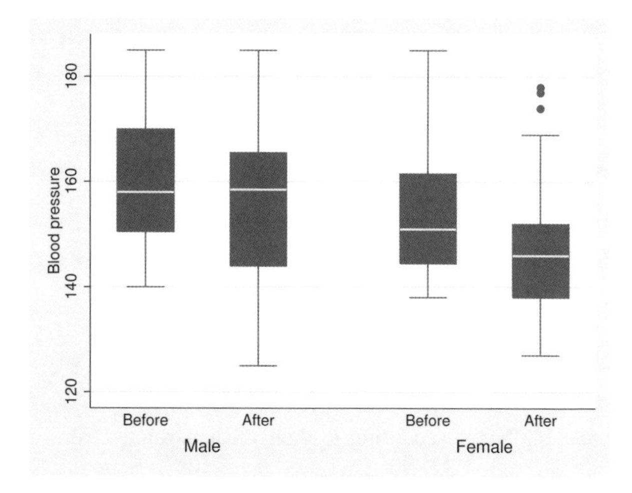

The two graphs are virtually identical. They differ in that

	multiple *yvars*	over() groups
boxes different colors	yes	no
boxes identified via ...	legend	axis label

Option ascategory will cause multiple *yvars* to be presented as if they were the first over() group, and option asyvars will cause the first over() group to be presented as if they were multiple *yvars*. Thus

```
. graph box bp, over(when) over(sex) asyvars
```

would produce the first chart and

```
. graph box bp_before bp_after, over(sex) ascategory
```

would produce the second.

How boxes are ordered

The default is to place the boxes in the order of the *yvars* and to order each over(*varname*) group according to the values of *varname*. Let us consider some examples:

graph box bp_before bp_after
  Boxes appear in the order specified, bp_before and bp_after.

graph box bp, over(when)
  Boxes are ordered according to the values of variable when.

  If variable when is a numeric, the lowest when number comes first, followed by the next lowest, and so on. This is true even if variable when has a value label. Say that when = 1 has been labeled "Before" and when = 2, labeled "After". The boxes will be in the order Before followed by After.

If variable when is a string, the boxes will be ordered by the sort order of the values of the variable (that is, alphabetically, but with capital letters placed before lowercase letters). If variable when contains "Before" and "After", the boxes will be in the order After followed by Before.

graph box bp_before bp_after, over(sex)
Boxes appear in the order specified, bp_before and bp_after, and are repeated for each sex, which will be ordered as explained above.

graph box bp_before bp_after, over(sex) over(agegrp)
Boxes appear in the order specified, bp_before and bp_after, repeated for sex ordered on the values of variable sex, repeated for agegrp ordered on the values of variable agegrp.

Reordering the boxes

There are two ways you may wish to reorder the boxes:

1. You want to control the order in which the elements of each over() group appear. String variable when might contain "After" and "Before", but you want the boxes to appear in the order Before and After.
2. You wish to order the boxes according to their median values. You wish to draw the graph

```
. graph box wage, over(industry)
```

and you want the industries ordered by wage.

We will consider each of these desires separately.

Putting the boxes in a prespecified order

You have drawn the graph

```
. graph box bp, over(when) over(sex)
```

Variable when is a string containing "Before" and "After". You wish the boxes to be in that order.

To do that, you create a new numeric variable that orders the group as you would like:

```
. generate order = 1 if when=="Before"
. replace  order = 2 if when=="After"
```

You may name the variable and create it however you wish, but be sure that there is a one-to-one correspondence between the new variable and the over() group's values. You then specify over()'s sort(*varname*) option:

```
. graph box bp, over(when, sort(order)) over(sex)
```

If you want to reverse the order, you may specify the descending suboption:

```
. graph box bp, over(when, sort(order) descending) over(sex)
```

Putting the boxes in median order

You have drawn the graph

```
. graph hbox wage, over(industry)
```

and now wish to put the boxes in median order, lowest first. You type

```
. graph hbox wage, over( industry, sort(1) )
```

If you wanted the largest first, you would type

```
. graph hbox wage, over(industry, sort(1) descending)
```

The `1` in `sort(1)` refers to the first (and here only) *yvar*. If you had multiple *yvars*, you might type

```
. graph hbox wage benefits, over( industry, sort(1) )
```

and you would have a chart showing `wage` and `benefits` sorted on `wage`. If you typed

```
. graph hbox wage benefits, over( industry, sort(2) )
```

the graph would be sorted on `benefits`.

Use with by()

`graph box` and `graph hbox` may be used with `by()`, but in general, you will want to use `over()` in preference to `by()`. Box charts are explicitly categorical and do an excellent job of presenting summary statistics for multiple groups in one chart.

A good use of `by()`, however, is when the graph would otherwise be long. Consider the graph

```
. use http://www.stata-press.com/data/r12/nlsw88, clear
(NLSW, 1988 extract)
. graph hbox wage, over(ind) over(union)
```

In the above graph, there are 12 industry categories and two union categories, resulting in 24 separate boxes. The graph, presented at normal size, would be virtually unreadable. One way around that problem would be to make the graph longer than usual,

```
. graph hbox wage, over(ind) over(union) ysize(7)
```

See *Charts with many categories* in [G-2] **graph bar** for more information about that solution. The other solution would be to introduce union as a `by()` category rather than an `over()` category:

```
. graph hbox wage, over(ind) by(union)
```

Below we do precisely that, adding some extra options to produce a good-looking chart:

```
. graph hbox wage, over(ind, sort(1)) nooutside
        ytitle("")
        by(
                union,
                title("Hourly wage, 1988, woman aged 34-46", span)
                subtitle(" ")
                note("Source:  1988 data from NLS, U.S. Dept. of Labor,
                      Bureau of Labor Statistics", span)
        )
```

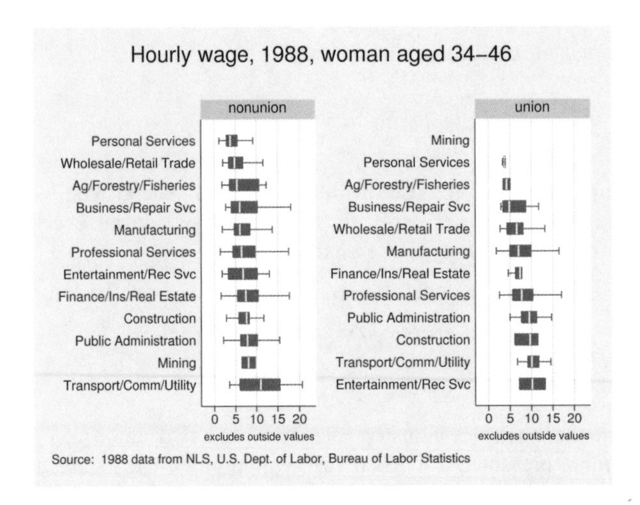

The title options were specified inside the `by()` so that they would not be applied to each graph separately; see [G-3] ***by_option***.

History

Box plots have been used in geography and climatology, under the name "dispersion diagrams", since at least 1933; see Crowe (1933). His figure 1 shows all the data points, medians, quartiles, and octiles by month for monthly rainfalls for Glasgow, 1868–1917. His figure 2, a map of Europe with several climatic stations, shows monthly medians, quartiles, and octiles.

Methods and formulas

For a description of box plots, see Cleveland (1993, 25–27).

Summary statistics are obtained from `summarize`; see [R] **summarize**.

The upper and lower adjacent values are as defined by Tukey (1977):

Let x represent a variable for which adjacent values are being calculated. Define $x_{(i)}$ as the ith ordered value of x, and define $x_{[25]}$ and $x_{[75]}$ as the 25th and 75th percentiles.

Define U as $x_{[75]} + \frac{3}{2}(x_{[75]} - x_{[25]})$. The upper adjacent value is defined as x_i, such that $x_{(i)} \leq U$ and $x_{(i+1)} > U$.

Define L as $x_{[25]} - \frac{3}{2}(x_{[75]} - x_{[25]})$. The lower adjacent value is defined as x_i, such that $x_{(i)} \geq L$ and $x_{(i-1)} < L$.

References

Chambers, J. M., W. S. Cleveland, B. Kleiner, and P. A. Tukey. 1983. *Graphical Methods for Data Analysis.* Belmont, CA: Wadsworth.

Cleveland, W. S. 1993. *Visualizing Data.* Summit, NJ: Hobart.

——. 1994. *The Elements of Graphing Data.* Rev. ed. Summit, NJ: Hobart.

Cox, N. J. 2005. Stata tip 24: Axis labels on two or more levels. *Stata Journal* 5: 469.

——. 2009. Speaking Stata: Creating and varying box plots. *Stata Journal* 9: 478–496.

Crowe, P. R. 1933. The analysis of rainfall probability. A graphical method and its application to European data. *Scottish Geographical Magazine* 49: 73–91.

Nash, J. C. 1996. gr19: Misleading or confusing boxplots. *Stata Technical Bulletin* 29: 14–17. Reprinted in *Stata Technical Bulletin Reprints*, vol. 5, pp. 60–64. College Station, TX: Stata Press.

Tukey, J. W. 1977. *Exploratory Data Analysis.* Reading, MA: Addison–Wesley.

Also see

[G-2] **graph bar** — Bar charts

[R] **lv** — Letter-value displays

[R] **summarize** — Summary statistics

Title

[G-2] graph combine — Combine multiple graphs

Syntax

`graph combine` *name* [*name* ...] [, *options*]

name	Description
simplename	name of graph in memory
name.gph	name of graph stored on disk
"*name*"	name of graph stored on disk

options	Description
`colfirst`	display down columns
`rows(`*#*`)` | `cols(`*#*`)`	display in # rows or # columns
`holes(`*numlist*`)`	positions to leave blank
`iscale(`[*]*#*`)`	size of text and markers
`altshrink`	alternate scaling of text, etc.
`imargin(`*marginstyle*`)`	margins for individual graphs
`ycommon`	give y axes common scales
`xcommon`	give x axes common scales
title_options	titles to appear on combined graph
region_options	outlining, shading, aspect ratio
`commonscheme`	put graphs on common scheme
`scheme(`*schemename*`)`	overall look
`nodraw`	suppress display of combined graph
`name(`*name*`, ...)`	specify name for combined graph
`saving(`*filename*`, ...)`	save combined graph in file

See [G-4] ***marginstyle***, [G-3] ***title_options***, [G-3] ***region_options***, [G-3] ***nodraw_option***, [G-3] ***name_option***, and [G-3] ***saving_option***.

Description

`graph combine` arrays separately drawn graphs into one.

Options

`colfirst`, `rows(`*#*`)`, `cols(`*#*`)`, and `holes(`*numlist*`)` specify how the resulting graphs are arrayed. These are the same options described in [G-3] ***by_option***.

`iscale(#)` and `iscale(*#)` specify a size adjustment (multiplier) to be used to scale the text and markers used in the individual graphs.

By default, `iscale()` gets smaller and smaller the larger is G, the number of graphs being combined. The default is parameterized as a multiplier f(G)—$0 < f(G) < 1$, $f'(G) < 0$—that is used to multiply `msize()`, $\{$ `y` | `x` $\}$`label(,labsize())`, etc., in the individual graphs.

If you specify `iscale(#)`, the number you specify is substituted for f(G). `iscale(1)` means that text and markers should appear the same size that they were originally. `iscale(.5)` displays text and markers at half that size. We recommend that you specify a number between 0 and 1, but you are free to specify numbers larger than 1.

If you specify `iscale(*#)`, the number you specify is multiplied by f(G), and that product is used to scale the text and markers. `iscale(*1)` is the default. `iscale(*1.2)` means that text and markers should appear at 20% larger than `graph combine` would ordinarily choose. `iscale(*.8)` would make them 20% smaller.

`altshrink` specifies an alternate method of determining the size of text, markers, line thicknesses, and line patterns. The size of everything drawn on each graph is as though the graph were drawn at full size, but at the aspect ratio of the combined individual graph, and then the individual graph and everything on it were shrunk to the size shown in the combined graph.

`imargin(`*marginstyle*`)` specifies margins to be put around the individual graphs. See [G-4] ***marginstyle***.

`ycommon` and `xcommon` specify that the individual graphs previously drawn by `graph twoway`, and for which the `by()` option was not specified, be put on common y or x axis scales. See *Combining twoway graphs* under *Remarks* below.

These options have no effect when applied to the categorical axes of `bar`, `box`, and `dot` graphs. Also, when `twoway` graphs are combined with `bar`, `box`, and `dot` graphs, the options affect only those graphs of the same type as the first graph combined.

title_options allow you to specify titles, subtitles, notes, and captions to be placed on the combined graph; see [G-3] ***title_options***.

region_options allow you to control the aspect ratio, size, etc., of the combined graph; see [G-3] ***region_options***. Important among these options are `ysize(#)` and `xsize(#)`, which specify the overall size of the resulting graph. It is sometimes desirable to make the combined graph wider or longer than usual.

`commonscheme` and `scheme(`*schemename*`)` are for use when combining graphs that use different schemes. By default, each subgraph will be drawn according to its own scheme.

`commonscheme` specifies that all subgraphs be drawn using the same scheme and, by default, that scheme will be your default scheme; see [G-4] **schemes intro**.

`scheme(`*schemename*`)` specifies that the *schemename* be used instead; see [G-3] ***scheme_option***.

`nodraw` causes the combined graph to be constructed but not displayed; see [G-3] ***nodraw_option***.

`name(`*name*[`, replace`]`)` specifies the name of the resulting combined graph. `name(Graph, replace)` is the default. See [G-3] ***name_option***.

`saving(`*filename*[`, asis replace`]`)` specifies that the combined graph be saved as *filename*. If *filename* is specified without an extension, `.gph` is assumed. `asis` specifies that the graph be saved in as-is format. `replace` specifies that, if the file already exists, it is okay to replace it. See [G-3] ***saving_option***.

Remarks

Remarks are presented under the following headings:

Typical use

We have previously drawn

```
. use http://www.stata-press.com/data/r12/uslifeexp
(U.S. life expectancy, 1900-1999)
. line le_male   year, saving(male)
. line le_female year, saving(female)
```

We now wish to combine these two graphs:

```
. gr combine male.gph female.gph
```

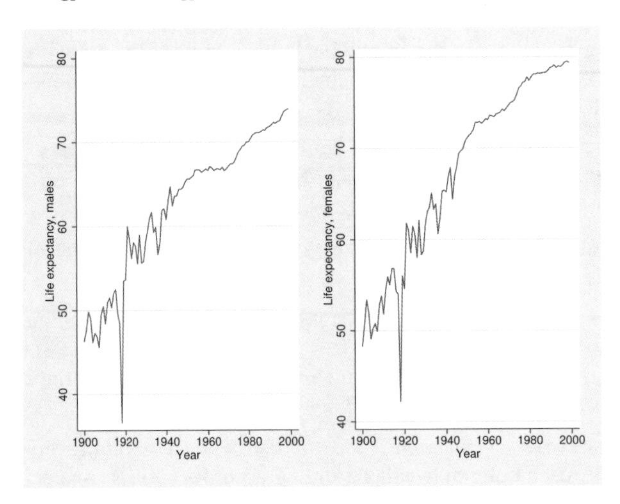

This graph would look better combined into one column and if we specified `iscale(1)` to prevent the font from shrinking:

```
. gr combine male.gph female.gph, col(1) iscale(1)
```

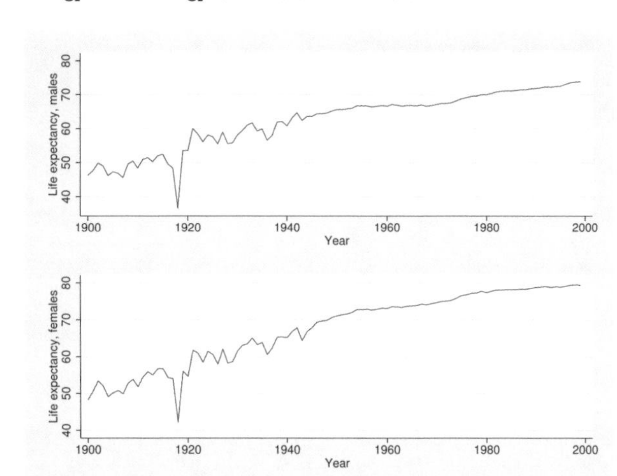

Typical use with memory graphs

In both the above examples, we explicitly typed the .gph suffix on the ends of the filenames:

```
. gr combine male.gph female.gph
. gr combine male.gph female.gph, col(1) iscale(1)
```

We must do that, or we must enclose the filenames in quotes:

```
. gr combine "male" "female"
. gr combine "male" "female", col(1) iscale(1)
```

If we did neither, graph combine would assume that the graphs were stored in memory and would then have issued the error that the graphs could not be found. Had we wanted to do these examples by using memory graphs rather than disk files, we could have substituted name() for saving on the individual graphs

```
. use http://www.stata-press.com/data/r12/uslifeexp, clear
(U.S. life expectancy, 1990-1999)
. line le_male   year, name(male)
. line le_female year, name(female)
```

and then we could type the names without quotes on the graph combine commands:

```
. gr combine male female
. gr combine male female, col(1) iscale(1)
```

Combining twoway graphs

In the first example of *Typical use*, the y axis of the two graphs did not align: one had a minimum of 40, whereas the other was approximately 37. Option ycommon will put all twoway graphs on a common y scale.

```
. use http://www.stata-press.com/data/r12/uslifeexp, clear
(U.S. life expectancy, 1990-1999)
. line le_male   year, saving(male)
. line le_female year, saving(female)
. gr combine male.gph female.gph, ycommon
```

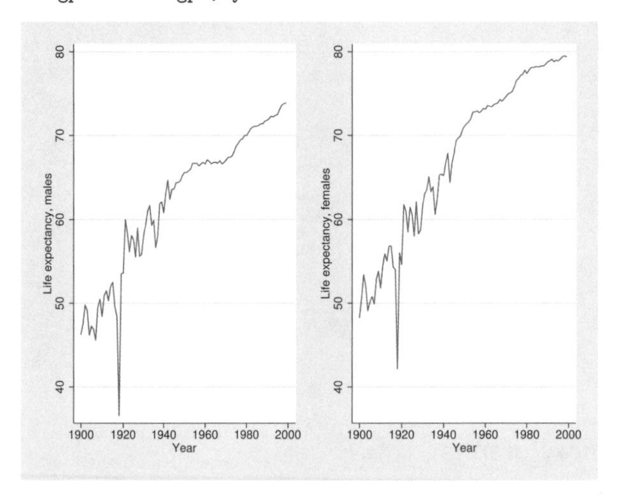

Advanced use

```
. use http://www.stata-press.com/data/r12/lifeexp, clear
(Life expectancy, 1998)
. generate loggnp = log10(gnppc)
. label var loggnp "Log base 10 of GNP per capita"
. scatter lexp loggnp,
        ysca(alt) xsca(alt)
        xlabel(, grid gmax)      saving(yx)
. twoway histogram lexp, fraction
        xsca(alt reverse) horiz  saving(hy)
. twoway histogram loggnp, fraction
        ysca(alt reverse)
        ylabel(,nogrid)
        xlabel(,grid gmax)       saving(hx)
. graph combine hy.gph yx.gph hx.gph,
        hole(3)
        imargin(0 0 0 0) graphregion(margin(l=22 r=22))
        title("Life expectancy at birth vs. GNP per capita")
        note("Source:  1998 data from The World Bank Group")
```

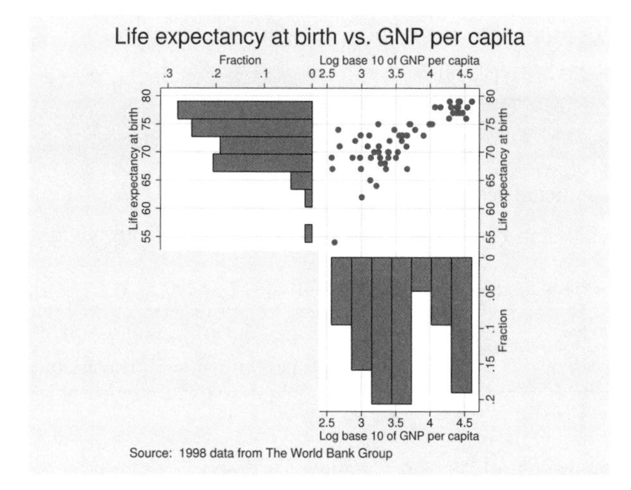

Note the specification of

```
imargin(0 0 0 0) graphregion(margin(l=22 r=22))
```

on the `graph combine` statement. Specifying `imargin()` pushes the graphs together by eliminating the margins around them. Specifying `graphregion(margin())` makes the graphs more square—to control the aspect ratio.

Controlling the aspect ratio of subgraphs

The above graph can be converted to look like this

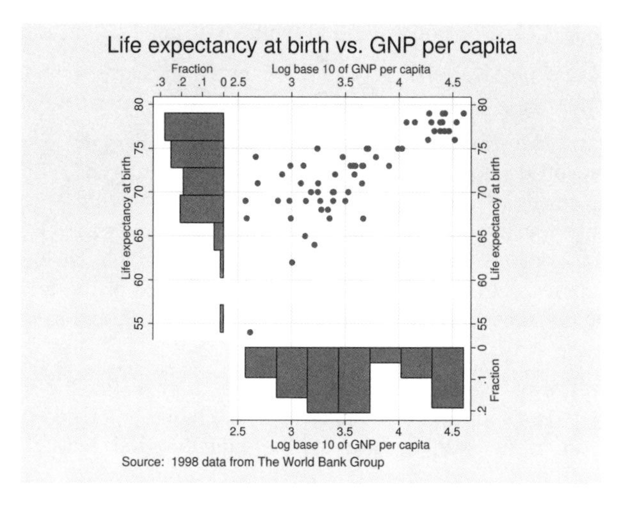

by adding `fysize(25)` to the drawing of the histogram for the x axis,

```
. twoway histogram loggnp, fraction
        ysca(alt reverse)
ylabel(0(.1).2, nogrid)
        xlabel(, grid gmax)       saving(hx)
        fysize(25)                                      ← new
```

and adding `fxsize(25)` to the drawing of the histogram for the y axis:

```
. twoway histogram lexp, fraction
        xsca(alt reverse) horiz
                                   saving(hy)
        fxsize(25)                                          ← new
```

The `graph combine` command remained unchanged.

The *forced_size_options* `fysize()` and `fxsize()` are allowed with any graph, their syntax being

forced_size_options	Description
`fysize(`*relativesize*`)`	use only percent of height available
`fxsize(`*relativesize*`)`	use only percent of width available

See [G-4] ***relativesize***.

There are three ways to control the aspect ratio of a graph:

1. Specify the *region_options* `ysize(#)` and `xsize(#)`; # is specified in inches.
2. Specify the *region_option* `graphregion(margin(`*marginstyle*`))`.
3. Specify the *forced_size_options* `fysize(`*relativesize*`)` and `fxsize(`*relativesize*`)`.

Now let us distinguish between

a. controlling the aspect ratio of the overall graph, and

b. controlling the aspect ratio of individual graphs in a combined graph.

For problem (a), methods (1) and (2) are best. We used method (2) when we constructed the overall combined graph above—we specified `graphregion(margin(l=22 r=22))`. Methods 1 and 2 are discussed under *Controlling the aspect ratio* in [G-3] ***region_options***.

For problem (b), method (1) will not work, and methods (2) and (3) do different things.

Method (1) controls the physical size at which the graph appears, so it indirectly controls the aspect ratio. `graph combine`, however, discards this physical-size information.

Method (2) is one way of controlling the aspect ratio of subgraphs. `graph combine` honors margins, assuming that you do not specify `graph combine`'s `imargin()` option, which overrides the original margin information. In any case, if you want the subgraph long and narrow, or short and wide, you need only specify the appropriate `graphregion(margin())` at the time you draw the subgraph. When you combine the resulting graph with other graphs, it will look exactly as you want it. The long-and-narrow or short-and-wide graph will appear in the array adjacent to all the other graphs. Each graph is allocated an equal-sized area in the array, and the oddly shaped graph is drawn into it.

Method (3) is the only way you can obtain unequally sized areas. For the combined graph above, you specified `graph combine`'s `imargin()` option and that alone precluded our use of method (2), but most importantly, you did not want an array of four equally sized areas:

1 histogram	2 scatter
3	4 histogram

We wanted

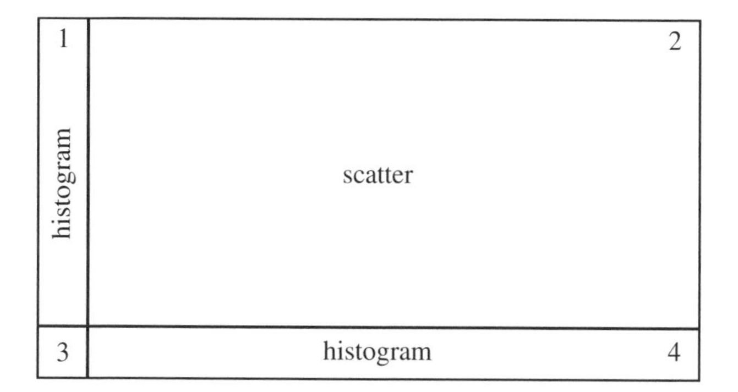

The *forced_size_options* allowed us to achieve that. You specify the *forced_size_options* `fysize()` and `fxsize()` with the commands that draw the subgraphs, not with `graph combine`. Inside the parentheses, you specify the percentage of the graph region to be used. Although you could use `fysize()` and `fxsize()` to control the aspect ratio in ordinary cases, there is no reason to do that. Use `fysize()` and `fxsize()` to control the aspect ratio when you are going to use `graph combine` and you want unequally sized areas or when you will be specifying `graph combine`'s `imargin()` option.

Also see

[G-2] **graph use** — Display graph stored on disk

[G-2] **graph save** — Save graph to disk

[G-3] ***saving_option*** — Option for saving graph to disk

[G-4] **concept: gph files** — Using gph files

Title

[G-2] graph copy — Copy graph in memory

Syntax

`graph copy [oldname] newname [, replace]`

If *oldname* is not specified, the name of the current graph is assumed.

Menu

Graphics > Manage graphs > Copy graph in memory

Description

`graph copy` makes a copy of a graph stored in memory under a new name.

Option

`replace` specifies that it is okay to replace *newname*, if it already exists.

Remarks

See [G-2] **graph manipulation** for an introduction to the graph manipulation commands.

`graph copy` is rarely used. Perhaps you have a graph displayed in the Graph window (known as the current graph), and you wish to experiment with changing its aspect ratio or scheme using the `graph display` command. Before starting your experiments, you make a copy of the original:

```
. graph copy backup
. graph display ...
```

Also see

[G-2] **graph manipulation** — Graph manipulation commands

[G-2] **graph rename** — Rename graph in memory

Title

[G-2] graph describe — Describe contents of graph in memory or on disk

Syntax

`graph describe` [*name*]

name	Description
simplename	name of graph in memory
filename.gph	name of graph on disk
"*filename*"	name of graph on disk

If *name* is not specified, the graph currently displayed in the Graph window is described.

Menu

Graphics > Manage graphs > Describe graph

Description

`graph describe` describes the contents of a graph in memory or a graph stored on disk.

Remarks

See [G-2] **graph manipulation** for an introduction to the graph manipulation commands.

`graph describe` describes the contents of a graph, which may be stored in memory or on disk. Without arguments, the graph stored in memory named `Graph` is described:

```
. use http://www.stata-press.com/data/r12/auto
(1978 Automobile Data)

. scatter mpg weight

. graph describe

Graph stored in memory

        name:  Graph
      format:  live
     created:  9 May 2011 14:26:12
      scheme:  default
        size:  4 x 5.5
    dta file:  auto.dta dated 13 Apr 2011 17:45
     command:  twoway scatter mpg weight
```

In the above, the size is reported as *ysize* × *xsize*, not the other way around.

When you type a name ending in .gph, the disk file is described:

```
. graph save myfile
. graph describe myfile.gph
myfile.gph stored on disk
        name:  myfile.gph
      format:  live
     created:  9 May 2011 14:26:12
      scheme:  default
        size:  4 x 5.5
    dta file:  auto.dta dated 13 Apr 2011 17:45
     command:  twoway scatter mpg weight
```

If the file is saved in asis format—see [G-4] **concept: gph files**—only the name and format are listed:

```
. graph save picture, asis
. graph describe picture.gph
picture.gph stored on disk
        name:  picture.gph
      format:  asis
```

Saved results

graph describe saves the following in r():

Macros

r(fn)	*filename* or *filename*.gph
r(ft)	"old", "asis", or "live"

and, if r(ft)=="live",

Macros

r(command)	command
r(family)	subcommand; twoway, matrix, bar, dot, box, or pie
r(command_date)	date on which command was run
r(command_time)	time at which command was run
r(scheme)	scheme name
r(ysize)	ysize() value
r(xsize)	xsize() value
r(dtafile)	.dta file in memory at command_time
r(dtafile_date)	.dta file date

Any of r(command), ..., r(dtafile_date) may be undefined, so refer to contents by using macro quoting.

Also see

[G-2] **graph manipulation** — Graph manipulation commands

[G-2] **graph dir** — List names of graphs in memory and on disk

Title

[G-2] graph dir — List names of graphs in memory and on disk

Syntax

`graph dir` [*pattern*] [, *options*]

where *pattern* is allowed by Stata's `strmatch()` function: * means that 0 or more characters go here, and ? means that exactly one character goes here; see `strmatch()` in [D] **functions**.

options	Description
`memory`	list only graphs stored in memory
`gph`	list only graphs stored on disk
`detail`	produce detailed listing

Description

`graph dir` lists the names of graphs stored in memory and stored on disk in the current directory.

Options

`memory` and `gph` restrict what is listed; `memory` lists only the names of graphs stored in memory and `gph` lists only the names of graphs stored on disk.

`detail` specifies that, in addition to the names, the commands that created the graphs be listed.

Remarks

See [G-2] **graph manipulation** for an introduction to the graph manipulation commands.

`graph dir` without options lists in column format the names of the graphs stored in memory and those stored on disk in the current directory.

```
. graph dir
           Graph       figure1.gph      large.gph     s7.gph
           dot.gph     figure2.gph      old.gph       yx_lines.gph
```

Graphs in memory are listed first, followed by graphs stored on disk. In the example above, we have only one graph in memory: `Graph`.

You may specify a pattern to restrict the files listed:

```
. graph dir fig*
           figure1.gph  figure2.gph
```

The `detail` option lists the names and the commands that drew the graphs:

```
. graph dir fig*, detail
  name          command
  ----------------------------------------------------------
  figure1.gph   matrix  h-tempjul, msy(p) name(myview)
  figure2.gph   twoway scatter mpg weight, saving(figure2)
  ----------------------------------------------------------
```

Saved results

`graph dir` returns in macro `r(list)` the names of the graphs.

Also see

[G-2] **graph manipulation** — Graph manipulation commands

[G-2] **graph describe** — Describe contents of graph in memory or on disk

Title

[G-2] graph display — Display graph stored in memory

Syntax

`graph display` [*name*] [, *options*]

If *name* is not specified, the name of the current graph—the graph displayed in the Graph window—is assumed.

options	Description
`ysize(#)`	change height of graph (in inches)
`xsize(#)`	change width of graph (in inches)
`margins(`*marginstyle*`)`	change outer margins
`scale(#)`	resize text, markers, and line widths
`scheme(`*schemename*`)`	change overall look

See [G-4] ***marginstyle***, [G-3] ***scale_option***, and [G-3] ***scheme_option***.

Menu

Graphics > Manage graphs > Make memory graph current

Description

`graph display` redisplays a graph stored in memory.

Options

`ysize(#)` and `xsize(#)` specify in inches the height and width of the entire graph (also known as the *available area*). The defaults are the original height and width of the graph. These two options can be used to change the aspect ratio; see *Changing the size and aspect ratio* under *Remarks* below.

`margins(`*marginstyle*`)` specifies the outer margins: the margins between the outer graph region and the inner graph region as shown in the diagram in [G-3] ***region_options***. See *Changing the margins and aspect ratio* under *Remarks* below, and see [G-4] ***marginstyle***.

`scale(#)` specifies a multiplier that affects the size of all text, markers, and line widths in a graph. `scale(1)` is the default, and `scale(1.2)` would make all text and markers 20% larger. See [G-3] ***scale_option***.

`scheme(`*schemename*`)` specifies the overall look of the graph. The default is the original scheme with which the graph was drawn. See *Changing the scheme* under *Remarks* below, and see [G-3] ***scheme_option***.

Remarks

See [G-2] **graph manipulation** for an introduction to the graph manipulation commands.

Remarks are presented under the following headings:

Changing the size and aspect ratio

Under *Controlling the aspect ratio* in [G-3] ***region_options***, we compared

```
. use http://www.stata-press.com/data/r12/auto
(1978 Automobile Data)
. scatter mpg weight
```

with

```
. scatter mpg weight, ysize(5)
```

We do not need to reconstruct the graph merely to change the `ysize()` or `xsize()`. We could start with some graph

```
. scatter mpg weight
```

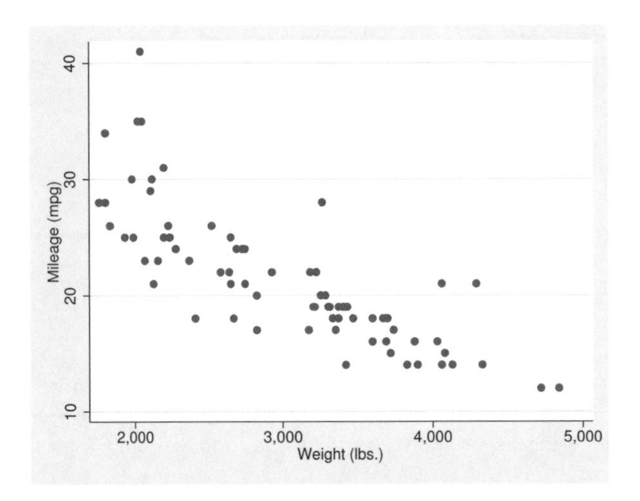

and then we could redisplay it with different `ysize()` and/or `xsize()` values:

```
. graph display, ysize(5)
```

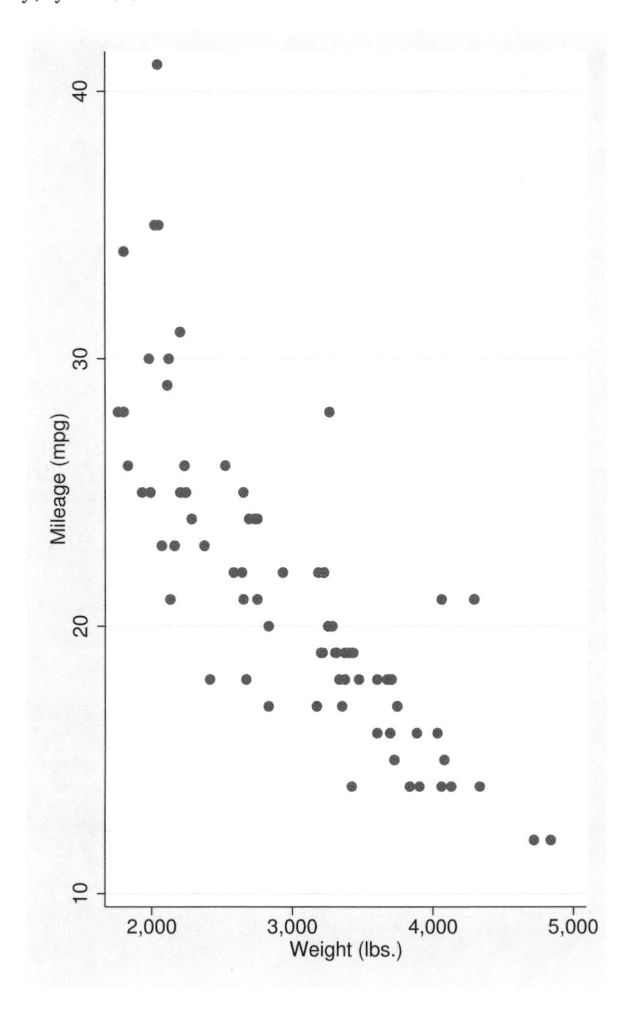

In this way we can quickly find the best `ysize()` and `xsize()` values. This works particularly well when the graph we have drawn required many options:

```
. use http://www.stata-press.com/data/r12/uslifeexp, clear
(U.S. life expectancy, 1900-1999)
. generate diff = le_wm - le_bm
. label var diff "Difference"
. line le_wm year, yaxis(1 2) xaxis(1 2)
  || line le_bm year
  || line diff  year
  || lfit diff  year
  ||,
     ylabel(0(5)20, axis(2) gmin angle(horizontal))
     ylabel(0 20(10)80,     gmax angle(horizontal))
     ytitle("", axis(2))
     xlabel(1918, axis(2)) xtitle("", axis(2))
     ytitle("Life expectancy at birth (years)")
     ylabel(, axis(2) grid)
     title("White and black life expectancy")
     subtitle("USA, 1900-1999")
     note("Source: National Vital Statistics, Vol 50, No. 6"
          "(1918 dip caused by 1918 Influenza Pandemic)")
     legend(label(1 "White males") label(2 "Black males"))
```

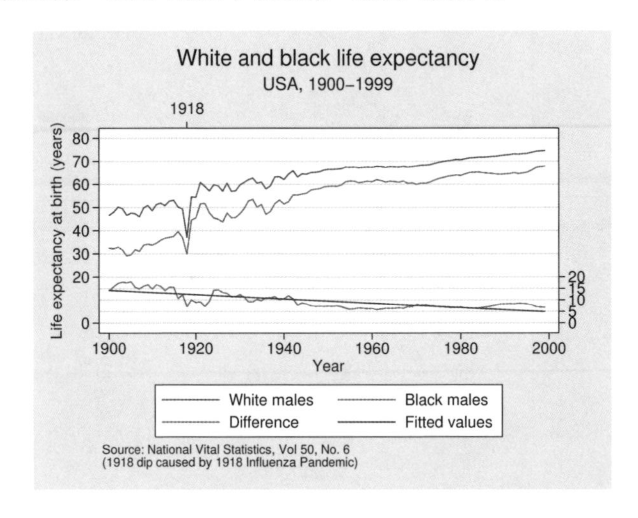

```
. graph display, ysize(5.25)
```

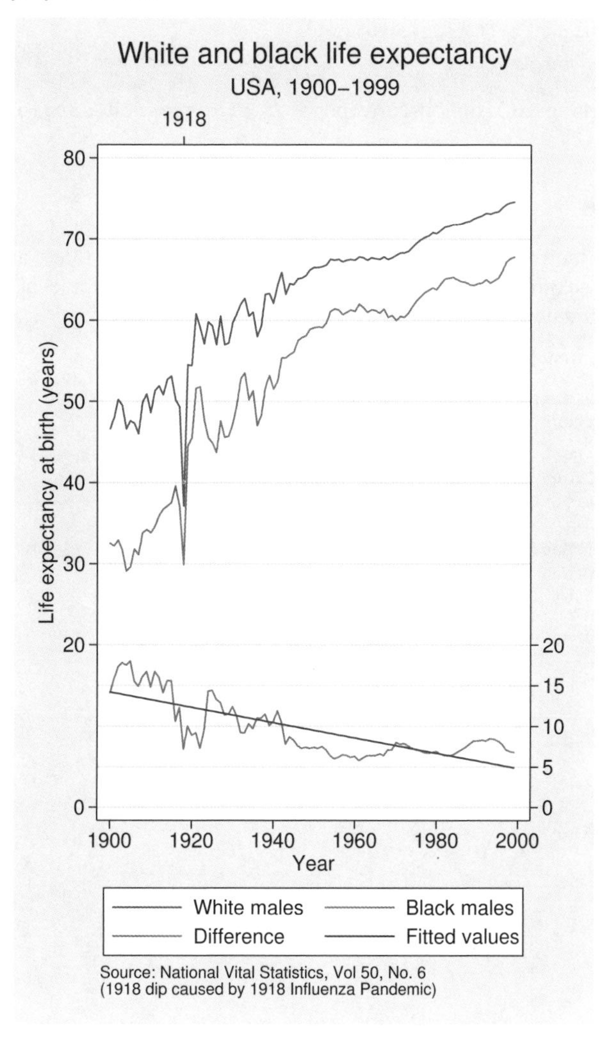

Also, we can change sizes of graphs we have previously drawn and stored on disk:

```
. graph use ...
. graph display, ysize(...) xsize(...)
```

You may not remember what `ysize()` and `xsize()` values were used (the defaults are `ysize(4)` and `xsize(5.5)`). Then use `graph describe` to describe the file; it reports the `ysize()` and `xsize()` values; see [G-2] **graph describe**.

Changing the margins and aspect ratio

We can change the size of a graph or change its margins to control the aspect ratio; this is discussed in *Controlling the aspect ratio* of [G-3] ***region_options***, which gives the example

```
scatter mpg weight, by(foreign, total graphregion(margin(l+10 r+10)))
```

This too can be done in two steps:

```
. scatter mpg weight, by(foreign, total)
. graph display, margins(l+10 r+10)
```

`graph display`'s `margin()` option corresponds to `graphregion(margin())` used at the time we construct graphs.

Changing the scheme

Schemes determine the overall look of a graph, such as where axes, titles, and legends are placed and the color of the background; see [G-4] **schemes intro**. Changing the scheme after a graph has been constructed sometimes works well and sometimes works poorly.

Here is an example in which it works well:

```
. use http://www.stata-press.com/data/r12/uslifeexp2, clear
(U.S. life expectancy, 1900-1940)

. line le year, sort
        title("Line plot")
        subtitle("Life expectancy at birth, U.S.")
        note("1")
        caption("Source: National Vital Statistics Report,
         Vol. 50 No. 6")
```

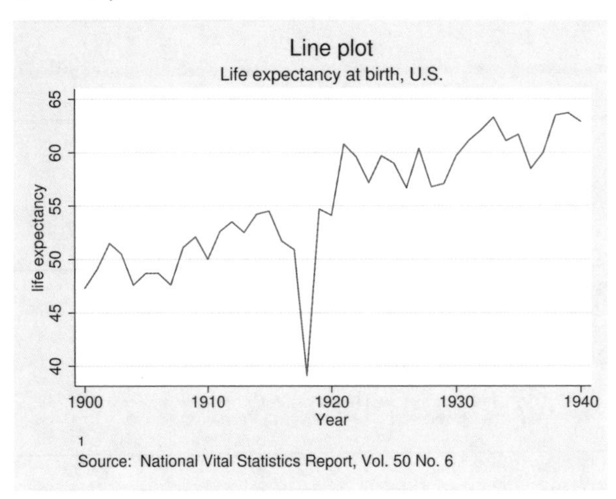

```
. graph display, scheme(economist)
```

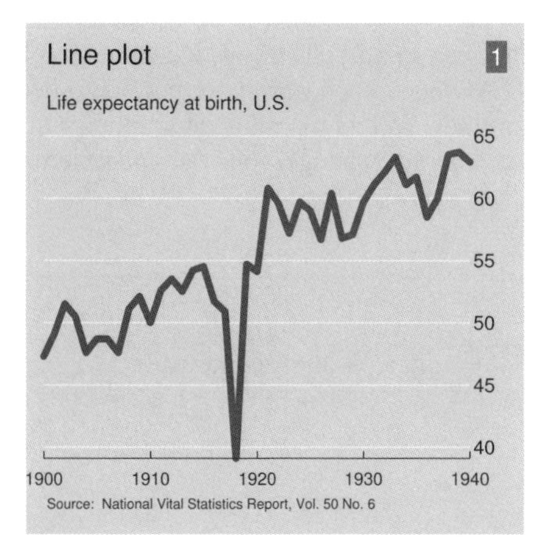

The above example works well because no options were specified to move from their default location things such as axes, titles, and legends, and no options were specified to override default colors. The issue is simple: if we draw a graph and say, "Move the title from its default location to over here", over here may be a terrible place for the title once we change schemes. Or if we override a color and make it magenta, magenta may clash terribly.

The above does not mean that the graph command need be simple. The example shown in *Changing the size and aspect ratio* above,

```
. line le_wm year, yaxis(1 2) xaxis(1 2)
  || line le_bm year
  || line diff  year
  || lfit diff  year
  ||,
     ylabel(0(5)20, axis(2) gmin angle(horizontal))
     ylabel(0 20(10)80,     gmax angle(horizontal))
     ytitle("", axis(2))
     xlabel(1918, axis(2)) xtitle("", axis(2))
     ytitle("Life expectancy at birth (years)")
     title("White and black life expectancy")
     subtitle("USA, 1900-1999")
     note("Source: National Vital Statistics, Vol 50, No. 6"
          "(1918 dip caused by 1918 Influenza Pandemic)")
     legend(label(1 "White males") label(2 "Black males"))
```

moves across schemes just fine, the only potential problem being our specification of `angle(horizontal)` for labeling the two y axes. That might not look good with some schemes.

If you are concerned about moving between schemes, when you specify options, specify style options in preference to options that directly control the outcome. For example, to have two sets of points with the same color, specify the `mstyle()` option rather than changing the color of one set to match the color you currently see of the other set.

There is another issue when moving between styles that have different background colors. Styles are said to have naturally white or naturally black background colors; see [G-4] **schemes intro**. When you move from one type of scheme to another, if the colors were not changed, colors that previously stood out would blend into the background and vice versa. To prevent this, `graph display` changes all the colors to be in accordance with the scheme, except that `graph display` does not change colors you specify by name (for example, you specify `mcolor(magenta)` or `mcolor("255 0 255")` to change the color of a symbol).

We recommend that you do not use `graph display` to change graphs from having naturally black to naturally white backgrounds. As long as you print in monochrome, `print` does an excellent job translating black to white backgrounds, so there is no need to change styles for that reason. If you are printing in color, we recommend that you change your default scheme to a naturally white scheme; see [G-2] **set scheme**.

Also see

[G-2] **graph manipulation** — Graph manipulation commands

Title

[G-2] graph dot — Dot charts (summary statistics)

Syntax

graph dot *yvars* [*if*] [*in*] [*weight*] [, *options*]

where *yvars* is

(asis) *varlist*

or is

[(*stat*)] *varname*	[[(*stat*)] ...]
[(*stat*)] *varlist*	[[(*stat*)] ...]
[(*stat*)] [*name*=]*varname*[...]	[[(*stat*)] ...]

where *stat* may be any of

mean median p1 p2 ...p99 sum count min max

or

any of the other *stats* defined in [D] **collapse**

mean is the default. p1 means the first percentile, p2 means the second percentile, and so on; p50 means the same as median. count means the number of nonmissing values of the specified variable.

options	Description
group_options	groups over which lines of dots are drawn
yvar_options	variables that are the dots
linelook_options	how the lines of dots look
legending_options	how *yvars* are labeled
axis_options	how numerical y axis is labeled
title_and_other_options	titles, added text, aspect ratio, etc.

Each is defined below.

group_options	Description
over(*varname*[, *over_subopts*])	categories; option may be repeated
nofill	omit empty categories
missing	keep missing value as category
allcategories	include all categories in the dataset

yvar_options	Description
ascategory	treat *yvars* as first over() group
asyvars	treat first over() group as *yvars*
percentages	show percentages within *yvars*
cw	calculate *yvar* statistics omitting missing values of any *yvar*

linelook_options	Description
outergap([*]#)	gap between top and first line and between last line and bottom
linegap(#)	gap between *yvar* lines; default is 0
marker(#, *marker_options*)	marker used for #th *yvar* line
pcycle(#)	marker styles before pstyles recycle
linetype(dot | line | rectangle)	type of line
ndots(#)	# of dots if linetype(dot); default is 100
dots(*marker_options*)	look if linetype(dot)
lines(*line_options*)	look if linetype(line)
rectangles(*area_options*)	look if linetype(rectangle)
rwidth(*relativesize*)	rectangle width if linetype(rectangle)
[no]extendline	whether line extends through plot region margins; extendline is usual default
lowextension(*relativesize*)	extend line through axis (advanced)
highextension(*relativesize*)	extend line through axis (advanced)

See [G-3] ***marker_options***, [G-3] ***line_options***, [G-3] ***area_options***, and [G-4] ***relativesize***.

legending_options	Description
legend_options	control of *yvar* legend
nolabel	use *yvar* names, not labels, in legend
yvaroptions(*over_subopts*)	*over_subopts* for *yvars*; seldom specified
showyvars	label *yvars* on x axis; seldom specified

See [G-3] ***legend_options***.

axis_options	Description
yalternate	put numerical y axis on right (top)
xalternate	put categorical x axis on top (right)
exclude0	do not force y axis to include 0
yreverse	reverse y axis
axis_scale_options	y-axis scaling and look
axis_label_options	y-axis labeling
ytitle(...)	y-axis titling

See [G-3] ***axis_scale_options***, [G-3] ***axis_label_options***, and [G-3] ***axis_title_options***.

title_and_other_options	Description
`text(...)`	add text on graph; x range $\left[0, 100\right]$
`yline(...)`	add y lines to graph
aspect_option	constrain aspect ratio of plot region
std_options	titles, graph size, saving to disk
`by(`*varlist*`, ...)`	repeat for subgroups

See [G-3] ***added_text_options***, [G-3] ***added_line_options***, [G-3] ***aspect_option***, [G-3] ***std_options***, and [G-3] ***by_option***.

The *over_subopts*—used in `over(`*varname*`,` *over_subopts*`)` and, on rare occasion, in `yvaroptions(`*over_subopts*`)`—are

over_subopts	Description
`relabel(`# "*text*" ...`)`	change axis labels
`label(`*cat_axis_label_options*`)`	rendition of labels
`axis(`*cat_axis_line_options*`)`	rendition of axis line
`gap(`[*]#`)`	gap between lines within `over()` category
`sort(`*varname*`)`	put lines in prespecified order
`sort(`#`)`	put lines in height order
`sort((`*stat*`)` *varname*`)`	put lines in derived order
`descending`	reverse default or specified line order

See [G-3] ***cat_axis_label_options*** and [G-3] ***cat_axis_line_options***.

`aweights`, `fweights`, and `pweights` are allowed; see [U] **11.1.6 weight** and see note concerning weights in [D] **collapse**.

Menu

Graphics > Dot chart

Description

`graph dot` draws horizontal dot charts. In a dot chart, the categorical axis is presented vertically, and the numerical axis is presented horizontally. Even so, the numerical axis is called the y axis, and the categorical axis is still called the x axis:

```
. graph dot (mean) numeric_var, over(cat_var)
```

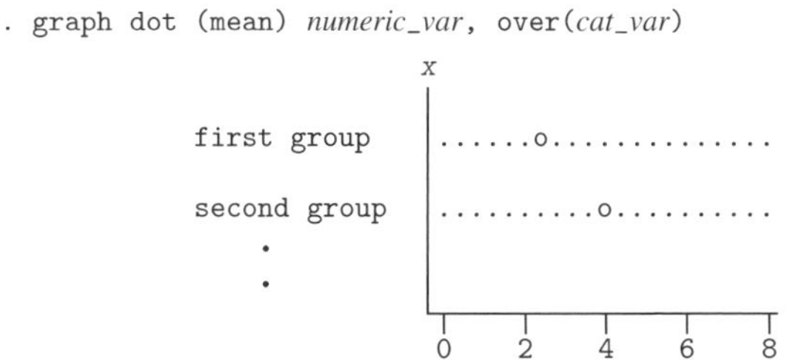

The syntax for dot charts is identical to that for bar charts; see [G-2] **graph bar**.

We use the following words to describe a dot chart:

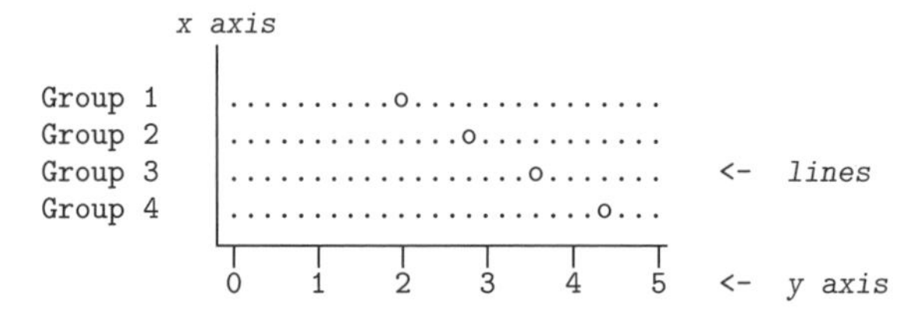

The above dot chart contains four *lines*. The words used to describe a line are

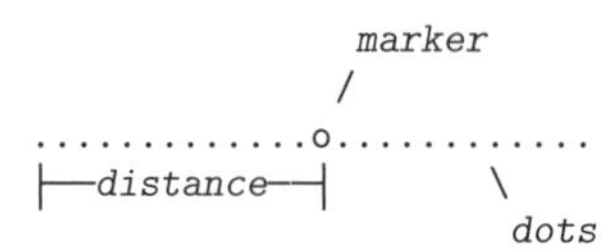

group_options

over(*varname*[, *over_subopts*]) specifies a categorical variable over which the *yvars* are to be repeated. *varname* may be string or numeric. Up to two over() options may be specified when multiple *yvars* are specified, and up to three over()s may be specified when one *yvar* is specified; options may be specified; see *Appendix: Examples of syntax* below.

nofill specifies that missing subcategories be omitted. For instance, consider

```
. graph dot (mean) y, over(division) over(region)
```

Say that one of the divisions has no data for one of the regions, either because there are no such observations or because y==. for such observations. In the resulting chart, the marker will be missing:

```
                   division 1  |..........o....
         region 1  division 2  |....o..........
                   division 3  |.......o.......

                   division 1  |..o............
         region 2  division 2  |...............
                   division 3  |.........o.....
                               |_______________
```

If you specify nofill, the missing category will be removed from the chart:

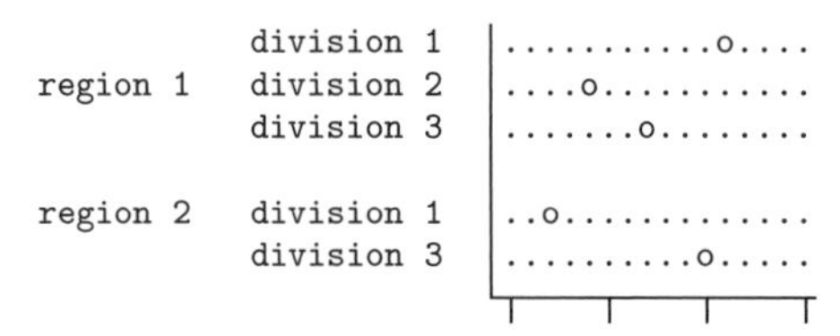

missing specifies that missing values of the over() variables be kept as their own categories, one for ., another for .a, etc. The default is to ignore such observations. An over() variable is considered to be missing if it is numeric and contains a missing value or if it is string and contains "".

`allcategories` specifies that all categories in the entire dataset be retained for the `over()` variables. When `if` or `in` is specified without `allcategories`, the graph is drawn, completely excluding any categories for the `over()` variables that do not occur in the specified subsample. With the `allcategories` option, categories that do not occur in the subsample still appear in the legend, but no markers are drawn where these categories would appear. Such behavior can be convenient when comparing graphs of subsamples that do not include completely common categories for all `over()` variables. This option has an effect only when `if` or `in` is specified or if there are missing values in the variables. `allcategories` may not be combined with `by()`.

yvar_options

`ascategory` specifies that the *yvars* be treated as the first `over()` group.

When you specify `ascategory`, results are the same as if you specified one *yvar* and introduced a new first `over()` variable. Anyplace you read in the documentation that something is done over the first `over()` category, or using the first `over()` category, it will be done over or using *yvars*.

Suppose that you specified

```
. graph dot y1 y2 y3, ascategory whatever_other_options
```

The results will be the same as if you typed

```
. graph dot y, over(newcategoryvariable) whatever_other_options
```

with a long rather than wide dataset in memory.

`asyvars` specifies that the first `over()` group be treated as *yvars*.

When you specify `asyvars`, results are the same as if you removed the first `over()` group and introduced multiple *yvars*. We said in most ways, not all ways, but let's ignore that for a moment. If you previously had *k yvars* and, in your first `over()` category, *G* groups, results will be the same as if you specified *k*G* yvars and removed the `over()`. Anyplace you read in the documentation that something is done over the *yvars* or using the *yvars*, it will be done over or using the first `over()` group.

Suppose that you specified

```
. graph dot y, over(group) asyvars whatever_other_options
```

Results will be the same as if you typed

```
. graph dot y1 y2 y3 ..., whatever_other_options
```

with a wide rather than long dataset in memory. Variables *y1*, *y2*, ..., are sometimes called the virtual *yvars*.

`percentages` specifies that marker positions be based on percentages that *yvar_i* represents of all the *yvars*. That is,

```
. graph dot (mean) inc_male inc_female
```

would produce a chart with the markers reflecting average income.

```
. graph dot (mean) inc_male inc_female, percentage
```

would produce a chart with the markers being located at $100 \times$ `inc_male`/(`inc_male` + `inc_female`) and $100 \times$ `inc_female`/(`inc_male` + `inc_female`).

If you have one *yvar* and want percentages calculated over the first `over()` group, specify the `asyvars` option. For instance,

```
. graph dot (mean) wage, over(i) over(j)
```

would produce a chart where marker positions reflect mean wages.

```
. graph dot (mean) wage, over(i) over(j) asyvars percentages
```

would produce a chart where marker positions are $100 \times (\text{mean}_{ij}/(\text{Sum}_i\ \text{mean}_{ij}))$

`cw` specifies casewise deletion. If `cw` is specified, observations for which any of the *yvars* are missing are ignored. The default is to calculate each statistic by using all the data possible.

linelook_options

`outergap(*#)` and `outergap(#)` specify the gap between the top of the graph to the beginning of the first line and the last line to the bottom of the graph.

`outergap(*#)` specifies that the default be modified. Specifying `outergap(*1.2)` increases the gap by 20%, and specifying `outergap(*.8)` reduces the gap by 20%.

`outergap(#)` specifies the gap as a percentage-of-bar-width units. `graph dot` is related to `graph bar`. Just remember that `outergap(50)` specifies a sizable but not excessive gap.

`linegap(#)` specifies the gap to be left between *yvar* lines. The default is `linegap(0)`, meaning that multiple *yvars* appear on the same line. For instance, typing

```
. graph dot y1 y2, over(group)
```

results in

```
group 1  |..x....o........
group 2  |........x..o....
group 3  |.......x.....o..
         +-----------------
          |    |    |    |
```

In the above, o represents the symbol for `y1` and x the symbol for `y2`. If you want to have separate lines for the separate *yvars*, specify `linegap(20)`:

```
. graph dot y1 y2, over(group) linegap(20)
group 1  |.......o........
         |..x.............
         |
group 2  |...........o....
         |........x.......
         |
group 3  |.............o..
         |.......x........
         +-----------------
          |    |    |    |
```

Specify a number smaller or larger than 20 to reduce or increase the distance between the `y1` and `y2` lines.

Alternatively, and generally preferred, is specifying option `ascategory`, which will result in

```
. graph dot y1 y2, over(group) ascategory
group 1  y1  |.......o........
         y2  |..o.............
             |
group 2  y1  |...........o....
         y2  |........o.......
             |
group 3  y1  |.............o..
         y2  |.......o........
             +-----------------
              |    |    |    |
```

linegap() affects only the *yvar* lines. If you want to change the gap for the first, second, or third over() groups, specify the *over_subopt* gap() inside the over() itself.

marker(*#*, *marker_options*) specifies the shape, size, color, etc., of the marker to be used to mark the value of the #th *yvar* variable. marker(1, ...) refers to the marker associated with the first *yvar*, marker(2, ...) refers to the marker associated with the second, and so on. A particularly useful *marker_option* is mcolor(*colorstyle*), which sets the color of the marker. For instance, you might specify marker(1, mcolor(green)) to make the marker associated with the first *yvar* green. See [G-4] ***colorstyle*** for a list of color choices, and see [G-3] ***marker_options*** for information on the other *marker_options*.

pcycle(*#*) specifies how many variables are to be plotted before the pstyle (see [G-4] ***pstyle***) of the markers for the next variable begins again at the pstyle of the first variable—p1dot (with the markers for the variable following that using p2dot and so on). Put another way, # specifies how quickly the look of markers is recycled when more than # variables are specified. The default for most schemes is pcycle(15).

linetype(dot), linetype(line), and linetype(rectangle) specify the style of the line.

linetype(dot) is the usual default. In this style, dots are used to fill the line around the marker:

........o........

linetype(line) specifies that a solid line be used to fill the line around the marker:

————o————

linetype(rectangle) specifies that a long "rectangle" (which looks more like two parallel lines) be used to fill the area around the marker:

========o========

ndots(*#*) and dots(*marker_options*) are relevant only in the linetype(dots) case.

ndots(*#*) specifies the number of dots to be used to fill the line. The default is ndots(100).

dots(*marker_options*) specifies the marker symbol, color, and size to be used as the dot symbol. The default is to use dots(msymbol(p)). See [G-3] ***marker_options***.

lines(*line_options*) is relevant only if linetype(line) is specified. It specifies the look of the line to be used; see [G-3] ***line_options***.

rectangles(*area_options*) and rwidth(*relativesize*) are relevant only if linetype(rectangle) is specified.

rectangles(*area_options*) specifies the look of the parallel lines (rectangle); see [G-3] ***area_options***.

rwidth(*relativesize*) specifies the width (height) of the rectangle (the distance between the parallel lines). The default is usually rwidth(.45); see [G-4] ***relativesize***.

noextendline and extendline are relevant in all cases. They specify whether the line (dots, a line, or a rectangle) is to extend through the plot region margin and touch the axes. The usual default is extendline, so noextendline is the option. See [G-3] ***region_options*** for a definition of the plot region.

lowextension(*relativesize*) and highextension(*relativesize*) are advanced options that specify the amount by which the line (dots, line or a rectangle) is extended through the axes. The usual defaults are lowextension(0) and highextension(0). See [G-4] ***relativesize***.

legending_options

legend_options allows you to control the legend. If more than one *yvar* is specified, a legend is produced. Otherwise, no legend is needed because the over() groups are labeled on the categorical x axis. See [G-3] ***legend_options***.

nolabel specifies that, in automatically constructing the legend, the variable names of the *yvars* be used in preference to "mean of *varname*" or "sum of *varname*", etc.

yvaroptions(*over_subopts*) allows you to specify *over_subopts* for the *yvars*. This is seldom specified.

showyvars specifies that, in addition to building a legend, the identities of the *yvars* be shown on the categorical x axis. If showyvars is specified, it is typical to also specify legend(off).

axis_options

yalternate and xalternate switch the side on which the axes appear. yalternate moves the numerical y axis from the bottom to the top; xalternate moves the categorical x axis from the left to the right. If your scheme by default puts the axes on the opposite sides, yalternate and xalternate reverse their actions.

exclude0 specifies that the numerical y axis need not be scaled to include 0.

yreverse specifies that the numerical y axis have its scale reversed so that it runs from maximum to minimum.

axis_scale_options specify how the numerical y axis is scaled and how it looks; see [G-3] ***axis_scale_options***. There you will also see option xscale() in addition to yscale(). Ignore xscale(), which is irrelevant for dot plots.

axis_label_options specify how the numerical y axis is to be labeled. The *axis_label_options* also allow you to add and suppress grid lines; see [G-3] ***axis_label_options***. There you will see that, in addition to options ylabel(), ytick(), ymlabel(), and ymtick(), options xlabel(), ..., xmtick() are allowed. Ignore the x*() options, which are irrelevant for dot charts.

ytitle() overrides the default title for the numerical y axis; see [G-3] ***axis_title_options***. There you will also find option xtitle() documented, which is irrelevant for dot charts.

title_and_other_options

text() adds text to a specified location on the graph; see [G-3] ***added_text_options***. The basic syntax of text() is

text($\#_y$ $\#_x$ "*text*")

text() is documented in terms of twoway graphs. When used with dot charts, the "numeric" x axis is scaled to run from 0 to 100.

yline() adds vertical lines at specified y values; see [G-3] ***added_line_options***. The xline() option, also documented there, is irrelevant for dot charts. If your interest is in adding grid lines, see [G-3] ***axis_label_options***.

aspect_option allows you to control the relationship between the height and width of a graph's plot region; see [G-3] ***aspect_option***.

std_options allow you to add titles, control the graph size, save the graph on disk, and much more; see [G-3] ***std_options***.

by(*varlist*, ...) draws separate plots within one graph; see [G-3] ***by_option***.

Suboptions for use with over( ) and yvaroptions( )

`relabel(`*# "text" ...*`)` specifies text to override the default category labeling. See the description of the `relabel()` option in [G-2] **graph bar** for more information about this very useful option.

`label(`*cat_axis_label_options*`)` determines other aspects of the look of the category labels on the x axis. Except for `label(labcolor())` and `label(labsize())`, these options are seldom specified; see [G-3] ***cat_axis_label_options***.

`axis(`*cat_axis_line_options*`)` specifies how the axis line is rendered. This is a seldom specified option. See [G-3] ***cat_axis_line_options***.

`gap(`*#*`)` and `gap(*`*#*`)` specify the gap between the lines in this `over()` group. `gap(`*#*`)` is specified in percentage-of-bar-width units. Just remember that `gap(50)` is a considerable, but not excessive width. `gap(*`*#*`)` allows modifying the default gap. `gap(*1.2)` would increase the gap by 20%, and `gap(*.8)` would decrease the gap by 20%.

`sort(`*varname*`)`, `sort(`*#*`)`, and `sort((`*stat*`)` *varname*`)` control how the lines are ordered. See *How bars are ordered* and *Reordering the bars* in [G-2] **graph bar**.

`sort(`*varname*`)` puts the lines in the order of *varname*.

`sort(`*#*`)` puts the markers in distance order. # refers to the *yvar* number on which the ordering should be performed.

`sort((`*stat*`)` *varname*`)` puts the lines in an order based on a calculated statistic.

`descending` specifies that the order of the lines—default or as specified by `sort()`—be reversed.

Remarks

Remarks are presented under the following headings:

Relationship between dot plots and horizontal bar charts

Despite appearances, `graph hbar` and `graph dot` are in fact the same command, meaning that concepts and options are the same:

```
. graph hbar y, over(group)
```

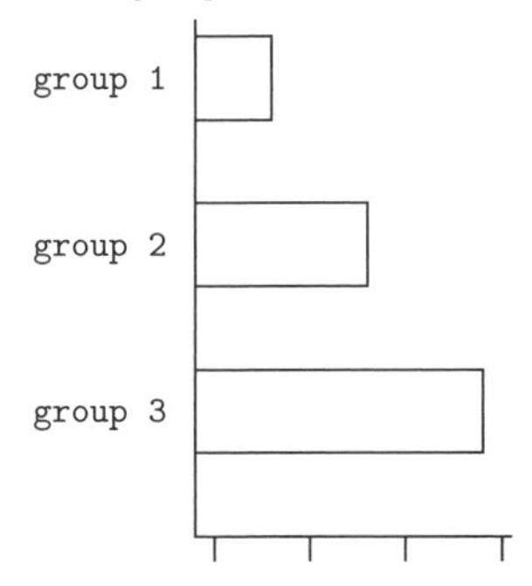

```
. graph dot y, over(group)

                    group 1 |...o...........
                    group 2 |........o......
                    group 3 |..............o.
                            +-----------------
```

There is only one substantive difference between the two commands: Given multiple *yvars*, `graph hbar` draws multiple bars:

```
. graph hbar y1 y2, over(group)
```

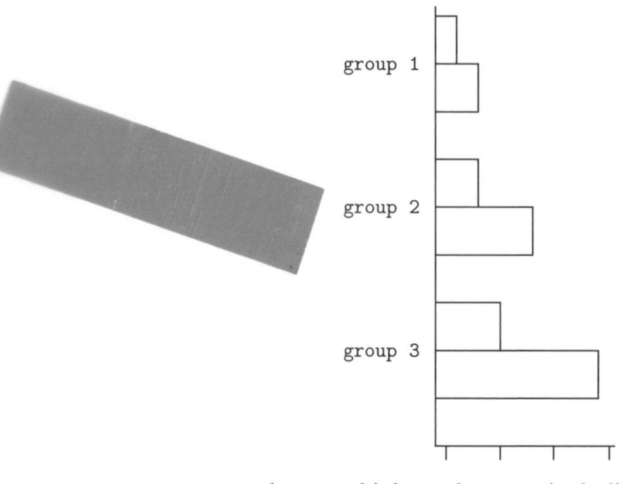

`graph dot` draws multiple markers on single lines:

```
. graph dot y1 y2, over(group)

                    group 1 |.x.o...........
                    group 2 |...x....o......
                    group 3 |.....x........o.
                            +-----------------
```

The way around this problem (if it is a problem) is to specify option `ascategory` or to specify option `linegap(#)`. Specifying `ascategory` is usually best.

Read about `graph hbar` in [G-2] **graph bar**.

Examples

Because `graph dot` and `graph hbar` are so related, the following examples should require little by way of explanation:

```
. use http://www.stata-press.com/data/r12/nlsw88
(NLSW, 1988 extract)

. graph dot wage, over(occ, sort(1))
        ytitle("")
        title("Average hourly wage, 1988, women aged 34-46", span)
        subtitle(" ")
        note("Source:  1988 data from NLS, U.S. Dept. of Labor,
              Bureau of Labor Statistics", span)
```

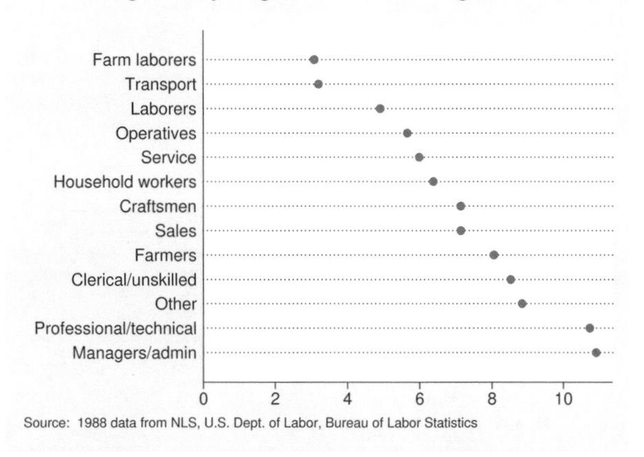


```
. graph dot (p10) wage (p90) wage,
        over(occ, sort(2))
        legend(label(1 "10th percentile") label(2 "90th percentile"))
        title("10th and 90th percentiles of hourly wage", span)
        subtitle("Women aged 34-46, 1988" " ", span)
        note("Source:  1988 data from NLS, U.S. Dept. of Labor,
              Bureau of Labor Statistics", span)
```

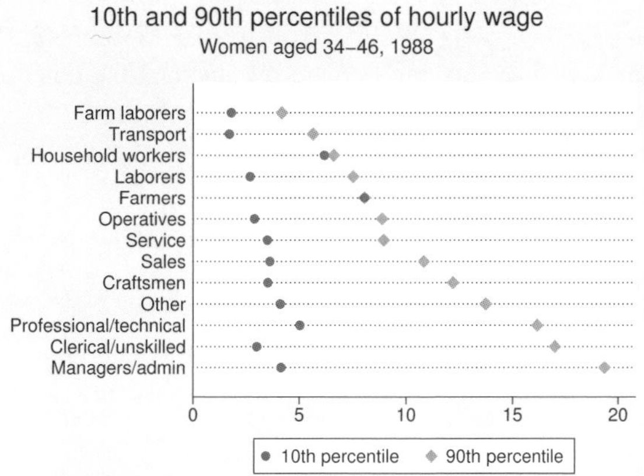

```
. graph dot (mean) wage,
        over(occ, sort(1))
        by(collgrad,
             title("Average hourly wage, 1988, women aged 34-46", span)
             subtitle(" ")
             note("Source:  1988 data from NLS, U.S. Dept. of Labor,
                   Bureau of Labor Statistics", span)
        )
```

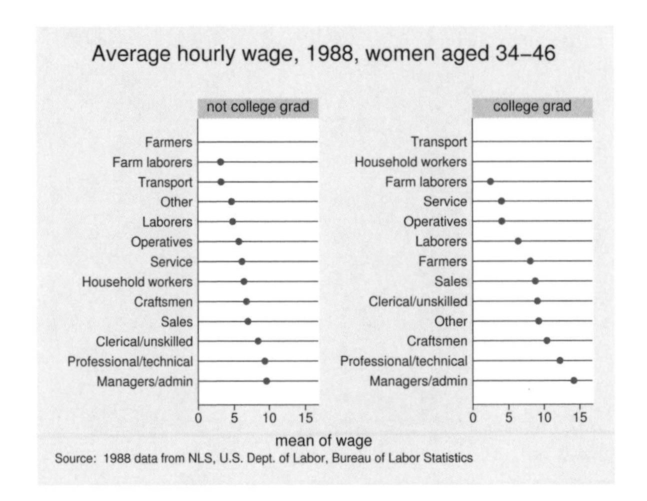

Appendix: Examples of syntax

Let us consider some `graph dot` commands and what they do:

`graph dot revenue`
One line showing average revenue.

`graph dot revenue profit`
One line with two markers, one showing average revenue and the other average profit.

`graph dot revenue, over(division)`
#_of_divisions lines, each with one marker showing average revenue for each division.

`graph dot revenue profit, over(division)`
#_of_divisions lines, each with two markers, one showing average revenue and the other average profit for each division.

`graph dot revenue, over(division) over(year)`
#_of_divisions × *#_of_years* lines, each with one marker showing average revenue for each division, repeated for each of the years. The grouping would look like this (assuming 3 divisions and 2 years):

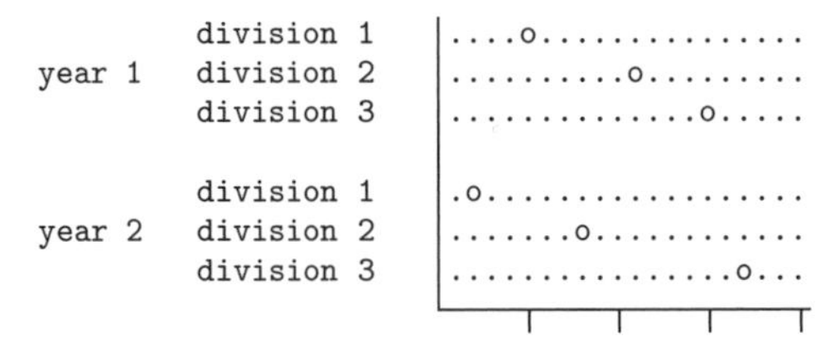

`graph dot revenue, over(year) over(division)`

Same as above, but ordered differently. In the previous example, we typed `over(division) over(year)`. This time, we reverse it:

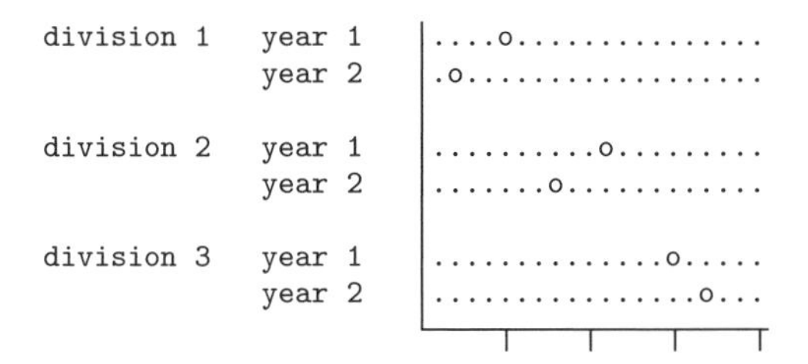

`graph dot revenue profit, over(division) over(year)`

#_of_divisions × *#_of_years* lines each with two markers, one showing average revenue and the other showing average profit for each division, repeated for each of the years.

`graph dot (sum) revenue profit, over(division) over(year)`

#_of_divisions × *#_of_years* lines each with two markers, the first showing the sum of revenue and the second showing the sum of profit for each division, repeated for each of the years.

`graph dot (median) revenue profit, over(division) over(year)`

#_of_divisions × *#_of_years* lines each with two markers showing the median of revenue and median of profit for each division, repeated for each of the years.

`graph dot (median) revenue (mean) profit, over(division) over(year)`

#_of_divisions × *#_of_years* lines each with two markers showing the median of revenue and mean of profit for each division, repeated for each of the years.

References

Cleveland, W. S. 1993. *Visualizing Data.* Summit, NJ: Hobart.

——. 1994. *The Elements of Graphing Data.* Rev. ed. Summit, NJ: Hobart.

Cox, N. J. 2008. Speaking Stata: Between tables and graphs. *Stata Journal* 8: 269–289.

Robbins, N. B. 2010. Trellis display. *Wiley Interdisciplinary Reviews: Computational Statistics* 2: 600–605.

Also see

[G-2] **graph bar** — Bar charts

[D] **collapse** — Make dataset of summary statistics

Title

[G-2] graph drop — Drop graphs from memory

Syntax

```
graph drop name [name ...]

graph drop _all
```

Menu

Graphics > Manage graphs > Drop graphs

Description

`graph drop` *name* drops (discards) the specified graphs from memory and closes any associated graph windows.

`graph drop _all` drops all graphs from memory and closes all associated graph windows.

Remarks

See [G-2] **graph manipulation** for an introduction to the graph manipulation commands.

Remarks are presented under the following headings:

Typical use
Relationship between graph drop _all and discard
Erasing graphs on disk

Typical use

Graphs contain the data they display, so when datasets are large, graphs can consume much memory. `graph drop` frees that memory. `Graph` is the name of a graph when you do not specify otherwise.

```
. graph twoway scatter faminc educ, ms(p)
. ...
. graph drop Graph
```

We often use graphs in memory to prepare the pieces for `graph combine`:

```
. graph ..., ... name(p1)
. graph ..., ... name(p2)
. graph ...   , ... name(p3)
. graph combine p1 p2 p3, ... saving(result, replace)
. graph drop _all
```

Relationship between graph drop _all and discard

The discard command performs graph drop _all and more:

1. discard eliminates prior estimation results and automatically loaded programs and thereby frees even more memory.
2. discard closes any open dialog boxes and thereby frees even more memory.

We nearly always type discard in preference to graph drop _all if only because discard has fewer characters. The exception to that is when we have fit a model and still plan on redisplaying prior results, performing tests on that model, or referring to _b[], _se[], etc.

See [P] **discard** for a description of the discard command.

Erasing graphs on disk

graph drop is not used to erase .gph files; instead, use Stata's standard erase command:

```
. erase matfile.gph
```

Also see

[G-2] **graph manipulation** — Graph manipulation commands

[D] **erase** — Erase a disk file

[P] **discard** — Drop automatically loaded programs

Title

[G-2] graph export — Export current graph

Syntax

```
graph export newfilename.suffix [, options]
```

options	Description
`name(`*windowname*`)`	name of Graph window to export
`as(`*fileformat*`)`	desired format of output
`replace`	*newfilename* may already exist
override_options	override defaults in conversion

If `as()` is not specified, the output format is determined by the suffix of *newfilename*`.`*suffix*:

suffix	Implied option	Output format
`.ps`	`as(ps)`	PostScript
`.eps`	`as(eps)`	EPS (Encapsulated PostScript)
`.wmf`	`as(wmf)`	Windows Metafile
`.emf`	`as(emf)`	Windows Enhanced Metafile
`.pdf`	`as(pdf)`	PDF
`.png`	`as(png)`	PNG (Portable Network Graphics)
`.tif`	`as(tif)`	TIFF
other		must specify `as()`

`ps`, `eps`, and `pdf` are available with all versions of Stata; `png` and `tif` are available for all versions of Stata except Stata(console) for Unix; and `wmf` and `emf` are available only with Stata for Windows.

override_options	Description
ps_options	when exporting to `ps`
eps_options	when exporting to `eps`
tif_options	when exporting to `tif`
png_options	when exporting to `png`

See [G-3] ***ps_options***, [G-3] ***eps_options***, [G-3] ***tif_options***, and [G-3] ***png_options***.

There are no *override_options* for the `pdf` format.

Description

`graph export` exports to a file the graph displayed in a Graph window.

Options

`name(`*windowname*`)` specifies which window to export from when exporting a graph. Omitting the `name()` option exports the topmost graph (Stata for Unix(GUI) users: see *Technical note for Stata for Unix(GUI) users*). The name for a window is displayed inside parentheses in the window title. For example, if the title for a Graph window is **Graph (MyGraph)**, the name for the window is **MyGraph**. If a graph is an **asis** or **graph7** graph where there is no name in the window title, specify `""` for *windowname*.

`as(`*fileformat*`)` specifies the file format to which the graph is to be exported. This option is rarely specified because, by default, `graph export` determines the format from the suffix of the file being created.

`replace` specifies that it is okay to replace *filename*`.`*suffix* if it already exists.

override_options modify how the graph is converted. See [G-3] ***ps_options***, [G-3] ***eps_options***, [G-3] ***tif_options***, and [G-3] ***png_options***. See also [G-2] **graph set** for permanently setting default values for the *override_options*.

Remarks

Graphs are exported by displaying them on the screen and then typing

```
. graph export filename.suffix
```

Remarks are presented under the following headings:

Exporting the graph displayed in a Graph window
Exporting a graph stored on disk
Exporting a graph stored in memory

If your interest is simply in printing a graph, see [G-2] **graph print**.

Exporting the graph displayed in a Graph window

There are three ways to export the graph displayed in a Graph window:

1. Right-click on the Graph window, select **Save Graph...**, and choose the appropriate **Save as type**.
2. Select **File > Save Graph...**, and choose the appropriate **Save as type**.
3. Type "`graph export` *filename*`.`*suffix*" in the Command window. Stata for Unix(GUI) users should use the `name()` option if there is more than one graph displayed to ensure that the correct graph is exported (see *Technical note for Stata for Unix(GUI) users*).

All three are equivalent. The advantage of `graph export` is that you can include it in do-files:

```
. graph ...                            (draw a graph)
. graph export filename.suffix  (and export it)
```

By default, `graph export` determines the output type by the *suffix*. If we wanted to create an Encapsulated PostScript file, we might type

```
. graph export figure57.eps
```

Exporting a graph stored on disk

To export a graph stored on disk, type

```
. graph use gph_filename
. graph export output_filename.suffix
```

Do not specify `graph use`'s `nodraw` option; see [G-2] **graph use**.

Stata for Unix(console) users: follow the instructions just given, even though you have no Graph window and cannot see what has just been "displayed". Use the graph, and then export it.

Exporting a graph stored in memory

To export a graph stored in memory but not currently displayed, type

```
. graph display name
. graph export filename.suffix
```

Do not specify `graph display`'s `nodraw` option; see [G-2] **graph display**.

Stata for Unix(console) users: follow the instructions just given, even though you have no Graph window and cannot see what has just been "displayed". Display the graph, and then export it.

❑ Technical note

Stata for Unix(GUI) users should note that X-Windows does not have a concept of a window z-order, which prevents Stata from determining which window is the topmost window. Instead, Stata determines which window is topmost based on which window has the focus. However, some window managers will set the focus to a window without bringing the window to the top. What Stata considers the topmost window may not appear topmost visually. For this reason, you should always use the `name()` option to ensure that the correct Graph window is exported.

❑

Also see

[G-3] ***eps_options*** — Options for exporting to Encapsulated PostScript

[G-3] ***png_options*** — Options for exporting to portable network graphics (PNG) format

[G-3] ***ps_options*** — Options for exporting or printing to PostScript

[G-3] ***tif_options*** — Options for exporting to tagged image file format (TIFF)

[G-2] **graph set** — Set graphics options

[G-2] **graph display** — Display graph stored in memory

[G-2] **graph use** — Display graph stored on disk

[G-2] **graph print** — Print a graph

Title

[G-2] graph manipulation — Graph manipulation commands

Syntax

Command	Description
`graph dir`	list names of graphs
`graph describe`	describe contents of graph
`graph drop`	discard graph stored in memory
`graph rename`	rename graph stored in memory
`graph copy`	copy graph stored in memory
`graph export`	export current graph
`graph use`	load graph on disk into memory and display it
`graph display`	redisplay graph stored in memory
`graph combine`	combine multiple graphs

See [G-2] **graph dir**, [G-2] **graph describe**, [G-2] **graph drop**, [G-2] **graph rename**, [G-2] **graph copy**, [G-2] **graph export**, [G-2] **graph use**, [G-2] **graph display**, and [G-2] **graph combine**.

Description

The graph manipulation commands manipulate graphs stored in memory or stored on disk.

Remarks

Remarks are presented under the following headings:

Overview of graphs in memory and graphs on disk
Summary of graph manipulation commands

Overview of graphs in memory and graphs on disk

Graphs are stored in memory and on disk. When you draw a graph, such as by typing

```
. graph twoway scatter mpg weight
```

the resulting graph is stored in memory, and, in particular, it is stored under the name `Graph`. Were you next to type

```
. graph matrix mpg weight displ
```

this new graph would replace the existing graph named `Graph`.

`Graph` is the default name used to record graphs in memory, and when you draw graphs, they replace what was previously recorded in `Graph`.

You can specify the `name()` option—see [G-3] ***name_option***—to record graphs under different names:

```
. graph twoway scatter mpg weight, name(scat)
```

Now there are two graphs in memory: `Graph`, containing a scatterplot matrix, and `scat`, containing a graph of `mpg` versus `weight`.

Graphs in memory are forgotten when you exit Stata, and they are forgotten at other times, too, such as when you type `clear` or `discard`; see [D] **drop** and [P] **discard**.

Graphs can be stored on disk, where they will reside permanently until you erase them. They are saved in files known as `.gph` files—files whose names end in `.gph`; see [G-4] **concept: gph files**.

You can save on disk the graph currently showing in the Graph window by typing

```
. graph save mygraph.gph
```

The result is to create a new file `mygraph.gph`; see [G-2] **graph save**. Or—see [G-3] ***saving_option***—you can save on disk graphs when you originally draw them:

```
. graph twoway scatter mpg weight, saving(mygraph.gph)
```

Either way, graphs saved on disk can be reloaded:

```
. graph use mygraph.gph
```

loads `mygraph.gph` into memory under the name—you guessed it—`Graph`. Of course, you could load it under a different name:

```
. graph use mygraph.gph, name(memcp)
```

Having brought this graph back into memory, you find that things are just as if you had drawn the graph for the first time. Anything you could do back then—such as combine the graph with other graphs or change its aspect ratio—you can do now. And, of course, after making any changes, you can save the result on disk, either replacing file `mygraph.gph` or saving it under a new name.

There is only one final, and minor, wrinkle: graphs on disk can be saved in either of two formats, known as `live` and `asis`. `live` is preferred and is the default, and what was said above applies only to `live`-format files. `asis` files are more like pictures—all you can do is admire them and make copies. To save a file in `asis` format, you type

```
. graph save ..., asis
```

or

```
. graph ..., ... saving(..., asis)
```

`asis` format is discussed in [G-4] **concept: gph files**.

There is a third format called `old`, which is like `asis`, except that it refers to graphs made by versions of Stata older than Stata 8. This is discussed in [G-4] **concept: gph files**, too.

Summary of graph manipulation commands

The graph manipulation commands help you manage your graphs, whether stored in memory or on disk. The commands are

`graph dir`
Lists the names under which graphs are stored, both in memory and on disk; see [G-2] **graph dir**.

`graph describe`
Provides details about a graph, whether stored in memory or on disk; see [G-2] **graph describe**.

`graph drop`
Eliminates from memory graphs stored there; see [G-2] **graph drop**.

`graph rename`
Changes the name of a graph stored in memory; see [G-2] **graph rename**.

`graph copy`
Makes a copy of a graph stored in memory; see [G-2] **graph copy**.

`graph export`
Exports the graph currently displayed in the Graph window to a file; see [G-2] **graph export**.

`graph use`
Copies a graph on disk into memory and displays it; see [G-2] **graph use**.

`graph display`
Redisplays a graph stored in memory; see [G-2] **graph display**.

`graph combine`
Combines graphs stored in memory or on disk; see [G-2] **graph combine**.

Also see

[G-2] **graph save** — Save graph to disk

[G-3] ***name_option*** — Option for naming graph in memory

[G-3] ***saving_option*** — Option for saving graph to disk

[G-4] **concept: gph files** — Using gph files

[D] **clear** — Clear memory

[D] **drop** — Eliminate variables or observations

[P] **discard** — Drop automatically loaded programs

Title

[G-2] graph matrix — Matrix graphs

Syntax

graph matrix *varlist* [*if*] [*in*] [*weight*] [, *options*]

options	Description
half	draw lower triangle only
marker_options	look of markers
marker_label_options	include labels on markers
jitter(*relativesize*)	perturb location of markers
jitterseed(*#*)	random-number seed for jitter()
diagonal(*stringlist*, ...)	override text on diagonal
scale(*#*)	overall size of symbols, labels, etc.
iscale([*]*#*)	size of symbols, labels, within plots
maxes(*axis_scale_options* *axis_label_options*)	labels, ticks, grids, log scales, etc.
axis_label_options	axis-by-axis control
by(*varlist*, ...)	repeat for subgroups
std_options	title, aspect ratio, saving to disk

All options allowed by graph twoway scatter are also allowed, but they are ignored.

half, diagonal(), scale(), and iscale() are *unique*; jitter() and jitterseed() are *rightmost* and maxes() is *merged-implicit*; see [G-4] **concept: repeated options**.

See [G-3] ***marker_options***, [G-3] ***marker_label_options***, [G-4] ***relativesize***, [G-3] ***scale_option***, [G-3] ***axis_scale_options***, [G-3] ***axis_label_options***, [G-3] ***by_option***, and [G-3] ***std_options***.

stringlist, ..., the argument allowed by diagonal(), is defined

[{ . | "*string*" }] [{ . | "*string*" } ...] [, *textbox_options*]

aweights, fweights, and pweights are allowed; see [U] **11.1.6 weight**. Weights affect the size of the markers. See *Weighted markers* in [G-2] **graph twoway scatter**.

Menu

Graphics > Scatterplot matrix

Description

`graph matrix` draws scatterplot matrices.

Options

`half` specifies that only the lower triangle of the scatterplot matrix be drawn.

marker_options specify the look of the markers used to designate the location of the points. The important *marker_options* are `msymbol()`, `mcolor()`, and `msize()`.

The default symbol used is `msymbol(O)`—solid circles. You specify `msymbol(Oh)` if you want hollow circles (a recommended alternative). If you have many observations, we recommend specifying `msymbol(p)`; see *Marker symbols and the number of observations* under *Remarks* below. See [G-4] ***symbolstyle*** for a list of marker symbol choices.

The default `mcolor()` is dictated by the scheme; see [G-4] **schemes intro**. See [G-4] ***colorstyle*** for a list of color choices.

Be careful specifying the `msize()` option. In `graph matrix`, the size of the markers varies with the number of variables specified; see option `iscale()` below. If you specify `msize()`, that will override the automatic scaling.

See [G-3] ***marker_options*** for more information on markers.

marker_label_options allow placing identifying labels on the points. To obtain this, you specify the *marker_label_option* `mlabel(`*varname*`)`; see [G-3] ***marker_label_options***. These options are of little use for scatterplot matrices because they make the graph seem too crowded.

`jitter(`*relativesize*`)` adds spherical random noise to the data before plotting. This is useful when plotting data that otherwise would result in points plotted on top of each other. See *Jittered markers* in [G-2] **graph twoway scatter** for an explanation of jittering.

`jitterseed(`*#*`)` specifies the seed for the random noise added by the `jitter()` option. # should be specified as a positive integer. Use this option to reproduce the same plotted points when the `jitter()` option is specified.

`diagonal(`[*stringlist*] [`,` *textbox_options*]`)` specifies text and its style to be displayed along the diagonal. This text serves to label the graphs (axes). By default, what appears along the diagonals are the variable labels of the variables of *varlist* or, if a variable has no variable label, its name. Typing

```
. graph matrix mpg weight displ, diag(. "Weight of car")
```

would change the text appearing in the cell corresponding to variable `weight`. We specified period (`.`) to leave the text in the first cell unchanged, and we did not bother to type a third string or a period, so we left the third element unchanged, too.

You may specify *textbox_options* following *stringlist* (which may itself be omitted) and a comma. These options will modify the style in which the text is presented but are of little use here. We recommend that you do not specify `diagonal(,size())` to override the default sizing of the text. By default, the size of text varies with the number of variables specified; see option `iscale()` below. Specifying `diagonal(,size())` will override the automatic size scaling. See [G-3] ***textbox_options*** for more information on textboxes.

`scale(`*#*`)` specifies a multiplier that affects the size of all text and markers in a graph. `scale(1)` is the default, and `scale(1.2)` would make all text and markers 20% larger.
See [G-3] ***scale_option***.

`iscale(#)` and `iscale(*#)` specify an adjustment (multiplier) to be used to scale the markers, the text appearing along the diagonals, and the labels and ticks appearing on the axes.

By default, `iscale()` gets smaller and smaller the larger *n* is, the number of variables specified in *varlist*. The default is parameterized as a multiplier f(*n*)—$0 < f(n) < 1$, $f'(n) < 0$—that is used as a multiplier for `msize()`, `diagonal(,size())`, `maxes(labsize())`, and `maxes(tlength())`.

If you specify `iscale(#)`, the number you specify is substituted for f(*n*). We recommend that you specify a number between 0 and 1, but you are free to specify numbers larger than 1.

If you specify `iscale(*#)`, the number you specify is multiplied by f(*n*), and that product is used to scale text. Here you should specify #>0; #>1 merely means you want the text to be bigger than `graph matrix` would otherwise choose.

`maxes(`*axis_scale_options axis_label_options*`)` affect the scaling and look of the axes. This is a case where you specify options within options.

Consider the *axis_scale_options* {`y`|`x`}`scale(log)`, which produces logarithmic scales. Type `maxes(yscale(log) xscale(log))` to draw the scatterplot matrix by using log scales. Remember to specify both `xscale(log)` and `yscale(log)`, unless you really want just the y axis or just the x axis logged.

Or consider the *axis_label_options* {`y`|`x`}`label(,grid)`, which adds grid lines. Specify `maxes(ylabel(,grid))` to add grid lines across, `maxes(xlabel(,grid))` to add grid lines vertically, and both options to add grid lines in both directions. When using both, you can specify the `maxes()` option twice—`maxes(ylabel(,grid)) maxes(xlabel(,grid))`—or once combined—`maxes(ylabel(,grid) xlabel(,grid))`—it makes no difference because `maxes()` is *merged-implicit*; see [G-4] **concept: repeated options**.

See [G-3] ***axis_scale_options*** and [G-3] ***axis_label_options*** for the suboptions that may appear inside `maxes()`. In reading those entries, ignore the `axis(#)` suboption; `graph matrix` will ignore it if you specify it.

axis_label_options allow you to assert axis-by-axis control over the labeling. Do not confuse this with `maxes(`*axis_label_options*`)`, which specifies options that affect all the axes. *axis_label_options* specified outside the `maxes()` option specify options that affect just one of the axes. *axis_label_options* can be repeated for each axis.

When you specify *axis_label_options* outside `maxes()`, you must specify the axis-label suboption `axis(#)`. For instance, you might type

```
. graph matrix mpg weight displ, ylabel(0(5)40, axis(1))
```

The effect of that would be to label the specified values on the first y axis (the one appearing on the far right). The axes are numbered as follows:

		x `axis(2)`		x `axis(4)`		
		v1/v2	v1/v3	v1/v4	v1/v5	y `axis(1)`
y `axis(2)`	v2/v1		v2/v3	v2/v4	v2/v5	
	v3/v1	v3/v2		v3/v4	v3/v5	y `axis(3)`
y `axis(4)`	v4/v1	v4/v2	v4/v3		v4/v5	
	v5/v1	v5/v2	v5/v3	v5/v4		y `axis(5)`
	x `axis(1)`		x `axis(3)`		x `axis(5)`	

and if `half` is specified, the numbering scheme is

y `axis(2)`	v2/v1				
y `axis(3)`	v3/v1	v3/v2			
y `axis(4)`	v4/v1	v4/v2	v4/v3		
y `axis(5)`	v5/v1	v5/v2	v5/v3	v5/v4	
	x `axis(1)`	x `axis(2)`	x `axis(3)`	x `axis(4)`	x `axis(5)`

See [G-3] ***axis_label_options***; remember to specify the `axis(#)` suboption, and do not specify the `graph matrix` option `maxes()`.

`by(`*varlist*`, ...)` allows drawing multiple graphs for each subgroup of the data. See *Use with by()* under *Remarks* below, and see [G-3] ***by_option***.

std_options allow you to specify titles (see *Adding titles* under *Remarks* below, and see [G-3] ***title_options***), control the aspect ratio and background shading (see [G-3] ***region_options***), control the overall look of the graph (see [G-3] ***scheme_option***), and save the graph to disk (see [G-3] ***saving_option***).

See [G-3] ***std_options*** for an overview of the standard options.

Remarks

Remarks are presented under the following headings:

Typical use

`graph matrix` provides an excellent alternative to correlation matrices (see [R] **correlate**) as a quick way to examine the relationships among variables:

```
. use http://www.stata-press.com/data/r12/lifeexp
(Life expectancy, 1998)
. graph matrix popgrowth-safewater
```

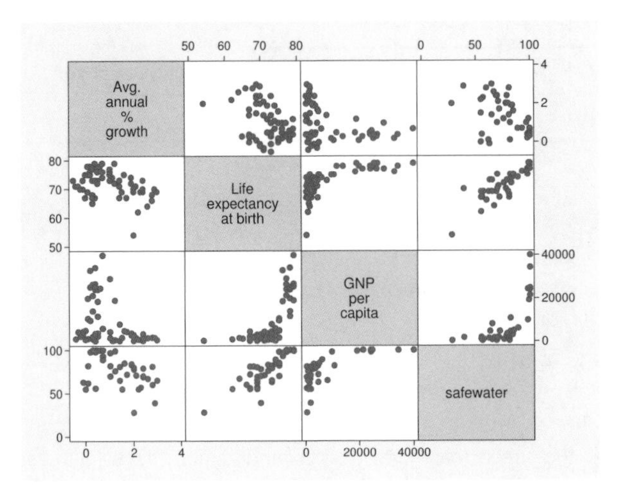

Seeing the above graph, we are tempted to transform gnppc into log units:

```
. generate lgnppc = ln(gnppc)
. graph matrix popgr lexp lgnp safe
```

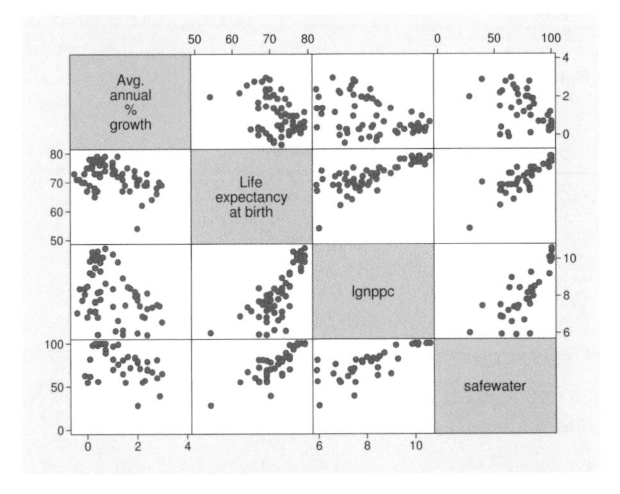

Some people prefer showing just half the matrix, moving the "dependent" variable to the end of the list:

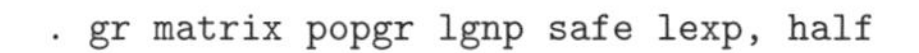

```
. gr matrix popgr lgnp safe lexp, half
```

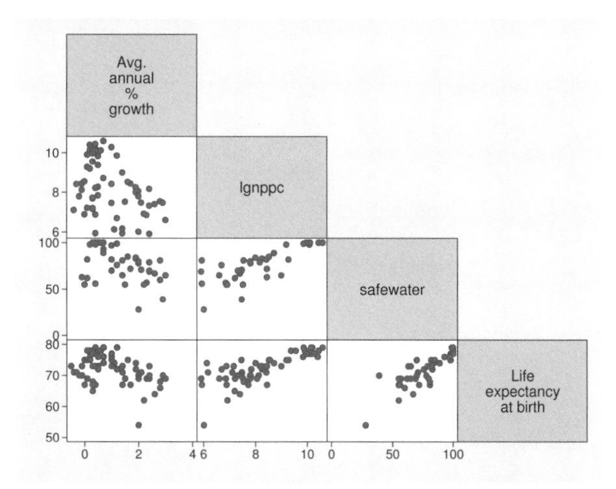

Marker symbols and the number of observations

The `msymbol()` option—abbreviation `ms()`—allows us to control the marker symbol used; see [G-3] ***marker_options***. Hollow symbols sometimes work better as the number of observations increases:

```
. use http://www.stata-press.com/data/r12/auto, clear
(1978 Automobile Data)
. gr mat mpg price weight length, ms(Oh)
```

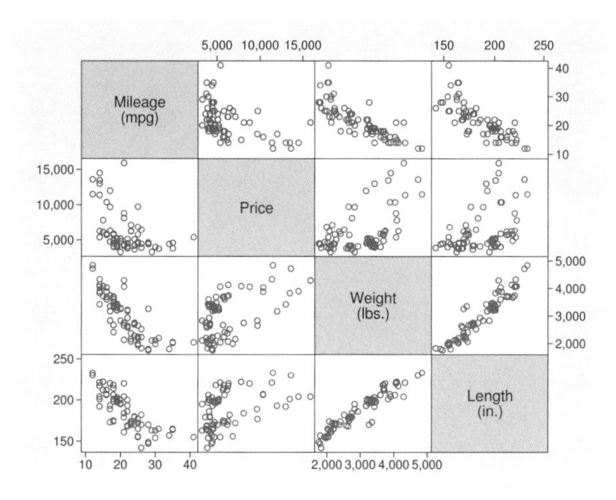

Points work best when there are many data:

```
. use http://www.stata-press.com/data/r12/citytemp, clear
(City Temperature Data)
. gr mat heatdd-tempjuly, ms(p)
```

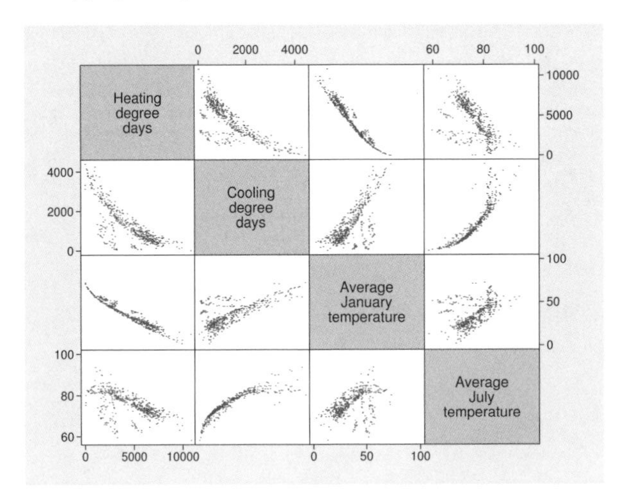

Controlling the axes labeling

By default, approximately three values are labeled and ticked on the y and x axes. When graphing only a few variables, increasing this often works well:

```
. use http://www.stata-press.com/data/r12/citytemp, clear
(City Temperature Data)
. gr mat heatdd-tempjuly, ms(p) maxes(ylab(#4) xlab(#4))
```

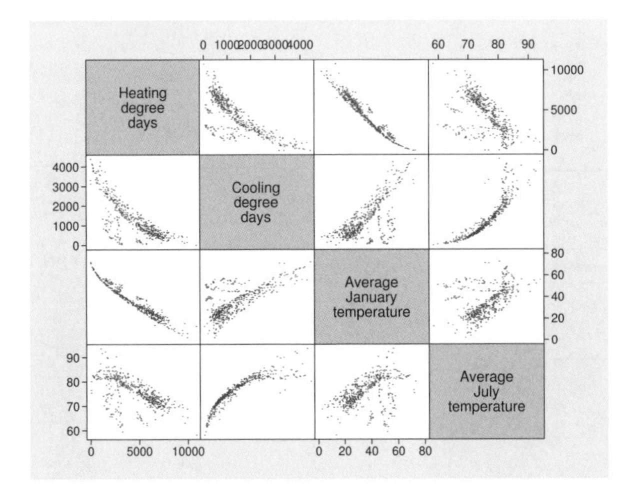

Specifying #4 does not guarantee four labels; it specifies that approximately four labels be used; see [G-3] ***axis_label_options***. Also see *axis_label_options* under *Options* above for instructions on controlling the axes individually.

Adding grid lines

To add horizontal grid lines, specify `maxes(ylab(,grid))`, and to add vertical grid lines, specify `maxes(xlab(,grid))`. Below we do both and specify that four values be labeled:

```
. use http://www.stata-press.com/data/r12/lifeexp, clear
(Life expectancy, 1998)
. generate lgnppc = ln(gnppc)
. graph matrix popgr lexp lgnp safe, maxes(ylab(#4, grid) xlab(#4, grid))
```

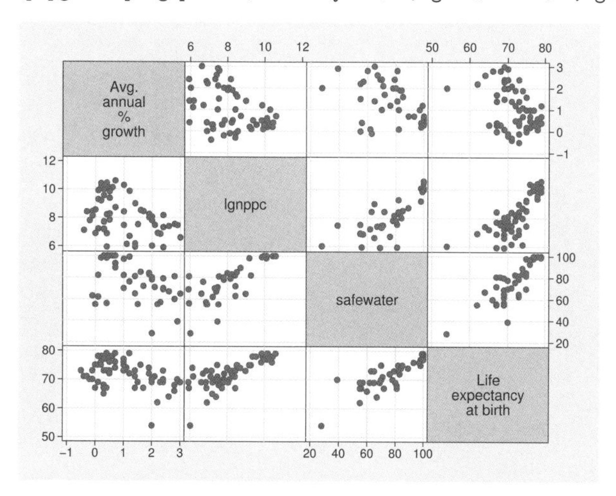

Adding titles

The standard title options may be used with `graph matrix`:

```
. use http://www.stata-press.com/data/r12/lifeexp, clear
(Life expectancy, 1998)
. generate lgnppc = ln(gnppc)
. label var lgnppc "ln GNP per capita"
. graph matrix popgr lexp lgnp safe, maxes(ylab(#4, grid) xlab(#4, grid))
                     subtitle("Summary of 1998 life-expectancy data")
                     note("Source:  The World Bank Group")
```

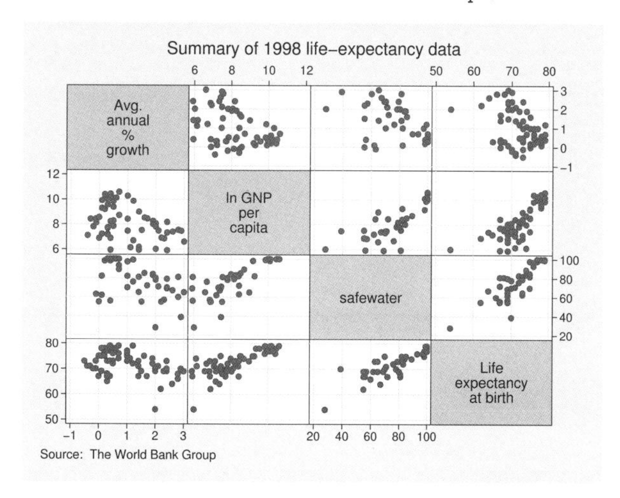

Use with by()

graph matrix may be used with by():

```
. use http://www.stata-press.com/data/r12/auto, clear
(1978 Automobile Data)
. gr matrix mpg weight displ, by(foreign) xsize(5)
```

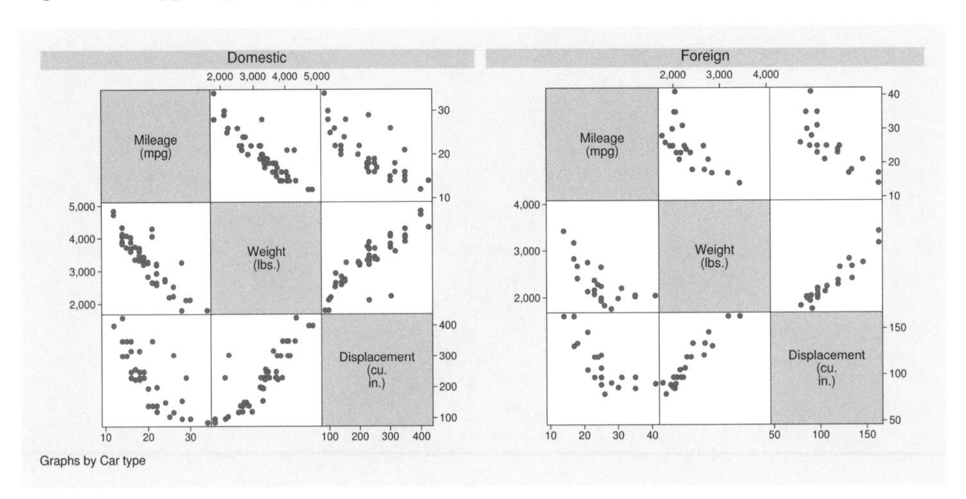

See [G-3] ***by_option***.

History

The origin of the scatterplot matrix is unknown, although early written discussions may be found in Hartigan (1975), Tukey and Tukey (1981), and Chambers et al. (1983). The scatterplot matrix has also been called the *draftman's display* and *pairwise scatterplot*. Regardless of the name used, we believe that the first "canned" implementation was by Becker and Chambers in a system called S—see Becker and Chambers (1984)—although S predates 1984. We also believe that Stata provided the second implementation, in 1985.

References

Basford, K. E., and J. W. Tukey. 1998. *Graphical Analysis of Multiresponse Data.* Boca Raton, FL: Chapman & Hall/CRC.

Becker, R. A., and J. M. Chambers. 1984. *S: An Interactive Environment for Data Analysis and Graphics.* Belmont, CA: Wadsworth.

Chambers, J. M., W. S. Cleveland, B. Kleiner, and P. A. Tukey. 1983. *Graphical Methods for Data Analysis.* Belmont, CA: Wadsworth.

Hartigan, J. A. 1975. Printer graphics for clustering. *Journal of Statistical Computation and Simulation* 4: 187–213.

Tukey, P. A., and J. W. Tukey. 1981. Preparation; prechosen sequences of views. In *Interpreting Multivariate Data*, ed. V. Barnett, 189–213. Chichester, UK: Wiley.

Also see

[G-2] **graph** — The graph command

[G-2] **graph twoway scatter** — Twoway scatterplots

Title

[G-2] graph other — Other graphics commands

Syntax

Distributional diagnostic plots:

Command	Description
`histogram`	histograms
`symplot`	symmetry plots
`quantile`	quantile plots
`qnorm`	quantile–normal plots
`pnorm`	normal probability plots, standardized
`qchi`	chi-squared quantile plots
`pchi`	chi-squared probability plots
`qqplot`	quantile–quantile plots
`gladder`	ladder-of-powers plots
`qladder`	ladder-of-powers quantiles
`spikeplot`	spikeplots and rootograms
`dotplot`	means or medians by group
`sunflower`	density-distribution sunflower plots

See [R] **histogram**, [R] **diagnostic plots**, [R] **ladder**, [R] **spikeplot**, and [R] **sunflower**.

Smoothing and densities:

Command	Description
`kdensity`	kernel density estimation, univariate
`lowess`	lowess smoothing
`lpoly`	local polynomial smoothing

See [R] **kdensity**, [R] **lowess**, and [R] **lpoly**.

Regression diagnostics:

Command	Description
`avplot`	added-variable (leverage) plots
`cprplot`	component-plus-residual plots
`lvr2plot`	L-R (leverage versus squared residual) plots
`rvfplot`	residual-versus-fitted plots
`rvpplot`	residual-versus-predicted plots

See [R] **regress postestimation**.

Time series:

Command	Description
`ac`	correlograms
`pac`	partial correlograms
`pergram`	periodograms
`cumsp`	spectral distribution plots, cumulative
`xcorr`	cross-correlograms for bivariate time series
`wntestb`	Bartlett's periodogram-based white-noise test

See [TS] **corrgram**, [TS] **pergram**, [TS] **cumsp**, [TS] **xcorr**, and [TS] **wntestb**.

Vector autoregressive (VAR, SVAR, VECM) models:

Command	Description
`fcast graph`	`var`, `svar`, and `vec` forecasts
`varstable`	eigenvalues of the companion matrix after `var` and `svar`
`vecstable`	eigenvalues of the companion matrix after `vec`
`irf graph`	impulse–response functions (IRFs) and forecast-error variance decompositions (FEVDs)
`irf ograph`	overlaid IRFs and FEVDs
`irf cgraph`	combined IRFs and FEVDs

See [TS] **fcast graph**, [TS] **varstable**, [TS] **vecstable**, [TS] **irf graph**, [TS] **irf ograph**, and [TS] **irf cgraph**.

Longitudinal data/panel data:

Command	Description
`xtline`	panel-data line plots

See [XT] **xtline**.

Survival analysis:

Command	Description
`sts graph`	survivor and cumulative-hazard functions
`strate`	failure rates and cumulative hazard comparisons
`ltable`	life tables
`stci`	means and percentiles of survival time, with CIs
`stphplot`	log-log plots
`stcoxkm`	Kaplan–Meier observed survival curves
`estat phtest`	verify proportional-hazards assumption
`stcurve`	survivor, hazard, cumulative hazard, or cumulative incidence function

See [ST] **sts graph**, [ST] **strate**, [ST] **ltable**, [ST] **stci**, [ST] **stcox PH-assumption tests**, and [ST] **stcurve**.

ROC analysis:

Command	Description
`roctab`	ROC curve
`rocplot`	parametric ROC curve
`roccomp`	multiple ROC curves, compared
`rocregplot`	marginal and covariate-specific ROC curves
`lroc`	ROC curve after `logistic`, `logit`, `probit`, and `ivprobit`
`lsens`	sensitivity and specificity versus probability cutoff

See [R] **roctab**, [R] **rocfit postestimation**, [R] **roccomp**, [R] **rocregplot**, and [R] **logistic postestimation**.

Multivariate analysis:

Command	Description
`biplot`	biplot
`cluster dendrogram`	dendrograms for hierarchical cluster analysis
`screeplot`	scree plot of eigenvalues
`scoreplot`	factor or component score plot
`loadingplot`	factor or component loading plot
`procoverlay`	Procrustes overlay plot
`cabiplot`	correspondence analysis biplot
`caprojection`	correspondence analysis dimension projection plot
`mcaplot`	plot of category coordinates
`mcaprojection`	MCA dimension projection plot
`mdsconfig`	multidimensional scaling configuration plot
`mdsshepard`	multidimensional scaling Shepard plot

See [MV] **biplot**, [MV] **cluster dendrogram**, [MV] **screeplot**, [MV] **scoreplot**, [MV] **procrustes postestimation**, [MV] **ca postestimation**, and [MV] **mds postestimation**.

Quality-control charts:

Command	Description
`cusum`	cusum plots
`cchart`	c charts
`pchart`	p charts
`rchart`	r charts
`xchart`	$\overline{X}$ charts
`shewhart`	$\overline{X}$ charts, vertically aligned
`serrbar`	standard error bar charts

See [R] **cusum**, [R] **qc**, and [R] **serrbar**.

Other statistical graphs:

Command	Description
`marginsplot`	graph of results from `margins` (profile plots, etc.)
`tabodds`	odds-of-failure versus categories
`pkexamine`	summarize pharmacokinetic data

See [R] **marginsplot**, [ST] **epitab**, and [R] **pkexamine**.

Description

In addition to `graph`, there are many other commands that draw graphs. They are listed above.

Remarks

The other graph commands are implemented in terms of `graph`, which provides the following capabilities:

Command	Description
`graph bar`	bar charts
`graph pie`	pie charts
`graph dot`	dot charts
`graph matrix`	scatterplot matrices
`graph twoway`	twoway (y-x) graphs, including
`graph twoway scatter`	scatterplots
`graph twoway line`	line plots
`graph twoway function`	function plots
`graph twoway histogram`	histograms
`graph twoway *`	more

See [G-2] **graph bar**, [G-2] **graph pie**, [G-2] **graph dot**, [G-2] **graph matrix**, and [G-2] **graph twoway**.

Also see

[G-1] **graph intro** — Introduction to graphics

Title

[G-2] graph pie — Pie charts

Syntax

Slices as totals or percentages of each variable

graph pie *varlist* [*if*] [*in*] [*weight*] [, *options*]

Slices as totals or percentages within over() *categories*

graph pie *varname* [*if*] [*in*] [*weight*], over(*varname*) [*options*]

Slices as frequencies within over() *categories*

graph pie [*if*] [*in*] [*weight*], over(*varname*) [*options*]

options	Description
* over(*varname*)	slices are distinct values of *varname*
missing	do not ignore missing values of *varname*
allcategories	include all categories in the dataset
cw	casewise treatment of missing values
noclockwise	counterclockwise pie chart
angle0(#)	angle of first slice; default is angle(90)
sort	put slices in size order
sort(*varname*)	put slices in *varname* order
descending	reverse default or specified order
pie(...)	look of slice, including explosion
plabel(...)	labels to appear on the slice
ptext(...)	text to appear on the pie
intensity([*]#)	color intensity of slices
line(*line_options*)	outline of slices
legend(...)	legend explaining slices
std_options	titles, saving to disk
by(*varlist*, ...)	repeat for subgroups

* over(*varname*) is required in syntaxes 2 and 3.

See [G-3] ***line_options***, [G-3] ***legend_options***, [G-3] ***std_options***, and [G-3] ***by_option***.

The syntax of the pie() option is

pie({*numlist* | _all} [, *pie_subopts*])

pie_subopts	Description
explode	explode slice by *relativesize* = 3.8
explode(*relativesize*)	explode slice by *relativesize*
color(*colorstyle*)	color of slice

See [G-4] ***relativesize*** and [G-4] ***colorstyle***.

The syntax of the plabel() option is

plabel({# | _all} {sum | percent | name | "*text*"} [, *plabel_subopts*])

plabel_subopts	Description
format(%*fmt*)	display format for sum or percent
gap(*relativesize*)	additional radial distance
textbox_options	look of label

See [G-4] ***relativesize*** and [G-3] ***textbox_options***.

The syntax for the ptext() option is

ptext($\#_a \#_r$ "*text*" ["*text*" ...] [$\#_a$ $\#_r$ ...] [, *ptext_subopts*])

ptext_subopts	Description
textbox_options	look of added text

See [G-3] ***textbox_options***.

aweights, fweights, and pweights are allowed; see [U] **11.1.6 weight**.

Menu

Graphics > Pie chart

Description

graph pie draws pie charts.

graph pie has three modes of operation. The first corresponds to the specification of two or more variables:

```
. graph pie div1_revenue div2_revenue div3_revenue
```

Three pie slices are drawn, the first corresponding to the sum of variable div1_revenue, the second to the sum of div2_revenue, and the third to the sum of div3_revenue.

The second mode of operation corresponds to the specification of one variable and the over() option:

```
. graph pie revenue, over(division)
```

Pie slices are drawn for each value of variable `division`; the first slice corresponds to the sum of revenue for the first division, the second to the sum of revenue for the second division, and so on.

The third mode of operation corresponds to the specification of `over()` with no variables:

```
. graph pie, over(popgroup)
```

Pie slices are drawn for each value of variable `popgroup`; the slices correspond to the number of observations in each group.

Options

`over(`*varname*`)` specifies a categorical variable to correspond to the pie slices. *varname* may be string or numeric.

`missing` is for use with `over()`; it specifies that missing values of *varname* not be ignored. Instead, separate slices are to be formed for *varname*`==.`, *varname*`==.a`, ..., or *varname*`==""`.

`allcategories` specifies that all categories in the entire dataset be retained for the `over()` variables. When `if` or `in` is specified without `allcategories`, the graph is drawn, completely excluding any categories for the `over()` variables that do not occur in the specified subsample. With the `allcategories` option, categories that do not occur in the subsample still appear in the legend, and zero-sized slices are drawn where these categories would appear. Such behavior can be convenient when comparing graphs of subsamples that do not include completely common categories for all `over()` variables. This option has an effect only when `if` or `in` is specified or if there are missing values in the variables. `allcategories` may not be combined with `by()`.

`cw` specifies casewise deletion and is for use when `over()` is not specified. `cw` specifies that, in calculating the sums, observations be ignored for which any of the variables in *varlist* contain missing values. The default is to calculate sums for each variable by using all nonmissing observations.

`noclockwise` and `angle0(`*#*`)` specify how the slices are oriented on the pie. The default is to start at 12 o'clock (known as `angle(90)`) and to proceed clockwise.

`noclockwise` causes slices to be placed counterclockwise.

`angle0(`*#*`)` specifies the angle at which the first slice is to appear. Angles are recorded in degrees and measured in the usual mathematical way: counterclockwise from the horizontal.

`sort`, `sort(`*varname*`)`, and `descending` specify how the slices are to be ordered. The default is to put the slices in the order specified; see *How slices are ordered* under *Remarks* below.

`sort` specifies that the smallest slice be put first, followed by the next largest, etc. See *Ordering slices by size* under *Remarks* below.

`sort(`*varname*`)` specifies that the slices be put in (ascending) order of *varname*. See *Reordering the slices* under *Remarks* below.

`descending`, which may be specified whether or not `sort` or `sort(`*varname*`)` is specified, reverses the order.

`pie({`*numlist* | `_all}`, *pie_subopts*`)` specifies the look of a slice or of a set of slices. This option allows you to "explode" (offset) one or more slices of the pie and to control the color of the slices. Examples include

```
. graph pie ..., ... pie(2, explode)
. graph pie ..., ... pie(2, explode color(red))
. graph pie ..., ... pie(2, explode color(red)) pie(5, explode)
```

numlist specifies the slices; see [U] **11.1.8 numlist**. The slices (after any sorting) are referred to as slice 1, slice 2, etc. `pie(1 ...)` would change the look of the first slice. `pie(2 ...)` would change the look of the second slice. `pie(1 2 3 ...)` would change the look of the first through third slices, as would `pie(1/3 ...)`. The `pie()` option may be specified more than once to specify a different look for different slices. You may also specify `pie(_all ...)` to specify a common characteristic for all slices.

The *pie_subopts* are `explode`, `explode(`*relativesize*`)`, and `color(`*colorstyle*`)`.

`explode` and `explode(`*relativesize*`)` specify that the slice be offset. Specifying `explode` is equivalent to specifying `explode(3.8)`. `explode(`*relativesize*`)` specifies by how much (measured radially) the slice is to be offset; see [G-4] ***relativesize***.

`color(`*colorstyle*`)` sets the color of the slice. See [G-4] ***colorstyle*** for a list of color choices.

`plabel({`*#*`|_all} {sum|percent|name|"`*text*`"},` *plabel_subopts*`)` specifies labels to appear on the slice. Slices may be labeled with their sum, their percentage of the overall sum, their identity, or with text you specify. The default is that no labels appear. Think of the syntax of `plabel()` as

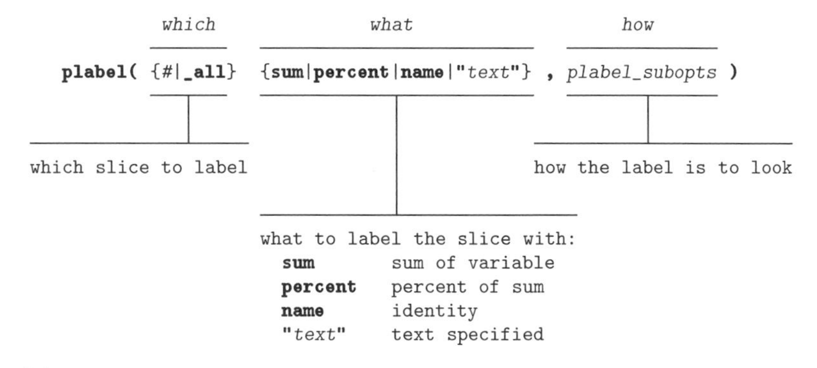

Thus you might type

```
. graph pie ..., ... plabel(_all sum)
. graph pie ..., ... plabel(_all percent)
. graph pie ..., ... plabel(1 "New appropriation")
```

The `plabel()` option may appear more than once, so you might also type

```
. graph pie ..., ... plabel(1 "New appropriation") plabel(2 "old")
```

If you choose to label the slices with their identities, you will probably also want to suppress the legend:

```
. graph pie ..., ... plabel(_all name) legend(off)
```

The *plabel_subopts* are `format(`*%fmt*`)`, `gap(`*relativesize*`)`, and *textbox_options*.

`format(`*%fmt*`)` specifies the display format to be used to format the number when `sum` or `percent` is chosen; see [D] **format**.

`gap(`*relativesize*`)` specifies a radial distance from the origin by which the usual location of the label is to be adjusted. `gap(0)` is the default. `gap(`*#*`)`, *#*$<$ 0, moves the text inward. `gap(`*#*`)`, *#*$>$ 0, moves the text outward. See [G-4] ***relativesize***.

textbox_options specify the size, color, etc., of the text; see [G-3] ***textbox_options***.

`ptext(`$\#_a$ $\#_r$ `"`*text*`"` [`"`*text*`"` ...] [$\#_a$ $\#_r$ ...] `,` *ptext_subopts*`)` specifies additional text to appear on the pie. The position of the text is specified by the polar coordinates $\#_a$ and $\#_r$. $\#_a$ specifies the angle in degrees, and $\#_r$ specifies the distance from the origin in relative-size units; see [G-4] ***relativesize***.

`intensity(`*#*`)` and `intensity(*`*#*`)` specify the intensity of the color used to fill the slices. `intensity(`*#*`)` specifies the intensity, and `intensity(*`*#*`)` specifies the intensity relative to the default.

Specify `intensity(*`*#*`)`, $\# < 1$, to attenuate the interior color and specify `intensity(*`*#*`)`, $\# > 1$, to amplify it.

Specify `intensity(0)` if you do not want the slice filled at all.

`line(`*line_options*`)` specifies the look of the line used to outline the slices. See [G-3] ***line_options***, but ignore option `lpattern()`, which is not allowed for pie charts.

`legend()` allows you to control the legend. See [G-3] ***legend_options***.

std_options allow you to add titles, save the graph on disk, and more; see [G-3] ***std_options***.

`by(`*varlist*`, ...)` draws separate pies within one graph; see [G-3] ***by_option*** and see *Use with by()* under *Remarks* below.

Remarks

Remarks are presented under the following headings:

Typical use
Data are summed
Data may be long rather than wide
How slices are ordered
Ordering slices by size
Reordering the slices
Use with by()
History

Typical use

We have been told that the expenditures for XYZ Corp. are $12 million in sales, $14 million in marketing, $2 million in research, and $8 million in development:

```
. input sales marketing research development

    sales  marketing   research  develop~t
 1. 12 14 2 8
 2. end

. label var sales "Sales"

. label var market "Marketing"

. label var research "Research"

. label var develop  "Development"
```

```
. graph pie sales marketing research development,
        plabel(_all name, size(*1.5) color(white))       (Note 1)
        legend(off)                                      (Note 2)
        plotregion(lstyle(none))                         (Note 3)
        title("Expenditures, XYZ Corp.")
        subtitle("2002")
        note("Source:  2002 Financial Report (fictional data)")
```

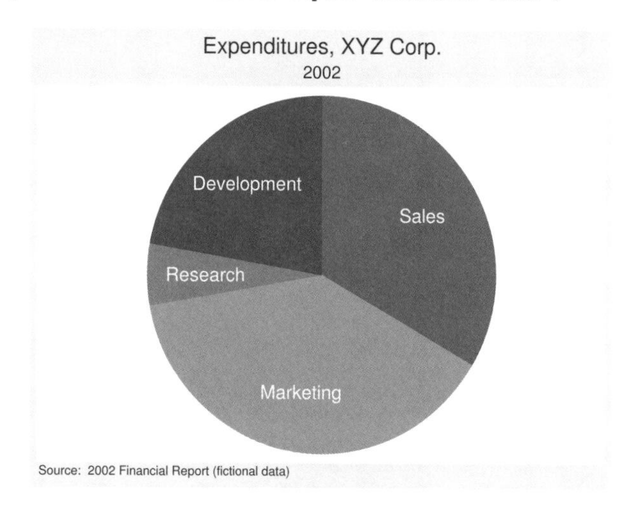

Notes:

1. We specified `plabel(_all name)` to put the division names on the slices. We specified `plabel()`'s textbox-option `size(*1.5)` to make the text 50% larger than usual. We specified `plabel()`'s textbox-option `color(white)` to make the text white. See [G-3] ***textbox_options***.
2. We specified the legend-option `legend(off)` to keep the division names from being repeated in a key at the bottom of the graph; see [G-3] ***legend_options***.
3. We specified the region-option `plotregion(lstyle(none))` to prevent a border from being drawn around the plot area; see [G-3] ***region_options***.

Data are summed

Rather than having the above summary data, we have

```
. list
```

	qtr	sales	marketing	research	development
1.	1	3	4.5	.3	1
2.	2	4	3	.5	2
3.	3	4	4	.6	2
4.	4	2	2.5	.6	3

The sums of these data are the same as the totals in the previous section. The same `graph pie` command

```
. graph pie sales marketing research development, ...
```

will result in the same chart.

Data may be long rather than wide

Rather than having the quarterly data in wide form, we have it in the long form:

```
. list, sepby(qtr)

     +--------------------------+
     | qtr      division   cost |
     |--------------------------|
  1. |   1   Development      1 |
  2. |   1     Marketing    4.5 |
  3. |   1      Research     .3 |
  4. |   1         Sales      3 |
     |--------------------------|
  5. |   2   Development      2 |
  6. |   2     Marketing      3 |
  7. |   2      Research     .5 |
  8. |   2         Sales      4 |
     |--------------------------|
  9. |   3   Development      2 |
 10. |   3     Marketing      4 |
 11. |   3      Research     .6 |
 12. |   3         Sales      3 |
     |--------------------------|
 13. |   4   Development      3 |
 14. |   4     Marketing    2.5 |
 15. |   4      Research     .6 |
 16. |   4         Sales      2 |
     +--------------------------+
```

Here rather than typing

```
. graph pie sales marketing research development, ...
```

we type

```
. graph pie cost, over(division) ...
```

For example,

```
. graph pie cost, over(division),
        plabel(_all name, size(*1.5) color(white))
        legend(off)
        plotregion(lstyle(none))
        title("Expenditures, XYZ Corp.")
        subtitle("2002")
        note("Source:  2002 Financial Report (fictional data)")
```

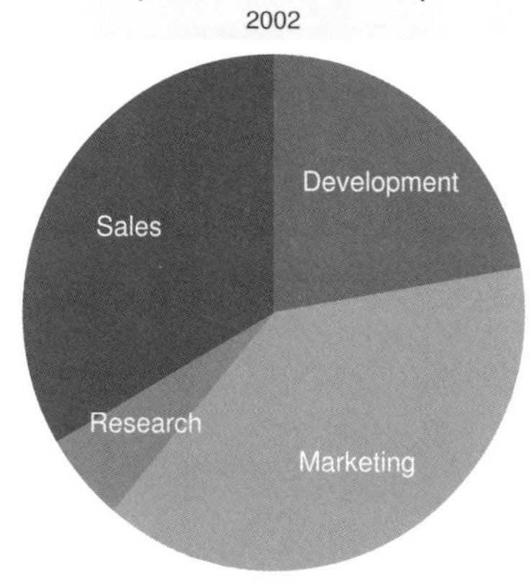

This is the same pie chart as the one drawn previously, except for the order in which the divisions are presented.

How slices are ordered

When we type

```
. graph pie sales marketing research development, ...
```

the slices are presented in the order we specify. When we type

```
. graph pie cost, over(division) ...
```

the slices are presented in the order implied by variable division. If division is numeric, slices are presented in ascending order of division. If division is string, slices are presented in alphabetical order (except that all capital letters occur before lowercase letters).

Ordering slices by size

Regardless of whether we type

```
. graph pie sales marketing research development, ...
```

or

```
. graph pie cost, over(division) ...
```

if we add the `sort` option, slices will be presented in the order of the size, smallest first:

```
. graph pie sales marketing research development, sort ...
. graph pie cost, over(division) sort ...
```

If we also specify the descending option, the largest slice will be presented first:

```
. graph pie sales marketing research development, sort descending ...
. graph pie cost, over(division) sort descending ...
```

Reordering the slices

If we wish to force a particular order, then if we type

```
. graph pie sales marketing research development, ...
```

specify the variables in the desired order. If we type

```
. graph pie cost, over(division) ...
```

then create a numeric variable that has a one-to-one correspondence with the order in which we wish the divisions to appear. For instance, we might type

```
. generate order     = 1 if division=="Sales"
. replace order = 2 if division=="Marketing"
. replace order = 3 if division=="Research"
. replace order = 4 if division=="Development"
```

then type

```
. graph pie cost, over(division) sort(order) ...
```

Use with by()

We have two years of data on XYZ Corp.:

```
. list
```

	year	sales	marketing	research	development
1.	2002	12	14	2	8
2.	2003	15	17.5	8.5	10

```
. graph pie sales marketing research development,
        plabel(_all name, size(*1.5) color(white))
        by(year,
                legend(off)
                title("Expenditures, XYZ Corp.")
                note("Source:  2002 Financial Report (fictional data)")
        )
```

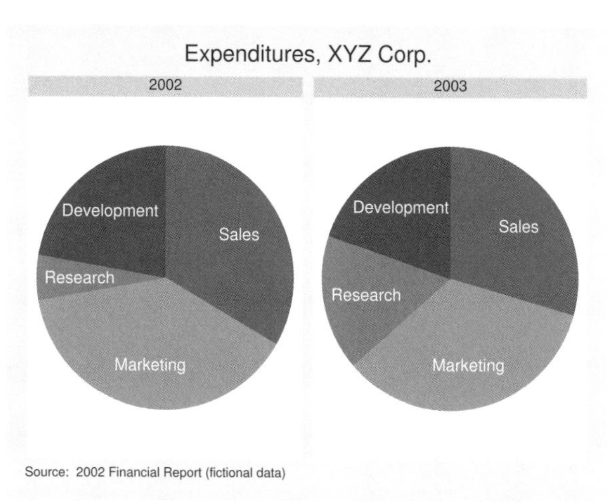

History

The first pie chart is credited to William Playfair (1801). See Beniger and Robyn (1978), Funkhouser (1937, 283–285), or Tufte (2001, 44–45) for more historical details.

William Playfair (1759–1823) was born in Liff, Scotland. He had a varied life with many highs and lows. He participated in the storming of the Bastille, made several engineering inventions, and did path-breaking work in statistical graphics, devising bar charts and pie charts. Playfair also was involved in some shady business ventures and had to shift base from time to time. His brother John (1748–1819) was a distinguished mathematician still remembered for his discussion of Euclidean geometry and his contributions to geology.

References

Beniger, J. R., and D. L. Robyn. 1978. Quantitative graphics in statistics: A brief history. *American Statistician* 32: 1–11.

Funkhouser, H. G. 1937. Historical development of the graphical representation of statistical data. *Osiris* 3: 269–404.

Playfair, W. 1801. *The Statistical Breviary: Shewing, on a Principle Entirely New, the Resources of Every State and Kingdom in Europe to Which is Added, a Similar Exhibition of the Ruling Powers of Hindoostan.* London: Wallis.

——. 2005. *The Commercial and Political Atlas and Statistical Breviary.* Cambridge University Press: Cambridge.

Spence, I., and H. Wainer. 2001. William Playfair. In *Statisticians of the Centuries*, ed. C. C. Heyde and E. Seneta, 105–110. New York: Springer.

Tufte, E. R. 2001. *The Visual Display of Quantitative Information.* 2nd ed. Cheshire, CT: Graphics Press.

Also see

[G-2] **graph** — The graph command

[G-2] **graph bar** — Bar charts

Title

[G-2] graph play — Apply edits from a recording on current graph

Syntax

`graph play` *recordingname*

Description

`graph play` applies edits that were previously recorded using the Graph Recorder to the current graph.

Remarks

Edits made in the Graph Editor (see [G-1] **graph editor**) can be saved as a recording and the edits subsequently played on another graph. In addition to being played from the Graph Editor, these recordings can be played on the currently active graph using the command `graph play` *recordingname*.

If you have previously created a recording named `xyz`, you can replay the edits from that recording on your currently active graph by typing

```
. graph play xyz
```

To learn about creating recordings, see *Graph Recorder* in [G-1] **graph editor**.

Also see

[G-1] **graph editor** — Graph Editor

[G-3] ***play_option*** — Option for playing graph recordings

Title

[G-2] graph print — Print a graph

Syntax

`graph print [, name(windowname) pr_options]`

where *pr_options* are defined in [G-3] ***pr_options***.

Description

`graph print` prints the graph displayed in a Graph window.

Stata for Unix users must do some setup before using `graph print` for the first time; see *Appendix: Setting up Stata for Unix to print graphs* below.

Options

`name(`*windowname*`)` specifies which window to print when printing a graph. The default is for Stata to print the topmost graph (Unix(GUI) users: see the technical note in *Appendix: Setting up Stata for Unix to print graphs*).

The window name is located inside parentheses in the window title. For example, if the title for a Graph window is *Graph (MyGraph)*, the name for the window is *MyGraph*. If a graph is an `asis` or `graph7` graph, where there is no name in the window title, then specify `""` for *windowname*.

pr_options modify how the graph is printed. See [G-3] ***pr_options***.

Default values for the options may be set using [G-2] **graph set**.

Remarks

Graphs are printed by displaying them on the screen and then typing

```
. graph print
```

Remarks are presented under the following headings:

Also see [G-2] **set printcolor**. By default, if the graph being printed has a black background, it is printed in monochrome.

In addition to printing graphs, Stata can export graphs in PostScript, Encapsulated PostScript (EPS), Portable Network Graphics (PNG), TIFF, Windows Metafile (WMF), and Windows Enhanced Metafile (EMF); see [G-2] **graph export**.

Printing the graph displayed in a Graph window

There are three ways to print the graph displayed in a Graph window:

1. Right-click in the Graph window, and select **Print...**.
2. Select **File > Print Graph...**.
3. Type "`graph print`" in the Command window. Unix(GUI) users should use the `name()` option if there is more than one graph displayed to ensure that the correct graph is printed (see the technical note in *Appendix: Setting up Stata for Unix to print graphs*).

All are equivalent. The advantage of `graph print` is that you may include it in do-files:

```
. graph ...                          (draw a graph)
. graph print                        (and print it)
```

Printing a graph stored on disk

To print a graph stored on disk, type

```
. graph use filename
. graph print
```

Do not specify `graph use`'s `nodraw` option; see [G-2] **graph use**.

Stata for Unix(console) users: follow the instructions just given, even though you have no Graph window and cannot see what has just been "displayed". Use the graph, and then print it.

Printing a graph stored in memory

To print a graph stored in memory but not currently displayed, type

```
. graph display name
. graph print
```

Do not specify `graph display`'s `nodraw` option; see [G-2] **graph display**.

Stata for Unix(console) users: follow the instructions just given, even though you have no Graph window and cannot see what has just been "displayed". Display the graph, and then print it.

Appendix: Setting up Stata for Unix to print graphs

Before you can print graphs, you must tell Stata the command you ordinarily use to print PostScript files. By default, Stata assumes that the command is

```
$ lpr < filename
```

That command may be correct for you. If, on the other hand, you usually type something like

```
$ lpr -Plexmark filename
```

you need to tell Stata that by typing

```
. printer define prn ps "lpr -Plexmark @"
```

Type an `@` where you ordinarily would type the filename. If you want the command to be "`lpr -Plexmark < @`", type

```
. printer define prn ps "lpr -Plexmark < @"
```

Stata assumes that the printer you specify understands PostScript format.

❑ Technical note

Unix(GUI) users: X-Windows does not have the concept of a window z-order, which prevents Stata from determining which window is the topmost window. Instead, Stata determines which window is topmost based on which window has the focus. However, some window managers will set the focus to a window without bringing the window to the top. What Stata considers the topmost window may not appear topmost visually. For this reason, you should always use the `name()` option to ensure that the correct Graph window is printed.

❑

Also see

[G-3] ***pr_options*** — Options for use with graph print

[G-2] **set printcolor** — Set how colors are treated when graphs are printed

[G-2] **graph display** — Display graph stored in memory

[G-2] **graph use** — Display graph stored on disk

[G-2] **graph export** — Export current graph

[G-2] **graph set** — Set graphics options

Title

[G-2] graph query — List available schemes and styles

Syntax

```
graph query, schemes
```

```
graph query
```

`graph query` *stylename*

Menu

Graphics > Manage graphs > Query styles and schemes

Description

`graph query, schemes` lists the available schemes.

`graph query` without options lists the available styles.

`graph query` *stylename* lists the styles available within *stylename*.

Remarks

This manual may not be—probably is not—complete. Schemes and styles can be added by StataCorp via updates (see [R] **update**), by other users and traded over the Internet (see [R] **net** and [R] **ssc**), and by you.

Schemes define how graphs look (see [G-4] **schemes intro**), and styles define the features that are available to you (see [G-4] ***symbolstyle*** or [G-4] ***linestyle***).

To find out which schemes are installed on your computer, type

```
. graph query, schemes
```

See [G-4] **schemes intro** for information on schemes and how to use them.

To find out which styles are installed on your computer, type

```
. graph query
```

Many styles will be listed. How you use those styles is described in this manual. For instance, one of the styles that will be listed is *symbolstyle*. See [G-4] ***symbolstyle*** for more information on symbol styles. To find out which symbol styles are available to you, type

```
. graph query symbolstyle
```

All styles end in "*style*", and you may omit typing that part:

```
. graph query symbol
```

Also see

[G-4] **schemes intro** — Introduction to schemes

[G-2] **palette** — Display palettes of available selections

[G-4] ***addedlinestyle*** — Choices for overall look of added lines

[G-4] ***alignmentstyle*** — Choices for vertical alignment of text

[G-4] ***anglestyle*** — Choices for the angle at which text is displayed

[G-4] ***areastyle*** — Choices for look of regions

[G-4] ***axisstyle*** — Choices for overall look of axes

[G-4] ***bystyle*** — Choices for look of by-graphs

[G-4] ***clockposstyle*** — Choices for location: Direction from central point

[G-4] ***colorstyle*** — Choices for color

[G-4] ***compassdirstyle*** — Choices for location

[G-4] ***connectstyle*** — Choices for how points are connected

[G-4] ***gridstyle*** — Choices for overall look of grid lines

[G-4] ***intensitystyle*** — Choices for the intensity of a color

[G-4] ***justificationstyle*** — Choices for how text is justified

[G-4] ***legendstyle*** — Choices for look of legends

[G-4] ***linepatternstyle*** — Choices for whether lines are solid, dashed, etc.

[G-4] ***linestyle*** — Choices for overall look of lines

[G-4] ***linewidthstyle*** — Choices for thickness of lines

[G-4] ***marginstyle*** — Choices for size of margins

[G-4] ***markerlabelstyle*** — Choices for overall look of marker labels

[G-4] ***markersizestyle*** — Choices for the size of markers

[G-4] ***markerstyle*** — Choices for overall look of markers

[G-4] ***orientationstyle*** — Choices for orientation of textboxes

[G-4] ***plotregionstyle*** — Choices for overall look of plot regions

[G-4] ***pstyle*** — Choices for overall look of plot

[G-4] ***ringposstyle*** — Choices for location: Distance from plot region

[G-4] ***shadestyle*** — Choices for overall look of filled areas

[G-4] ***symbolstyle*** — Choices for the shape of markers

[G-4] ***textboxstyle*** — Choices for the overall look of text including border

[G-4] ***textsizestyle*** — Choices for the size of text

[G-4] ***textstyle*** — Choices for the overall look of text

[G-4] ***ticksetstyle*** — Choices for overall look of axis ticks

[G-4] ***tickstyle*** — Choices for the overall look of axis ticks and axis tick labels

Title

[G-2] graph rename — Rename graph in memory

Syntax

`graph rename` [*oldname*] *newname* [, `replace`]

If *oldname* is not specified, the name of the current graph is assumed.

Menu

Graphics > Manage graphs > Rename graph in memory

Description

`graph rename` changes the name of a graph stored in memory.

Option

`replace` specifies that it is okay to replace *newname* if it already exists.

Remarks

See [G-2] **graph manipulation** for an introduction to the graph manipulation commands.

`graph rename` is most commonly used to rename the current graph—the graph currently displayed in the Graph window—when creating the pieces for `graph combine`:

```
. graph ..., ...
. graph rename p1
. graph ..., ...
. graph rename p2
. graph combine p1 p2, ...
```

Also see

[G-2] **graph manipulation** — Graph manipulation commands

[G-2] **graph copy** — Copy graph in memory

Title

[G-2] graph save — Save graph to disk

Syntax

`graph save` [*graphname*] *filename* [`, asis replace`]

Description

`graph save` saves the specified graph to disk. If *graphname* is not specified, the graph currently displayed is saved to disk in Stata's `.gph` format.

If *filename* is specified without an extension, `.gph` is assumed.

Options

`asis` specifies that the graph be frozen and saved as is. The alternative—and the default if `asis` is not specified—is live format. In live format, the graph can be edited in future sessions, and the overall look of the graph continues to be controlled by the chosen scheme (see [G-4] **schemes intro**).

Say that you type

```
. scatter yvar xvar, ...
. graph save mygraph
```

which will create file `mygraph.gph`. Suppose that you send the file to a colleague. The way the graph will appear on your colleague's computer might be different from how it appeared on yours. Perhaps you display titles on the top, and your colleague has set his scheme to display titles on the bottom. Or perhaps your colleague prefers y axes on the right rather than on the left. It will still be the same graph, but it might look different.

Or perhaps you just file away `mygraph.gph` for use later. If the file is stored in the default live format, you can come back to it and change the way it looks by specifying a different scheme, and you can edit it.

If, on the other hand, you specify `asis`, the graph will forever look just as it looked the instant it was saved. You cannot edit it, and you cannot change the scheme. If you send the as-is graph to colleagues, they will see it exactly in the form that you see it.

Whether a graph is saved as-is or live makes no difference for printing. As-is graphs usually require fewer bytes to store, and they generally display more quickly, but that is all.

`replace` specifies that the file may be replaced if it already exists.

Remarks

You may instead specify that the graph be saved at the instant you draw it by specifying the `saving(`*filename*[`, asis replace`]`)` option; see [G-3] ***saving_option***.

Also see

[G-3] ***saving_option*** — Option for saving graph to disk

[G-2] **graph export** — Export current graph

[G-4] **concept: gph files** — Using gph files

[G-2] **graph manipulation** — Graph manipulation commands

Title

[G-2] graph set — Set graphics options

Syntax

Manage graph print settings

`graph set print` [*setopt setval*]

Manage graph export settings

`graph set` [*exporttype*] [*setopt setval*]

where *exporttype* is the export file type and may be one of

`ps` | `eps`

and *setopt* is the option to set with the setting *setval*.

Manage Graph window font settings

`graph set window fontface` {*fontname* | `default`}
`graph set window fontfacemono` {*fontname* | `default`}
`graph set window fontfacesans` {*fontname* | `default`}
`graph set window fontfaceserif` {*fontname* | `default`}
`graph set window fontfacesymbol` {*fontname* | `default`}

Description

`graph set` without options lists the current graphics font, print, and export settings for all *exporttype*s. `graph set` with `window`, `print`, or *exporttype* lists the current settings for the Graph window, for printing, or for the specified *exporttype*, respectively.

`graph set print` allows you to change the print settings for graphics.

`graph set` *exporttype* allows you to change the graphics export settings for export file type *exporttype*.

`graph set window fontface*` allows you to change the Graph window font settings. (To change font settings for graphs exported to PostScript or Encapsulated PostScript files, use `graph set` { `ps` | `eps` } `fontface*`; see [G-3] ***ps_options*** or [G-3] ***eps_options***.) If *fontname* contains spaces, enclose it in double quotes. If you specify `default` for any of the `fontface*` settings, the default setting will be restored.

Remarks

Remarks are presented under the following headings:

Overview
Setting defaults

Overview

`graph set` allows you to permanently set the primary font face used in the Graph window as well as the font faces to be used for the four Stata "font faces" supported by the graph SMCL tags `{stMono}`, `{stSans}`, `{stSerif}`, and `{stSymbol}`. See [G-4] ***text*** for more details on these SMCL tags.

`graph set` also allows you to permanently set any of the options supported by `graph print` (see [G-2] **graph print**) or by the specific export file types provided by `graph export` (see [G-2] **graph export**).

To find out more about the `graph set print` *setopt* options and their associated values (*setval*), see [G-3] ***pr_options***.

Some graphics file types supported by `graph export` (see [G-2] **graph export**) have options that can be set. The file types that allow option settings and their associated *exporttype*s are

exporttype	Description	Available settings
`ps`	PostScript	[G-3] ***ps_options***
`eps`	Encapsulated PostScript	[G-3] ***eps_options***

Setting defaults

If you always want the Graph window to use Times New Roman as its default font, you could type

```
. graph set window fontface "Times New Roman"
```

Later, you could type

```
. graph set window fontface default
```

to restore the factory setting.

To change the font used by `{stMono}` in the Graph window, you could type

```
. graph set window fontfacemono "Lucida Console"
```

and to reset it, you could type

```
. graph set window fontfacemono default
```

You can list the current graph settings by typing

```
. graph set
```

Also see

[G-2] **graph print** — Print a graph

[G-2] **graph export** — Export current graph

[G-4] ***text*** — Text in graphs

[G-3] ***pr_options*** — Options for use with graph print

[G-3] ***ps_options*** — Options for exporting or printing to PostScript

[G-3] ***eps_options*** — Options for exporting to Encapsulated PostScript

Title

[G-2] graph twoway — Twoway graphs

Syntax

[graph] twoway *plot* [*if*] [*in*] [, *twoway_options*]

where the syntax of *plot* is

[(] *plottype varlist* ..., *options* [)] [||]

plottype	Description
`scatter`	scatterplot
`line`	line plot
`connected`	connected-line plot
`scatteri`	`scatter` with immediate arguments
`area`	line plot with shading
`bar`	bar plot
`spike`	spike plot
`dropline`	dropline plot
`dot`	dot plot
`rarea`	range plot with area shading
`rbar`	range plot with bars
`rspike`	range plot with spikes
`rcap`	range plot with capped spikes
`rcapsym`	range plot with spikes capped with symbols
`rscatter`	range plot with markers
`rline`	range plot with lines
`rconnected`	range plot with lines and markers
`pcspike`	paired-coordinate plot with spikes
`pccapsym`	paired-coordinate plot with spikes capped with symbols
`pcarrow`	paired-coordinate plot with arrows
`pcbarrow`	paired-coordinate plot with arrows having two heads
`pcscatter`	paired-coordinate plot with markers
`pci`	`pcspike` with immediate arguments
`pcarrowi`	`pcarrow` with immediate arguments
`tsline`	time-series plot
`tsrline`	time-series range plot
`contour`	contour plot with filled areas
`contourline`	contour lines plot

`mband`	median-band line plot
`mspline`	spline line plot
`lowess`	LOWESS line plot
`lfit`	linear prediction plot
`qfit`	quadratic prediction plot
`fpfit`	fractional polynomial plot
`lfitci`	linear prediction plot with CIs
`qfitci`	quadratic prediction plot with CIs
`fpfitci`	fractional polynomial plot with CIs
`function`	line plot of function
`histogram`	histogram plot
`kdensity`	kernel density plot
`lpoly`	local polynomial smooth plot
`lpolyci`	local polynomial smooth plot with CIs

For each of the above, see [G] **graph twoway** *plottype*, where you substitute for *plottype* a word from the left column.

twoway_options are as defined in [G-3] ***twoway_options***.

The leading `graph` is optional. If the first (or only) *plot* is `scatter`, you may omit `twoway` as well, and then the syntax is

`scatter` ... [, *scatter_options*] [|| *plot* [*plot* [...]]]

and the same applies to `line`. The other *plottypes* must be preceded by `twoway`.

Regardless of how the command is specified, *twoway_options* may be specified among the *scatter_options*, *line_options*, etc., and they will be treated just as if they were specified among the *twoway_options* of the `graph twoway` command.

Menu

Graphics > Twoway graph (scatter, line, etc.)

Description

`twoway` is a family of plots, all of which fit on numeric y and x scales.

Remarks

Remarks are presented under the following headings:

Definition
Syntax
Multiple if and in restrictions
twoway and plot options

Definition

Twoway graphs show the relationship between numeric data. Say that we have data on life expectancy in the United States between 1900 and 1940:

```
. use http://www.stata-press.com/data/r12/uslifeexp2
(U.S. life expectancy, 1900-1940)
. list in 1/8
```

	year	le
1.	1900	47.3
2.	1901	49.1
3.	1902	51.5
4.	1903	50.5
5.	1904	47.6
6.	1905	48.7
7.	1906	48.7
8.	1907	47.6

We could graph these data as a twoway scatterplot,

```
. twoway scatter le year
```

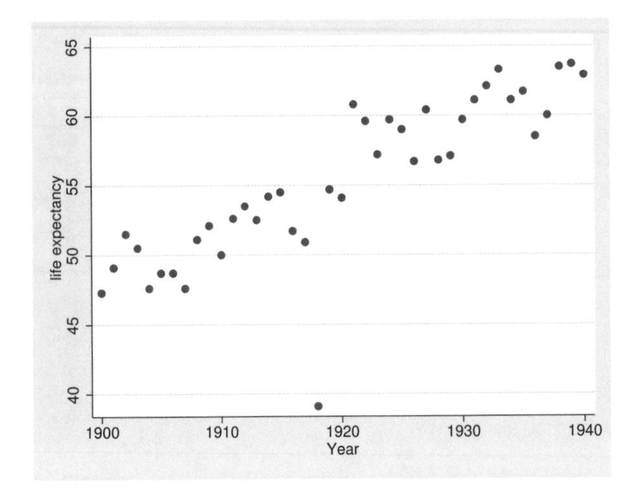

or we could graph these data as a twoway line plot,

```
. twoway line le year
```

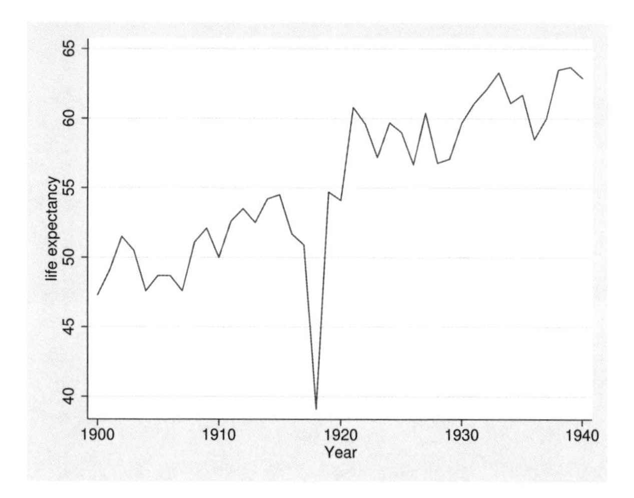

or we could graph these data as a twoway connected plot, marking both the points and connecting them with straight lines,

```
. twoway connected le year
```

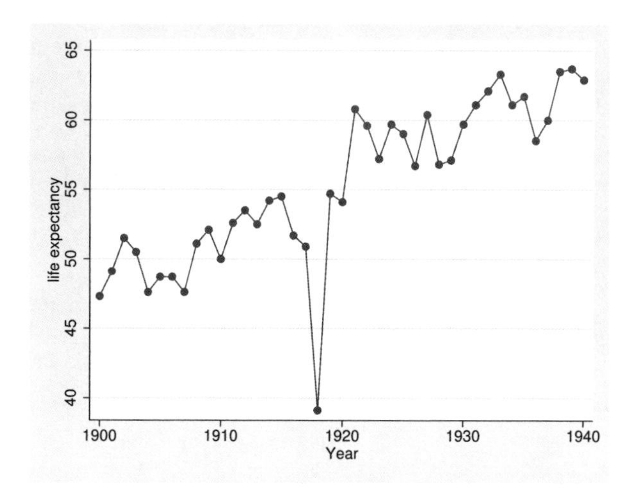

or we could graph these data as a scatterplot and put on top of that the prediction from a linear regression of `le` on `year`,

```
. twoway (scatter le year) (lfit le year)
```

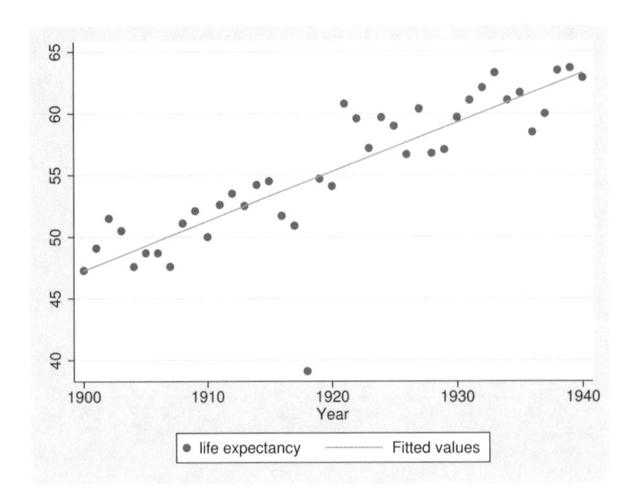

or we could graph these data in many other ways.

These all are examples of twoway graphs. What distinguishes a twoway graph is that it fits onto numeric y and x axes.

Each of what we produced above is called a *graph*. What appeared in the graphs are called *plots*. In the first graph, the plottype was a scatter; in the second, the plottype was a line; in the third, the plottype was connected; and in the fourth, there were two plots: a scatter combined with a line plot of a linear fit.

`twoway` provides many different plottypes. Some, such as `scatter` and `line`, simply render the data in different ways. Others, such as `lfit`, transform the data and render that. And still others, such as `function`, actually make up data to be rendered. This last class makes it easy to overlay $y = x$ lines or $y = f(x)$ functions on your graphs.

By the way, in case you are wondering, there are no errors in the above data. In 1918, there was an outbreak of influenza known as the 1918 Influenza Pandemic, which in the United States, was the worst epidemic ever known and which killed more citizens than all combat deaths of the 20th century.

Syntax

If we want to graph `y1` versus `x` and `y2` versus `x`, the formal way to type this is

```
. graph twoway (scatter y1 x) (scatter y2 x)
```

If we wanted `y1` versus `x` plotted with solid circles and `y2` versus `x` plotted with hollow circles, formally we would type

```
. graph twoway (scatter y1 x, ms(O)) (scatter y2 x, ms(Oh))
```

If we wanted `y1` versus `x` plotted with solid circles and wanted a line graph for `y2` versus `x`, formally we would type

```
. graph twoway (scatter y1 x, ms(O)) (line y2 x, sort)
```

The `sort` option is included under the assumption that the data are not already sorted by `x`.

We have shown the formal way to type each of our requests, but few people would type that. First, most users omit the graph:

```
. twoway (scatter y1 x) (scatter y2 x)
. twoway (scatter y1 x, ms(O)) (scatter y2 x, ms(Oh))
. twoway (scatter y1 x, ms(O)) (line y2 x, sort)
```

Second, most people use the ||-separator notation rather than the ()-binding notation:

```
. twoway scatter y1 x || scatter y2 x
. twoway scatter y1 x, ms(O) || scatter y2 x, ms(Oh)
. twoway scatter y1 x, ms(O) || line y2 x, sort
```

Third, most people now omit the twoway:

```
. scatter y1 x || scatter y2 x
. scatter y1 x, ms(O) || scatter y2 x, ms(Oh)
. scatter y1 x, ms(O) || line y2 x, sort
```

And finally, most people quickly realize that scatter allows us to plot more than one y variable against the same x variable:

```
. scatter y1 y2 x
. scatter y1 y2 x, ms(O Oh)
. scatter y1 x, ms(O) || line y2 x, sort
```

The third example did not change: in that example, we are combining a scatterplot and a line plot. Actually, in this particular case, there is a way we can combine that, too:

```
. scatter y1 y2 x, ms(O i) connect(. l)
```

That we can combine scatter and line just happens to be an oddity of the examples we picked. It is important to understand that there is nothing wrong with any of the above ways of typing our request, and sometimes the wordier syntaxes are the only way to obtain what we want. If we wanted to graph y1 versus x1 and y2 versus x2, the only way to type that is

```
. scatter y1 x1 || scatter y2 x2
```

or to type the equivalent in one of the wordier syntaxes above it. We have to do this because scatter (see [G-2] **graph twoway scatter**) draws a scatterplot against one x variable. Therefore, if we want two different x variables, we need two different scatters.

In any case, we will often refer to the graph twoway command, even though, when we give the command, we will seldom type the graph, and mostly, we will not type the twoway either.

Multiple if and in restrictions

Each *plot* may have its own if *exp* and in *range* restrictions:

```
. twoway (scatter mpg weight if foreign, msymbol(O))
         (scatter mpg weight if !foreign, msymbol(Oh))
```

Multiple *plots* in one graph twoway command draw one graph with multiple things plotted in it. The above will produce a scatter of mpg versus weight for foreign cars (making the points with solid circles) and a scatter of mpg versus weight for domestic cars (using hollow circles).

Also, the graph twoway command itself can have if *exp* and in *range* restrictions:

```
. twoway (scatter mpg weight if foreign, msymbol(O))
         (scatter mpg weight if !foreign, msymbol(Oh)) if mpg>20
```

The if mpg>20 restriction will apply to both scatters.

We have chosen to show these two examples with the ()-binding notation because it makes the scope of each `if` *exp* so clear. In ||-separator notation, the commands would read

```
. twoway scatter mpg weight if foreign, msymbol(O) ||
         scatter mpg weight if !foreign, msymbol(Oh)
```

and

```
. twoway scatter mpg weight if foreign, msymbol(O) ||
         scatter mpg weight if !foreign, msymbol(Oh) || if mpg>20
```

or even

```
. scatter mpg weight if foreign, msymbol(O) ||
         scatter mpg weight if !foreign, msymbol(Oh)
```

and

```
. scatter mpg weight if foreign, msymbol(O) ||
         scatter mpg weight if !foreign, msymbol(Oh) || if mpg>20
```

We may specify `graph twoway` restrictions only, of course:

```
. twoway (scatter mpg weight) (lfit mpg weight) if !foreign
. scatter mpg weight || lfit mpg weight || if !foreign
```

twoway and plot options

`graph twoway` allows options, and the individual *plots* allow options. For instance, `graph twoway` allows the `saving()` option, and `scatter` (see [G-2] **graph twoway scatter**) allows the `msymbol()` option, which specifies the marker symbol to be used. Nevertheless, we do not have to keep track of which option belongs to which. If we type

```
. scatter mpg weight, saving(mygraph) msymbol(Oh)
```

the results will be the same as if we more formally typed

```
. twoway (scatter mpg weight, msymbol(Oh)), saving(mygraph)
```

Similarly, we could type

```
. scatter mpg weight, msymbol(Oh) || lfit mpg weight, saving(mygraph)
```

or

```
. scatter mpg weight, msymbol(Oh) saving(mygraph) || lfit mpg weight
```

and, either way, the results would be the same as if we typed

```
. twoway (scatter mpg weight, msymbol(Oh))
         (lfit mpg weight), saving(mygraph)
```

We may specify a `graph twoway` option "too deeply", but we cannot go the other way. The following is an error:

```
. scatter mpg weight || lfit mpg weight ||, msymbol(Oh) saving(mygraph)
```

It is an error because we specified a `scatter` option where only a `graph twoway` option may be specified, and given what we typed, there is insufficient information for `graph twoway` to determine for which *plot* we meant the `msymbol()` option. Even when there is sufficient information (say that option `msymbol()` were not allowed by `lfit`), it would still be an error. `graph twoway` can reach in and pull out its options, but it cannot take from its options and distribute them back to the individual *plots*.

Title

[G-2] graph twoway area — Twoway line plot with area shading

Syntax

`twoway area` *yvar xvar* [*if*] [*in*] [, *options*]

options	Description
`vertical`	vertical area plot; the default
`horizontal`	horizontal area plot
`cmissing(y|n)`	missing values do not force gaps in area; default is `cmissing(y)`
`base(#)`	value to drop to; default is 0
`nodropbase`	programmer's option
`sort`	sort by *xvar*; recommended
area_options	change look of shaded areas
axis_choice_options	associate plot with alternative axis
twoway_options	titles, legends, axes, added lines and text, by, regions, name, aspect ratio, etc.

See [G-3] ***area_options***, [G-3] ***axis_choice_options***, and [G-3] ***twoway_options***.

Option `base()` is *rightmost*; `vertical`, `horizontal`, `nodropbase`, and `sort` are *unique*; see [G-4] **concept: repeated options**.

Menu

Graphics > Twoway graph (scatter, line, etc.)

Description

`twoway area` displays (y,x) connected by straight lines and shaded underneath.

Options

`vertical` and `horizontal` specify either a vertical or a horizontal area plot. `vertical` is the default. If `horizontal` is specified, the values recorded in *yvar* are treated as x values, and the values recorded in *xvar* are treated as y values. That is, to make horizontal plots, do not switch the order of the two variables specified.

In the `vertical` case, shading at each *xvar* value extends up or down from 0 according to the corresponding *yvar* values. If 0 is not in the range of the y axis, shading extends up or down to the x axis.

In the `horizontal` case, shading at each *xvar* value extends left or right from 0 according to the corresponding *yvar* values. If 0 is not in the range of the x axis, shading extends left or right to the y axis.

`cmissing(y|n)` specifies whether missing values are to be ignored when drawing the area or if they are to create breaks in the area. The default is `cmissing(y)`, meaning that they are ignored. Consider the following data:

	y1	y2	x
1.	1	2	1
2.	3	5	2
3.	5	4	3
4.	.	.	.
5.	6	7	5
6.	11	12	8

Say that you graph these data by using `twoway area y1 y2 x`. Do you want a break in the area between 3 and 5? If so, you type

```
. twoway area y1 y2 x, cmissing(n)
```

and two areas will be drawn, one for the observations before the missing values at observation 4 and one for the observations after the missing values.

If you omit the option (or type `cmissing(y)`), the data are treated as if they contained

	y1	y2	x
1.	1	2	1
2.	3	5	2
3.	5	4	3
4.	6	7	5
5.	11	12	8

meaning that one contiguous area will be drawn over the range (1,8).

`base(#)` specifies the value from which the shading should extend. The default is `base(0)`, and in the above description of options `vertical` and `horizontal`, this default was assumed.

`nodropbase` is a programmer's option and is an alternative to `base()`. It specifies that rather than the enclosed area dropping to `base(#)`—or `base(0)`—it drops to the line formed by (y_1,x_1) and (y_N,x_N), where (y_1,x_1) are the y and x values in the first observation being plotted and (y_N,x_N) are the values in the last observation being plotted.

`sort` specifies that the data be sorted by *xvar* before plotting.

area_options set the look of the shaded areas. The most important of these options is `color(`*colorstyle*`)`, which specifies the color of both the area and its outline; see [G-4] ***colorstyle*** for a list of color choices. See [G-3] ***area_options*** for information on the other *area_options*.

axis_choice_options associate the plot with a particular y or x axis on the graph; see [G-3] ***axis_choice_options***.

twoway_options are a set of common options supported by all `twoway` graphs. These options allow you to title graphs, name graphs, control axes and legends, add lines and text, set aspect ratios, create graphs over `by()` groups, and change some advanced settings. See [G-3] ***twoway_options***.

Remarks

Remarks are presented under the following headings:

Typical use
Advanced use
Cautions

Typical use

We have quarterly data recording the U.S. GNP in constant 1996 dollars:

```
. use http://www.stata-press.com/data/r12/gnp96
. list in 1/5
```

	date	gnp96
1.	1967q1	3631.6
2.	1967q2	3644.5
3.	1967q3	3672
4.	1967q4	3703.1
5.	1968q1	3757.5

In our opinion, the area under a curve should be shaded only if the area is meaningful:

```
. use http://www.stata-press.com/data/r12/gnp96, clear
. twoway area d.gnp96 date
```

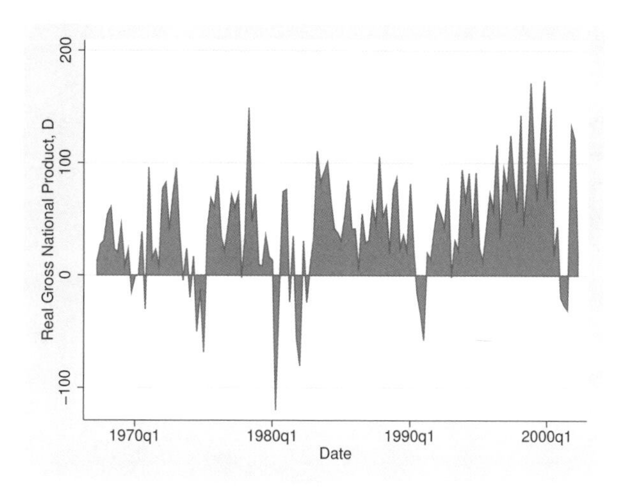

Advanced use

Here is the same graph, but greatly improved with some advanced options:

```
. twoway area d.gnp96 date, xlabel(36(8)164, angle(90))
        ylabel(-100(50)200, angle(0))
        ytitle("Billions of 1996 Dollars")
        xtitle("")
        subtitle("Change in U.S. GNP", position(11))
        note("Source: U.S. Department of Commerce,
                       Bureau of Economic Analysis")
```

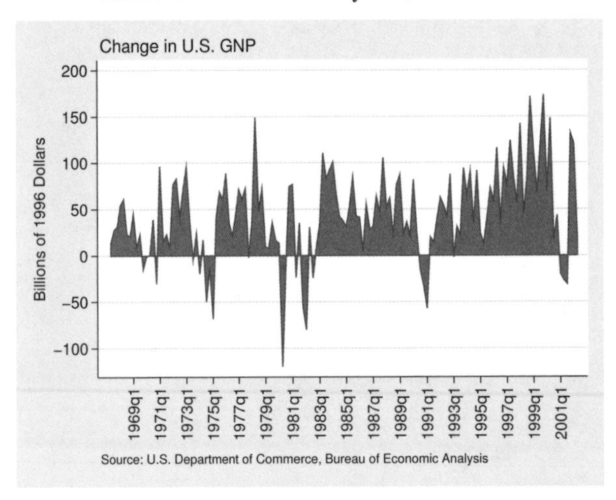

Cautions

Be sure that the data are in the order of *xvar*, or specify area's sort option. If you do neither, you will get something that looks like modern art:

```
. use http://www.stata-press.com/data/r12/gnp96, clear
. generate d = d.gnp96
. generate u = runiform()
. sort u                                  (put in random order)
. twoway area d date
```

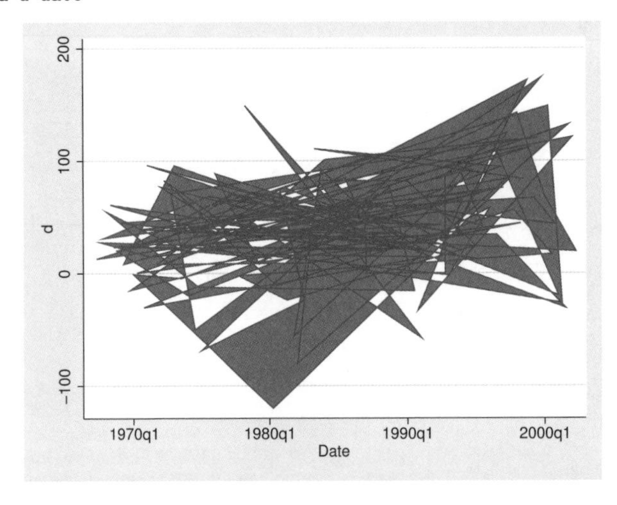

Also see

[G-2] **graph twoway scatter** — Twoway scatterplots

[G-2] **graph twoway dot** — Twoway dot plots

[G-2] **graph twoway dropline** — Twoway dropped-line plots

[G-2] **graph twoway histogram** — Histogram plots

[G-2] **graph twoway spike** — Twoway spike plots

[G-2] **graph bar** — Bar charts

Title

[G-2] graph twoway bar — Twoway bar plots

Syntax

`twoway bar` *yvar xvar* [*if*] [*in*] [, *options*]

options	Description
`vertical`	vertical bar plot; the default
`horizontal`	horizontal bar plot
`base(#)`	value to drop to; default is 0
`barwidth(#)`	width of bar in *xvar* units
barlook_options	change look of bars
axis_choice_options	associate plot with alternative axis
twoway_options	titles, legends, axes, added lines and text, by, regions, name, aspect ratio, etc.

See [G-3] ***barlook_options***, [G-3] ***axis_choice_options***, and [G-3] ***twoway_options***.

Options `base()` and `barwidth()` are *rightmost*, and `vertical` and `horizontal` are *unique*; see [G-4] **concept: repeated options**.

Menu

Graphics > Twoway graph (scatter, line, etc.)

Description

`twoway bar` displays numeric (y,x) data as bars. `twoway bar` is useful for drawing bar plots of time-series data or other equally spaced data and is useful as a programming tool. For finely spaced data, also see [G-2] **graph twoway spike**.

Also see [G-2] **graph bar** for traditional bar charts and [G-2] **graph twoway histogram** for histograms.

Options

`vertical` and `horizontal` specify either a vertical or a horizontal bar plot. `vertical` is the default. If `horizontal` is specified, the values recorded in *yvar* are treated as x values, and the values recorded in *xvar* are treated as y values. That is, to make horizontal plots, do not switch the order of the two variables specified.

In the `vertical` case, bars are drawn at the specified *xvar* values and extend up or down from 0 according to the corresponding *yvar* values. If 0 is not in the range of the y axis, bars extend up or down to the x axis.

In the `horizontal` case, bars are drawn at the specified *xvar* values and extend left or right from 0 according to the corresponding *yvar* values. If 0 is not in the range of the x axis, bars extend left or right to the y axis.

`base(#)` specifies the value from which the bar should extend. The default is `base(0)`, and in the above description of options `vertical` and `horizontal`, this default was assumed.

`barwidth(#)` specifies the width of the bar in *xvar* units. The default is `width(1)`. When a bar is plotted, it is centered at x, so half the width extends below x and half above.

barlook_options set the look of the bars. The most important of these options is `color(`*colorstyle*`)`, which specifies the color of the bars; see [G-4] ***colorstyle*** for a list of color choices. See [G-3] ***barlook_options*** for information on the other *barlook_options*.

axis_choice_options associate the plot with a particular y or x axis on the graph; see [G-3] ***axis_choice_options***.

twoway_options are a set of common options supported by all `twoway` graphs. These options allow you to title graphs, name graphs, control axes and legends, add lines and text, set aspect ratios, create graphs over `by()` groups, and change some advanced settings. See [G-3] ***twoway_options***.

Remarks

Remarks are presented under the following headings:

Typical use
Advanced use: Overlaying
Advanced use: Population pyramid
Cautions

Typical use

We have daily data recording the values for the S&P 500 in 2001:

```
. use http://www.stata-press.com/data/r12/sp500
(S&P 500)

. list date close change in 1/5
```

	date	close	change
1.	02jan2001	1283.27	.
2.	03jan2001	1347.56	64.29004
3.	04jan2001	1333.34	-14.22009
4.	05jan2001	1298.35	-34.98999
5.	08jan2001	1295.86	-2.48999

We will use the first 57 observations from these data:

```
. twoway bar change date in 1/57
```

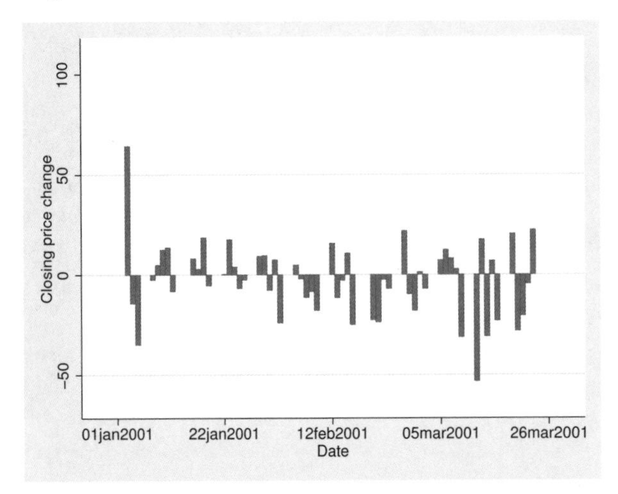

We get a different visual effect if we reduce the width of the bars from 1 day to .6 days:

```
. twoway bar change date in 1/57, barw(.6)
```

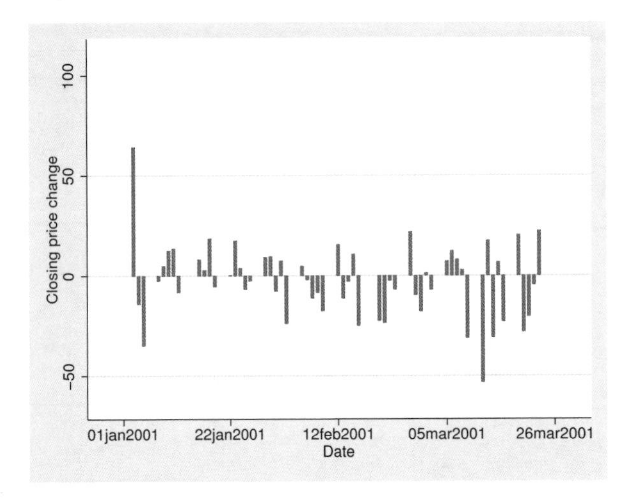

Advanced use: Overlaying

The useful thing about `twoway bar` is that it can be combined with other `twoway` plottypes (see [G-2] **graph twoway**):

```
. twoway line close date || bar change date || in 1/52
```

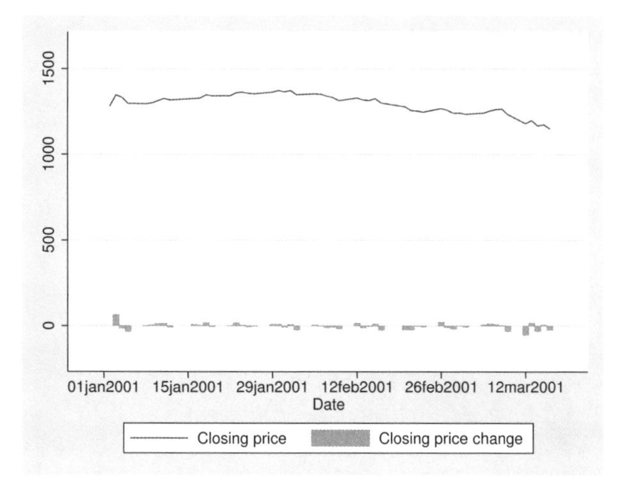

We can improve this graph by typing

```
. twoway
        line close date, yaxis(1)
  ||
        bar change date, yaxis(2)
  ||
  in 1/52,
        ysca(axis(1) r(1000 1400)) ylab(1200(50)1400, axis(1))
        ysca(axis(2) r(-50 300)) ylab(-50 0 50, axis(2))
                ytick(-50(25)50, axis(2) grid)
        legend(off)
        xtitle("Date")
        title("S&P 500")
        subtitle("January - March 2001")
        note("Source:  Yahoo!Finance and Commodity Systems, Inc.")
        yline(1150, axis(1) lstyle(foreground))
```

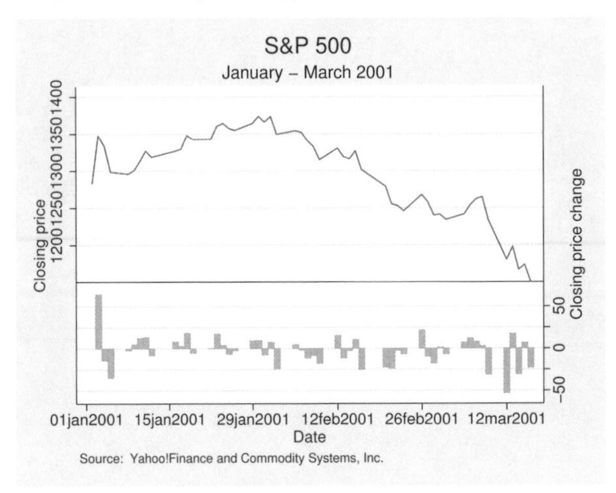

Notice the use of

```
yline(1150, axis(1) lstyle(foreground))
```

The 1150 put the horizontal line at $y = 1150$; `axis(1)` stated that y should be interpreted according to the left y axis; and `lstyle(foreground)` specified that the line be drawn in the foreground style.

Instead of `lstyle(foreground)` we could have coded `lcolor(white)` or `lcolor(black)`, depending on whether we were producing a graph for display on the screen or for printing in the manuals. Specifying `lstyle(foreground)` got us around the problem of having to use a different option for the help files and the manual.

Advanced use: Population pyramid

We have the following aggregate data from the U.S. 2000 Census recording total population by age and sex. From this, we produce a population pyramid:

```
. use http://www.stata-press.com/data/r12/pop2000, clear
. list agegrp maletotal femtotal
```

	agegrp	maletotal	femtotal
1.	Under 5	9,810,733	9,365,065
2.	5 to 9	10,523,277	10,026,228
3.	10 to 14	10,520,197	10,007,875
4.	15 to 19	10,391,004	9,828,886
5.	20 to 24	9,687,814	9,276,187
6.	25 to 29	9,798,760	9,582,576
7.	30 to 34	10,321,769	10,188,619
8.	35 to 39	11,318,696	11,387,968
9.	40 to 44	11,129,102	11,312,761
10.	45 to 49	9,889,506	10,202,898
11.	50 to 54	8,607,724	8,977,824
12.	55 to 59	6,508,729	6,960,508
13.	60 to 64	5,136,627	5,668,820
14.	65 to 69	4,400,362	5,133,183
15.	70 to 74	3,902,912	4,954,529
16.	75 to 79	3,044,456	4,371,357
17.	80 to 84	1,834,897	3,110,470

```
. replace maletotal = -maletotal/1e+6
. replace femtotal = femtotal/1e+6
. twoway
        bar maletotal agegrp, horizontal xvarlab(Males)
  ||
        bar  femtotal agegrp, horizontal xvarlab(Females)
  ||
   , ylabel(1(1)17, angle(horizontal) valuelabel labsize(*.8))
     xtitle("Population in millions") ytitle("")
     xlabel(-10 "10" -7.5 "7.5" -5 "5" -2.5 "2.5" 2.5 5 7.5 10)
     legend(label(1 Males) label(2 Females))
     title("US Male and Female Population by Age")
     subtitle("Year 2000")
     note("Source:  U.S. Census Bureau, Census 2000, Tables 1, 2 and 3", span)
```

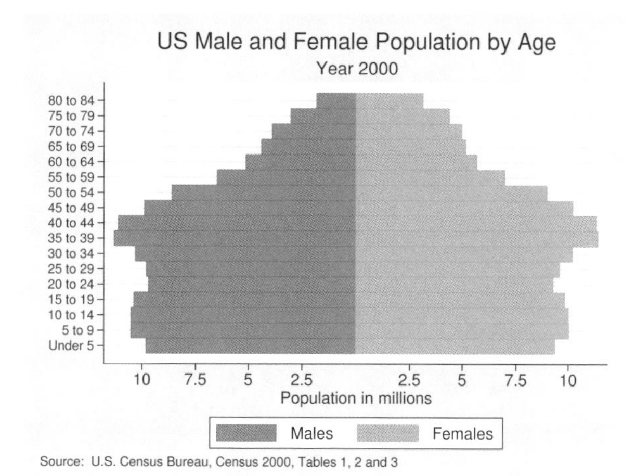

At its heart, the above graph is simple: we turned the bars sideways and changed the male total to be negative. Our first attempt at the above was simply

```
. use http://www.stata-press.com/data/r12/pop2000, clear
. replace maletotal = -maletotal
. twoway bar maletotal agegrp, horizontal ||
         bar  femtotal agegrp, horizontal
```

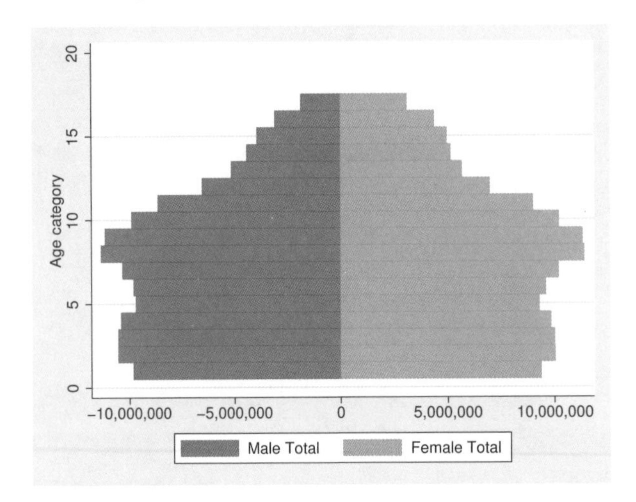

From there, we divided the population totals by 1 million and added options.

`xlabel(-10 "10" -7.5 "7.5" -5 "5" -2.5 "2.5" 2.5 5 7.5 10)` was a clever way to disguise that the bars for males extended in the negative direction. We said to label the values −10, −7.5, −5, −2.5, 2.5, 5, 7.5, and 10, but then we substituted text for the negative numbers to make it appear that they were positive. See [G-3] ***axis_label_options***.

Using the `span` suboption to `note()` aligned the text on the left side of the graph rather than on the plot region. See [G-3] ***textbox_options***.

For another rendition of the pyramid, we tried

```
. use http://www.stata-press.com/data/r12/pop2000, clear
. replace maletotal = -maletotal/1e+6
. replace femtotal = femtotal/1e+6
. generate zero = 0
. twoway
        bar maletotal agegrp, horizontal xvarlab(Males)
  ||
        bar  femtotal agegrp, horizontal xvarlab(Females)
  ||
        sc  agegrp zero     , mlabel(agegrp) mlabcolor(black) msymbol(i)
  ||
       , xtitle("Population in millions") ytitle("")
  plotregion(style(none))                                        (note 1)
  ysca(noline) ylabel(none)                                      (note 2)
  xsca(noline titlegap(-3.5))                                    (note 3)
  xlabel(-12 "12" -10 "10" -8 "8" -6 "6" -4 "4" 4(2)12 , tlength(0)
                                                        grid gmin gmax)
  legend(label(1 Males) label(2 Females)) legend(order(1 2))
  title("US Male and Female Population by Age, 2000")
  note("Source:  U.S. Census Bureau, Census 2000, Tables 1, 2 and 3")
```

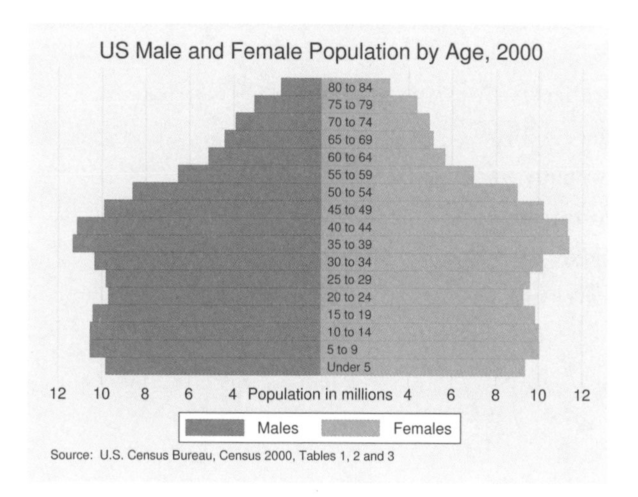

In the above rendition, we moved the labels from the x axis to inside the bars by overlaying a `scatter` on top of the bars. The points of the scatter we plotted at $y =$ `agegrp` and $x = 0$, and rather than showing the markers, we displayed marker labels containing the desired labelings. See [G-3] ***marker_label_options***.

We also played the following tricks:

1. `plotregion(style(none))` suppressed outlining the plot region; see [G-3] ***region_options***.

2. `ysca(noline)` suppressed drawing the y axis—see [G-3] ***axis_scale_options***—and `ylabel(none)` suppressed labeling it—see [G-3] ***axis_label_options***.

3. `xsca(noline titlegap(-3.5))` suppressed drawing the x axis and moved the x-axis title up to be in between its labels; see [G-3] ***axis_scale_options***.

Cautions

You must extend the scale of the axis, if that is necessary. Consider using twoway bar to produce a histogram (ignoring the better alternative of using twoway histogram; see [G-2] **graph twoway histogram**). Assume that you have already aggregated data of the form

x	frequency
1	400
2	800
3	3,000
4	1,800
5	1,100

which you enter into Stata to make variables x and frequency. You type

```
. twoway bar frequency x
```

to make a histogram-style bar chart. The y axis will be scaled to go between 400 and 3,000 (labeled at 500, 1,000, ..., 3,000), and the shortest bar will have zero height. You need to type

```
. twoway bar frequency x, ysca(r(0))
```

Also see

[G-2] **graph twoway scatter** — Twoway scatterplots

[G-2] **graph twoway dot** — Twoway dot plots

[G-2] **graph twoway dropline** — Twoway dropped-line plots

[G-2] **graph twoway histogram** — Histogram plots

[G-2] **graph twoway spike** — Twoway spike plots

[G-2] **graph bar** — Bar charts

Title

[G-2] graph twoway connected — Twoway connected plots

Syntax

`twoway connected` *varlist* [*if*] [*in*] [*weight*] [, *scatter_options*]

where *varlist* is

y_1 [y_2[...]] x

`aweight`s, `fweight`s, and `pweight`s are allowed; see [U] **11.1.6 weight**.

Menu

Graphics > Twoway graph (scatter, line, etc.)

Description

`twoway connected` draws connected-line plots. In a connected-line plot, the markers are displayed and the points are connected.

`connected` is a *plottype* as defined in [G-2] **graph twoway**. Thus the syntax for `connected` is

```
. graph twoway connected ...
. twoway connected ...
```

Being a plottype, `connected` may be combined with other plottypes in the `twoway` family (see [G-2] **graph twoway**), as in,

```
. twoway (connected ...) (scatter ...) (lfit ...) ...
```

Options

scatter_options are any of the options allowed by the `graph twoway scatter` command; see [G-2] **graph twoway scatter**.

Remarks

`connected` is, in fact, `scatter`, the difference being that by default the points are connected:

Default `connect()` option: `connect(l ...)`

Thus you get the same results by typing

```
. twoway connected yvar xvar
```

as typing

```
. scatter yvar xvar, connect(l)
```

You can just as easily turn `connected` into `scatter`: Typing

```
. scatter yvar xvar
```

is the same as typing

```
. twoway connected yvar xvar, connect(none)
```

Also see

[G-2] **graph twoway scatter** — Twoway scatterplots

Title

[G-2] graph twoway contour — Twoway contour plot with area shading

Syntax

`twoway contour` *z y x* [*if*] [*in*] [, *options*]

options	Description
`ccuts(`*numlist*`)`	list of values for contour lines or cuts
`levels(`*#*`)`	number of contour levels
`minmax`	include minimum and maximum of *z* in levels
`crule(`*crule*`)`	rule for creating contour-level colors
`scolor(`*colorstyle*`)`	starting color for contour rule
`ecolor(`*colorstyle*`)`	ending color for contour rule
`ccolors(`*colorstylelist*`)`	list of colors for contour levels
`interp(`*interpmethod*`)`	interpolation method if (*z*, *y*, *x*) does not fill a regular grid
twoway_options	titles, legends, axes, added lines and text, by, regions, name, aspect ratio, etc.

See [G-4] ***colorstyle*** and [G-3] ***twoway_options***.

crule	Description
`hue`	use equally spaced hues between `scolor()` and `ecolor()`; the default
`chue`	use equally spaced hues between `scolor()` and `ecolor()`; unlike `hue`, it uses 360 + `hue` of the `ecolor()` if the hue of the `ecolor()` is less than the hue of the `scolor()`
`intensity`	use equally spaced intensities with `ecolor()` as the base; `scolor()` is ignored
`linear`	use equally spaced interpolations of the RGB values between `scolor()` and `ecolor()`

interpmethod	Description
`thinplatespline`	thin-plate-spline interpolation; the default
`shepard`	Shepard interpolation
`none`	no interpolation; plot data as is

Menu

Graphics > Twoway graph (scatter, line, etc.)

Description

`twoway contour` displays z as filled contours in (y, x).

Options

`ccuts()`, `levels()`, and `minmax` determine how many contours are created and the values of those contours.

An alternative way of controlling the contour values is using the standard axis-label options available through the `zlabel()` option; see [G-3] ***axis_label_options***. Even when `ccuts()` or `levels()` are specified, you can further control the appearance of the contour labels using the `zlabel()` option.

`ccuts(`*numlist*`)` specifies the z values for the contour lines. Contour lines are drawn at each value of *numlist* and color- or shade-filled levels are created for each area between the lines and for the areas below the minimum and above the maximum.

`levels(`*#*`)` specifies the number of filled contour levels to create; $\# - 1$ contour cuts will be created.

`minmax` is a modifier of `levels()` and specifies that the minimum and maximum values of z be included in the cuts.

`ccuts()` and `levels()` are different ways of specifying the contour cuts and may not be combined.

`crule()`, `scolor()`, `ecolor()`, and `ccolors()` determine the colors that are used for each filled contour level.

`crule(`*crule*`)` specifies the rule used to set the colors for the contour levels. Valid *crule*s are `hue`, `chue`, `intensity`, and `linear`. The default is `crule(hue)`.

`scolor(`*colorstyle*`)` specifies the starting color for the rule. See [G-4] ***colorstyle***.

`ecolor(`*colorstyle*`)` specifies the ending color for the rule. See [G-4] ***colorstyle***.

`ccolors(`*colorstylelist*`)` specifies a list of *colorstyle*s for the area of each contour level. If RGB, CMYK, HSV, or intensity-adjusted (for example, `red*.3`) colorstyle is specified, they should be placed in quotes. Examples of valid `ccolors()` options include `ccolors(red green magenta)` and `ccolors(red "55 132 22" ".3 .9 .3 hsv" blue)`. See [G-4] ***colorstyle***.

`interp(`*interpmethod*`)` specifies the interpolation method to use if z, y, and x do not fill a regular grid. Variables z, y, and x fill a regular grid if for every combination of nonmissing (y, x), there is at least one nonmissing z corresponding to the pair in the dataset. For example, the following dataset forms a 2×2 grid.

```
. input z y x
       z y x
  1.   1 1 1
  2.   2 4 1
  3.   3 4 1
  4.   1 1 2
  5.   1 4 2
  6.   end
```

If there is more than one z value corresponding to a pair of (y, x), the smallest z value is used in plotting. In the above example, there are two z values corresponding to pair $(4, 1)$, and the smallest value, 2, is used.

```
. input z y x
       z y x
  1.   1 1 1
  2.   2 2 1
  3.   1 1 2
  4.   end
```

does not fill a regular grid because there is no z value corresponding to the pair $(2,2)$.

twoway_options are any of the options documented in [G-3] ***twoway_options***. These include options for titling the graph (see [G-3] ***title_options***); for saving the graph to disk (see [G-3] ***saving_option***); for controlling the labeling and look of the axes (see [G-3] ***axis_options***); for controlling the look, contents, position, and organization of the legend (see [G-3] ***legend_options***); for adding lines (see [G-3] ***added_line_options***) and text (see [G-3] ***added_text_options***); and for controlling other aspects of the graph's appearance (see [G-3] ***twoway_options***).

Remarks

Remarks are presented under the following headings:

Controlling the number of contours and their values
Controlling the colors of the contour areas
Choose the interpolation method

Controlling the number of contours and their values

We could draw a contour plot with default values by typing

```
. use http://www.stata-press.com/data/r12/sandstone
(Subsea elevation of Lamont sandstone in an area of Ohio)
. twoway contour depth northing easting
```

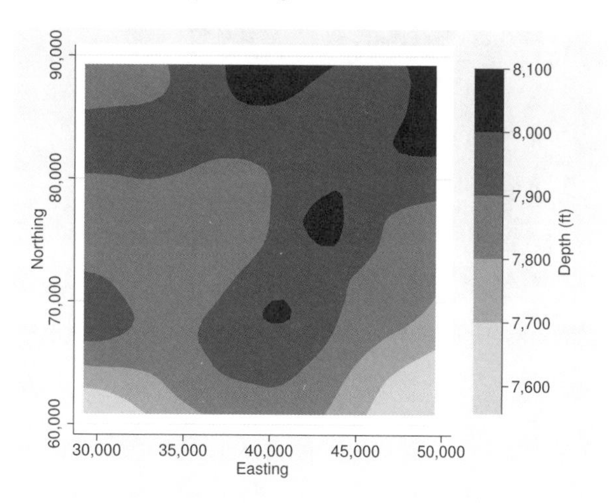

We could add the `levels()` option to the above command to create # − 1 equally spaced contours between `min(depth)` and `max(depth)`.

```
. twoway contour depth northing easting, levels(10)
```

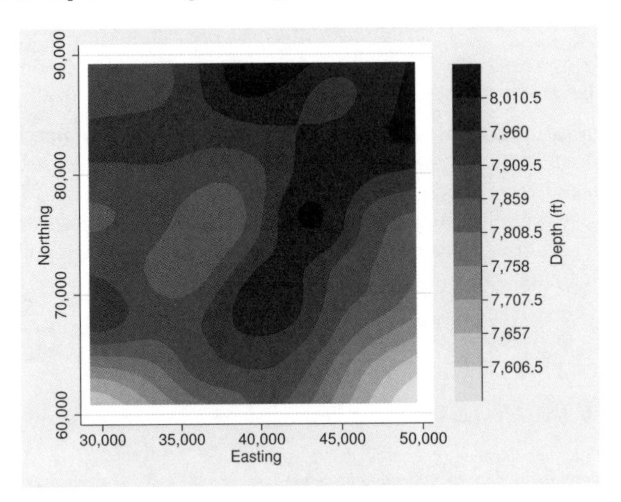

We could use the `ccuts()` option to draw a contour plot with 7 levels determined by 6 cuts at 7500, 7600, 7700, 7800, 7900, and 8000. `ccuts()` gives you the finest control over creating contour levels.

```
. twoway contour depth northing easting, ccuts(7500(100)8000)
```

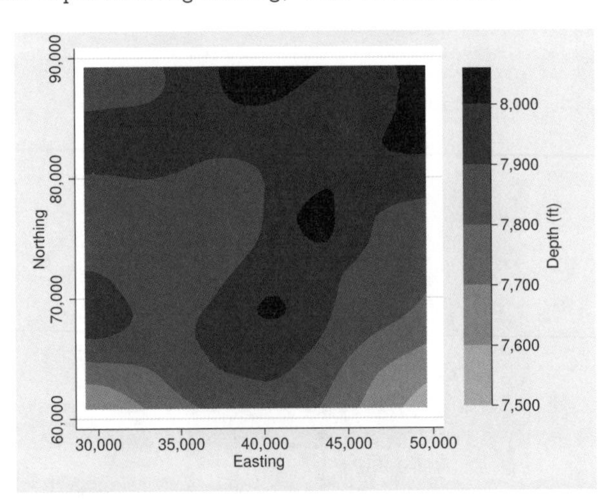

`zlabel()` controls the axis on the contour legend. When `ccuts()` and `levels()` are not specified, `zlabel()` also controls the number and value of contours. To obtain about 7 nicely spaced cuts, specify `zlabel(#7)`:

```
. twoway contour depth northing easting, zlabel(#7)
```

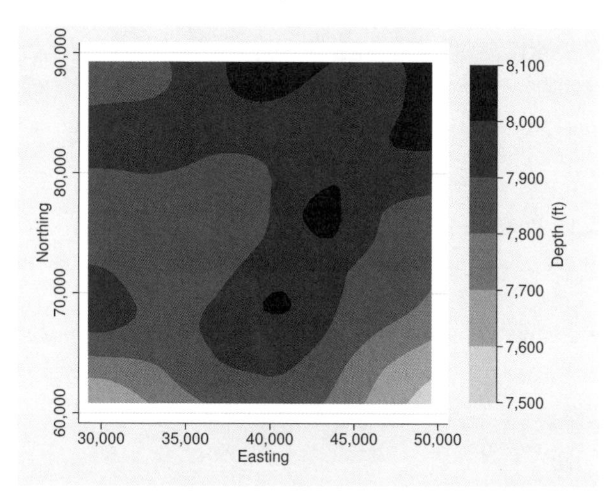

With either `levels()` or `ccuts()`, `zlabel()` becomes an option that only affects the labels of the contour legend. The contour legend can label different values than the actual contour cuts. The legend can have more (or fewer) ticks than the number of contour levels. See [G-3] ***axis_label_options*** for details.

We now specify the `twoway contour` command with the `levels()` and `zlabel()` options and the `format()` suboption to draw a 10-level contour plot with 7 labels on the contour legend. The labels' display format is `%9.1f`.

```
. twoway contour depth northing easting, levels(10) zlabel(#7, format(%9.1f))
```

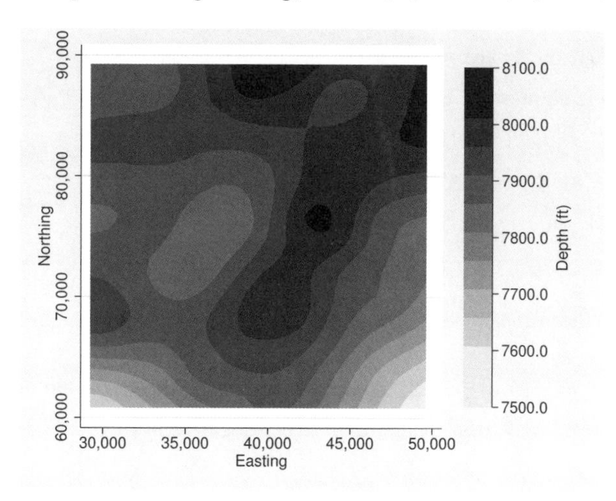

Controlling the colors of the contour areas

crule(), scolor(), and ecolor() control the colors for each contour level. Typing

```
. twoway contour depth northing easting, level(10) scolor(green) ecolor(red)
```

draws a 10-level contour plot with starting color green and ending color red. Because the hue of green is 120 and the hue of red is 0, the hues of levels are moving downward under the default crule(hue). Hence you will see yellow, but not blue and purple.

For the above example, you can use crule(chue) if you want hues of the levels to move up:

```
. twoway contour depth northing easting, level(10) crule(chue) scolor(green)
> ecolor(red)
```

Now you will see blue and purple as the hue varies from 120 to 360 $(0 + 360)$, but not yellow.

ccolors() specifies a list of colors to be used for each contour level.

```
. twoway contour depth northing easting, levels(5)
> ccolors(red green magenta blue yellow)
```

Choose the interpolation method

If z, y, and x do not fill a regular grid, the missing z values on grid points (y, x) need to be interpolated.

Thin-plate-spline interpolation uses a weight vector $(\mathbf{w}_i)$ obtained from solving a dimension $n+3$ linear equation system, where n is the number of unique pairs (y, x) with nonmissing z values in the dataset. Then the z value on a pair (y, x) can be interpolated by

$$z = w_1 \times f(y - y_1, x - x_1) + \cdots + w_n \times f(y - y_n, x - x_n) + w_{n+1} + w_{n+2} \times x + w_{n+3} \times y$$

where $f(y, x) = \sqrt{y^2 + x^2}$. interp(thinplatespline) is the default.

Shepard interpolation obtains the z value on a pair (y, x) from

$$z = (z_1 \times f(y - y_1, x - x_1) + \cdots + z_n \times \frac{f(y - y_n, x - x_n)}{\sum}$$

where $\sum$ is

$$\sum = f(y - y_1, x - x_1) + \cdots + f(y - y_n, x - x_n)$$

and $f(y, x) = 1/(x^2 + y^2)$. You specify interp(shepard) to use this method.

For the detailed formulas of thin-plate-spline and Shepard interpolation, see Press et al. (2007, 140–144).

Thin-plate-spline interpolation needs to solve a dimension $n + 3$ linear system, where n is the number of unique pairs (y, x) with nonmissing z value in the dataset. It becomes expensive when n becomes large. A rule-of-thumb number for choosing the thin-plate-spline method is n 1000.

Shepard interpolation is usually not as good as thin-plate-spline interpolation but is faster.

Method none plots data as is without any interpolation. Any grid cell with edge points containing a missing z value will be displayed using background color. If the dataset (z, y, x) is dense (that is, there are few missing grid points), interp(none) may be adequate.

Reference

Press, W. H., S. A. Teukolsky, W. T. Vetterling, and B. P. Flannery. 2007. *Numerical Recipes in C: The Art of Scientific Computing*. 3rd ed. Cambridge: Cambridge University Press.

Also see

[G-2] **graph twoway contourline** — Twoway contour-line plot

[G-2] **graph twoway area** — Twoway line plot with area shading

[G-2] **graph twoway rarea** — Range plot with area shading

Title

[G-2] graph twoway contourline — Twoway contour-line plot

Syntax

`twoway contourline` *z y x* [*if*] [*in*] [, *options*]

options	Description
`ccuts(`*numlist*`)`	list of values for contour lines or cuts
`levels(`*#*`)`	number of contour levels
`minmax`	include contour lines for minimum and maximum of *z*
`format(`*%fmt*`)`	display format for `ccuts()` or `levels()`
`colorlines`	display contour lines in different colors
`crule(`*crule*`)`	rule for creating contour-line colors
`scolor(`*colorstyle*`)`	starting color for contour rule
`ecolor(`*colorstyle*`)`	ending color for contour rule
`ccolors(`*colorstylelist*`)`	list of colors for contour lines
`clwidths(`*linewidthstylelist*`)`	list of widths for contour lines
`reversekey`	reverse the order of the keys in contour-line legend
`interp(`*interpmethod*`)`	interpolation method if (*z*, *y*, *x*) does not fill a regular grid
twoway_options	titles, legends, axes, added lines and text, by, regions, name, aspect ratio, etc.

See [G-4] ***colorstyle***, [G-4] ***linewidthstyle***, and [G-3] ***twoway_options***.

crule	Description
`hue`	use equally spaced hues between `scolor()` and `ecolor()`; the default
`chue`	use equally spaced hues between `scolor()` and `ecolor()`; unlike `hue`, it uses 360 + `hue` of the `ecolor()` if the hue of the `ecolor()` is less than the hue of the `scolor()`
`intensity`	use equally spaced intensities with `ecolor()` as the base; `scolor()` is ignored
`linear`	use equally spaced interpolations of the RGB values between `scolor()` and `ecolor()`

interpmethod	Description
`thinplatespline`	thin-plate-spline interpolation; the default
`shepard`	Shepard interpolation
`none`	no interpolation; plot data as is

Menu

Graphics > Twoway graph (scatter, line, etc.)

Description

`twoway contourline` displays *z* as contour lines in (*y*, *x*).

Options

`ccuts()`, `levels()`, `minmax`, and `format()` determine how many contours are created and the values of those contours.

`ccuts(`*numlist*`)` specifies the *z* values for the contour lines. Contour lines are drawn at each value of *numlist*.

`levels(`*#*`)` specifies the number of contour lines to create; # − 1 contour lines will be created.

`minmax` is a modifier of `levels()` and specifies that contour lines be drawn for the minimum and maximum values of *z*. By default, lines are drawn only for the cut values implied by levels, not the full range of *z*.

`format(%`*fmt*`)` specifies the display format used to create the labels in the contour legend for the contour lines.

`ccuts()` and `levels()` are different ways of specifying the contour cuts and may not be combined.

`colorlines`, `crule()`, `scolor()`, `ecolor()`, `ccolors()`, and `clwidths()` determine the colors and width that are used for each contour line.

`colorlines` specifies that the contour lines be drawn in different colors. Unless the `ccolors()` option is specified, the colors are determined by `crule()`.

`crule(`*crule*`)` specifies the rule used to set the colors for the contour lines. Valid *crule*s are `hue`, `chue`, `intensity`, and `linear`. The default is `crule(hue)`.

`scolor(`*colorstyle*`)` specifies the starting color for the rule. See [G-4] ***colorstyle***.

`ecolor(`*colorstyle*`)` specifies the ending color for the rule. See [G-4] ***colorstyle***.

`ccolors(`*colorstylelist*`)` specifies a list of *colorstyle*s for each contour line. If RGB, CMYK, HSV, or intensity-adjusted (for example, `red*.3`) colorstyle is specified, they should be placed in quotes. Examples of valid `ccolors()` options include `ccolors(red green magenta)` and `ccolors(red "55 132 22" ".3 .9 .3 hsv" blue)`. See [G-4] ***colorstyle***.

`clwidths(`*linewidthstylelist*`)` specifies a list of *linewidthstyle*s, one for each contour line. See [G-4] ***linewidthstyle***.

`reversekey` specifies that the order of the keys in the contour-line legend be reversed. By default, the keys are ordered from top to bottom, starting with the key for the highest values of *z*. See *plegend_option* in [G-3] ***legend_options***.

`interp(`*interpmethod*`)` specifies the interpolation method to use if *z*, *y*, and *x* do not fill a regular grid. Variables *z*, *y*, and *x* fill a regular grid if for every combination of nonmissing (*y*, *x*), there is at least one nonmissing *z* corresponding to the pair in the dataset. For example, the following dataset forms a 2×2 grid.

```
. input z y x
        z y x
  1.    1 1 1
  2.    2 4 1
  3.    3 4 1
  4.    1 1 2
  5.    1 4 2
  6.    end
```

If there is more than one z value corresponding to a pair of (y, x), the smallest z value is used in plotting. In the above example, there are two z values corresponding to pair (4, 1), and the smallest value, 2, is used.

```
. input z y x
        z y x
  1.    1 1 1
  2.    2 2 1
  3.    1 1 2
  4.    end
```

does not fill a regular grid because there is no z value corresponding to the pair $(2, 2)$.

twoway_options are any of the options documented in [G-3] ***twoway_options***. These include options for titling the graph (see [G-3] ***title_options***); for saving the graph to disk (see [G-3] ***saving_option***); for controlling the labeling and look of the axes (see [G-3] ***axis_options***); for controlling the look, contents, position, and organization of the legend (see [G-3] ***legend_options***); for adding lines (see [G-3] ***added_line_options***) and text (see [G-3] ***added_text_options***); and for controlling other aspects of the graph's appearance (see [G-3] ***twoway_options***).

Remarks

Remarks are presented under the following headings:

Controlling the number of contour lines and their values
Controlling the colors of the contour lines
Choose the interpolation method

Controlling the number of contour lines and their values

We could draw a contour-line plot with default values by typing

```
. use http://www.stata-press.com/data/r12/sandstone
(Subsea elevation of Lamont sandstone in an area of Ohio)
. twoway contourline depth northing easting
```

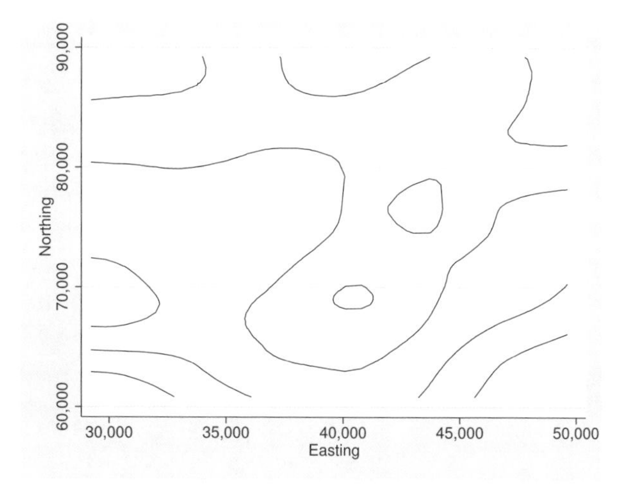

We add the `colorlines` option to display the values of cuts in the contour legend. We also include the `levels()` option to create # − 1 contour lines equally spaced between `min(depth)` and `max(depth)`.

```
. twoway contourline depth northing easting, colorlines levels(10)
```

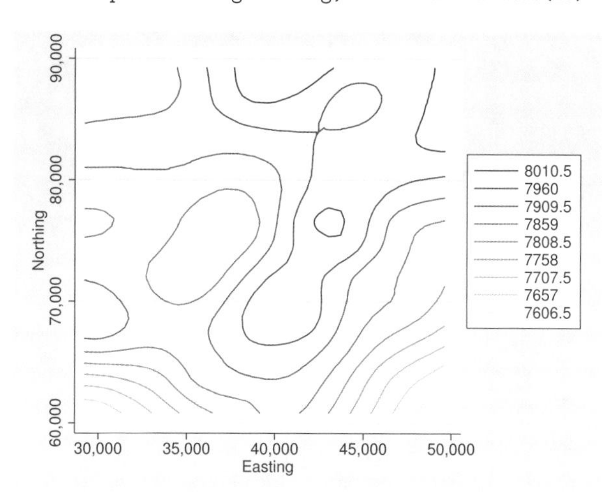

The ccuts() option gives you the finest control over creating contour lines. Here we use it to draw a contour-line plot with 6 cuts at 7500, 7600, 7700, 7800, 7900, and 8000.

```
. twoway contourline depth northing easting, colorlines ccuts(7500(100)8000)
```

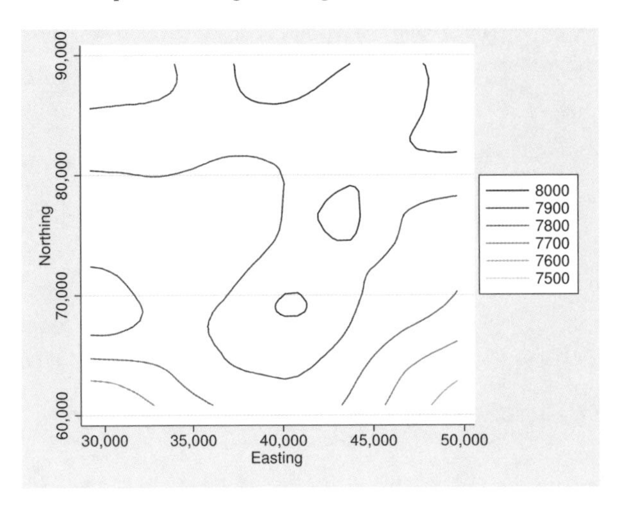

Controlling the colors of the contour lines

crule(), scolor(), and ecolor() control the colors for each contour line. Typing

```
. twoway contourline depth northing easting, level(10) format(%9.1f)
> colorlines scolor(green) ecolor(red)
```

draws a contour-line plot with lines of 9 equally spaced z values between min(depth) and max(depth). The starting color for lines is green and the ending color for lines is red. Also, the legend labels' display format is %9.1f.

ccolors() specifies a list of colors to be used for each contour line.

```
. twoway contourline depth northing easting, ccuts(7600(100)8000)
> colorlines ccolors(red green magenta blue yellow)
```

Choose the interpolation method

See *Choose the interpolation method* in [G-2] **graph twoway contour**.

Also see

[G-2] **graph twoway contour** — Twoway contour plot with area shading

[G-2] **graph twoway line** — Twoway line plots

[G-2] **graph twoway connected** — Twoway connected plots

Title

[G-2] graph twoway dot — Twoway dot plots

Syntax

`twoway dot` *yvar xvar* [*if*] [*in*] [, *options*]

options	Description
`vertical`	vertical bar plot; the default
`horizontal`	horizontal bar plot
`dotextend(yes|no)`	dots extend beyond point
`base(#)`	value to drop to if `dotextend(no)`
`ndots(#)`	# of dots in full span of y or x
`dstyle(`*markerstyle*`)`	overall marker style of dots
`dsymbol(`*symbolstyle*`)`	marker symbol for dots
`dcolor(`*colorstyle*`)`	fill and outline color for dots
`dfcolor(`*colorstyle*`)`	fill color for dots
`dsize(`*markersizestyle*`)`	size of dots
`dlstyle(`*linestyle*`)`	overall outline style of dots
`dlcolor(`*colorstyle*`)`	outline color for dots
`dlwidth(`*linewidthstyle*`)`	thickness of outline for dots
scatter_options	any options other than *connect_options* documented in [G-2] **graph twoway scatter**

See [G-4] ***markerstyle***, [G-4] ***symbolstyle***, [G-4] ***colorstyle***, [G-4] ***markersizestyle***, [G-4] ***linestyle***, and [G-4] ***linewidthstyle***.

All options are *rightmost*, except `vertical` and `horizontal`, which are *unique*; see [G-4] **concept: repeated options**.

Menu

Graphics > Twoway graph (scatter, line, etc.)

Description

`twoway dot` displays numeric (y,x) data as dot plots. Also see [G-2] **graph dot** to create dot plots of categorical variables. `twoway dot` is useful in programming contexts.

Options

`vertical` and `horizontal` specify either a vertical or a horizontal dot plot. `vertical` is the default. If `horizontal` is specified, the values recorded in *yvar* are treated as x values, and the values recorded in *xvar* are treated as y values. That is, to make horizontal plots, do not switch the order of the two variables specified.

In the `vertical` case, dots are drawn at the specified *xvar* values and extend up and down.

In the `horizontal` case, lines are drawn at the specified *xvar* values and extend left and right.

`dotextend(yes | no)` determines whether the dots extend beyond the y value (or x value if `horizontal` is specified). `dotextend(yes)` is the default.

`base(#)` is relevant only if `dotextend(no)` is also specified. `base()` specifies the value from which the dots are to extend. The default is `base(0)`.

`ndots(#)` specifies the number of dots across a line; `ndots(75)` is the default. Depending on printer/screen resolution, using fewer or more dots can make the graph look better.

`dstyle(`*markerstyle*`)` specifies the overall look of the markers used to create the dots, including their shape and color. The other options listed below allow you to change their attributes, but `dstyle()` provides the starting point.

You need not specify `dstyle()` just because there is something you want to change. You specify `dstyle()` when another style exists that is exactly what you desire or when another style would allow you to specify fewer changes to obtain what you want.

See [G-4] ***markerstyle*** for a list of available marker styles.

`dsymbol(`*symbolstyle*`)` specifies the shape of the marker used for the dot. See [G-4] ***symbolstyle*** for a list of symbol choices, although it really makes little sense to change the shape of dots; else why would it be called a dot plot?

`dcolor(`*colorstyle*`)` specifies the color of the symbol used for the dot. See [G-4] ***colorstyle*** for a list of color choices.

`dfcolor(`*colorstyle*`)`, `dsize(`*markersizestyle*`)`, `dlstyle(`*linestyle*`)`, `dlcolor(`*colorstyle*`)`, and `dlwidth(`*linewidthstyle*`)` are rarely (never) specified options. They control, respectively, the fill color, size, outline style, outline color, outline width, and, if you are really using dots, dots are affected by none of these things. For these options to be useful, you must also specify `dsymbol()`; as we said earlier, why then would it be called a dot plot? In any case, see [G-4] ***colorstyle***, [G-4] ***markersizestyle***, [G-4] ***linestyle***, and [G-4] ***linewidthstyle*** for a list of choices.

scatter_options refer to any of the options allowed by `scatter`, and most especially the *marker_options*, which control how the marker (not the dot) appears. *connect_options*, even if specified, are ignored. See [G-2] **graph twoway scatter**.

Remarks

`twoway dot` is of little, if any use. We cannot think of a use for it, but perhaps someday, somewhere, someone will. We have nothing against the dot plot used with categorical data—see [G-2] **graph dot** for a useful command—but using the dot plot in a twoway context would be bizarre. It is nonetheless included for logical completeness.

In [G-2] **graph twoway bar**, we graphed the change in the value for the S&P 500. Here are a few of that data graphed as a dot plot:

```
. use http://www.stata-press.com/data/r12/sp500
(S&P 500)
. twoway dot change date in 1/45
```

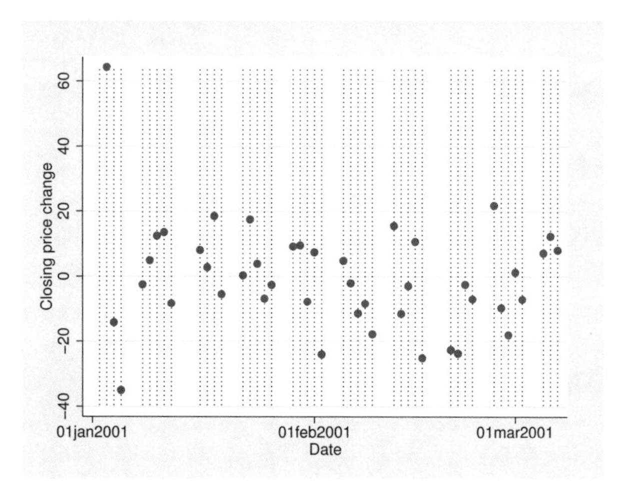

Dot plots are usually presented horizontally,

```
. twoway dot change date in 1/45, horizontal
```

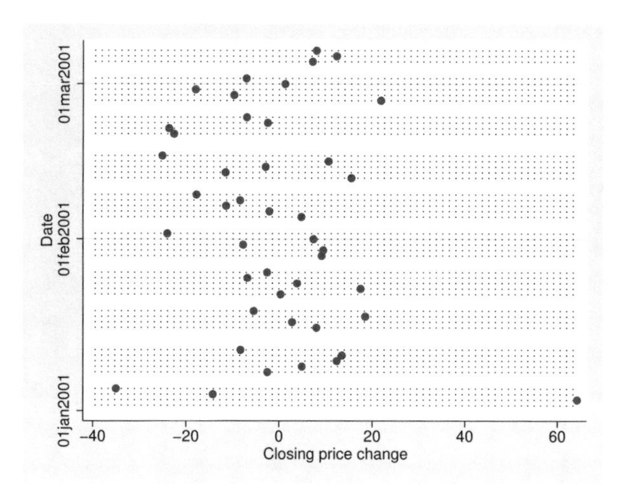

and below we specify the `dotextend(n)` option to prevent the dots from extending across the range of x:

```
. twoway dot change date in 1/45, horizontal dotext(n)
```

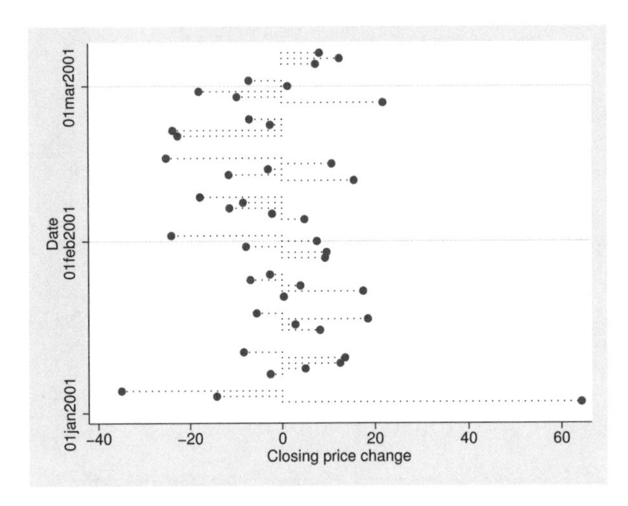

Reference

Cox, N. J. 2008. Speaking Stata: Between tables and graphs. *Stata Journal* 8: 269–289.

Also see

[G-2] **graph twoway scatter** — Twoway scatterplots

[G-2] **graph dot** — Dot charts (summary statistics)

Title

[G-2] graph twoway dropline — Twoway dropped-line plots

Syntax

`twoway dropline` *yvar xvar* [*if*] [*in*] [, *options*]

options	Description
`vertical`	vertical dropped-line plot; the default
`horizontal`	horizontal dropped-line plot
`base(#)`	value to drop to; default is 0
marker_options	change look of markers (color, size, etc.)
marker_label_options	add marker labels; change look or position
line_options	change look of dropped lines
axis_choice_options	associate plot with alternative axis
twoway_options	titles, legends, axes, added lines and text, by, regions, name, aspect ratio, etc.

See [G-3] ***marker_options***, [G-3] ***marker_label_options***, [G-3] ***line_options***, [G-3] ***axis_choice_options***, and [G-3] ***twoway_options***.

All explicit options are *rightmost*, except `vertical` and `horizontal`, which are *unique*; see [G-4] **concept: repeated options**.

Menu

Graphics > Twoway graph (scatter, line, etc.)

Description

`twoway dropline` displays numeric (y,x) data as dropped lines capped with a marker. `twoway dropline` is useful for drawing plots in which the numbers vary around zero.

Options

`vertical` and `horizontal` specify either a vertical or a horizontal dropped-line plot. `vertical` is the default. If `horizontal` is specified, the values recorded in *yvar* are treated as x values, and the values recorded in *xvar* are treated as y values. That is, to make horizontal plots, do not switch the order of the two variables specified.

In the `vertical` case, dropped lines are drawn at the specified *xvar* values and extend up or down from 0 according to the corresponding *yvar* values. If 0 is not in the range of the y axis, lines extend up or down to the x axis.

In the `horizontal` case, dropped lines are drawn at the specified *xvar* values and extend left or right from 0 according to the corresponding *yvar* values. If 0 is not in the range of the x axis, lines extend left or right to the y axis.

`base(#)` specifies the value from which the lines should extend. The default is `base(0)`, and in the above description of options `vertical` and `horizontal`, this default was assumed.

marker_options specify the look of markers plotted at the data points. This look includes the marker symbol and its size, color, and outline; see [G-3] ***marker_options***.

marker_label_options specify if and how the markers are to be labeled; see [G-3] ***marker_label_options***.

line_options specify the look of the dropped lines, including pattern, width, and color; see [G-3] ***line_options***.

axis_choice_options associate the plot with a particular y or x axis on the graph; see [G-3] ***axis_choice_options***.

twoway_options are a set of common options supported by all `twoway` graphs. These options allow you to title graphs, name graphs, control axes and legends, add lines and text, set aspect ratios, create graphs over `by()` groups, and change some advanced settings. See [G-3] ***twoway_options***.

Remarks

Remarks are presented under the following headings:

Typical use
Advanced use
Cautions

Typical use

We have daily data recording the values for the S&P 500 in 2001:

```
. use http://www.stata-press.com/data/r12/sp500
(S&P 500)

. list date close change in 1/5
```

	date	close	change
1.	02jan2001	1283.27	.
2.	03jan2001	1347.56	64.29004
3.	04jan2001	1333.34	-14.22009
4.	05jan2001	1298.35	-34.98999
5.	08jan2001	1295.86	-2.48999

In [G-2] **graph twoway bar**, we graphed the first 57 observations of these data by using bars. Here is the same graph presented as dropped lines:

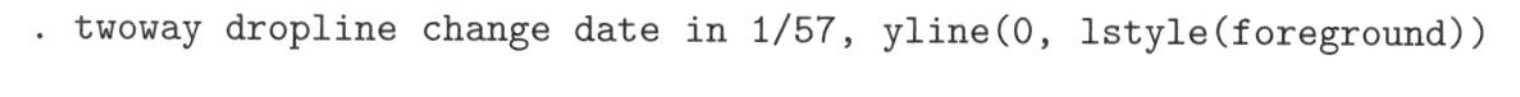

```
. twoway dropline change date in 1/57, yline(0, lstyle(foreground))
```

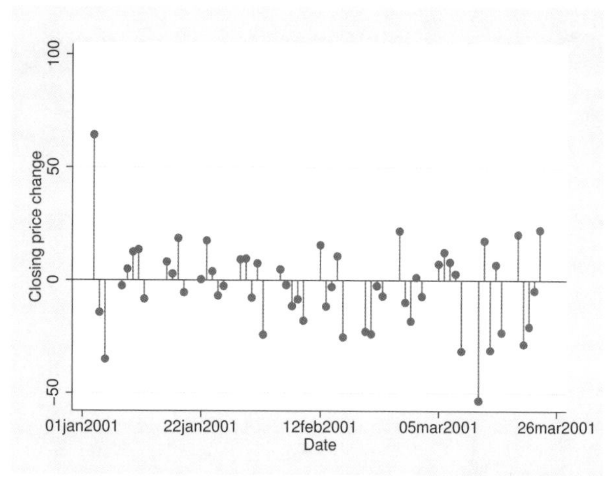

In the above, we specified `yline(0)` to add a line across the graph at 0, and then we specified `yline(, lstyle(foreground))` so that the line would have the same color as the foreground. We could have instead specified `yline(, lcolor())`. For an explanation of why we chose `lstyle()` over `foreground()`, see *Advanced use: Overlaying* in [G-2] **graph twoway bar**.

Advanced use

Dropped-line plots work especially well when the points are labeled. For instance,

```
. use http://www.stata-press.com/data/r12/lifeexp, clear
(Life expectancy, 1998)

. keep if region==3
(58 observations deleted)

. generate lngnp = ln(gnppc)

. quietly regress le lngnp

. predict r, resid
```

```
. twoway dropline r gnp,
        yline(0, lstyle(foreground)) mlabel(country) mlabpos(9)
        ylab(-6(1)6)
        subtitle("Regression of life expectancy on ln(gnp)"
                 "Residuals:" " ", pos(11))
        note("Residuals in years; positive values indicate"
             "longer than predicted life expectancy")
```

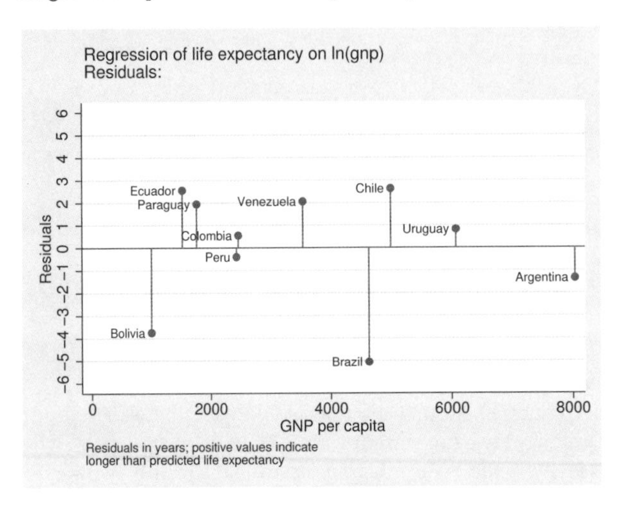

Cautions

See *Cautions* in [G-2] **graph twoway bar**, which applies equally to `twoway dropline`.

Also see

[G-2] **graph twoway scatter** — Twoway scatterplots

[G-2] **graph twoway spike** — Twoway spike plots

Title

[G-2] graph twoway fpfit — Twoway fractional-polynomial prediction plots

Syntax

`twoway fpfit` *yvar xvar* [*if*] [*in*] [*weight*] [, *options*]

options	Description
`estcmd(`*estcmd*`)`	estimation command; default is `regress`
`estopts(`*fracpoly_options*`)`	options for `fracpoly` *estcmd*
`predopts(`*predict_options*`)`	options for `predict`
cline_options	change look of predicted line
axis_choice_options	associate plot with alternative axis
twoway_options	titles, legends, axes, added lines and text, by, regions, name, aspect ratio, etc.

See [G-3] ***cline_options***, [G-3] ***axis_choice_options***, and [G-3] ***twoway_options***.

Options `estcmd()`, `estopts()`, and `predopts()` are *unique*; see [G-4] **concept: repeated options**.

`aweight`s, `fweight`s, and `pweight`s are allowed. Weights, if specified, affect estimation but not how the weighted results are plotted. See [U] **11.1.6 weight**.

Menu

Graphics > Twoway graph (scatter, line, etc.)

Description

`twoway fpfit` calculates the prediction for *yvar* from estimation of a fractional polynomial of *xvar* and plots the resulting curve.

Options

`estcmd(`*estcmd*`)` specifies the estimation command to be used; `estcmd(regress)` is the default.

`estopts(`*fracpoly_options*`)` specifies options to be passed along to `fracpoly` to estimate the fractional polynomial regression from which the curve will be predicted; see [R] **fracpoly**.

`predopts(`*predict_options*`)` specifies options to be passed along to `predict` to obtain the predictions after estimation by `fracpoly: regress`; see [R] **regress postestimation**. `predopts()` may be used only with `estcmd(regress)`. Predictions in all cases are calculated at all the *xvar* values in the data.

cline_options specify how the prediction line is rendered; see [G-3] ***cline_options***.

axis_choice_options associate the plot with a particular y or x axis on the graph; see [G-3] ***axis_choice_options***.

twoway_options are a set of common options supported by all `twoway` graphs. These options allow you to title graphs, name graphs, control axes and legends, add lines and text, set aspect ratios, create graphs over `by()` groups, and change some advanced settings. See [G-3] ***twoway_options***.

Remarks

Remarks are presented under the following headings:

Typical use
Cautions
Use with by()

Typical use

`twoway fpfit` is nearly always used in conjunction with other `twoway` plottypes, such as

```
. use http://www.stata-press.com/data/r12/auto
(1978 Automobile Data)
. scatter mpg weight || fpfit mpg weight
```

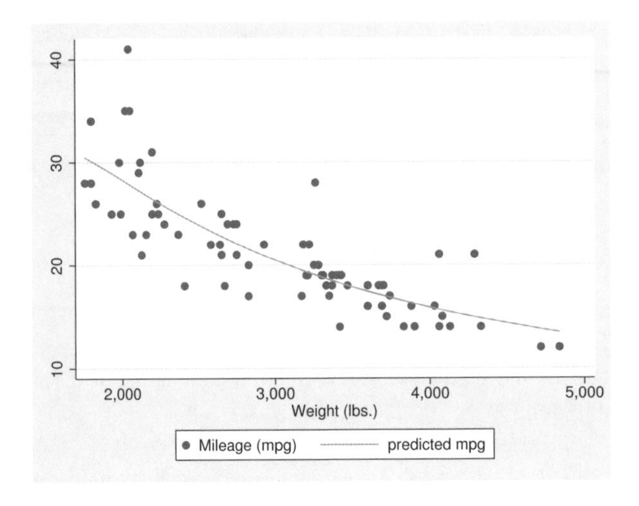

Results are visually the same as typing

```
. fracpoly regress mpg weight
. predict fitted
. scatter mpg weight || line fitted weight
```

Cautions

Do not use `twoway fpfit` when specifying the *axis_scale_options* `yscale(log)` or `xscale(log)` to create log scales. Typing

```
. scatter mpg weight, xscale(log) || fpfit mpg weight
```

will produce a curve that will be fit from a fractional polynomial regression of `mpg` on `weight` rather than `log(weight)`.

Use with by()

fpfit may be used with by() (as can all the twoway plot commands):

```
. scatter mpg weight || fpfit mpg weight ||, by(foreign, total row(1))
```

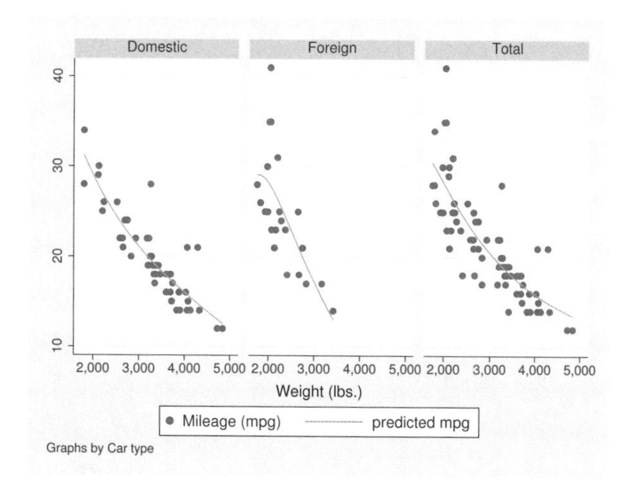

Also see

[G-2] **graph twoway line** — Twoway line plots

[G-2] **graph twoway fpfitci** — Twoway fractional-polynomial prediction plots with CIs

[G-2] **graph twoway lfit** — Twoway linear prediction plots

[G-2] **graph twoway qfit** — Twoway quadratic prediction plots

[G-2] **graph twoway mband** — Twoway median-band plots

[G-2] **graph twoway mspline** — Twoway median-spline plots

[R] **fracpoly** — Fractional polynomial regression

Title

[G-2] graph twoway fpfitci — Twoway fractional-polynomial prediction plots with CIs

Syntax

`twoway fpfitci` *yvar xvar* [*if*] [*in*] [*weight*] [, *options*]

options	Description
fpfit_options	estimation command and options
`level(#)`	set confidence level; default is `level(95)`
`nofit`	prevent plotting the prediction
`fitplot(`*plottype*`)`	how to plot fit; default is `fitplot(line)`
`ciplot(`*plottype*`)`	how to plot CIs; default is `ciplot(rarea)`
fcline_options	change look of predicted line
fitarea_options	change look of CI
axis_choice_options	associate plot with alternative axis
twoway_options	titles, legends, axes, added lines and text, by, regions, name, aspect ratio, etc.

See [G-2] **graph twoway fpfit**, [G-3] ***fcline_options***, [G-3] ***fitarea_options***, [G-3] ***axis_choice_options***, and [G-3] ***twoway_options***.

Option `level()` is *rightmost*; `nofit`, `fitplot()`, and `ciplot()` are *unique*; see [G-4] **concept: repeated options**.

`aweights`, `fweights`, and `pweights` are allowed. Weights, if specified, affect estimation but not how the weighted results are plotted. See [U] **11.1.6 weight**.

Menu

Graphics > Twoway graph (scatter, line, etc.)

Description

`twoway fpfitci` calculates the prediction for *yvar* from estimation of a fractional polynomial of *xvar* and plots the resulting curve along with the confidence interval of the mean.

Options

fpfit_options refers to any of the options of `graph twoway fpfit`; see [G-2] **graph twoway fpfit**. These options are seldom specified.

`level(#)` specifies the confidence level, as a percentage, for the confidence intervals. The default is `level(95)` or as set by `set level`; see [U] **20.7 Specifying the width of confidence intervals**.

`nofit` prevents the prediction from being plotted.

`fitplot(`*plottype*`)` is seldom specified. It specifies how the prediction is to be plotted. The default is `fitplot(line)`, meaning that the prediction will be plotted by `graph twoway line`. See [G-2] **graph twoway** for a list of *plottype* choices. You may choose any plottypes that expect one y variable and one x variable.

`ciplot(`*plottype*`)` specifies how the confidence interval is to be plotted. The default is `ciplot(rarea)`, meaning that the prediction will be plotted by `graph twoway rarea`.

A reasonable alternative is `ciplot(rline)`, which will substitute lines around the prediction for shading. See [G-2] **graph twoway** for a list of *plottype* choices. You may choose any plottypes that expect two y variables and one x variable.

fcline_options specify how the prediction line is rendered; see [G-3] ***fcline_options***. If you specify `fitplot()`, then rather than using *fcline_options*, you should select options that affect the specified *plottype* from the options in `scatter`; see [G-2] **graph twoway scatter**.

fitarea_options specify how the confidence interval is rendered; see [G-3] ***fitarea_options***. If you specify `ciplot()`, then rather than using *fitarea_options*, you should specify whatever is appropriate.

axis_choice_options associate the plot with a particular y or x axis on the graph; see [G-3] ***axis_choice_options***.

twoway_options are a set of common options supported by all `twoway` graphs. These options allow you to title graphs, name graphs, control axes and legends, add lines and text, set aspect ratios, create graphs over `by()` groups, and change some advanced settings. See [G-3] ***twoway_options***.

Remarks

Remarks are presented under the following headings:

Typical use
Advanced use
Cautions
Use with by()

Typical use

`twoway fpfitci` by default draws the confidence interval of the predicted mean:

```
. use http://www.stata-press.com/data/r12/auto
(1978 Automobile Data)
. twoway fpfitci mpg weight
```

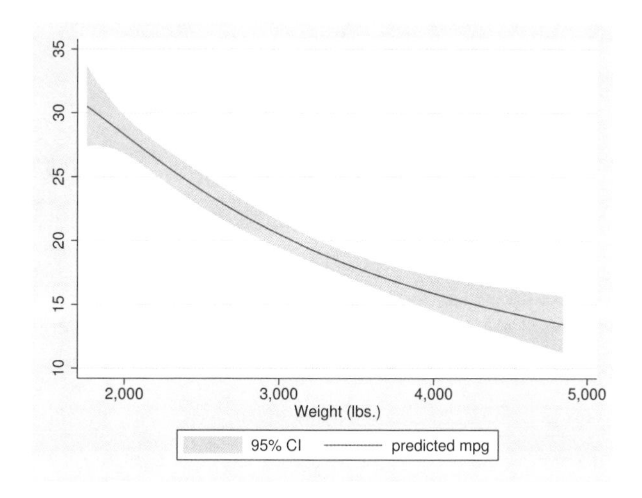

If you specify the ciplot(rline) option, the confidence interval will be designated by lines rather than shading:

```
. twoway fpfitci mpg weight, ciplot(rline)
```

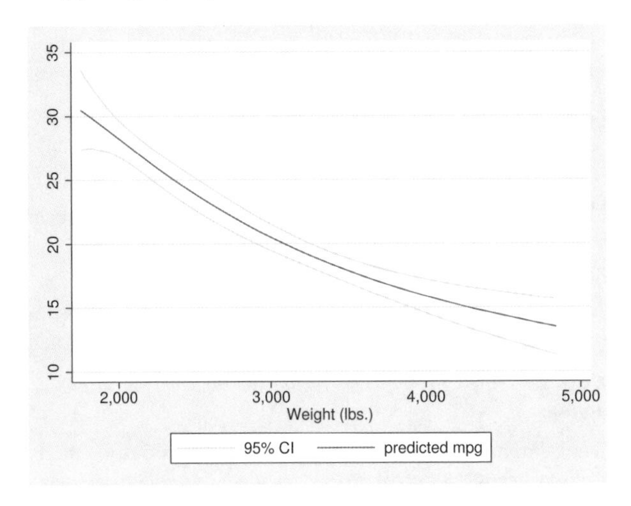

Advanced use

fpfitci can be usefully overlaid with other plots:

```
. use http://www.stata-press.com/data/r12/auto, clear
(1978 Automobile Data)
. twoway fpfitci mpg weight || scatter mpg weight
```

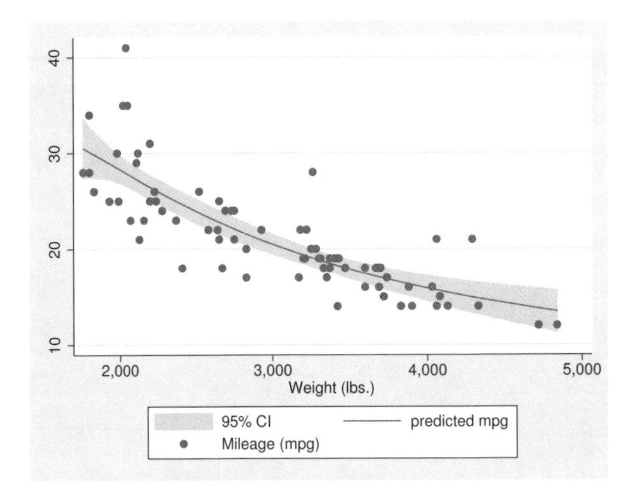

In the above graph, the shaded area corresponds to the 95% confidence interval for the mean.

It is of great importance to note that we typed

```
. twoway fpfitci ... || scatter ...
```

and not

```
. twoway scatter ... || fpfitci ...
```

Had we drawn the scatter diagram first, the confidence interval would have covered up most of the points.

Cautions

Do not use `twoway fpfitci` when specifying the *axis_scale_options* `yscale(log)` or `xscale(log)` to create log scales. Typing

```
. twoway fpfitci mpg weight || scatter mpg weight ||, xscale(log)
```

will produce a curve that will be fit from a fractional polynomial regression of `mpg` on `weight` rather than `log(weight)`.

See *Cautions* in [G-2] **graph twoway lfitci**.

Use with by()

`fpfitci` may be used with `by()` (as can all the `twoway` plot commands):

```
. twoway fpfitci  mpg weight  ||
         scatter  mpg weight  ||
  , by(foreign, total row(1))
```

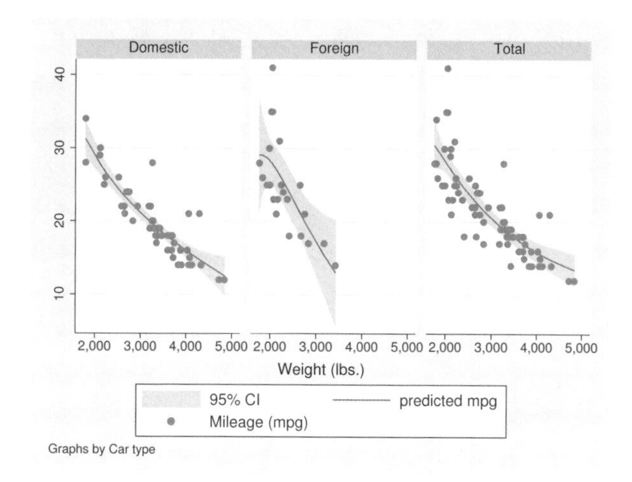

Also see

[G-2] **graph twoway lfitci** — Twoway linear prediction plots with CIs

[G-2] **graph twoway qfitci** — Twoway quadratic prediction plots with CIs

[G-2] **graph twoway fpfit** — Twoway fractional-polynomial prediction plots

[R] **fracpoly** — Fractional polynomial regression

Title

[G-2] graph twoway function — Twoway line plot of function

Syntax

`twoway function` [[`y`] `=`] *f*(`x`) [*if*] [*in*] [, *options*]

options	Description
`range(`*# #*`)`	plot over `x` = # to #
`range(`*varname*`)`	plot over `x` = min to max of *varname*
`n(`*#*`)`	evaluate at # points; default is 300
`droplines(`*numlist*`)`	draw lines to axis at specified `x` values
`base(`*#*`)`	base value for `dropline()`; default is 0
`horizontal`	draw plot horizontally
`yvarformat(`*%fmt*`)`	display format for `y`
`xvarformat(`*%fmt*`)`	display format for `x`
cline_options	change look of plotted line
axis_choice_options	associate plot with alternative axis
twoway_options	titles, legends, axes, added lines and text, by, regions, name, aspect ratio, etc.

See [G-3] ***cline_options***, [G-3] ***axis_choice_options***, and [G-3] ***twoway_options***.

All explicit options are *rightmost*, except `horizontal`, which is *unique*; see [G-4] **concept: repeated options**.

`if` *exp* and `in` *range* play no role unless option `range(`*varname*`)` is specified.

In the above syntax diagram, *f*(`x`) stands for an *exp*ression in terms of `x`.

Menu

Graphics > Twoway graph (scatter, line, etc.)

Description

`twoway function` plots `y` = *f*(`x`), where *f*(`x`) is some function of `x`. That is, you type

```
. twoway function y=sqrt(x)
```

It makes no difference whether `y` and `x` are variables in your data.

Options

range(*# #*) and range(*varname*) specify the range of values for x. In the first syntax, range() is a pair of numbers identifying the minimum and maximum. In the second syntax, range() is a variable name, and the range used will be obtained from the minimum and maximum values of the variable. If range() is not specified, range(0 1) is assumed.

n(*#*) specifies the number of points at which $f(\mathtt{x})$ is to be evaluated. The default is n(300).

droplines(*numlist*) adds dropped lines from the function down to, or up to, the axis (or y = base() if base() is specified) at each x value specified in *numlist*.

base(*#*) specifies the base for the droplines(). The default is base(0). This option does not affect the range of the axes, so you may also want to specify the *axis_scale_option* yscale(range(*#*)) as well; see [G-3] ***axis_scale_options***.

horizontal specifies that the roles of y and x be interchanged and that the graph be plotted horizontally rather than vertically (that the plotted function be reflected along the identity line).

yvarformat(*%fmt*) and xvarformat(*%fmt*) specify the display format to be used for y and x. These formats are used when labeling the axes; see [G-3] ***axis_label_options***.

cline_options specify how the function line is rendered; see [G-3] ***cline_options***.

axis_choice_options associate the plot with a particular y or x axis on the graph; see [G-3] ***axis_choice_options***.

twoway_options are a set of common options supported by all twoway graphs. These options allow you to title graphs, name graphs, control axes and legends, add lines and text, set aspect ratios, create graphs over by() groups, and change some advanced settings. See [G-3] ***twoway_options***.

Remarks

Remarks are presented under the following headings:

Typical use
Advanced use 1
Advanced use 2

Typical use

You wish to plot the function y = exp(−x/6)sin(x) over the range 0 to 4π:

```
. twoway function y=exp(-x/6)*sin(x), range(0 12.57)
```

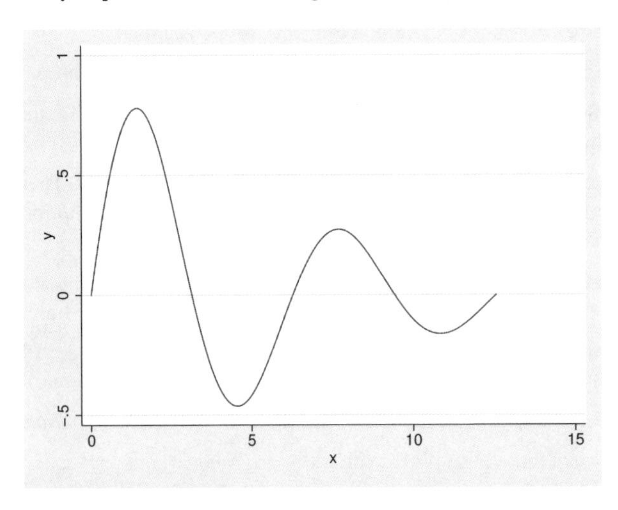

A better rendition of the graph above is

```
. twoway function y=exp(-x/6)*sin(x), range(0 12.57)
        yline(0, lstyle(foreground))
        xlabel(0 3.14 "{&pi}" 6.28 "2{&pi}" 9.42 "3{&pi}" 12.57 "4{&pi}")
        plotregion(style(none))
        xsca(noline)
```

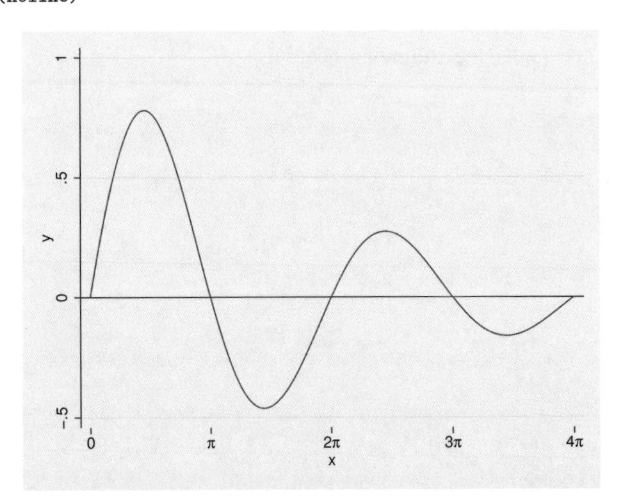

`yline(0, lstyle(foreground))` added a line at y = 0; `lstyle(foreground)` gave the line the same style as used for the axes. See [G-3] ***added_line_options***.

`xlabel(0 3.14 "{&pi}" 6.28 "2{&pi}" 9.42 "3{&pi}" 12.57 "4{&pi}")` labeled the x axis with the numeric values given; see [G-3] ***axis_label_options***.

`plotregion(style(none))` suppressed the border around the plot region; see [G-3] ***region_options***.

`xsca(noline)` suppressed the drawing of the x-axis line; see [G-3] ***axis_scale_options***.

Advanced use 1

The following graph appears in many introductory textbooks:

```
. twoway
      function y=normalden(x), range(-4 -1.96) color(gs12) recast(area)
   || function y=normalden(x), range(1.96 4)   color(gs12) recast(area)
   || function y=normalden(x), range(-4 4) lstyle(foreground)
   ||,
      plotregion(style(none))
      ysca(off) xsca(noline)
      legend(off)
      xlabel(-4 "-4 sd" -3 "-3 sd" -2 "-2 sd" -1 "-1 sd" 0 "mean"
              1  "1 sd"  2  "2 sd"  3  "3 sd"  4  "4 sd"
      , grid gmin gmax)
      xtitle("")
```

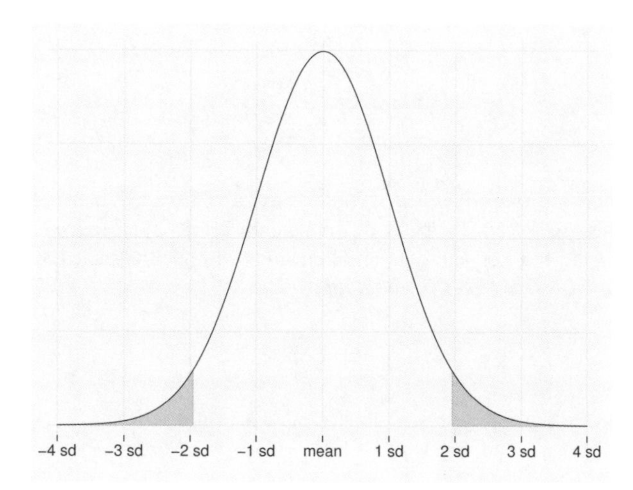

We drew the graph in three parts: the shaded area on the left, the shaded area on the right, and then the overall function. To obtain the shaded areas, we used the *advanced_option* `recast(area)` so that, rather than the function being plotted by `graph twoway line`, it was plotted by `graph twoway area`; see [G-3] ***advanced_options*** and [G-2] **graph twoway area**. Concerning the overall function, we drew it last so that its darker foreground-colored line would not get covered up by the shaded areas.

Advanced use 2

`function` plots may be overlaid with other `twoway` plots. For instance, `function` is one way to add $y = x$ lines to a plot:

```
. use http://www.stata-press.com/data/r12/sp500, clear
(S&P 500)
. scatter open close, msize(*.25) mcolor(*.6) ||
  function y=x, range(close) yvarlab("y=x") clwidth(*1.5)
```

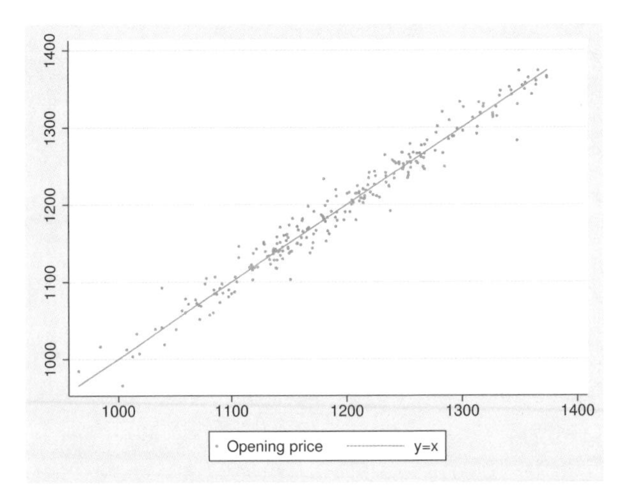

In the above, we specified the *advanced_option* `yvarlab("y=x")` so that the variable label of y would be treated as "y=x" in the construction of the legend; see [G-3] ***advanced_options***. We specified `msize(*.25)` to make the marker symbols smaller, and we specified `mcolor(*.6)` to make them dimmer; see [G-4] ***relativesize*** and [G-4] ***colorstyle***.

Reference

Cox, N. J. 2004. Stata tip 15: Function graphs on the fly. *Stata Journal* 4: 488–489.

Also see

[G-2] **graph twoway line** — Twoway line plots

Title

[G-2] graph twoway histogram — Histogram plots

Syntax

`twoway histogram` *varname* [*if*] [*in*] [*weight*]

[, [*discrete_options* | *continuous_options*] *common_options*]

discrete_options	Description
`discrete`	specify that data are discrete
`width(#)`	width of bins in *varname* units
`start(#)`	theoretical minimum value

continuous_options	Description
`bin(#)`	# of bins
`width(#)`	width of bins in *varname* units
`start(#)`	lower limit of first bin

common_options	Description
`density`	draw as density; the default
`fraction`	draw as fractions
`frequency`	draw as frequencies
`percent`	draw as percents
`vertical`	vertical bars; the default
`horizontal`	horizontal bars
`gap(#)`	reduce width of bars, $0 \leq \# < 100$
barlook_options	change look of bars
axis_choice_options	associate plot with alternative axis
twoway_options	titles, legends, axes, added lines and text, by, regions, name, aspect ratio, etc.

See [G-3] ***barlook_options***, [G-3] ***axis_choice_options***, and [G-3] ***twoway_options***.

`fweight`s are allowed; see [U] **11.1.6 weight**.

Menu

Graphics > Twoway graph (scatter, line, etc.)

Description

`twoway histogram` draws histograms of *varname*. Also see [R] **histogram** for an easier-to-use alternative.

Options for use in the discrete case

`discrete` specifies that *varname* is discrete and that each unique value of *varname* be given its own bin (bar of histogram).

`width(#)` is rarely specified in the discrete case; it specifies the width of the bins. The default is `width(`*d*`)`, where *d* is the observed minimum difference between the unique values of *varname*.

Specify `width()` if you are concerned that your data are sparse. For example, *varname* could in theory take on the values 1, 2, 3, ..., 9, but because of sparseness, perhaps only the values 2, 4, 7, and 8 are observed. Here the default width calculation would produce `width(2)`, and you would want to specify `width(1)`.

`start(#)` is also rarely specified in the discrete case; it specifies the theoretical minimum value of *varname*. The default is `start(`*m*`)`, where *m* is the observed minimum value.

As with `width()`, specify `start()` when you are concerned about sparseness. In the previous example, you would also want to specify `start(1)`. `start()` does nothing more than add white space to the left side of the graph.

`start()`, if specified, must be less than or equal to *m*, or an error will be issued.

Options for use in the continuous case

`bin(#)` and `width(#)` are alternatives that specify how the data are to be aggregated into bins. `bin()` specifies the number of bins (from which the width can be derived), and `width()` specifies the bin width (from which the number of bins can be derived).

If neither option is specified, the results are the same as if `bin(`*k*`)` were specified, where

$$k = \min\left(\sqrt{N}, 10 \times \frac{\ln(N)}{\ln(10)}\right)$$

and where *N* is the number of nonmissing observations of *varname*.

`start(#)` specifies the theoretical minimum of *varname*. The default is `start(`*m*`)`, where *m* is the observed minimum value of *varname*.

Specify `start()` when you are concerned about sparse data. For instance, you might know that *varname* can go down to 0, but you are concerned that 0 may not be observed.

`start()`, if specified, must be less than or equal to *m*, or an error will be issued.

Options for use in both the discrete and continuous cases

`density`, `fraction`, `frequency`, and `percent` are alternatives that specify whether you want the histogram scaled to density, fractional, or frequency units, or percentages. `density` is the default.

`density` scales the height of the bars so that the sum of their areas equals 1.

`fraction` scales the height of the bars so that the sum of their heights equals 1.

`frequency` scales the height of the bars so that each bar's height is equal to the number of observations in the category, and thus the sum of the heights is equal to the total number of nonmissing observations of *varname*.

`percent` scales the height of the bars so that the sum of their heights equals 100.

`vertical` and `horizontal` specify whether the bars are to be drawn vertically (the default) or horizontally.

`gap(#)` specifies that the bar width be reduced by # percent. `gap(0)` is the default; `histogram` sets the width so that adjacent bars just touch. If you wanted gaps between the bars, you would specify, for instance, `gap(5)`.

Also see [G-2] **graph twoway rbar** for other ways to set the display width of the bars. Histograms are actually drawn using `twoway rbar` with a restriction that 0 be included in the bars; `twoway histogram` will accept any options allowed by `twoway rbar`.

barlook_options set the look of the bars. The most important of these options is `color(`*colorstyle*`)`, which specifies the color of the bars; see [G-4] ***colorstyle*** for a list of color choices. See [G-3] ***barlook_options*** for information on the other *barlook_options*.

axis_choice_options associate the plot with a particular y or x axis on the graph; see [G-3] ***axis_choice_options***.

twoway_options are a set of common options supported by all `twoway` graphs. These options allow you to title graphs, name graphs, control axes and legends, add lines and text, set aspect ratios, create graphs over `by()` groups, and change some advanced settings. See [G-3] ***twoway_options***.

Remarks

Remarks are presented under the following headings:

Relationship between graph twoway histogram and histogram

`graph twoway histogram`—documented here—and `histogram`—documented in [R] **histogram**—are almost the same command. `histogram` has the advantages that

1. it allows overlaying of a normal density or a kernel estimate of the density;
2. if a density estimate is overlaid, it scales the density to reflect the scaling of the bars.

`histogram` is implemented in terms of `graph twoway histogram`.

Typical use

When you do not specify otherwise, `graph twoway histogram` assumes that the variable is continuous:

```
. use http://www.stata-press.com/data/r12/lifeexp
(Life expectancy, 1998)
. twoway histogram le
```

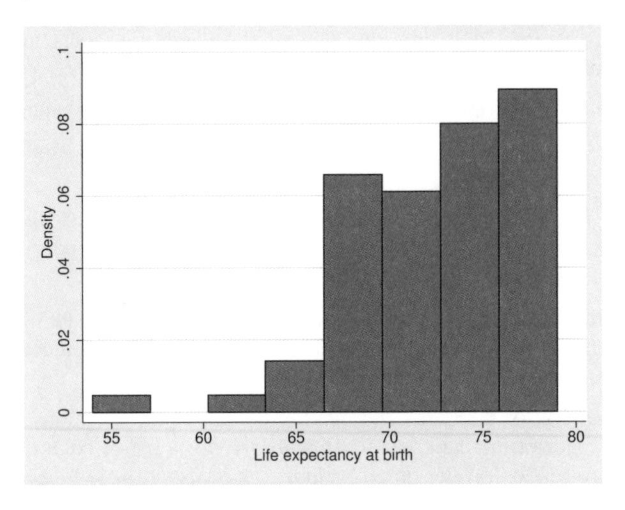

Even with a continuous variable, you may specify the `discrete` option to see the individual values:

```
. twoway histogram le, discrete
```

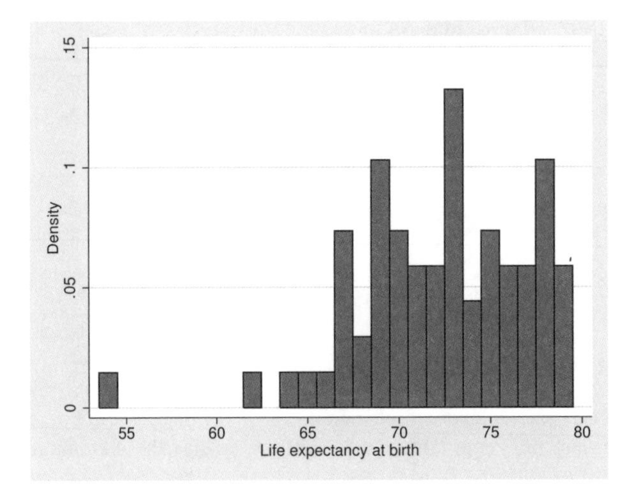

Use with by()

`graph twoway histogram` may be used with `by()`:

```
. use http://www.stata-press.com/data/r12/lifeexp, clear
(Life expectancy, 1998)
. twoway histogram le, discrete by(region, total)
```

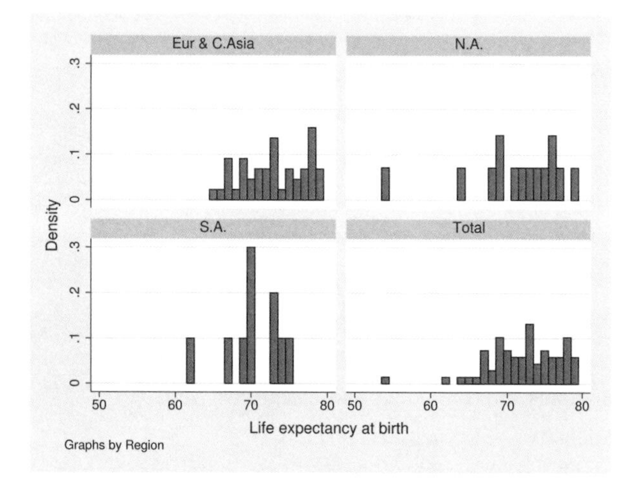

Here specifying `frequency` is a good way to show both the distribution and the overall contribution to the total:

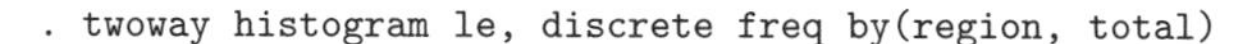

```
. twoway histogram le, discrete freq by(region, total)
```

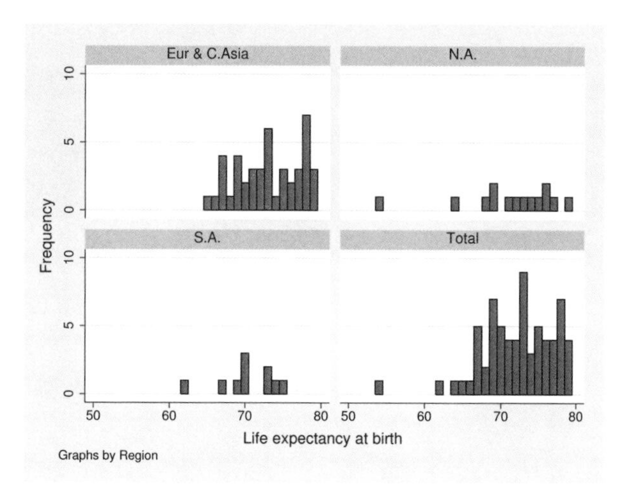

The height of the bars reflects the number of countries. Here—and in all the above examples—we would do better by obtaining population data on the countries and then typing

```
. twoway histogram le [fw=pop], discrete freq by(region, total)
```

so that bar height reflected total population.

History

According to Beniger and Robyn (1978, 4), although A. M. Guerry published a histogram in 1833, the word "histogram" was first used by Karl Pearson in 1895.

References

Beniger, J. R., and D. L. Robyn. 1978. Quantitative graphics in statistics: A brief history. *American Statistician* 32: 1–11.

Cox, N. J. 2005. Speaking Stata: Density probability plots. *Stata Journal* 5: 259–273.

——. 2007. Software Updates: Speaking Stata: Density probability plots. *Stata Journal* 7: 593.

Guerry, A.-M. 1833. *Essai sur la Statistique Morale de la France*. Paris: Crochard.

Harrison, D. A. 2005. Stata tip 20: Generating histogram bin variables. *Stata Journal* 5: 280–281.

Pearson, K. 1895. Contributions to the mathematical theory of evolution—II. Skew variation in homogeneous material. *Philosophical Transactions of the Royal Society of London, Series A* 186: 343–414.

Also see

[R] **histogram** — Histograms for continuous and categorical variables

[G-2] **graph twoway kdensity** — Kernel density plots

Title

[G-2] graph twoway kdensity — Kernel density plots

Syntax

`twoway kdensity` *varname* [*if*] [*in*] [*weight*] [, *options*]

options	Description
`bwidth(#)`	smoothing parameter
`kernel(kernel)`	specify kernel function; default is `kernel(epanechnikov)`
`range(# #)`	range for plot, minimum and maximum
`range(varname)`	range for plot obtained from *varname*
`n(#)`	number of points to evaluate
`area(#)`	rescaling parameter
`horizontal`	graph horizontally
cline_options	change look of the line
axis_choice_options	associate plot with alternative axis
twoway_options	titles, legends, axes, added lines and text, by, regions, name, aspect ratio, etc.

See [G-3] ***cline_options***, [G-3] ***axis_choice_options***, and [G-3] ***twoway_options***.

kernel	Description
`epanechnikov`	Epanechnikov kernel function; the default
`epan2`	alternative Epanechnikov kernel function
`biweight`	biweight kernel function
`cosine`	cosine kernel function
`gaussian`	Gaussian kernel function
`parzen`	Parzen kernel function
`rectangle`	rectangular kernel function
`triangle`	triangular kernel function

`fweights` and `aweights` are allowed; see [U] **11.1.6 weight**.

Menu

Graphics > Twoway graph (scatter, line, etc.)

Description

`graph twoway kdensity` plots a kernel density estimate for *varname* using `graph twoway line`; see [G-2] **graph twoway line**.

Options

`bwidth(#)` and `kernel(kernel)` specify how the kernel density estimate is to be obtained and are in fact the same options as those specified with the command `kdensity`; see [R] **kdensity**.

`bwidth(#)` specifies the smoothing parameter.

`kernel(kernel)` specify the kernel-weight function to be used. The default is `kernel(epanechnikov)`.

See [R] **kdensity** for more information about these options.

All the other `graph twoway kdensity` options modify how the result is displayed, not how it is obtained.

`range(# #)` and `range(varname)` specify the range of values at which the kernel density estimates are to be plotted. The default is `range(m M)`, where *m* and *M* are the minimum and maximum of the *varname* specified on the `graph twoway kdensity` command.

`range(# #)` specifies a pair of numbers to be used as the minimum and maximum.

`range(varname)` specifies another variable for which its minimum and maximum are to be used.

`n(#)` specifies the number of points at which the estimate is evaluated. The default is `n(300)`.

`area(#)` specifies a multiplier by which the density estimates are adjusted before being plotted. The default is `area(1)`. `area()` is useful when overlaying a density estimate on top of a histogram that is itself not scaled as a density. For instance, if you wished to scale the density estimate as a frequency, `area()` would be specified as the total number of nonmissing observations.

`horizontal` specifies that the result be plotted horizontally (i.e, reflected along the identity line).

cline_options specify how the density line is rendered and its appearance; [G-3] ***cline_options***.

axis_choice_options associate the plot with a particular y or x axis on the graph; see [G-3] ***axis_choice_options***.

twoway_options are a set of common options supported by all `twoway` graphs. These options allow you to title graphs, name graphs, control axes and legends, add lines and text, set aspect ratios, create graphs over `by()` groups, and change some advanced settings. See [G-3] ***twoway_options***.

Remarks

`graph twoway kdensity` *varname* uses the `kdensity` command to obtain an estimate of the density of *varname* and uses `graph twoway line` to plot the result.

Remarks are presented under the following headings:

Typical use
Use with by()

Typical use

The density estimate is often graphed on top of the histogram:

```
. use http://www.stata-press.com/data/r12/lifeexp
(Life expectancy, 1998)
. twoway histogram lexp, color(*.5) || kdensity lexp
```

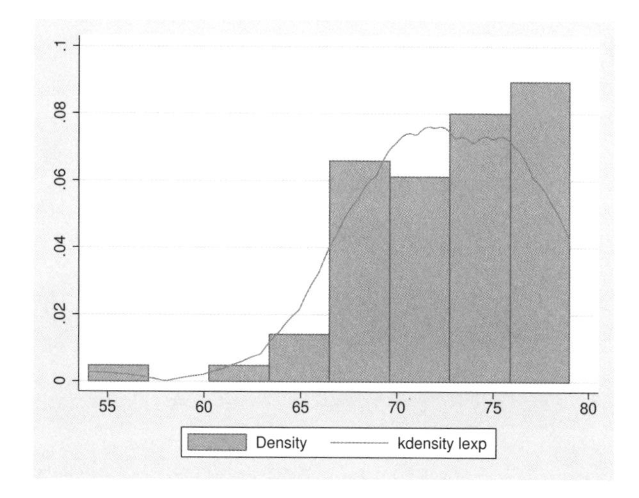

Notice the use of `graph twoway histogram`'s `color(*.5)` option to dim the bars and make the line stand out; see [G-4] ***colorstyle***.

Notice also the y and x axis titles: "Density/kdensity lexp" and "Life expectancy at birth/x". The "kdensity lexp" and "x" were contributed by the `twoway kdensity`. When you overlay graphs, you nearly always need to respecify the axis titles using the *axis_title_options* `ytitle()` and `xtitle()`; see [G-3] ***axis_title_options***.

Use with by()

`graph twoway kdensity` may be used with `by()`:

```
. use http://www.stata-press.com/data/r12/lifeexp, clear
(Life expectancy, 1998)
. twoway histogram lexp, color(*.5) || kdensity lexp ||, by(region)
```

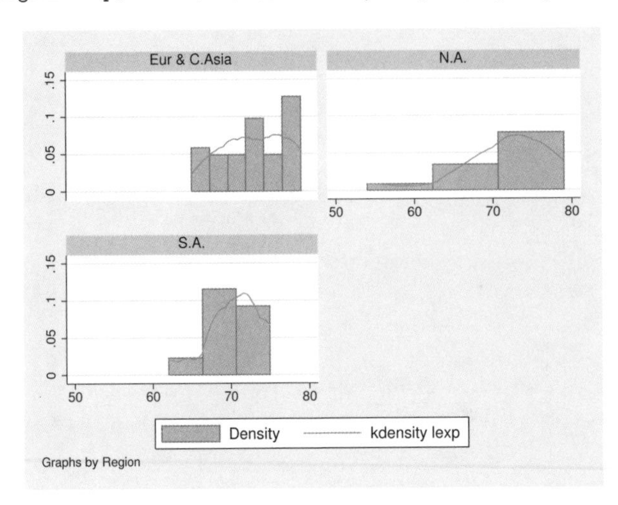

References

Cox, N. J. 2005. Speaking Stata: Density probability plots. *Stata Journal* 5: 259–273.

——. 2007. Software Updates: Speaking Stata: Density probability plots. *Stata Journal* 7: 593.

Also see

[R] **kdensity** — Univariate kernel density estimation

[G-2] **graph twoway histogram** — Histogram plots

Title

[G-2] graph twoway lfit — Twoway linear prediction plots

Syntax

`twoway lfit` *yvar xvar* [*if*] [*in*] [*weight*] [, *options*]

options	Description
`range(`*# #*`)`	range over which predictions calculated
`n(`*#*`)`	number of prediction points
`atobs`	calculate predictions at *xvar*
`estopts(`*regress_options*`)`	options for `regress`
`predopts(`*predict_options*`)`	options for `predict`
cline_options	change look of predicted line
axis_choice_options	associate plot with alternative axis
twoway_options	titles, legends, axes, added lines and text, by, regions, name, aspect ratio, etc.

See [G-3] ***cline_options***, [G-3] ***axis_choice_options***, and [G-3] ***twoway_options***.

All options are *rightmost*; see [G-4] **concept: repeated options**.

yvar and *xvar* may contain time-series operators; see [U] **11.4.4 Time-series varlists**.

`aweight`s, `fweight`s, and `pweight`s are allowed. Weights, if specified, affect estimation but not how the weighted results are plotted. See [U] **11.1.6 weight**.

Menu

Graphics > Twoway graph (scatter, line, etc.)

Description

`twoway lfit` calculates the prediction for *yvar* from a linear regression of *yvar* on *xvar* and plots the resulting line.

Options

`range(`*# #*`)` specifies the x range over which predictions are to be calculated. The default is `range(. .)`, meaning the minimum and maximum values of *xvar*. `range(0 10)` would make the range 0 to 10, `range(. 10)` would make the range the minimum to 10, and `range(0 .)` would make the range 0 to the maximum.

`n(`*#*`)` specifies the number of points at which predictions over `range()` are to be calculated. The default is `n(3)`.

`atobs` is an alternative to `n()`. It specifies that the predictions be calculated at the *xvar* values. `atobs` is the default if `predopts()` is specified and any statistic other than the `xb` is requested.

`estopts(`*regress_options*`)` specifies options to be passed along to `regress` to estimate the linear regression from which the line will be predicted; see [R] **regress**. If this option is specified, `estopts(nocons)` is also often specified.

`predopts(`*predict_options*`)` specifies options to be passed along to `predict` to obtain the predictions after estimation by `regress`; see [R] **regress postestimation**.

cline_options specify how the prediction line is rendered; see [G-3] ***cline_options***.

axis_choice_options associate the plot with a particular y or x axis on the graph; see [G-3] ***axis_choice_options***.

twoway_options are a set of common options supported by all `twoway` graphs. These options allow you to title graphs, name graphs, control axes and legends, add lines and text, set aspect ratios, create graphs over `by()` groups, and change some advanced settings. See [G-3] ***twoway_options***.

Remarks

Remarks are presented under the following headings:

Typical use
Cautions
Use with by()

Typical use

`twoway lfit` is nearly always used in conjunction with other `twoway` plottypes, such as

```
. use http://www.stata-press.com/data/r12/auto
(1978 Automobile Data)
. scatter mpg weight || lfit mpg weight
```

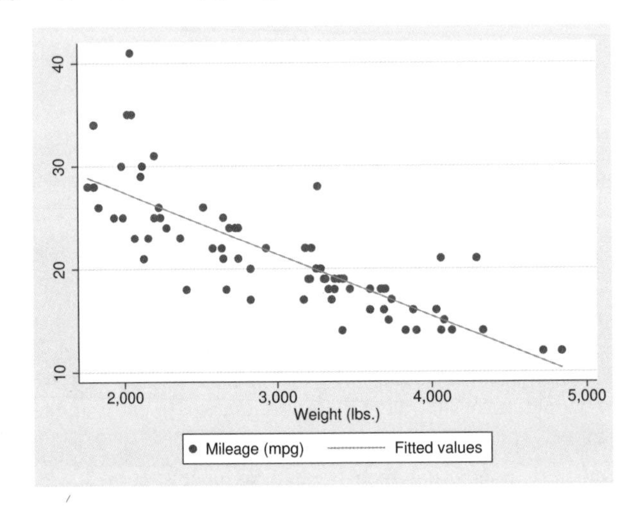

Results are visually the same as typing

```
. regress mpg weight
. predict fitted
. scatter mpg weight || line fitted weight
```

Cautions

Do not use `twoway lfit` when specifying the *axis_scale_options* `yscale(log)` or `xscale(log)` to create log scales. Typing

```
. scatter mpg weight, xscale(log) || lfit mpg weight
```

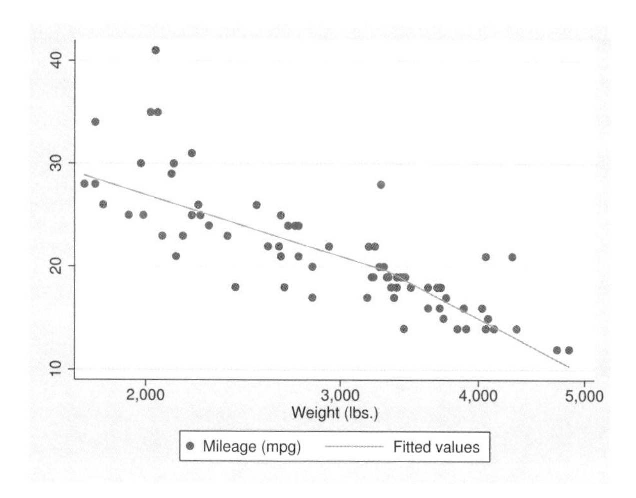

The line is not straight because the regression estimated for the prediction was for `mpg` on `weight`, not `mpg` on `log(weight)`. (The default for `n()` is 3 so that, if you make this mistake, you will spot it.)

Use with by()

`lfit` may be used with `by()` (as can all the `twoway` plot commands):

```
. scatter mpg weight || lfit mpg weight ||, by(foreign, total row(1))
```

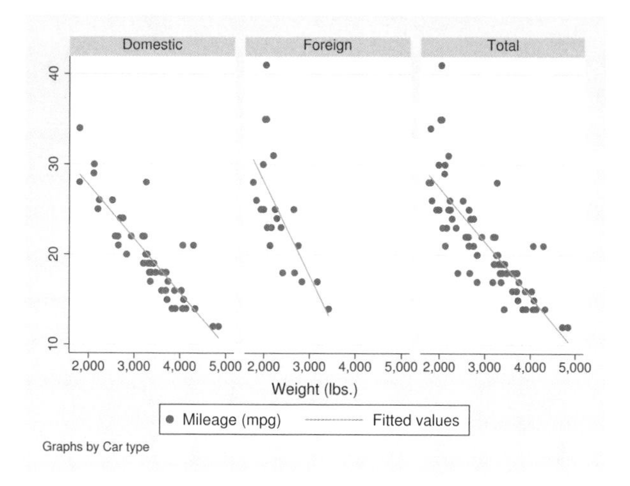

Also see

[G-2] **graph twoway line** — Twoway line plots

[G-2] **graph twoway qfit** — Twoway quadratic prediction plots

[G-2] **graph twoway fpfit** — Twoway fractional-polynomial prediction plots

[G-2] **graph twoway mband** — Twoway median-band plots

[G-2] **graph twoway mspline** — Twoway median-spline plots

[G-2] **graph twoway lfitci** — Twoway linear prediction plots with CIs

[R] **regress** — Linear regression

Title

[G-2] graph twoway lfitci — Twoway linear prediction plots with CIs

Syntax

`twoway lfitci` *yvar xvar* [*if*] [*in*] [*weight*] [, *options*]

options	Description
`stdp`	CIs from SE of prediction; the default
`stdf`	CIs from SE of forecast
`stdr`	CIs from SE of residual; seldom specified
`level(`*#*`)`	set confidence level; default is `level(95)`
`range(`*# #*`)`	range over which predictions are calculated
`n(`*#*`)`	number of prediction points
`atobs`	calculate predictions at *xvar*
`estopts(`*regress_options*`)`	options for `regress`
`predopts(`*predict_options*`)`	options for `predict`
`nofit`	do not plot the prediction
`fitplot(`*plottype*`)`	how to plot fit; default is `fitplot(line)`
`ciplot(`*plottype*`)`	how to plot CIs; default is `ciplot(rarea)`
fcline_options	change look of predicted line
fitarea_options	change look of CI
axis_choice_options	associate plot with alternative axis
twoway_options	titles, legends, axes, added lines and text, by, regions, name, aspect ratio, etc.

See [G-3] ***fcline_options***, [G-3] ***fitarea_options***, [G-3] ***axis_choice_options***, and [G-3] ***twoway_options***.

Options `range()`, `estopts()`, `predopts()`, `n()`, and `level()` are *rightmost*; `atobs`, `nofit`, `fitplot()`, `ciplot()`, `stdp`, `stdf`, and `stdr` are *unique*; see [G-4] **concept: repeated options**.

yvar and *xvar* may contain time-series operators; see [U] **11.4.4 Time-series varlists**.

`aweight`s, `fweight`s, and `pweight`s are allowed. Weights, if specified, affect estimation but not how the weighted results are plotted. See [U] **11.1.6 weight**.

Menu

Graphics > Twoway graph (scatter, line, etc.)

Description

`twoway lfitci` calculates the prediction for *yvar* from a linear regression of *yvar* on *xvar* and plots the resulting line, along with a confidence interval.

Options

stdp, stdf, and stdr determine the basis for the confidence interval. stdp is the default.

stdp specifies that the confidence interval be the confidence interval of the mean.

stdf specifies that the confidence interval be the confidence interval for an individual forecast, which includes both the uncertainty of the mean prediction and the residual.

stdr specifies that the confidence interval be based only on the standard error of the residual.

level(#) specifies the confidence level, as a percentage, for the confidence intervals. The default is level(95) or as set by set level; see [U] **20.7 Specifying the width of confidence intervals**.

range(# #) specifies the x range over which predictions are calculated. The default is range(. .), meaning the minimum and maximum values of *xvar*. range(0 10) would make the range 0 to 10, range(. 10) would make the range the minimum to 10, and range(0 .) would make the range 0 to the maximum.

n(#) specifies the number of points at which the predictions and the CI over range() are to be calculated. The default is n(100).

atobs is an alternative to n() and specifies that the predictions be calculated at the *xvar* values. atobs is the default if predopts() is specified and any statistic other than the xb is requested.

estopts(*regress_options*) specifies options to be passed along to regress to estimate the linear regression from which the line will be predicted; see [R] **regress**. If this option is specified, also commonly specified is estopts(nocons).

predopts(*predict_options*) specifies options to be passed along to predict to obtain the predictions after estimation by regress; see [R] **regress postestimation**.

nofit prevents the prediction from being plotted.

fitplot(*plottype*), which is seldom used, specifies how the prediction is to be plotted. The default is fitplot(line), meaning that the prediction will be plotted by graph twoway line. See [G-2] **graph twoway** for a list of *plottype* choices. You may choose any that expect one y and one x variable.

ciplot(*plottype*) specifies how the confidence interval is to be plotted. The default is ciplot(rarea), meaning that the prediction will be plotted by graph twoway rarea.

A reasonable alternative is ciplot(rline), which will substitute lines around the prediction for shading. See [G-2] **graph twoway** for a list of *plottype* choices. You may choose any that expect two y variables and one x variable.

fcline_options specify how the prediction line is rendered; see [G-3] ***fcline_options***. If you specify fitplot(), then rather than using *fcline_options*, you should select options that affect the specified *plottype* from the options in scatter; see [G-2] **graph twoway scatter**.

fitarea_options specify how the confidence interval is rendered; see [G-3] ***fitarea_options***. If you specify ciplot(), then rather than using *fitarea_options*, you should specify whatever is appropriate.

axis_choice_options associate the plot with a particular y or x axis on the graph; see [G-3] ***axis_choice_options***.

twoway_options are a set of common options supported by all twoway graphs. These options allow you to title graphs, name graphs, control axes and legends, add lines and text, set aspect ratios, create graphs over by() groups, and change some advanced settings. See [G-3] ***twoway_options***.

Remarks

Remarks are presented under the following headings:

Typical use

`twoway lfitci` by default draws the confidence interval of the predicted mean:

```
. use http://www.stata-press.com/data/r12/auto
(1978 Automobile Data)
. twoway lfitci mpg weight
```

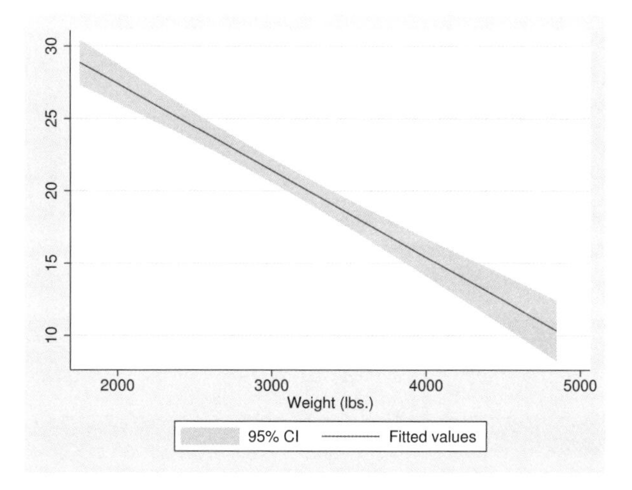

If you specify the `ciplot(rline)` option, then rather than being shaded, the confidence interval will be designated by lines:

```
. twoway lfitci mpg weight, ciplot(rline)
```

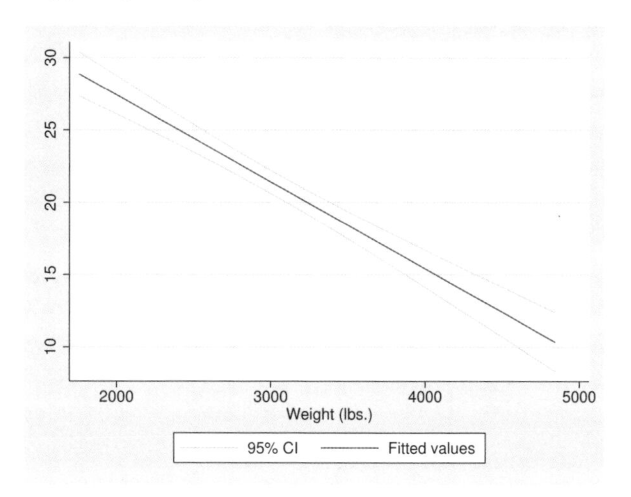

Advanced use

`lfitci` can be usefully overlaid with other plots:

```
. use http://www.stata-press.com/data/r12/auto, clear
(1978 Automobile Data)
. twoway lfitci mpg weight, stdf || scatter mpg weight
```

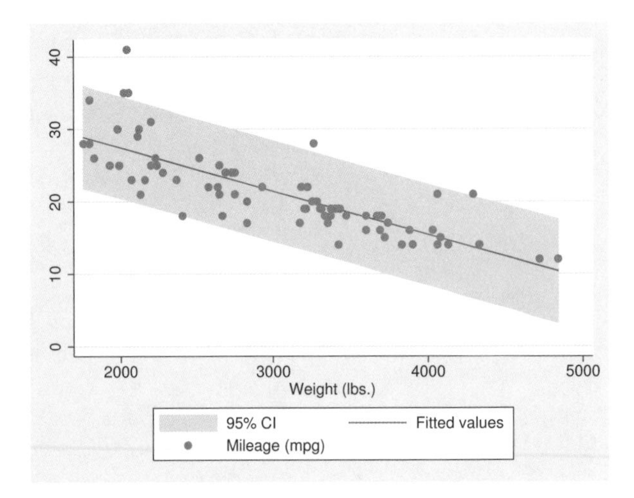

In the above example, we specified `stdf` to obtain a confidence interval based on the standard error of the forecast rather than the standard error of the mean. This is more useful for identifying outliers.

We typed

```
. twoway lfitci ... || scatter ...
```

and not

```
. twoway scatter ... || lfitci ...
```

Had we drawn the scatter diagram first, the confidence interval would have covered up most of the points.

Cautions

Do not use `twoway lfitci` when specifying the *axis_scale_options* `yscale(log)` or `xscale(log)` to create log scales. Typing

```
. twoway lfitci mpg weight, stdf || scatter mpg weight ||, xscale(log)
```

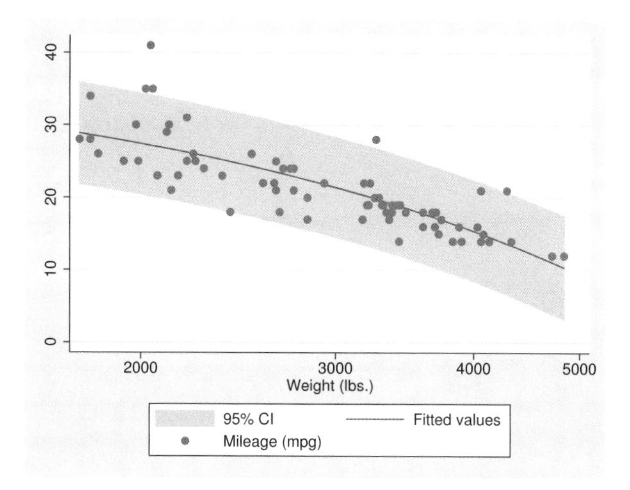

The result may look pretty, but if you think about it, it is not what you want. The prediction line is not straight because the regression estimated for the prediction was for `mpg` on `weight`, not for `mpg` on `log(weight)`.

Use with by()

`lfitci` may be used with `by()` (as can all the `twoway` plot commands):

```
. twoway lfitci  mpg weight, stdf ||
        scatter mpg weight       ||
  , by(foreign, total row(1))
```

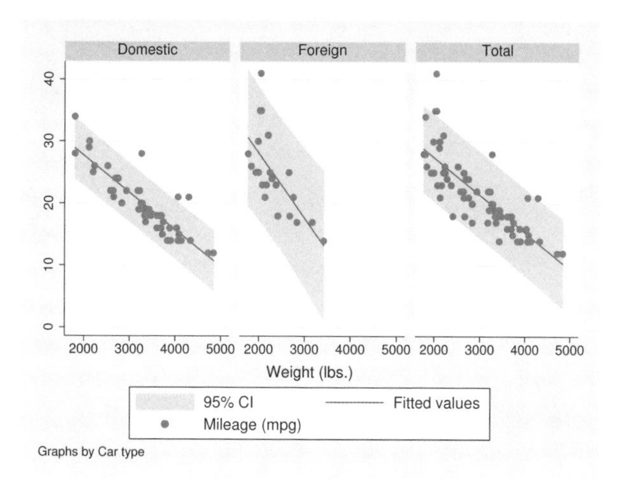

Also see

[G-2] **graph twoway qfitci** — Twoway quadratic prediction plots with CIs

[G-2] **graph twoway fpfitci** — Twoway fractional-polynomial prediction plots with CIs

[G-2] **graph twoway lfit** — Twoway linear prediction plots

[R] **regress** — Linear regression

Title

[G-2] graph twoway line — Twoway line plots

Syntax

[<u>tw</u>oway] `line` *varlist* [*if*] [*in*] [, *options*]

where *varlist* is

y_1 [y_2[...]] x

options	Description
connect_options	change look of lines or connecting method
axis_choice_options	associate plot with alternative axis
twoway_options	titles, legends, axes, added lines and text, by, regions, name, aspect ratio, etc.

connect_options discusses options for one y versus one x; see *connect_options* in [G-2] **graph twoway scatter** when plotting multiple ys against one x.

Menu

Graphics > Twoway graph (scatter, line, etc.)

Description

`line` draws line plots.

`line` is a command and a *plottype* as defined in [G-2] **graph twoway**. Thus the syntax for `line` is

```
. graph twoway line ...
. twoway line ...
. line ...
```

Being a plottype, `line` may be combined with other plottypes in the `twoway` family (see [G-2] **graph twoway**), as in

```
. twoway (line ...) (scatter ...) (lfit ...) ...
```

which can equivalently be written

```
. line ... || scatter ... || lfit ... || ...
```

Options

connect_options specify how the points forming the line are connected and the look of the lines, including pattern, width, and color; see [G-3] ***connect_options***.

[G-3] ***connect_options*** discusses options for one *y* versus one *x*, see *connect_options* in [G-2] **graph twoway scatter** when plotting multiple *y*s against one *x*.

axis_choice_options associate the plot with a particular *y* or *x* axis on the graph; see [G-3] ***axis_choice_options***.

twoway_options are a set of common options supported by all `twoway` graphs. These options allow you to title graphs, name graphs, control axes and legends, add lines and text, set aspect ratios, create graphs over `by()` groups, and change some advanced settings. See [G-3] ***twoway_options***.

Remarks

Remarks are presented under the following headings:

Oneway equivalency of line and scatter
Typical use
Advanced use
Cautions

Oneway equivalency of line and scatter

`line` is similar to `scatter`, the differences being that by default the marker symbols are not displayed and the points are connected:

Default `msymbol()` option: `msymbol(none ...)`

Default `connect()` option: `connect(l ...)`

Thus you get the same results typing

```
. line yvar xvar
```

as typing

```
. scatter yvar xvar, msymbol(none) connect(l)
```

You can use `scatter` in place of `line`, but you may not use `line` in place of `scatter`. Typing

```
. line yvar xvar, msymbol(O) connect(none)
```

will not achieve the same results as

```
. scatter yvar xvar
```

because `line`, while it allows you to specify the *marker_option* `msymbol()`, ignores its setting.

Typical use

`line` draws line charts:

```
. use http://www.stata-press.com/data/r12/uslifeexp
(U.S. life expectancy, 1900-1999)
. line le year
```

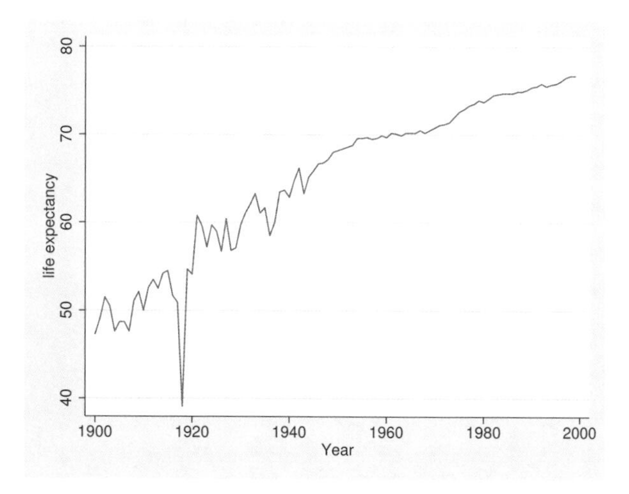

Line charts work well with time-series data. With other datasets, lines are often used to show predicted values and confidence intervals:

```
. use http://www.stata-press.com/data/r12/auto, clear
(1978 Automobile Data)
. quietly regress mpg weight
. predict hat
. predict stdf, stdf
. generate lo = hat - 1.96*stdf
. generate hi = hat + 1.96*stdf
. scatter mpg weight || line hat lo hi weight, pstyle(p2 p3 p3) sort
```

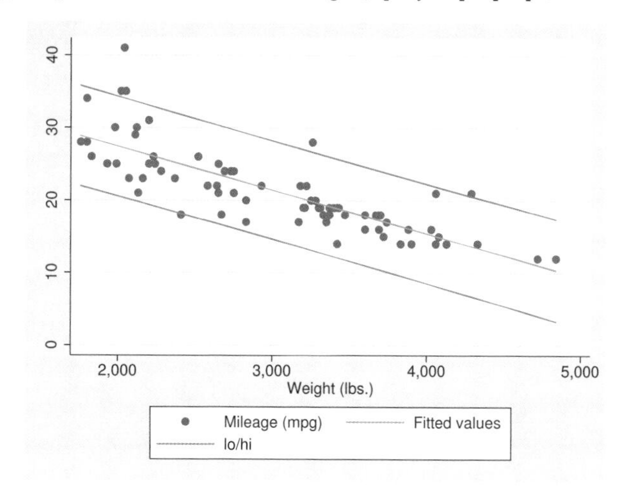

Do not forget to include the sort option when the data are not in the order of the x variable, as they are not above. We also included pstyle(p2 p3 p3) to give the lower and upper confidence limit lines the same look; see *Appendix: Styles and composite styles* under *Remarks* in [G-2] **graph twoway scatter**.

Because line is scatter, we can use any of the options allowed by scatter. Below we return to the U.S. life expectancy data and graph black and white male life expectancies, along with the difference, specifying many options to create an informative and visually pleasing graph:

```
. use http://www.stata-press.com/data/r12/uslifeexp, clear
(U.S. life expectancy, 1900-1999)
. generate diff = le_wm - le_bm
. label var diff "Difference"
.     line le_wm year, yaxis(1 2) xaxis(1 2)
   || line le_bm year
   || line diff  year
   || lfit diff  year
   ||,
      ylabel(0(5)20, axis(2) gmin angle(horizontal))
      ylabel(0 20(10)80,     gmax angle(horizontal))
      ytitle("", axis(2))
      xlabel(1918, axis(2)) xtitle("", axis(2))
      ylabel(, axis(2) grid)
      ytitle("Life expectancy at birth (years)")
      title("White and black life expectancy")
      subtitle("USA, 1900-1999")
      note("Source: National Vital Statistics, Vol 50, No. 6"
           "(1918 dip caused by 1918 Influenza Pandemic)")
```

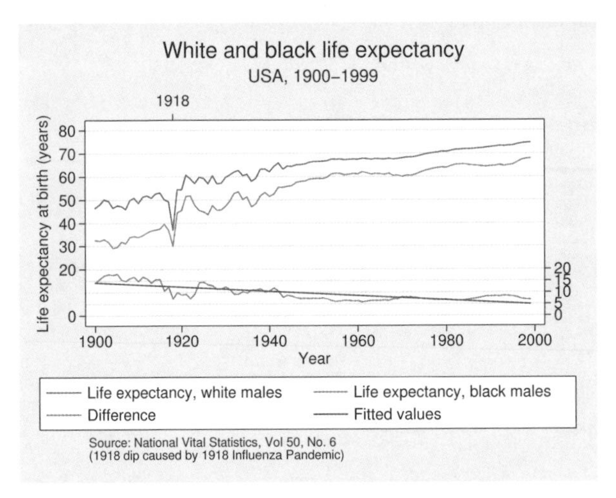

See [G-2] **graph twoway scatter**.

Advanced use

The above graph would look better if we shortened the descriptive text used in the keys. Below we add

```
legend(label(1 "White males") label(2 "Black males"))
```

to our previous command:

```
.    line le_wm year, yaxis(1 2) xaxis(1 2)
  || line le_bm year
  || line diff  year
  || lfit diff  year
  ||,
     ylabel(0(5)20, axis(2) gmin angle(horizontal))
     ylabel(0 20(10)80,     gmax angle(horizontal))
     ytitle("", axis(2))
     xlabel(1918, axis(2)) xtitle("", axis(2))
     ylabel(, axis(2) grid)
     ytitle("Life expectancy at birth (years)")
     title("White and black life expectancy")
     subtitle("USA, 1900-1999")
     note("Source: National Vital Statistics, Vol 50, No. 6"
          "(1918 dip caused by 1918 Influenza Pandemic)")
     legend(label(1 "White males") label(2 "Black males"))
```

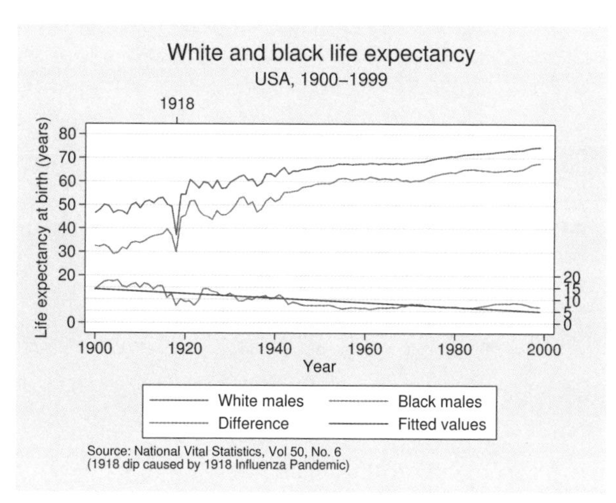

We might also consider moving the legend to the right of the graph, which we can do by adding

```
legend(col(1) pos(3))
```

resulting in

```
.    line le_wm year, yaxis(1 2) xaxis(1 2)
  || line le_bm year
  || line diff  year
  || lfit diff  year
  ||,
     ylabel(0(5)20, axis(2) gmin angle(horizontal))
     ylabel(0 20(10)80,     gmax angle(horizontal))
     ytitle("", axis(2))
     xlabel(1918, axis(2)) xtitle("", axis(2))
     ylabel(, axis(2) grid)
     ytitle("Life expectancy at birth (years)")
     title("White and black life expectancy")
     subtitle("USA, 1900-1999")
     note("Source: National Vital Statistics, Vol 50, No. 6"
          "(1918 dip caused by 1918 Influenza Pandemic)")
     legend(label(1 "White males") label(2 "Black males"))
     legend(col(1) pos(3))
```

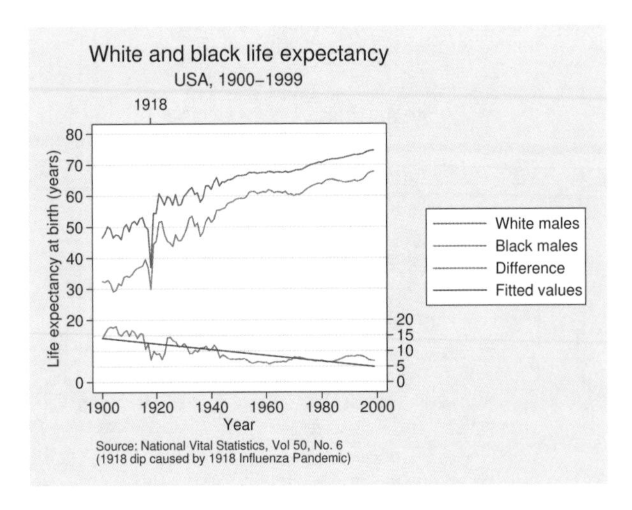

See [G-3] ***legend_options*** for more information about dealing with legends.

Cautions

Be sure that the data are in the order of the x variable, or specify `line`'s `sort` option. If you do neither, you will get something that looks like the scribblings of a child:

```
. use http://www.stata-press.com/data/r12/auto, clear
(1978 Automobile Data)
. line mpg weight
```

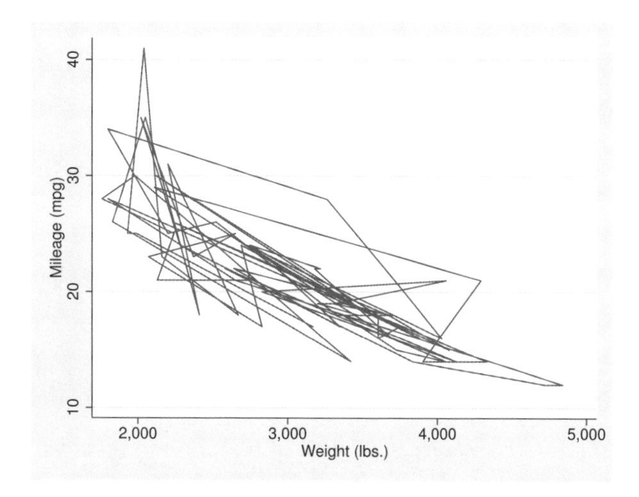

Also see

[G-2] **graph twoway scatter** — Twoway scatterplots

[G-2] **graph twoway fpfit** — Twoway fractional-polynomial prediction plots

[G-2] **graph twoway lfit** — Twoway linear prediction plots

[G-2] **graph twoway mband** — Twoway median-band plots

[G-2] **graph twoway mspline** — Twoway median-spline plots

[G-2] **graph twoway qfit** — Twoway quadratic prediction plots

Title

[G-2] graph twoway lowess — Local linear smooth plots

Syntax

`twoway lowess` *yvar xvar* [*if*] [*in*] [, *options*]

options	Description
`bwidth(#)`	smoothing parameter
`mean`	use running-mean smoothing
`noweight`	use unweighted smoothing
`logit`	transform the smooth to logits
`adjust`	adjust smooth's mean to equal *yvar*'s mean
cline_options	change look of the line
axis_choice_options	associate plot with alternative axis
twoway_options	titles, legends, axes, added lines and text, by, regions, name, aspect ratio, etc.

See [G-3] ***cline_options***, [G-3] ***axis_choice_options***, and [G-3] ***twoway_options***.

Menu

Graphics > Twoway graph (scatter, line, etc.)

Description

`graph twoway lowess` plots a lowess smooth of *yvar* on *xvar* using `graph twoway line`; see [G-2] **graph twoway line**.

Options

`bwidth(#)` specifies the bandwidth. `bwidth(.8)` is the default. Centered subsets of *N**`bwidth()` observations, *N* = number of observations, are used for calculating smoothed values for each point in the data except for endpoints, where smaller, uncentered subsets are used. The greater the `bwidth()`, the greater the smoothing.

`mean` specifies running-mean smoothing; the default is running-line least-squares smoothing.

`noweight` prevents the use of Cleveland's (1979) tricube weighting function; the default is to use the weighting function.

`logit` transforms the smoothed *yvar* into logits.

`adjust` adjusts by multiplication the mean of the smoothed *yvar* to equal the mean of *yvar*. This is useful when smoothing binary (0/1) data.

cline_options specify how the lowess line is rendered and its appearance; see [G-3] ***cline_options***.

axis_choice_options associate the plot with a particular y or x axis on the graph; see [G-3] ***axis_choice_options***.

twoway_options are a set of common options supported by all `twoway` graphs. These options allow you to title graphs, name graphs, control axes and legends, add lines and text, set aspect ratios, create graphs over `by()` groups, and change some advanced settings. See [G-3] ***twoway_options***.

Remarks

`graph twoway lowess` *yvar xvar* uses the `lowess` command—see [R] **lowess**—to obtain a local linear smooth of *yvar* on *xvar* and uses `graph twoway line` to plot the result.

Remarks are presented under the following headings:

Typical use
Use with by( )

Typical use

The local linear smooth is often graphed on top of the data, possibly with other regression lines:

```
. use http://www.stata-press.com/data/r12/auto
(1978 Automobile Data)

. twoway scatter mpg weight, mcolor(*.6) ||
         lfit    mpg weight    ||
         lowess  mpg weight
```

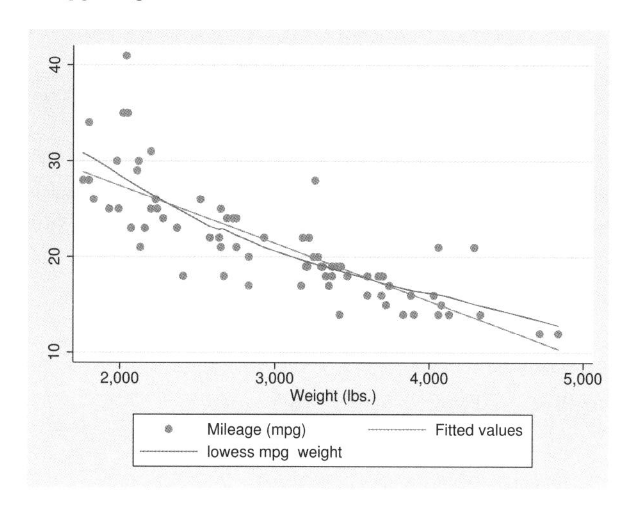

Notice our use of `mcolor(*.6)` to dim the points and thus make the lines stand out; see [G-4] ***colorstyle***.

Notice also the y-axis title: "Mileage (mpg)/Fitted values/lowess mpg weight". The "Fitted values" was contributed by `twoway lfit` and "lowess mpg weight" by `twoway lowess`. When you overlay graphs, you nearly always need to respecify the axis titles using the *axis_title_options* `ytitle()` and `xtitle()`; see [G-3] ***axis_title_options***.

Use with by()

`graph twoway lowess` may be used with `by()`:

```
. use http://www.stata-press.com/data/r12/auto, clear
(1978 Automobile Data)
. twoway scatter mpg weight, mcolor(*.6) ||
         lfit   mpg weight ||
         lowess mpg weight ||, by(foreign)
```

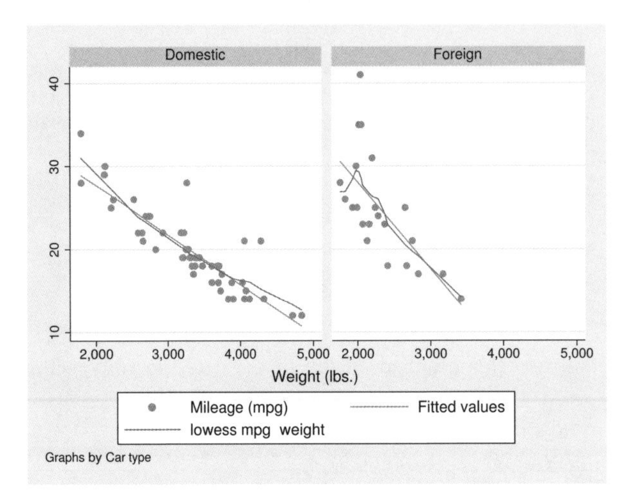

References

Cleveland, W. S. 1979. Robust locally weighted regression and smoothing scatterplots. *Journal of the American Statistical Association* 74: 829–836.

Cox, N. J. 2005. Speaking Stata: Smoothing in various directions. *Stata Journal* 5: 574–593.

——. 2010. Software Updates: Speaking Stata: Smoothing in various directions. *Stata Journal* 10: 164.

Royston, P., and N. J. Cox. 2005. A multivariable scatterplot smoother. *Stata Journal* 5: 405–412.

Also see

[R] **lowess** — Lowess smoothing

[G-2] **graph twoway mspline** — Twoway median-spline plots

Title

[G-2] graph twoway lpoly — Local polynomial smooth plots

Syntax

`twoway lpoly` *yvar xvar* [*if*] [*in*] [*weight*] [, *options*]

options	Description
`kernel(`*kernel*`)`	kernel function; default is `kernel(epanechnikov)`
`bwidth(`*#*`)`	kernel bandwidth
`degree(`*#*`)`	degree of the polynomial smooth; default is `degree(0)`
`n(`*#*`)`	obtain the smooth at # points; default is $\min(N, 50)$
cline_options	change look of the line
axis_choice_options	associate plot with alternative axis
twoway_options	titles, legends, axes, added lines and text, by, regions, name, aspect ratio, etc.

See [G-3] ***cline_options***, [G-3] ***axis_choice_options***, and [G-3] ***twoway_options***.

kernel	Description
`epanechnikov`	Epanechnikov kernel function; the default
`epan2`	alternative Epanechnikov kernel function
`biweight`	biweight kernel function
`cosine`	cosine trace kernel function
`gaussian`	Gaussian kernel function
`parzen`	Parzen kernel function
`rectangle`	rectangle kernel function
`triangle`	triangle kernel function

`fweight`s and `aweight`s are allowed; see [U] **11.1.6 weight**.

Menu

Graphics > Twoway graph (scatter, line, etc.)

Description

`graph twoway lpoly` plots a local polynomial smooth of *yvar* on *xvar*.

Options

`kernel(`*kernel*`)` specifies the kernel function for use in calculating the weighted local polynomial estimate. The default is `kernel(epanechnikov)`. See [R] **kdensity** for more information on this option.

`bwidth(#)` specifies the half-width of the kernel, the width of the smoothing window around each point. If `bwidth()` is not specified, a rule-of-thumb bandwidth estimator is calculated and used; see [R] **lpoly**.

`degree(#)` specifies the degree of the polynomial to be used in the smoothing. The default is `degree(0)`, meaning local mean smoothing.

`n(#)` specifies the number of points at which the smooth is to be calculated. The default is $\min(N, 50)$, where N is the number of observations.

cline_options specify how the line is rendered and its appearance; see [G-3] ***cline_options***.

axis_choice_options associate the plot with a particular y or x axis on the graph; see [G-3] ***axis_choice_options***.

twoway_options are a set of common options supported by all `twoway` graphs. These options allow you to title graphs, name graphs, control axes and legends, add lines and text, set aspect ratios, create graphs over `by()` groups, and change some advanced settings. See [G-3] ***twoway_options***.

Remarks

`graph twoway lpoly` *yvar xvar* uses the `lpoly` command—see [R] **lpoly**—to obtain a local polynomial smooth of *yvar* on *xvar* and uses `graph twoway line` to plot the result.

Remarks are presented under the following headings:

Typical use
Use with by()

Typical use

The local polynomial smooth is often graphed on top of the data, possibly with other smoothers or regression lines:

```
. use http://www.stata-press.com/data/r12/auto
(1978 Automobile Data)

. twoway scatter weight length, mcolor(*.6) ||
         lpoly weight length                ||
         lowess weight length
```

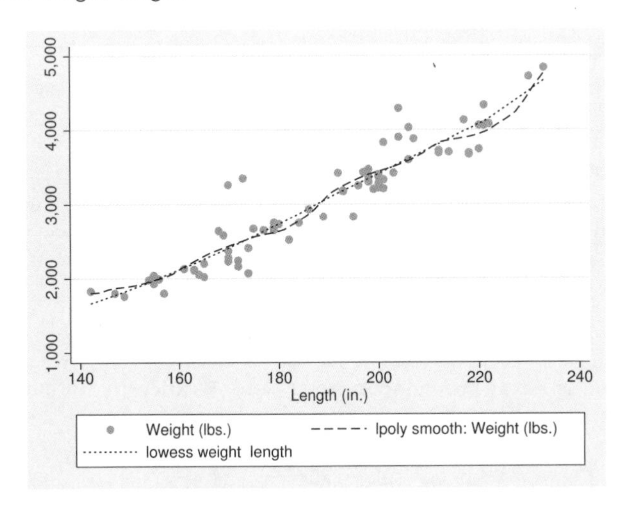

We used `mcolor(*.6)` to dim the points and thus make the lines stand out; see [G-4] ***colorstyle***.

Notice the y-axis title: "Mileage (mpg)/lpoly smooth: Mileage (mpg)/lowess mpg weight". The "lpoly smooth: Mileage (mpg)" was contributed by **twoway lpoly** and "lowess mpg weight" by **twoway lowess**. When you overlay graphs, you nearly always need to respecify the axis titles by using the *axis_title_options* `ytitle()` and `xtitle()`; see [G-3] ***axis_title_options***.

Use with by()

`graph twoway lpoly` may be used with `by()`:

```
. use http://www.stata-press.com/data/r12/auto, clear
(1978 Automobile Data)
. twoway scatter weight length, mcolor(*.6) ||
           lpoly weight length,             ||
     , by(foreign)
```

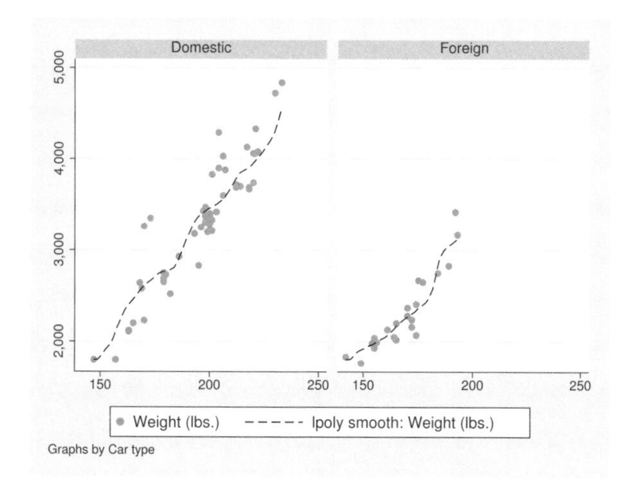

References

Cox, N. J. 2005. Speaking Stata: Smoothing in various directions. *Stata Journal* 5: 574–593.

——. 2010. Software Updates: Speaking Stata: Smoothing in various directions. *Stata Journal* 10: 164.

Also see

[R] **lpoly** — Kernel-weighted local polynomial smoothing

[G-2] **graph twoway lpolyci** — Local polynomial smooth plots with CIs

Title

[G-2] graph twoway lpolyci — Local polynomial smooth plots with CIs

Syntax

`twoway lpolyci` *yvar xvar* [*if*] [*in*] [*weight*] [, *options*]

options	Description
`kernel(`*kernel*`)`	kernel function; default is `kernel(epanechnikov)`
`bwidth(`#`)`	kernel bandwidth
`degree(`#`)`	degree of the polynomial smooth; default is `degree(0)`
`n(`#`)`	obtain the smooth at # points; default is min$(N, 50)$
`level(`#`)`	set confidence level; default is `level(95)`
`pwidth(`#`)`	pilot bandwidth for standard error calculation
`var(`#`)`	estimate of the constant conditional variance
`nofit`	do not plot the smooth
`fitplot(`*plottype*`)`	how to plot the smooth; default is `fitplot(line)`
`ciplot(`*plottype*`)`	how to plot CIs; default is `ciplot(rarea)`
fcline_options	change look of the smoothed line
fitarea_options	change look of CI
axis_choice_options	associate plot with alternative axis
twoway_options	titles, legends, axes, added lines and text, by, regions, name, aspect ratio, etc.

See [G-3] ***fcline_options***, [G-3] ***fitarea_options***, [G-3] ***axis_choice_options***, and [G-3] ***twoway_options***.

kernel	Description
`epanechnikov`	Epanechnikov kernel function; the default
`epan2`	alternative Epanechnikov kernel function
`biweight`	biweight kernel function
`cosine`	cosine trace kernel function
`gaussian`	Gaussian kernel function
`parzen`	Parzen kernel function
`rectangle`	rectangle kernel function
`triangle`	triangle kernel function

`fweight`s and `aweight`s are allowed; see [U] **11.1.6 weight**.

Menu

Graphics > Twoway graph (scatter, line, etc.)

Description

`graph twoway lpolyci` plots a local polynomial smooth of *yvar* on *xvar* by using `graph twoway line` (see [G-2] **graph twoway line**), along with a confidence interval by using `graph twoway rarea` (see [G-2] **graph twoway rarea**).

Options

`kernel(`*kernel*`)` specifies the kernel function for use in calculating the weighted local polynomial estimate. The default is `kernel(epanechnikov)`. See [R] **kdensity** for more information on this option.

`bwidth(`*#*`)` specifies the half-width of the kernel, the width of the smoothing window around each point. If `bwidth()` is not specified, a rule-of-thumb bandwidth estimator is calculated and used; see [R] **lpoly**.

`degree(`*#*`)` specifies the degree of the polynomial to be used in the smoothing. The default is `degree(0)`, meaning local mean smoothing.

`n(`*#*`)` specifies the number of points at which the smooth is to be evaluated. The default is $\min(N, 50)$, where N is the number of observations.

`level(`*#*`)` specifies the confidence level, as a percentage, for confidence intervals. The default is `level(95)` or as set by `set level`; see [U] **20.7 Specifying the width of confidence intervals**.

`pwidth(`*#*`)` specifies the pilot bandwidth to be used for standard error computations. The default is chosen to be 1.5 times the value of the rule-of-thumb bandwidth selector.

`var(`*#*`)` specifies an estimate of a constant conditional variance required for standard error computation. By default, the conditional variance at each smoothing point is estimated by the normalized weighted residual sum of squares obtained from locally fitting a polynomial of order $p + 2$, where p is the degree specified in `degree()`.

`nofit` prevents the smooth from being plotted.

`fitplot(`*plottype*`)` specifies how the prediction is to be plotted. The default is `fitplot(line)`, meaning that the smooth will be plotted by `graph twoway line`. See [G-2] **graph twoway** for a list of *plottype* choices. You may choose any that expects one y and one x variable. `fitplot()` is seldom used.

`ciplot(`*plottype*`)` specifies how the confidence interval is to be plotted. The default is `ciplot(rarea)`, meaning that the confidence bounds will be plotted by `graph twoway rarea`.

A reasonable alternative is `ciplot(rline)`, which will substitute lines around the smooth for shading. See [G-2] **graph twoway** for a list of *plottype* choices. You may choose any that expects two y variables and one x variable.

fcline_options specify how the lpoly line is rendered and its appearance; see [G-3] ***fcline_options***.

fitarea_options specify how the confidence interval is rendered; see [G-3] ***fitarea_options***. If you specify `ciplot()`, you should specify whatever is appropriate instead of using *fitarea_options*.

axis_choice_options associate the plot with a particular y or x axis on the graph; see [G-3] ***axis_choice_options***.

twoway_options are a set of common options supported by all `twoway` graphs. These options allow you to title graphs, name graphs, control axes and legends, add lines and text, set aspect ratios, create graphs over `by()` groups, and change some advanced settings. See [G-3] ***twoway_options***.

Remarks

`graph twoway lpolyci` *yvar xvar* uses the `lpoly` command—see [R] **lpoly**—to obtain a local polynomial smooth of *yvar* on *xvar* and confidence intervals and uses `graph twoway line` and `graph twoway rarea` to plot results.

Remarks are presented under the following headings:

Typical use

`graph twoway lpolyci` can be used to overlay the confidence bands obtained from different local polynomial smooths. For example, for local mean and local cubic polynomial smooths:

```
. use http://www.stata-press.com/data/r12/auto
(1978 Automobile Data)
. twoway lpolyci weight length, nofit                    ||
         lpolyci weight length, degree(3) nofit
                              ciplot(rline) pstyle(ci2) ||
         scatter weight length, msymbol(o)
```

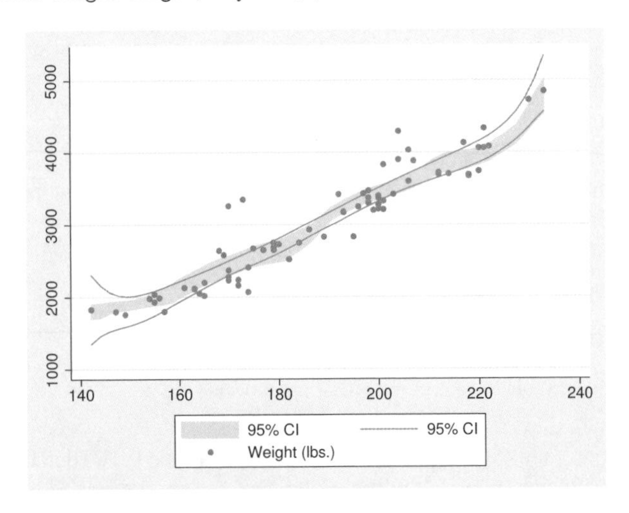

The plotted area corresponds to the confidence bands for the local mean smooth and lines correspond to confidence intervals for the local cubic smooth.

When you overlay graphs, you nearly always need to respecify the axis titles by using the *axis_title_options* `ytitle()` and `xtitle()`; see [G-3] ***axis_title_options***.

Use with by()

`graph twoway lpolyci` may be used with `by()`:

```
. use http://www.stata-press.com/data/r12/auto, clear
(1978 Automobile Data)
. twoway lpolyci weight length                       ||
             scatter weight length, msymbol(o) ||
     , by(foreign)
```

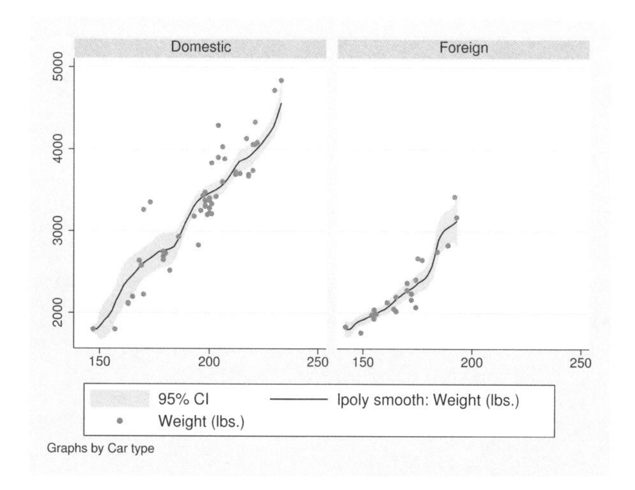

Also see

[R] **lpoly** — Kernel-weighted local polynomial smoothing

[G-2] **graph twoway lpolyci** — Local polynomial smooth plots with CIs

Title

[G-2] graph twoway mband — Twoway median-band plots

Syntax

`twoway mband` *yvar xvar* [*if*] [*in*] [, *options*]

options	Description
`bands(#)`	number of bands
cline_options	change look of the line
axis_choice_options	associate plot with alternative axis
twoway_options	titles, legends, axes, added lines and text, by, regions, name, aspect ratio, etc.

See [G-3] ***cline_options***, [G-3] ***axis_choice_options***, and [G-3] ***twoway_options***.

All options are *rightmost*; see [G-4] **concept: repeated options**.

Menu

Graphics > Twoway graph (scatter, line, etc.)

Description

`twoway mband` calculates cross medians and then graphs the cross medians as a line plot.

Options

`bands(#)` specifies the number of bands on which the calculation is to be based. The default is $\max(10, \text{round}(10 \times \log10(N)))$, where N is the number of observations.

In a median-band plot, the x axis is divided into # equal-width intervals and then the median of y and the median of x are calculated in each interval. It is these cross medians that `mband` graphs as a line plot.

cline_options specify how the median-band line is rendered and its appearance; see [G-3] ***cline_options***.

axis_choice_options associate the plot with a particular y or x axis on the graph; see [G-3] ***axis_choice_options***.

twoway_options are a set of common options supported by all `twoway` graphs. These options allow you to title graphs, name graphs, control axes and legends, add lines and text, set aspect ratios, create graphs over `by()` groups, and change some advanced settings. See [G-3] ***twoway_options***.

Remarks

Remarks are presented under the following headings:

Typical use
Use with by( )

Typical use

Median bands provide a convenient but crude way to show the tendency in the relationship between y and x:

```
. use http://www.stata-press.com/data/r12/auto
(1978 Automobile Data)
. scatter mpg weight, msize(*.5) || mband mpg weight
```

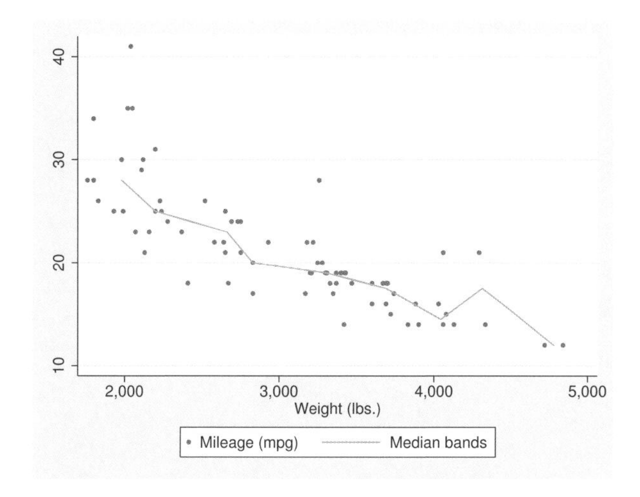

The important part of the above is "`mband mpg weight`". On the `scatter`, we specified `msize(*.5)` to make the marker symbols half their normal size; see [G-4] ***relativesize***.

Use with by( )

`mband` may be used with `by()` (as can all the `twoway` plot commands):

```
. scatter mpg weight, ms(oh) ||
  mband mpg weight ||, by(foreign, total row(1))
```

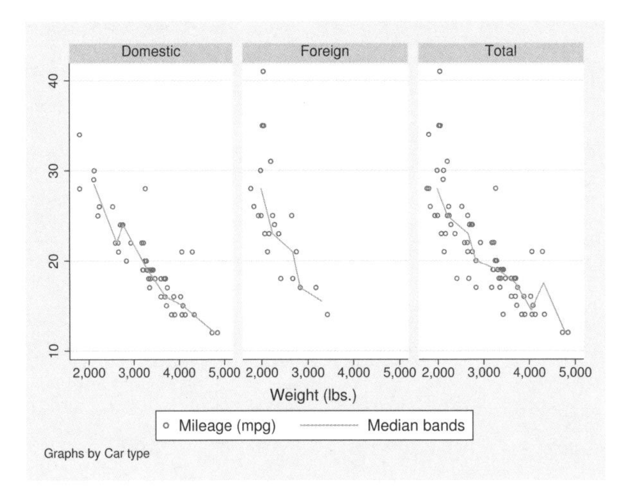

In the above graph, we specified `ms(oh)` so as to use hollow symbols; see [G-4] ***symbolstyle***.

Also see

[G-2] **graph twoway line** — Twoway line plots

[G-2] **graph twoway mspline** — Twoway median-spline plots

[G-2] **graph twoway lfit** — Twoway linear prediction plots

[G-2] **graph twoway qfit** — Twoway quadratic prediction plots

[G-2] **graph twoway fpfit** — Twoway fractional-polynomial prediction plots

Title

[G-2] graph twoway mspline — Twoway median-spline plots

Syntax

`twoway mspline` *yvar xvar* [*if*] [*in*] [, *options*]

options	Description
`bands(#)`	number of cross-median knots
`n(#)`	number of points between knots
cline_options	change look of the line
axis_choice_options	associate plot with alternative axis
twoway_options	titles, legends, axes, added lines and text, by, regions, name, aspect ratio, etc.

See [G-3] ***cline_options***, [G-3] ***axis_choice_options***, and [G-3] ***twoway_options***.

All options are *rightmost*; see [G-4] **concept: repeated options**.

Menu

Graphics > Twoway graph (scatter, line, etc.)

Description

`twoway mspline` calculates cross medians and then uses the cross medians as knots to fit a cubic spline. The resulting spline is graphed as a line plot.

Options

`bands(#)` specifies the number of bands for which cross medians should be calculated. The default is $\max\{\min(b_1, b_2), b_3\}$, where b_1 is $\text{round}\{10 * \text{log10}(N)\}$, b_2 is $\text{round}(\sqrt{N})$, b_3 is $\min(2, N)$, and N is the number of observations.

The x axis is divided into # equal-width intervals and then the median of y and the median of x are calculated in each interval. It is these cross medians to which a cubic spline is then fit.

`n(#)` specifies the number of points between the knots for which the cubic spline should be evaluated. `n(10)` is the default. `n()` does not affect the result that is calculated, but it does affect how smooth the result appears.

cline_options specify how the median-spline line is rendered and its appearance; see [G-3] ***cline_options***.

axis_choice_options associate the plot with a particular y or x axis on the graph; see [G-3] ***axis_choice_options***.

twoway_options are a set of common options supported by all `twoway` graphs. These options allow you to title graphs, name graphs, control axes and legends, add lines and text, set aspect ratios, create graphs over `by()` groups, and change some advanced settings. See [G-3] ***twoway_options***.

Remarks

Remarks are presented under the following headings:

Typical use
Cautions
Use with by( )

Typical use

Median splines provide a convenient way to show the relationship between y and x:

```
. use http://www.stata-press.com/data/r12/auto
(1978 Automobile Data)
. scatter mpg weight, msize(*.5) || mspline mpg weight
```

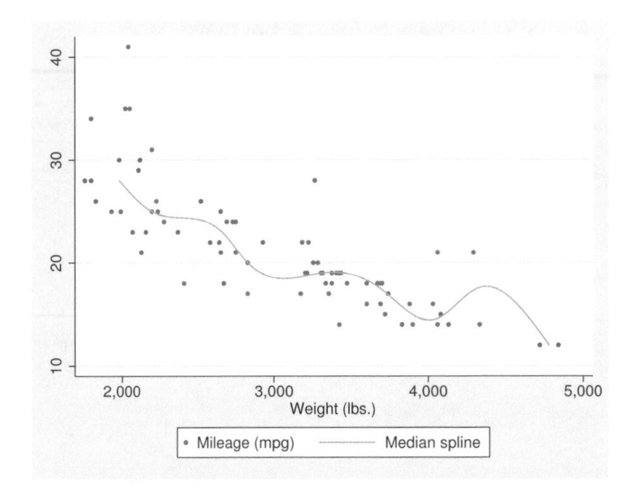

The important part of the above command is "`mspline mpg weight`". On the `scatter`, we specified `msize(*.5)` to make the marker symbols half their normal size; see [G-4] ***relativesize***.

Cautions

The graph shown above illustrates a common problem with this technique: it tracks wiggles that may not be real and can introduce wiggles if too many bands are chosen. An improved version of the graph above would be

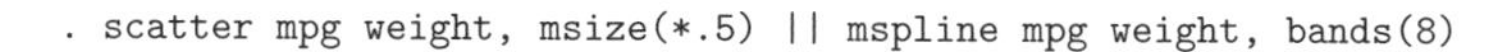

```
. scatter mpg weight, msize(*.5) || mspline mpg weight, bands(8)
```

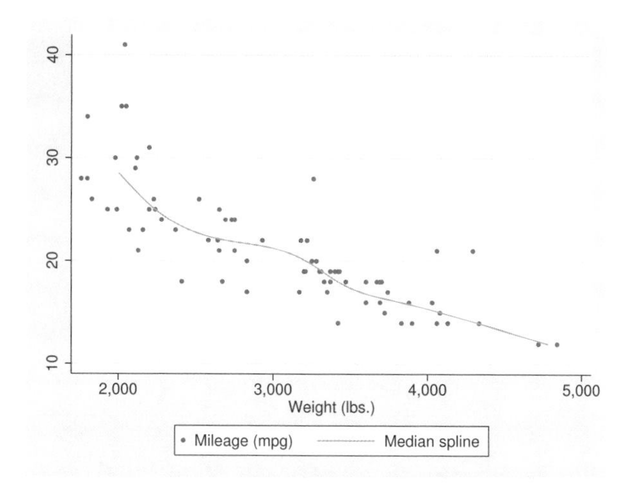

Use with by()

mspline may be used with by() (as can all the twoway plot commands):

```
. scatter mpg weight, msize(*.5) ||
  mspline mpg weight, bands(8)    ||, by(foreign, total row(1))
```

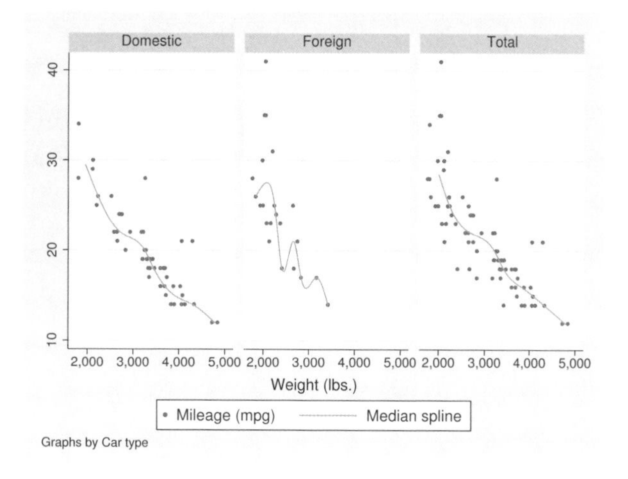

Also see

[G-2] **graph twoway line** — Twoway line plots

[G-2] **graph twoway mband** — Twoway median-band plots

[G-2] **graph twoway lfit** — Twoway linear prediction plots

[G-2] **graph twoway qfit** — Twoway quadratic prediction plots

[G-2] **graph twoway fpfit** — Twoway fractional-polynomial prediction plots

Title

[G-2] graph twoway pcarrow — Paired-coordinate plot with arrows

Syntax

Directional arrows

`twoway pcarrow` *y1var x1var y2var x2var* [*if*] [*in*] [, *options*]

Bidirectional arrows

`twoway pcbarrow` *y1var x1var y2var x2var* [*if*] [*in*] [, *options*]

options	Description
`mstyle(`*markerstyle*`)`	overall style of arrowhead
`msize(`*markersizestyle*`)`	size of arrowhead
`mangle(`*anglestyle*`)`	angle of arrowhead
`barbsize(`*markersizestyle*`)`	size of filled portion of arrowhead
`mcolor(`*colorstyle*`)`	color of arrowhead, inside and out
`mfcolor(`*colorstyle*`)`	arrowhead "fill" color
`mlcolor(`*colorstyle*`)`	arrowhead outline color
`mlwidth(`*linewidthstyle*`)`	arrowhead outline thickness
`mlstyle(`*linestyle*`)`	thickness and color
line_options	change look of arrow shaft lines
marker_label_options	add marker labels; change look or position
`headlabel`	label head of arrow, not tail
`vertical`	orient plot naturally; the default
`horizontal`	orient plot transposing y and x values
axis_choice_options	associate plot with alternative axis
twoway_options	titles, legends, axes, added lines and text, by regions, name, aspect ratio, etc.

See [G-4] ***markerstyle***, [G-4] ***markersizestyle***, [G-4] ***anglestyle***, [G-4] ***colorstyle***, [G-4] ***linewidthstyle***, [G-4] ***linestyle***, [G-3] ***line_options***, [G-3] ***marker_label_options***, [G-3] ***axis_choice_options***, and [G-3] ***twoway_options***.

Most options are *rightmost*, except *axis_choice_options*, `headlabel`, `vertical`, and `horizontal`, which are *unique*, and *twoway_options*, which are a mix of forms; see [G-4] **concept: repeated options**.

Menu

Graphics > Twoway graph (scatter, line, etc.)

Description

`twoway pcarrow` draws an arrow for each observation in the dataset. The arrow starts at the coordinate (*y1var*, *x1var*) and ends at the coordinate (*y2var*, *x2var*), with an arrowhead drawn at the ending coordinate.

`twoway pcbarrow` draws an arrowhead at each end; that is, it draws bidirectional arrows.

Options

`mstyle(`*markerstyle*`)` specifies the overall look of arrowheads, including their size, their color, etc. The other options allow you to change each attribute of the arrowhead, but `mstyle()` is the point from which they start.

You need not specify `mstyle()` just because you want to change the look of the arrowhead. In fact, most people seldom specify the `mstyle()` option. You specify `mstyle()` when another style exists that is exactly what you desire or when another style would allow you to specify fewer changes to obtain what you want.

`pcarrow` plots borrow their options and associated "look" from standard markers, so all its options begin with `m`. See [G-4] ***markerstyle*** for a list of available marker/arrowhead styles.

`msize(`*markersizestyle*`)` specifies the size of arrowheads. See [G-4] ***markersizestyle*** for a list of size choices.

`mangle(`*anglestyle*`)` specifies the angle that each side of an arrowhead forms with the arrow's line. For most schemes, the default angle is 28.64.

`barbsize(`*markersizestyle*`)` specifies the portion of the arrowhead that is to be filled. `barbsize(0)` specifies that just the lines for the arrowhead be drawn. When `barbsize()` is equal to `msize()`, the arrowhead is filled to a right angle with the arrow line. The effect of `barbsize()` is easier to see than to describe:

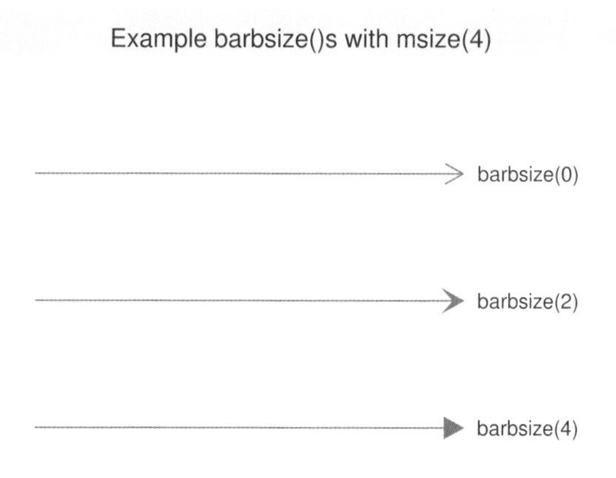

`mcolor(`*colorstyle*`)` specifies the color of the arrowhead. This option sets both the color of the line used to outline the arrowhead and the color of the inside the arrowhead. Also see options `mfcolor()` and `mlcolor()` below. See [G-4] ***colorstyle*** for a list of color choices.

`mfcolor(`*colorstyle*`)` specifies the color of the inside the arrowhead. See [G-4] ***colorstyle*** for a list of color choices.

`mlstyle(`*linestyle*`)`, `mlwidth(`*linewidthstyle*`)`, and `mlcolor(`*colorstyle*`)` specify the look of the line used to outline the arrowhead. See [G-4] **concept: lines**, but you cannot change the line pattern of an arrowhead.

line_options specify the look of the lines used to draw the shaft of the arrow, including pattern, width, and color; see [G-3] ***line_options***.

marker_label_options specify if and how the arrows are to be labeled. By default, the labels are placed at the tail of the arrow, the point defined by *y1var* and *x1var*. See [G-3] ***marker_label_options*** for options that change the look of the labels.

`headlabel` specifies that labels be drawn at the arrowhead, the (*y2var*,*x2var*) points rather than at the tail of the arrow, the (*y1var*,*x1var*) points. By default, when the `mlabel()` option is specified, labels are placed at the tail of the arrows; `headlabel` moves the labels from the tail to the head.

`vertical` and `horizontal` specify whether the y and x coordinates are to be swapped before plotting—`vertical` (the default) does not swap the coordinates, whereas `horizontal` does.

These options are rarely used when plotting only paired-coordinate data; they can, however, be used to good effect when combining paired-coordinate plots with range plots, such as `twoway rspike` or `twoway rbar`; see [G-2] **graph twoway rspike** and [G-2] **graph twoway rbar**.

axis_choice_options associate the plot with a particular y or x axis on the graph; see [G-3] ***axis_choice_options***.

twoway_options are a set of common options supported by all `twoway` graphs. These options allow you to title graphs, name graphs, control axes and legends, add lines and text, set aspect ratios, create graphs over `by()` groups, and change some advanced settings. See [G-3] ***twoway_options***.

Remarks

Remarks are presented under the following headings:

Basic use
Advanced use

Basic use

We have longitudinal data from 1968 and 1988 on the earnings and total experience of U.S. women by occupation. We will input data for two arrows, both originating at (0,0) and extending at right angles from each other, and plot them.

```
. input y1 x1 y2 x2
  1.     0  0  0  1
  2.     0  0  1  0
  3. end
. twoway pcarrow y1 x1 y2 x2
```

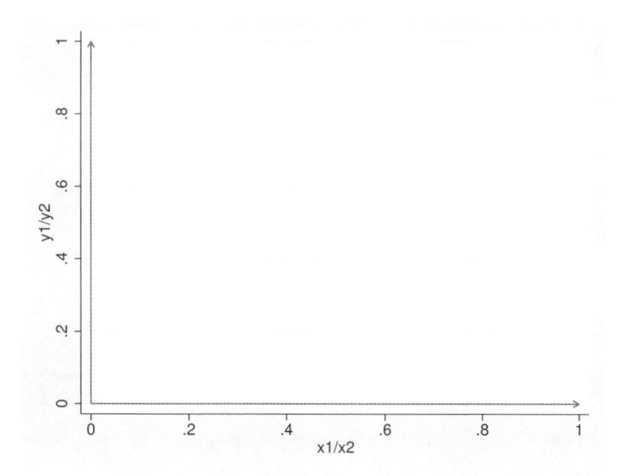

We could add labels to the heads of the arrows while also adding a little room in the plot region and constraining the plot region to be square:

```
. drop _all
. input y1 x1 y2 x2 str10 time   pos
  1.     0  0  0  1 "3 o'clock"    3
  2.     0  0  1  0 "12 o'clock"  12
  3. end
. twoway pcarrow y1 x1 y2 x2, aspect(1) mlabel(time) headlabel
                       mlabvposition(pos) plotregion(margin(vlarge))
```

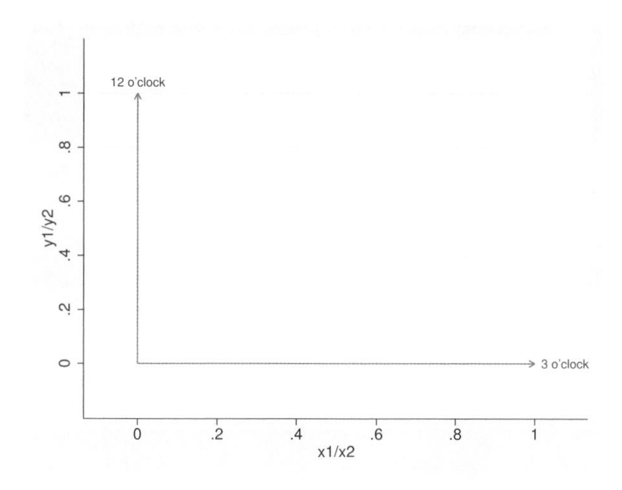

For examples of arrows in graphing multivariate results, see [MV] **biplot**.

Advanced use

As with many twoway plottypes, pcarrow and pcbarrow can be usefully combined with other twoway plottypes (see [G-2] **graph twoway**). Here a scatter plot is used to label ranges drawn by pcbarrow (though admittedly the ranges might better be represented using twoway rcap).

```
. use http://www.stata-press.com/data/r12/nlsw88, clear
. keep if occupation <= 8
. collapse (p05) p05=wage (p95) p95=wage (p50) p50=wage, by(occupation)
. gen mid = (p05 + p95) / 2
. gen dif = (p95 - p05)
. gsort -dif
. gen srt = _n
. twoway pcbarrow srt p05 srt p95 ||
        scatter  srt mid, msymbol(i) mlabel(occupation)
                          mlabpos(12) mlabcolor(black)
        plotregion(margin(t=5)) yscale(off)
        ylabel(, nogrid) legend(off)
        ytitle(Hourly wages)
        title("90 Percentile Range of US Women's Wages by Occupation")
        note("Source: National Longitudinal Survey of Young Women")
```

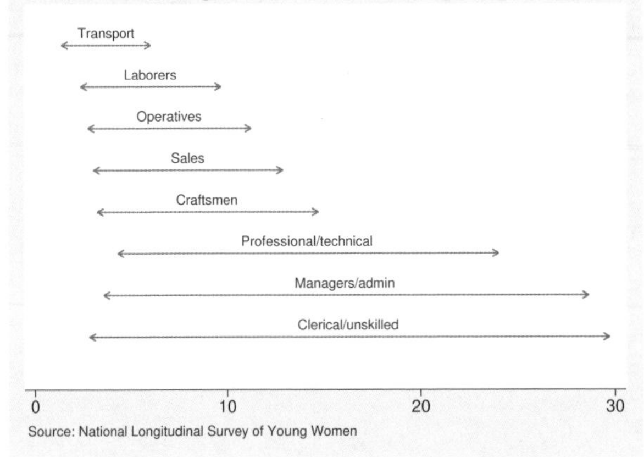

References

Cox, N. J. 2005. Stata tip 21: The arrows of outrageous fortune. *Stata Journal* 5: 282–284.

——. 2009. Speaking Stata: Paired, parallel, or profile plots for changes, correlations, and other comparisons. *Stata Journal* 9: 621–639.

Also see

[G-2] **graph twoway** — Twoway graphs

[G-2] **graph twoway pcarrowi** — Twoway pcarrow with immediate arguments

[G-2] **graph twoway pcspike** — Paired-coordinate plot with spikes

[G-2] **graph twoway pccapsym** — Paired-coordinate plot with spikes and marker symbols

[G-2] **graph twoway pcscatter** — Paired-coordinate plot with markers

[G-2] **graph twoway pci** — Twoway paired-coordinate plot with immediate arguments

Title

[G-2] graph twoway pcarrowi — Twoway pcarrow with immediate arguments

Syntax

`twoway pcarrowi` *immediate_values* [, *options*]

where *immediate_values* is one or more of

$\#_{y1}$ $\#_{x1}$ $\#_{y2}$ $\#_{x2}$ [($\#_{\text{clockposstyle}}$)] ["*text for label*"]

See [G-4] ***clockposstyle*** for a description of $\#_{\text{clockposstyle}}$.

Menu

Graphics > Twoway graph (scatter, line, etc.)

Description

`pcarrowi` is an immediate version of `twoway pcarrow`; see [U] **19 Immediate commands** and [G-2] **graph twoway pcarrow**. `pcarrowi` is intended for programmer use but can be useful interactively.

Options

options are as defined in [G-2] **graph twoway pcarrow**, with the following modifications:

If "*text for label*" is specified among any of the immediate arguments, option `mlabel()` is assumed.

If ($\#_{\text{clockposstyle}}$) is specified among any of the immediate arguments, option `mlabvposition()` is assumed.

Remarks

Immediate commands are commands that obtain data from numbers typed as arguments. Typing

```
. twoway pcarrowi 1.1 1.2 1.3 1.4  2.1 2.2 2.3 2.4, any_options
```

produces the same graph as typing

```
. clear
. input y1 x1 y2 x2
          y1         x1          y2          x2
  1. 1.1 1.2 1.3 1.4
  2. 2.1 2.2 2.3 2.4
  3. end
. twoway pcarrowi y x, any_options
```

`twoway pcarrowi` does not modify the data in memory.

`pcarrowi` is intended for programmer use but can be used interactively. In *Basic use* of [G-2] **graph twoway pcarrow**, we drew some simple clock hands from data that we input. We can draw the same graph by using `pcarrowi`.

```
. twoway pcarrowi 0  0  0  1  0  0  1  0
```

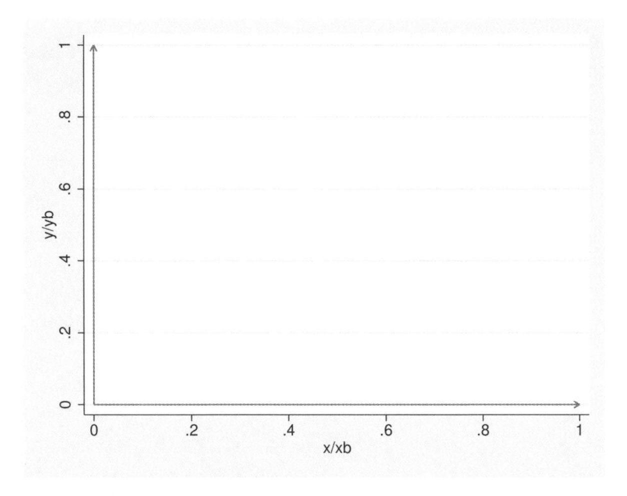

We can also draw the annotated second example,

```
. twoway pcarrowi 0  0  0  1  (3) "3 o'clock"
                  0  0  1  0 (12) "12 o'clock",
                  aspect(1) headlabel plotregion(margin(vlarge))
```

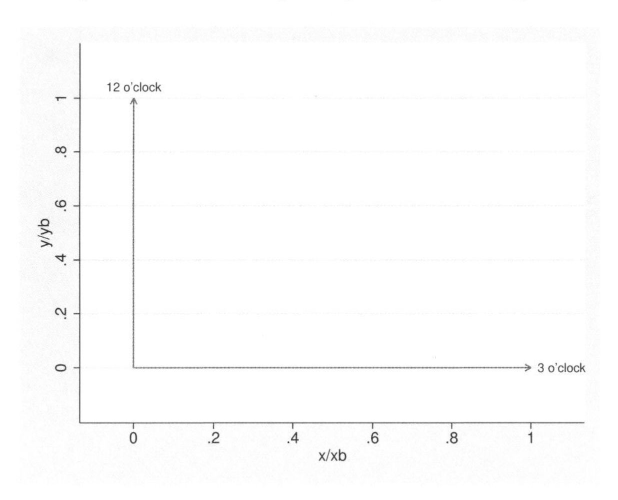

As another example, in [G-3] ***added_text_options***, we demonstrated the use of option `text()` to add text to a graph:

```
. twoway qfitci  mpg weight, stdf ||
         scatter mpg weight, ms(O)
                text(41 2040 "VW Diesel", place(e))
                text(28 3260 "Plymouth Arrow", place(e))
                text(35 2050 "Datsun 210 and Subaru", place(e))
```

Below we use pcarrowi to obtain similar results:

```
. twoway qfitci  mpg weight, stdf ||
         scatter mpg weight, ms(O) ||
         pcarrowi 41 2200 41 2060 (3) "VW Diesel"
                  28 3460 28 3280 (3) "Plymouth Arrow"
                  35 2250 35 2070 (3) "Datsun 210 and Subaru",
                  legend(order(1 2 3))
```

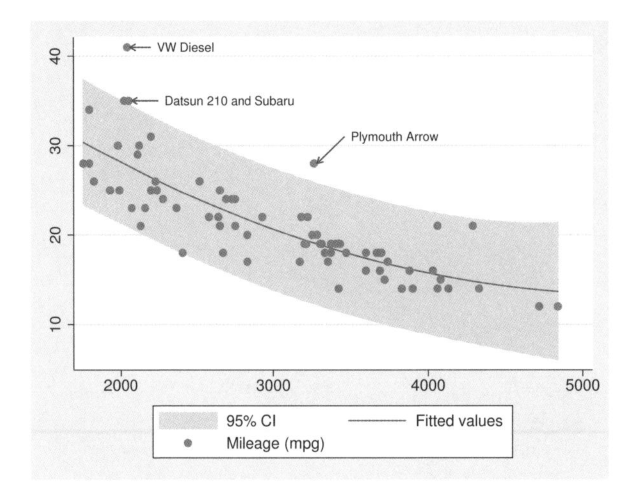

Also see

[G-2] **graph twoway scatteri** — Scatter with immediate arguments

[G-2] **graph twoway pcarrow** — Paired-coordinate plot with arrows

[G-2] **graph twoway** — Twoway graphs

Title

[G-2] graph twoway pccapsym — Paired-coordinate plot with spikes and marker symbols

Syntax

`twoway pccapsym` *y1var x1var y2var x2var* [*if*] [*in*] [, *options*]

options	Description
line_options	change look of spike lines
marker_options	change look of markers (color, size, etc.)
marker_label_options	add marker labels; change look or position
`headlabel`	label second coordinate, not first
`vertical`	orient plot naturally; the default
`horizontal`	orient plot transposing y and x values
axis_choice_options	associate plot with alternative axis
twoway_options	titles, legends, axes, added lines and text, by, regions, name, aspect ratio, etc.

See [G-3] ***line_options***, [G-3] ***marker_options***, [G-3] ***marker_label_options***, [G-3] ***axis_choice_options***, and [G-3] ***twoway_options***.

All explicit options are *rightmost*, except `headlabel`, `vertical`, and `horizontal`, which are *unique*; see [G-4] **concept: repeated options**.

Menu

Graphics > Twoway graph (scatter, line, etc.)

Description

A paired-coordinate capped-symbol plot draws a spike (or line) for each observation in the dataset and caps these spikes with a marker symbol at each end. The line starts at the coordinate (*y1var*, *x1var*) and ends at the coordinate (*y2var*, *x2var*), and both coordinates are designated with a marker.

Options

line_options specify the look of the lines used to draw the spikes, including pattern, width, and color; see [G-3] ***line_options***.

marker_options specify how the markers look, including shape, size, color, and outline; see [G-3] ***marker_options***. The same marker is used on both ends of the spikes.

marker_label_options specify if and how the markers are to be labeled; see [G-3] ***marker_label_options***.

`headlabel` specifies that labels be drawn on the markers of the (*y2var*, *x2var*) points rather than on the markers of the (*y1var*, *x1var*) points. By default, when the `mlabel()` option is specified, labels are placed on the points for the first two variables—*y1var* and *x1var*. `headlabel` moves the labels from these points to the points for the second two variables—*y2var* and *x2var*.

`vertical` and `horizontal` specify whether the y and x coordinates are to be swapped before plotting—`vertical` (the default) does not swap the coordinates, whereas `horizontal` does.

These options are rarely used when plotting only paired-coordinate data; they can, however, be used to good effect when combining paired-coordinate plots with range plots, such as `twoway rspike` or `twoway rbar`; see [G-2] **graph twoway rspike** and [G-2] **graph twoway rbar**.

axis_choice_options associate the plot with a particular y or x axis on the graph; see [G-3] ***axis_choice_options***.

twoway_options are a set of common options supported by all `twoway` graphs. These options allow you to title graphs, name graphs, control axes and legends, add lines and text, set aspect ratios, create graphs over `by()` groups, and change some advanced settings. See [G-3] ***twoway_options***.

Remarks

Remarks are presented under the following headings:

Basic use 1
Basic use 2

Basic use 1

We have longitudinal data from 1968 and 1988 on the earnings and total experience of U.S. women by occupation.

```
. use http://www.stata-press.com/data/r12/nlswide1
(National Longitudinal Survey.  Young Women 14-26 years of age in 1968)
. list occ wage68 ttl_exp68 wage88 ttl_exp88
```

	occ	wage68	ttl_e~68	wage88	ttl_e~88
1.	Professionals	6.121874	.860618	10.94776	14.11177
2.	Managers	5.426208	1.354167	11.53928	13.88886
3.	Sales	4.836701	.9896552	7.290306	12.62823
4.	Clerical/unskilled	4.088309	.640812	9.612672	11.08019
5.	Craftsmen	4.721373	1.091346	7.839769	12.64364
6.	Operatives	4.364782	.7959284	5.893025	11.99362
7.	Transport	1.987857	.5247414	3.200494	8.710394
8.	Laborers	3.724821	.775966	5.264415	10.56182
9.	Other	5.58524	.8278245	8.628641	12.78389

We graph a spike with symbols capping the end to show the movement from 1968 values to 1988 values for each observation (each occupation):

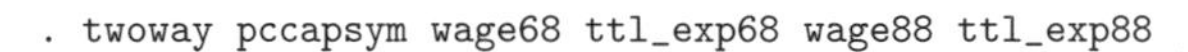

```
. twoway pccapsym wage68 ttl_exp68 wage88 ttl_exp88
```

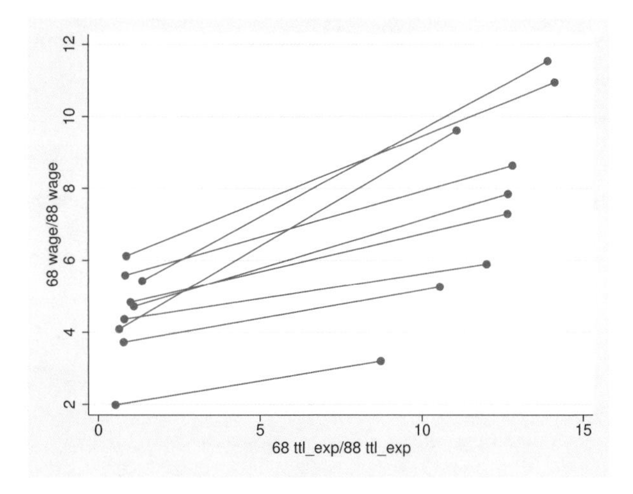

For a better presentation of these data, see *Advanced use* in [G-2] **graph twoway pcspike**; the comments there about combining plots apply equally well to `pccapsym` plots.

Basic use 2

We can draw both the edges and nodes of network diagrams by using `twoway pccapsym`.

```
. use http://www.stata-press.com/data/r12/network1
. twoway pccapsym y_c x_c y_l x_l
```

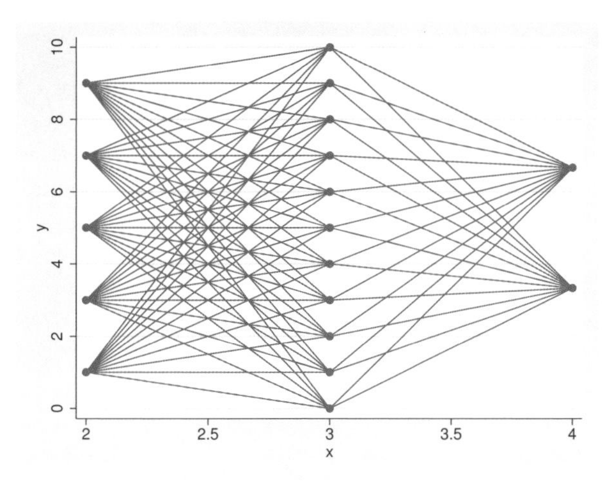

Again, a better presentation of these data can be found in [G-2] **graph twoway pcspike** under *Advanced use 2*.

Also see

[G-2] **graph twoway** — Twoway graphs

[G-2] **graph twoway line** — Twoway line plots

[G-2] **graph twoway rcapsym** — Range plot with spikes capped with marker symbols

[G-2] **graph twoway pcspike** — Paired-coordinate plot with spikes

[G-2] **graph twoway pcarrow** — Paired-coordinate plot with arrows

[G-2] **graph twoway pcscatter** — Paired-coordinate plot with markers

[G-2] **graph twoway pci** — Twoway paired-coordinate plot with immediate arguments

Title

[G-2] graph twoway pci — Twoway paired-coordinate plot with immediate arguments

Syntax

`twoway pci` *immediate_values* [, *options*]

where *immediate_values* is one or more of

$\#_{y1}$ $\#_{x1}$ $\#_{y2}$ $\#_{x2}$ [(#$_{\text{clockposstyle}}$)] ["*text for label*"]

See [G-4] ***clockposstyle*** for a description of #$_{\text{clockposstyle}}$.

Menu

Graphics > Twoway graph (scatter, line, etc.)

Description

`pci` is an immediate version of `twoway pcspike`; see [U] **19 Immediate commands** and [G-2] **graph twoway pcspike**. `pci` is intended for programmer use but can be useful interactively.

Options

options are as defined in [G-2] **graph twoway pcspike**, with the following modifications:

If "*text for label*" is specified among any of the immediate arguments, option `mlabel()` is assumed.

If (#$_{\text{clockposstyle}}$) is specified among any of the immediate arguments, option `mlabvposition()` is assumed.

Also see the *marker_options* defined in [G-2] **graph twoway pccapsym** if the `recast()` option is used to change the spikes into a paired-coordinate plot that plots markers.

Remarks

Immediate commands are commands that obtain data from numbers typed as arguments.

`twoway pci` does not modify the data in memory.

`pci` is intended for programmer use but can be used interactively. We can combine a `pci` plot with other `twoway` plots to produce a quick diagram.

```
. twoway function  y = -x^2, range(-2 2)        ||
        pci 0 1 0 -1                            ||
        pcarrowi 1.2 .5 0 0
```

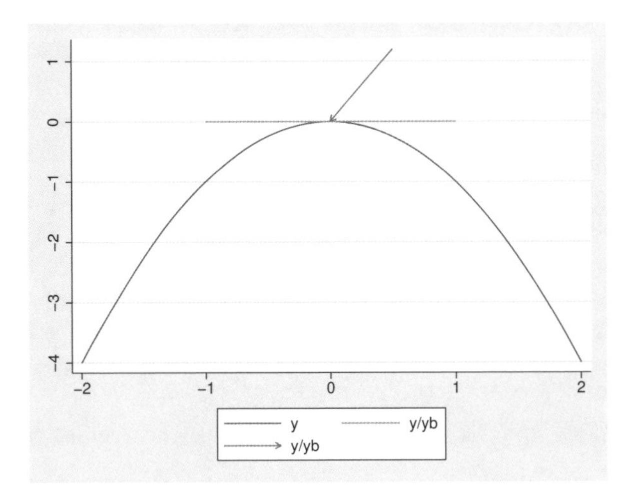

We can improve the annotation with

```
. twoway function  y = -x^2, range(-2 2)                          ||
        pci 0 1 0 -1 "Tangent", recast(pccapsym) msymbol(i) ||
        pcarrowi 1.2 .5 0.05 0 "Maximum at x=0",
        legend(off) title("Characteristics of y = -x{superscript:2}")
```

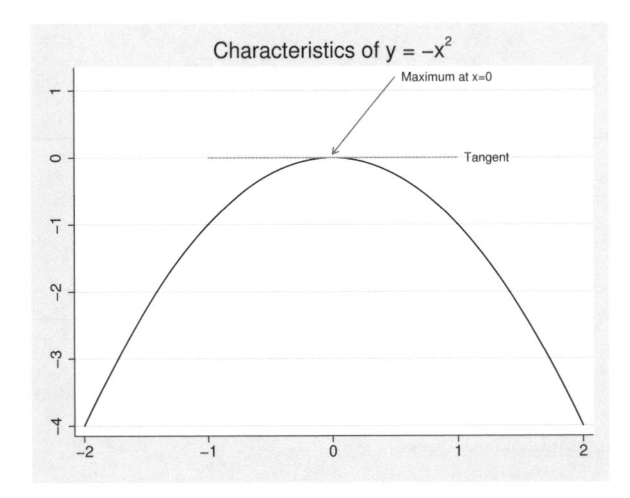

A slightly more whimsical example is

```
. twoway pci 2 0 2 6  4 0 4 6  0 2 6 2  0 4 6 4 ||
        scatteri 5 1  3 3, msize(ehuge) ms(X)  ||
        scatteri 5 5  1 5, msize(ehuge) ms(Oh) legend(off)
```

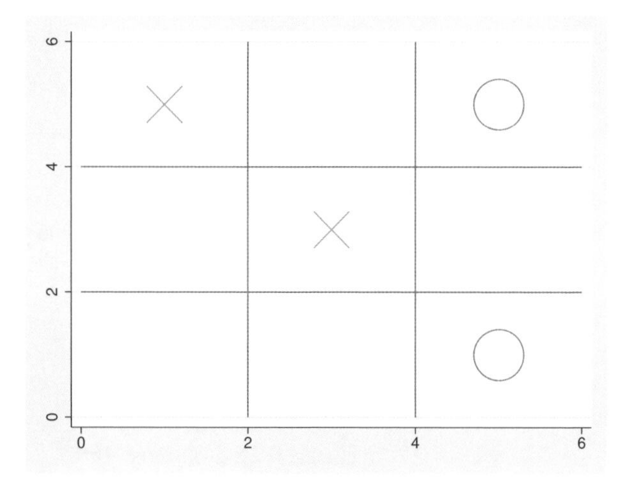

❑ Technical note

Programmers: Note carefully `twoway`'s *advanced_option* `recast()`; see [G-3] ***advanced_options***. It can be used to good effect, such as using `pci` to add marker labels.

❑

Also see

[G-2] **graph twoway scatteri** — Scatter with immediate arguments

[G-2] **graph twoway pcarrow** — Paired-coordinate plot with arrows

[G-2] **graph twoway** — Twoway graphs

Title

[G-2] graph twoway pcscatter — Paired-coordinate plot with markers

Syntax

`twoway pcscatter` *y1var x1var y2var x2var* [*if*] [*in*] [, *options*]

options	Description
marker_options	change look of markers (color, size, etc.)
marker_label_options	add marker labels; change look or position
`headlabel`	label second coordinate, not first
`vertical`	orient plot naturally; the default
`horizontal`	orient plot transposing y and x values
axis_choice_options	associate plot with alternative axis
twoway_options	titles, legends, axes, added lines and text, by, regions, name, aspect ratio, etc.

All explicit options are *unique*; see [G-4] **concept: repeated options**.

Menu

Graphics > Twoway graph (scatter, line, etc.)

Description

`twoway pcscatter` draws markers for each point designated by (*y1var*, *x1var*) and for each point designated by (*y2var*, *x2var*).

Options

marker_options specify how the markers look, including shape, size, color, and outline; see [G-3] ***marker_options***. The same marker is used for both sets of points.

marker_label_options specify if and how the markers are to be labeled; see [G-3] ***marker_label_options***.

`headlabel` specifies that labels be drawn on the markers of the (*y2var*, *x2var*) points rather than on the markers of the (*y1var*, *x1var*) points. By default, when the `mlabel()` option is specified, labels are placed on the points for the first two variables—*y1var* and *x1var*. `headlabel` moves the labels from these points to the points for the second two variables—*y2var* and *x2var*.

`vertical` and `horizontal` specify whether the y and x coordinates are to be swapped before plotting—`vertical` (the default) does not swap the coordinates, whereas `horizontal` does.

These options are rarely used when plotting only paired-coordinate data; they can, however, be used to good effect when combining paired-coordinate plots with range plots, such as `twoway rspike` or `twoway rbar`; see [G-2] **graph twoway rspike** and [G-2] **graph twoway rbar**.

axis_choice_options associate the plot with a particular y or x axis on the graph; see [G-3] ***axis_choice_options***.

twoway_options are a set of common options supported by all `twoway` graphs. These options allow you to title graphs, name graphs, control axes and legends, add lines and text, set aspect ratios, create graphs over `by()` groups, and change some advanced settings. See [G-3] ***twoway_options***.

Remarks

Visually, there is no difference between

```
. twoway pcscatter y1var x1var y2var x2var
```

and

```
. twoway scatter y1var x1var || scatter y2var x2var, pstyle(p1)
```

though in some cases the former is more convenient and better represents the conceptual structure of the data.

The two scatters are presented in the same overall style, meaning that the markers (symbol shape and color) are the same.

Also see

[G-2] **graph twoway** — Twoway graphs

[G-2] **graph twoway scatter** — Twoway scatterplots

[G-2] **graph twoway rscatter** — Range plot with markers

[G-2] **graph twoway pcarrow** — Paired-coordinate plot with arrows

[G-2] **graph twoway pccapsym** — Paired-coordinate plot with spikes and marker symbols

[G-2] **graph twoway pci** — Twoway paired-coordinate plot with immediate arguments

[G-2] **graph twoway pcspike** — Paired-coordinate plot with spikes

Title

[G-2] graph twoway pcspike — Paired-coordinate plot with spikes

Syntax

`twoway pcspike` *y1var x1var y2var x2var* [*if*] [*in*] [, *options*]

options	Description
line_options	change look of spike lines
`vertical`	orient plot naturally; the default
`horizontal`	orient plot transposing y and x values
axis_choice_options	associate plot with alternative axis
twoway_options	titles, legends, axes, added lines and text, by, regions, name, aspect ratio, etc.

See [G-3] ***line_options***, [G-3] ***axis_choice_options***, and [G-3] ***twoway_options***.

All explicit options are *rightmost*, except `vertical` and `horizontal`, which are *unique*; see [G-4] **concept: repeated options**.

Menu

Graphics > Twoway graph (scatter, line, etc.)

Description

A paired-coordinate spike plot draws a spike (or line) for each observation in the dataset. The line starts at the coordinate (*y1var*,*x1var*) and ends at the coordinate (*y2var*,*x2var*).

Options

line_options specify the look of the lines used to draw the spikes, including pattern, width, and color; see [G-3] ***line_options***.

`vertical` and `horizontal` specify whether the y and x coordinates are to be swapped before plotting—`vertical` (the default) does not swap the coordinates, whereas `horizontal` does.

These options are rarely used when plotting only paired-coordinate data; they can, however, be used to good effect when combining paired-coordinate plots with range plots, such as `twoway rspike` or `twoway rbar`; see [G-2] **graph twoway rspike** and [G-2] **graph twoway rbar**.

axis_choice_options associate the plot with a particular y or x axis on the graph; see [G-3] ***axis_choice_options***.

twoway_options are a set of common options supported by all `twoway` graphs. These options allow you to title graphs, name graphs, control axes and legends, add lines and text, set aspect ratios, create graphs over `by()` groups, and change some advanced settings. See [G-3] ***twoway_options***.

Remarks

Remarks are presented under the following headings:

Basic use
Advanced use
Advanced use 2

Basic use

We have longitudinal data from 1968 and 1988 on the earnings and total experience of U.S. women by occupation.

```
. use http://www.stata-press.com/data/r12/nlswide1
(National Longitudinal Survey. Young Women 14-26 years of age in 1968)
. list occ wage68 ttl_exp68 wage88 ttl_exp88
```

	occ	wage68	ttl_e~68	wage88	ttl_e~88
1.	Professionals	6.121874	.860618	10.94776	14.11177
2.	Managers	5.426208	1.354167	11.53928	13.88886
3.	Sales	4.836701	.9896552	7.290306	12.62823
4.	Clerical/unskilled	4.088309	.640812	9.612672	11.08019
5.	Craftsmen	4.721373	1.091346	7.839769	12.64364
6.	Operatives	4.364782	.7959284	5.893025	11.99362
7.	Transport	1.987857	.5247414	3.200494	8.710394
8.	Laborers	3.724821	.775966	5.264415	10.56182
9.	Other	5.58524	.8278245	8.628641	12.78389

We graph a spike showing the movement from 1968 values to 1988 values for each observation (each occupation):

```
. twoway pcspike wage68 ttl_exp68 wage88 ttl_exp88
```

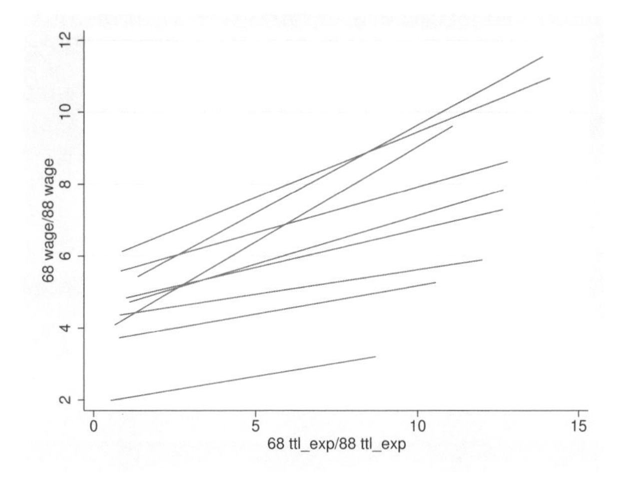

Advanced use

twoway pcspike can be usefully combined with other twoway plottypes (see [G-2] **graph twoway**). Here we add markers and labeled markers along with titles and such to improve the graph:

```
. twoway pcspike wage68 ttl_exp68 wage88 ttl_exp88        ||
         scatter wage68 ttl_exp68, msym(O)                ||
         scatter wage88 ttl_exp88, msym(O) pstyle(p4)
         mlabel(occ) xscale(range(17))
         title("Change in US Women's Experience and Earnings")
         subtitle("By Occupation -- 1968 to 1988")
         ytitle(Earnings) xtitle(Total experience)
         note("Source: National Longitudinal Survey of Young Women")
         legend(order(2 "1968" 3 "1988"))
```

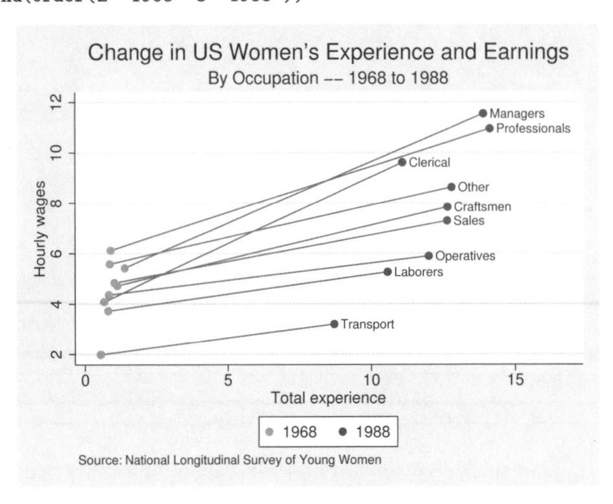

Advanced use 2

Drawing the edges of network diagrams is often easier with twoway pcspike than with other plottypes.

```
. use http://www.stata-press.com/data/r12/network1
. twoway pcspike y_c x_c y_l x_l
```

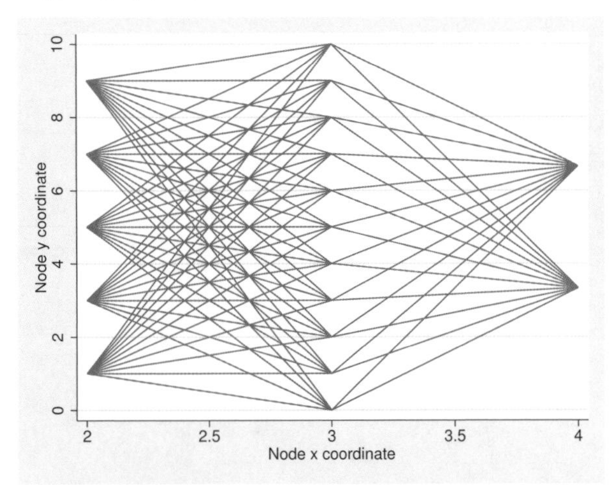

As with our first example, this graph can be made prettier by combining `twoway pcspike` with other plottypes.

```
. use http://www.stata-press.com/data/r12/network1a
. twoway pcspike y_c x_c y_l x_l, pstyle(p3)                          ||
              pcspike y_c x_c y_r x_r, pstyle(p4)                     ||
              scatter y_l x_l, pstyle(p3) msize(vlarge) msym(O)
                              mlabel(lab_l) mlabpos(9)                ||
              scatter y_c x_c, pstyle(p5) msize(vlarge) msym(O)       ||
              scatter y_r x_r, pstyle(p4) msize(vlarge) msym(O)
                              mlabel(lab_r) mlabpos(3)
            yscale(off) xscale(off) ylabels(, nogrid) legend(off)
            plotregion(margin(30 15 3 3))
```

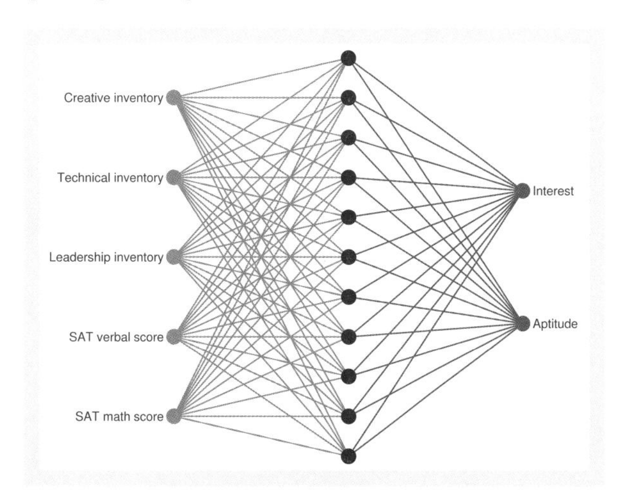

Reference

Cox, N. J. 2009. Speaking Stata: Paired, parallel, or profile plots for changes, correlations, and other comparisons. *Stata Journal* 9: 621–639.

Also see

[G-2] **graph twoway** — Twoway graphs

[G-2] **graph twoway line** — Twoway line plots

[G-2] **graph twoway rspike** — Range plot with spikes

[G-2] **graph twoway pccapsym** — Paired-coordinate plot with spikes and marker symbols

[G-2] **graph twoway pcarrow** — Paired-coordinate plot with arrows

[G-2] **graph twoway pcscatter** — Paired-coordinate plot with markers

[G-2] **graph twoway pci** — Twoway paired-coordinate plot with immediate arguments

Title

[G-2] graph twoway qfit — Twoway quadratic prediction plots

Syntax

`twoway qfit` *yvar xvar* [*if*] [*in*] [*weight*] [, *options*]

options	Description
`range(`*# #*`)`	range over which predictions calculated
`n(`*#*`)`	number of prediction points
`atobs`	calculate predictions at *xvar*
`estopts(`*regress_options*`)`	options for `regress`
`predopts(`*predict_options*`)`	options for `predict`
cline_options	change look of predicted line
axis_choice_options	associate plot with alternative axis
twoway_options	titles, legends, axes, added lines and text, by, regions, name, aspect ratio, etc.

See [G-3] ***cline_options***, [G-3] ***axis_choice_options***, and [G-3] ***twoway_options***.

All options are *rightmost*; see [G-4] **concept: repeated options**.

yvar and *xvar* may contain time-series operators; see [U] **11.4.4 Time-series varlists**.

`aweight`s, `fweight`s, and `pweight`s are allowed. Weights, if specified, affect estimation but not how the weighted results are plotted. See [U] **11.1.6 weight**.

Menu

Graphics > Twoway graph (scatter, line, etc.)

Description

`twoway qfit` calculates the prediction for *yvar* from a linear regression of *yvar* on *xvar* and $xvar^2$ and plots the resulting curve.

Options

`range(`*# #*`)` specifies the x range over which predictions are calculated. The default is `range(. .)`, meaning the minimum and maximum values of *xvar*. `range(0 10)` would make the range 0 to 10, `range(. 10)` would make the range the minimum to 10, and `range(0 .)` would make the range 0 to the maximum.

`n(`*#*`)` specifies the number of points at which predictions over `range()` are to be calculated. The default is `n(100)`.

`atobs` is an alternative to `n()`. It specifies that the predictions be calculated at the *xvar* values. `atobs` is the default if `predopts()` is specified and any statistic other than `xb` is requested.

`estopts(`*regress_options*`)` specifies options to be passed along to `regress` to estimate the linear regression from which the curve will be predicted; see [R] **regress**. If this option is specified, commonly specified is `estopts(nocons)`.

`predopts(`*predict_options*`)` specifies options to be passed along to `predict` to obtain the predictions after estimation by `regress`; see [R] **regress postestimation**.

cline_options specify how the prediction line is rendered; see [G-3] ***cline_options***.

axis_choice_options associate the plot with a particular y or x axis on the graph; see [G-3] ***axis_choice_options***.

twoway_options are a set of common options supported by all `twoway` graphs. These options allow you to title graphs, name graphs, control axes and legends, add lines and text, set aspect ratios, create graphs over `by()` groups, and change some advanced settings. See [G-3] ***twoway_options***.

Remarks

Remarks are presented under the following headings:

Typical use
Cautions
Use with by()

Typical use

`twoway qfit` is nearly always used in conjunction with other `twoway` plottypes, such as

```
. use http://www.stata-press.com/data/r12/auto
(1978 Automobile Data)
. scatter mpg weight || qfit mpg weight
```

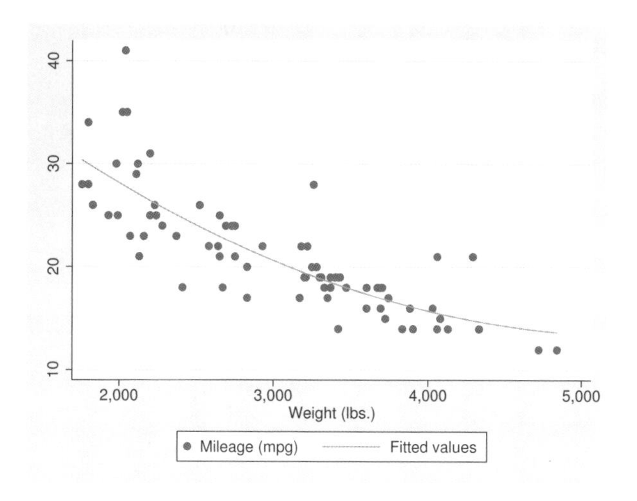

Results are visually the same as typing

```
. generate tempvar = weight^2
. regress mpg weight tempvar
. predict fitted
. scatter mpg weight || line fitted weight
```

Cautions

Do not use twoway qfit when specifying the *axis_scale_options* yscale(log) or xscale(log) to create log scales. Typing

```
. scatter mpg weight, xscale(log) || qfit mpg weight
```

produces something that is not a parabola because the regression estimated for the prediction was for mpg on weight and weight^2, not mpg on log(weight) and log(weight)^2.

Use with by()

qfit may be used with by() (as can all the twoway plot commands):

```
. scatter mpg weight || qfit mpg weight ||, by(foreign, total row(1))
```

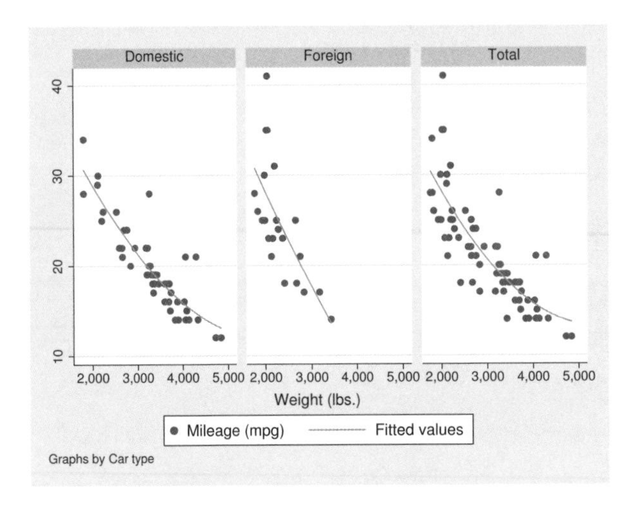

Also see

[G-2] **graph twoway line** — Twoway line plots

[G-2] **graph twoway lfit** — Twoway linear prediction plots

[G-2] **graph twoway fpfit** — Twoway fractional-polynomial prediction plots

[G-2] **graph twoway mband** — Twoway median-band plots

[G-2] **graph twoway mspline** — Twoway median-spline plots

[G-2] **graph twoway qfitci** — Twoway quadratic prediction plots with CIs

[R] **regress** — Linear regression

Title

[G-2] graph twoway qfitci — Twoway quadratic prediction plots with CIs

Syntax

`twoway qfitci` *yvar xvar* [*if*] [*in*] [*weight*] [, *options*]

options	Description
`stdp`	CIs from SE of prediction; the default
`stdf`	CIs from SE of forecast
`stdr`	CIs from SE of residual; seldom specified
`level(`*#*`)`	set confidence level; default is `level(95)`
`range(`*# #*`)`	range over which predictions are calculated
`n(`*#*`)`	number of prediction points
`atobs`	calculate predictions at *xvar*
`estopts(`*regress_options*`)`	options for `regress`
`predopts(`*predict_options*`)`	options for `predict`
`nofit`	do not plot the prediction
`fitplot(`*plottype*`)`	how to plot fit; default is `fitplot(line)`
`ciplot(`*plottype*`)`	how to plot CIs; default is `ciplot(rarea)`
fcline_options	change look of predicted line
fitarea_options	change look of CI
axis_choice_options	associate plot with alternative axis
twoway_options	titles, legends, axes, added lines and text, by, regions, name, aspect ratio, etc.

See [G-3] ***fcline_options***, [G-3] ***fitarea_options***, [G-3] ***axis_choice_options***; [G-3] ***twoway_options***.

Options `range()`, `estopts()` `predopts()`, `n()`, and `level()` are *rightmost*, and `atobs`, `nofit`, `fitplot()`, `ciplot()`, `stdp`, `stdf`, and `stdr` are *unique*; see [G-4] **concept: repeated options**.

yvar and *xvar* may contain time-series operators; see [U] **11.4.4 Time-series varlists**.

`aweight`s, `fweight`s, and `pweight`s are allowed. Weights, if specified, affect estimation but not how the weighted results are plotted. See [U] **11.1.6 weight**.

Menu

Graphics > Twoway graph (scatter, line, etc.)

Description

`twoway qfitci` calculates the prediction for *yvar* from a regression of *yvar* on *xvar* and $xvar^2$ and plots the resulting line along with a confidence interval.

Options

`stdp`, `stdf`, and `stdr` determine the basis for the confidence interval. `stdp` is the default.

`stdp` specifies that the confidence interval be the confidence interval of the mean.

`stdf` specifies that the confidence interval be the confidence interval for an individual forecast, which includes both the uncertainty of the mean prediction and the residual.

`stdr` specifies that the confidence interval be based only on the standard error of the residual.

`level(`*#*`)` specifies the confidence level, as a percentage, for the confidence intervals. The default is `level(95)` or as set by `set level`; see [U] **20.7 Specifying the width of confidence intervals**.

`range(`*# #*`)` specifies the x range over which predictions are calculated. The default is `range(. .)`, meaning the minimum and maximum values of *xvar*. `range(0 10)` would make the range 0 to 10, `range(. 10)` would make the range the minimum to 10, and `range(0 .)` would make the range 0 to the maximum.

`n(`*#*`)` specifies the number of points at which the predictions and the CI over `range()` are to be calculated. The default is `n(100)`.

`atobs` is an alternative to `n()` and specifies that the predictions be calculated at the *xvar* values. `atobs` is the default if `predopts()` is specified and any statistic other than the `xb` is requested.

`estopts(`*regress_options*`)` specifies options to be passed along to `regress` to estimate the linear regression from which the curve will be predicted; see [R] **regress**. If this option is specified, commonly specified is `estopts(nocons)`.

`predopts(`*predict_options*`)` specifies options to be passed along to `predict` to obtain the predictions after estimation by `regress`; see [R] **regress postestimation**.

`nofit` prevents the prediction from being plotted.

`fitplot(`*plottype*`)`, which is seldom used, specifies how the prediction is to be plotted. The default is `fitplot(line)`, meaning that the prediction will be plotted by `graph twoway line`. See [G-2] **graph twoway** for a list of *plottype* choices. You may choose any that expect one y and one x variable.

`ciplot(`*plottype*`)` specifies how the confidence interval is to be plotted. The default is `ciplot(rarea)`, meaning that the prediction will be plotted by `graph twoway rarea`.

A reasonable alternative is `ciplot(rline)`, which will substitute lines around the prediction for shading. See [G-2] **graph twoway** for a list of *plottype* choices. You may choose any that expect two y variables and one x variable.

fcline_options specify how the prediction line is rendered; see [G-3] ***fcline_options***. If you specify `fitplot()`, then rather than using *fcline_options*, you should select options that affect the specified *plottype* from the options in `scatter`; see [G-2] **graph twoway scatter**.

fitarea_options specify how the confidence interval is rendered; see [G-3] ***fitarea_options***. If you specify `ciplot()`, then rather than using *fitarea_options*, you should specify whatever is appropriate.

axis_choice_options associate the plot with a particular y or x axis on the graph; see [G-3] ***axis_choice_options***.

twoway_options are a set of common options supported by all `twoway` graphs. These options allow you to title graphs, name graphs, control axes and legends, add lines and text, set aspect ratios, create graphs over `by()` groups, and change some advanced settings. See [G-3] ***twoway_options***.

Remarks

Remarks are presented under the following headings:

Typical use
Advanced use
Cautions
Use with by()

Typical use

`twoway qfitci` by default draws the confidence interval of the predicted mean:

```
. use http://www.stata-press.com/data/r12/auto
(1978 Automobile Data)
. twoway qfitci mpg weight
```

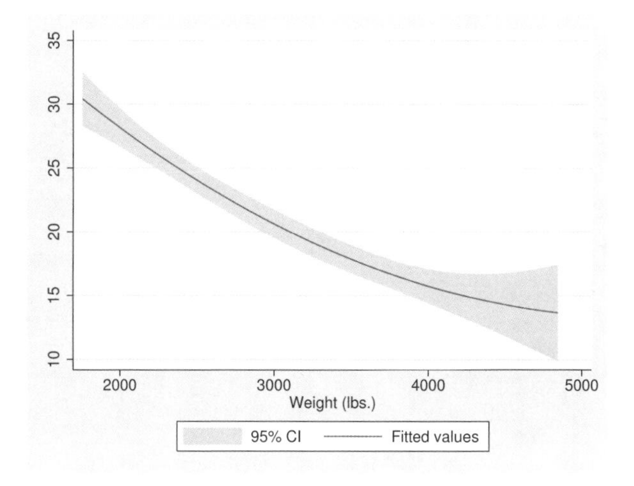

If you specify the `ciplot(rline)` option, rather than shading the confidence interval, it will be designated by lines:

```
. twoway qfitci mpg weight, ciplot(rline)
```

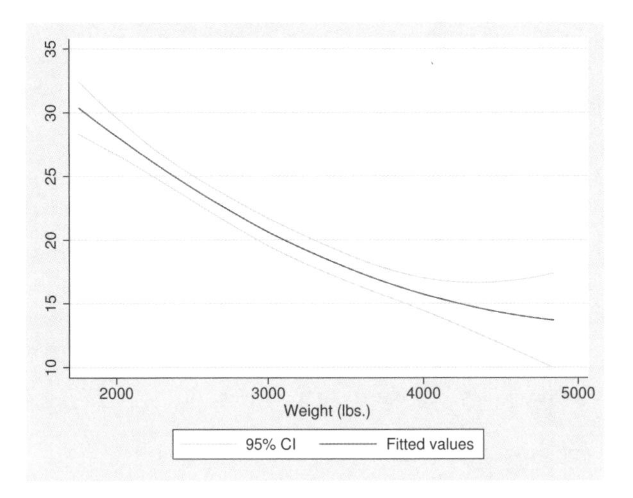

Advanced use

`qfitci` can be overlaid with other plots:

```
. use http://www.stata-press.com/data/r12/auto, clear
(1978 Automobile Data)
. twoway qfitci mpg weight, stdf || scatter mpg weight
```

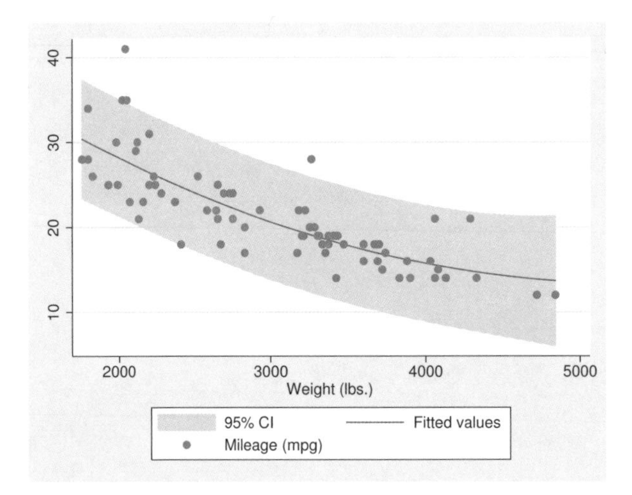

In the above command, we specified `stdf` to obtain a confidence interval based on the standard error of the forecast rather than the standard error of the mean. This is more useful for identifying outliers.

We typed

```
. twoway qfitci ... || scatter ...
```

and not

```
. twoway scatter ... || qfitci ...
```

Had we drawn the scatter diagram first, the confidence interval would have covered up most of the points.

Cautions

Do not use `twoway qfitci` when specifying the *axis_scale_options* `yscale(log)` or `xscale(log)` to create log scales. Typing

```
. twoway qfitci mpg weight, stdf || scatter mpg weight ||, xscale(log)
```

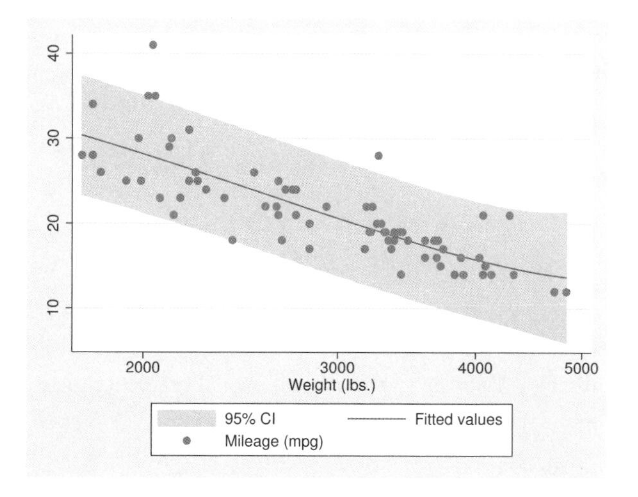

The result may look pretty but, if you think about it, it is not what you want. The prediction line is not a parabola because the regression estimated for the prediction was for `mpg` on `weight` and `weight^2`, not `mpg` on `log(weight)` and `log(weight)^2`.

Use with by()

qfitci may be used with by() (as can all the twoway plot commands):

```
. twoway qfitci  mpg weight, stdf ||
        scatter mpg weight       ||
         , by(foreign, total row(1))
```

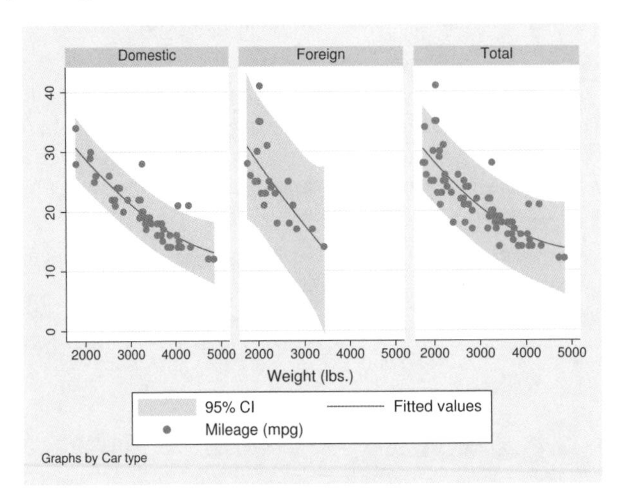

Also see

[G-2] **graph twoway lfitci** — Twoway linear prediction plots with CIs

[G-2] **graph twoway fpfitci** — Twoway fractional-polynomial prediction plots with CIs

[G-2] **graph twoway qfit** — Twoway quadratic prediction plots

[R] **regress** — Linear regression

Title

[G-2] graph twoway rarea — Range plot with area shading

Syntax

`twoway rarea` *y1var y2var xvar* [*if*] [*in*] [, *options*]

options	Description
`vertical`	vertical area plot; the default
`horizontal`	horizontal area plot
`cmissing(y|n)`	missing values do not force gaps in area; default is `cmissing(y)`
`sort`	sort by *xvar*; recommended
area_options	change look of shaded areas
axis_choice_options	associate plot with alternative axis
twoway_options	titles, legends, axes, added lines and text, by, regions, name, aspect ratio, etc.

See [G-3] ***area_options***, [G-3] ***axis_choice_options***, and [G-3] ***twoway_options***.

All explicit options are *unique*; see [G-4] **concept: repeated options**.

Menu

Graphics > Twoway graph (scatter, line, etc.)

Description

A range plot has two y variables, such as high and low daily stock prices or upper and lower 95% confidence limits.

`twoway rarea` plots range as a shaded area.

Also see [G-2] **graph twoway area** for area plots filled to the axis.

Options

`vertical` and `horizontal` specify whether the high and low y values are to be presented vertically (the default) or horizontally.

In the default `vertical` case, *y1var* and *y2var* record the minimum and maximum (or maximum and minimum) y values to be graphed against each *xvar* value.

If `horizontal` is specified, the values recorded in *y1var* and *y2var* are plotted in the x direction and *xvar* is treated as the y value.

`cmissing(y|n)` specifies whether missing values are to be ignored when drawing the area or if they are to create breaks in the area. The default is `cmissing(y)`, meaning that they are ignored. Consider the following data:

```
     +--------+
     |  y   x |
     |--------|
  1. |  1   1 |
  2. |  3   2 |
  3. |  5   3 |
  4. |  .   . |
  5. |  6   5 |
     |--------|
  6. | 11   8 |
     +--------+
```

Say that you graph these data by using `twoway rarea y x`. Do you want a break in the area between 3 and 5? If so, you type

```
. twoway rarea y x, cmissing(n)
```

and two areas will be drawn, one for the observations before the missing values at observation 4 and one for the observations after the missing values.

If you omit the option (or type `cmissing(y)`), the data are treated as if they contained

```
     +--------+
     |  y   x |
     |--------|
  1. |  1   1 |
  2. |  3   2 |
  3. |  5   3 |
  4. |  6   5 |
  5. | 11   8 |
     +--------+
```

meaning that one contiguous area will be drawn over the range (1,8).

`sort` specifies that the data be sorted by *xvar* before plotting.

area_options set the look of the shaded areas. The most important of these options is `color(`*colorstyle*`)`, which specifies the color of both the area and its outline; see [G-4] ***colorstyle*** for a list of color choices. See [G-3] ***area_options*** for information on the other *area_options*.

axis_choice_options associate the plot with a particular y or x axis on the graph; see [G-3] ***axis_choice_options***.

twoway_options are a set of common options supported by all `twoway` graphs. These options allow you to title graphs, name graphs, control axes and legends, add lines and text, set aspect ratios, create graphs over `by()` groups, and change some advanced settings. See [G-3] ***twoway_options***.

Remarks

Remarks are presented under the following headings:

Typical use
Advanced use
Cautions

Typical use

We have daily data recording the values for the S&P 500 in 2001:

```
. use http://www.stata-press.com/data/r12/sp500
(S&P 500)

. list date high low close in 1/5
```

	date	high	low	close
1.	02jan2001	1320.28	1276.05	1283.27
2.	03jan2001	1347.76	1274.62	1347.56
3.	04jan2001	1350.24	1329.14	1333.34
4.	05jan2001	1334.77	1294.95	1298.35
5.	08jan2001	1298.35	1276.29	1295.86

We will use the first 57 observations from these data:

```
. twoway rarea high low date in 1/57
```

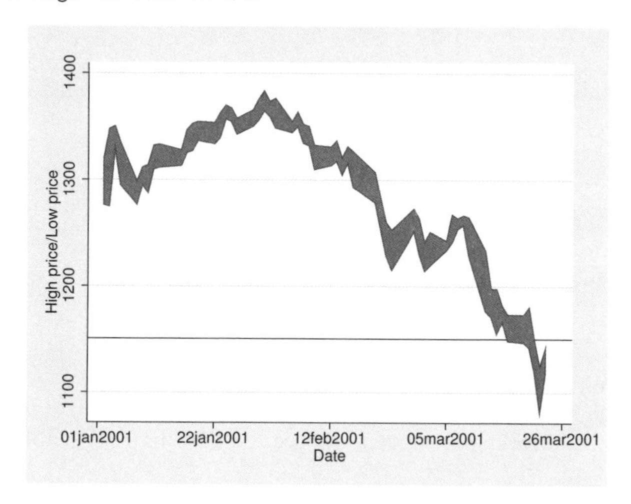

Advanced use

`rarea` works particularly well when the upper and lower limits are smooth functions and when the area is merely shaded rather than given an eye-catching color:

```
. use http://www.stata-press.com/data/r12/auto, clear
(1978 Automobile Data)

. quietly regress mpg weight

. predict hat

. predict s, stdf

. generate low = hat - 1.96*s

. generate hi  = hat + 1.96*s
```

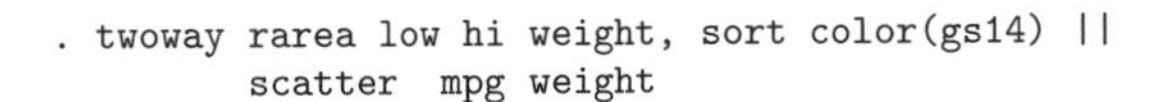

```
. twoway rarea low hi weight, sort color(gs14) ||
        scatter  mpg weight
```

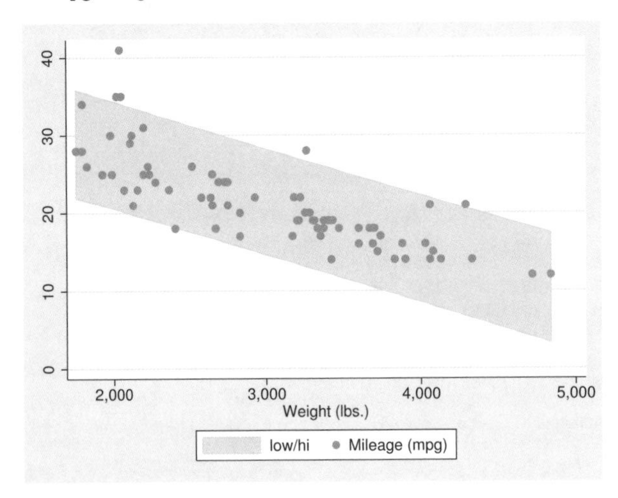

Notice the use of option `color()` to change the color of the shaded area. Also, we graphed the shaded area first and then the scatter. Typing

```
. twoway scatter ... || rarea ...
```

would not have produced the desired result because the shaded area would have covered up the scatterplot.

Also see [G-2] **graph twoway lfitci**.

Cautions

Be sure that the data are in the order of *xvar*, or specify `rarea`'s `sort` option. If you do neither, you will get something that looks like modern art; see *Cautions* in [G-2] **graph twoway area** for an example.

Also see

[G-2] **graph twoway area** — Twoway line plot with area shading

[G-2] **graph twoway rbar** — Range plot with bars

[G-2] **graph twoway rspike** — Range plot with spikes

[G-2] **graph twoway rcap** — Range plot with capped spikes

[G-2] **graph twoway rcapsym** — Range plot with spikes capped with marker symbols

[G-2] **graph twoway rline** — Range plot with lines

[G-2] **graph twoway rconnected** — Range plot with connected lines

[G-2] **graph twoway rscatter** — Range plot with markers

Title

[G-2] graph twoway rbar — Range plot with bars

Syntax

`twoway rbar` *y1var y2var xvar* [*if*] [*in*] [, *options*]

options	Description
`vertical`	vertical bars; the default
`horizontal`	horizontal bars
`barwidth(#)`	width of bar in *xvar* units
`mwidth`	use `msize()` rather than `barwidth()`
`msize(`*markersizestyle*`)`	width of bar in relative size units
barlook_options	change look of bars
axis_choice_options	associate plot with alternative axis
twoway_options	titles, legends, axes, added lines and text, by, regions, name, aspect ratio, etc.

See [G-4] ***markersizestyle***, [G-3] ***barlook_options***, [G-3] ***axis_choice_options***, and [G-3] ***twoway_options***.

Options `barwidth()`, `mwidth`, and `msize()` are *rightmost*, and `vertical` and `horizontal` are *unique*; see [G-4] **concept: repeated options**.

Menu

Graphics > Twoway graph (scatter, line, etc.)

Description

A range plot has two y variables, such as high and low daily stock prices or upper and lower 95% confidence limits.

`twoway rbar` plots a range, using bars to connect the high and low values.

Also see [G-2] **graph bar** for more traditional bar charts.

Options

`vertical` and `horizontal` specify whether the high and low y values are to be presented vertically (the default) or horizontally.

In the default `vertical` case, *y1var* and *y2var* record the minimum and maximum (or maximum and minimum) y values to be graphed against each *xvar* value.

If `horizontal` is specified, the values recorded in *y1var* and *y2var* are plotted in the x direction and *xvar* is treated as the y value.

`barwidth(#)` specifies the width of the bar in *xvar* units. The default is `barwidth(1)`. When a bar is plotted, it is centered at x, so half the width extends below x and half above.

`mwidth` and `msize(`*markersizestyle*`)` change how the width of the bars is specified. Usually, the width of the bars is determined by the `barwidth()` option documented below. If `mwidth` is specified, `barwidth()` becomes irrelevant and the bar width switches to being determined by `msize()`. This all has to do with the units in which the width of the bar is specified.

By default, bar widths are specified in the units of *xvar*, and if option `barwidth()` is not specified, the default width is 1 *xvar* unit.

`mwidth` specifies that you wish bar widths to be measured in relative size units; see [G-4] ***relativesize***. When you specify `mwidth`, the default changes from being 1 *xvar* unit to the default width of a marker symbol.

If you also specify `msize()`, the width of the bar is modified to be the relative size specified.

barlook_options set the look of the bars. The most important of these options is `color(`*colorstyle*`)`, which specifies the color of the bars; see [G-4] ***colorstyle*** for a list of color choices. See [G-3] ***barlook_options*** for information on the other *barlook_options*.

axis_choice_options associate the plot with a particular y or x axis on the graph; see [G-3] ***axis_choice_options***.

twoway_options are a set of common options supported by all `twoway` graphs. These options allow you to title graphs, name graphs, control axes and legends, add lines and text, set aspect ratios, create graphs over `by()` groups, and change some advanced settings. See [G-3] ***twoway_options***.

Remarks

Remarks are presented under the following headings:

Typical use
Advanced use

Typical use

We have daily data recording the values for the S&P 500 in 2001:

```
. use http://www.stata-press.com/data/r12/sp500
(S&P 500)

. list date high low close in 1/5
```

	date	high	low	close
1.	02jan2001	1320.28	1276.05	1283.27
2.	03jan2001	1347.76	1274.62	1347.56
3.	04jan2001	1350.24	1329.14	1333.34
4.	05jan2001	1334.77	1294.95	1298.35
5.	08jan2001	1298.35	1276.29	1295.86

We will use the first 57 observations from these data:

```
. twoway rbar high low date in 1/57, barwidth(.6)
```

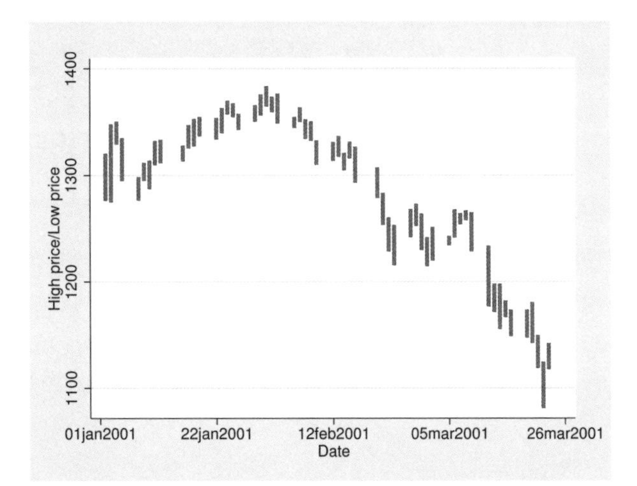

We specified `barwidth(.6)` to reduce the width of the bars. By default, bars are 1 x unit wide (meaning 1 day in our data). That default resulted in the bars touching. `barwidth(.6)` reduced the width of the bars to .6 days.

Advanced use

The useful thing about `twoway rbar` is that it can be combined with other `twoway` plottypes (see [G-2] **graph twoway**):

```
. twoway rbar high low date, barwidth(.6) color(gs7) ||
        line close date || in 1/57
```

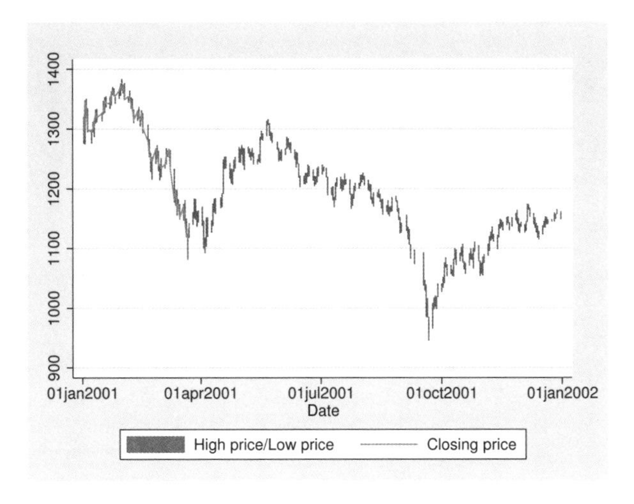

There are two things to note in the example above: our specification of `color(gs7)` and that we specified that the range bars be drawn first, followed by the line. We specified `color(gs7)` to tone down the bars: By default, the bars were too bright, making the line plot of close versus date all but invisible. Concerning the ordering, we typed

```
. twoway rbar high low date, barwidth(.6) color(gs7) ||
        line close date || in 1/57
```

so that the bars would be drawn first and then the line drawn over them. Had we specified

```
. twoway line close date ||
        rbar high low date, barwidth(.6) color(gs7) || in 1/57
```

the bars would have been placed on top of the line and thus would have occulted the line.

Reference

Kohler, U., and C. Brzinsky-Fay. 2005. Stata tip 25: Sequence index plots. *Stata Journal* 5: 601–602.

Also see

[G-2] **graph twoway bar** — Twoway bar plots

[G-2] **graph twoway rarea** — Range plot with area shading

[G-2] **graph twoway rspike** — Range plot with spikes

[G-2] **graph twoway rcap** — Range plot with capped spikes

[G-2] **graph twoway rcapsym** — Range plot with spikes capped with marker symbols

[G-2] **graph twoway rline** — Range plot with lines

[G-2] **graph twoway rconnected** — Range plot with connected lines

[G-2] **graph twoway rscatter** — Range plot with markers

Title

[G-2] graph twoway rcap — Range plot with capped spikes

Syntax

`twoway rcap` *y1var y2var xvar* [*if*] [*in*] [, *options*]

options	Description
`vertical`	vertical spikes; the default
`horizontal`	horizontal spikes
line_options	change look of spike and cap lines
`msize(`*markersizestyle*`)`	width of cap
axis_choice_options	associate plot with alternative axis
twoway_options	titles, legends, axes, added lines and text, by, regions, name, aspect ratio, etc.

See [G-3] ***line_options***, [G-4] ***markersizestyle***, [G-3] ***axis_choice_options***, and [G-3] ***twoway_options***.

All explicit options are *rightmost*, except `vertical` and `horizontal`, which are *unique*; see [G-4] **concept: repeated options**.

Menu

Graphics > Twoway graph (scatter, line, etc.)

Description

A range plot has two y variables, such as high and low daily stock prices or upper and lower 95% confidence limits.

`twoway rcap` plots a range, using capped spikes (I-beams) to connect the high and low values.

Options

`vertical` and `horizontal` specify whether the high and low y values are to be presented vertically (the default) or horizontally.

In the default `vertical` case, *y1var* and *y2var* record the minimum and maximum (or maximum and minimum) y values to be graphed against each *xvar* value.

If `horizontal` is specified, the values recorded in *y1var* and *y2var* are plotted in the x direction, and *xvar* is treated as the y value.

line_options specify the look of the lines used to draw the spikes and their caps, including pattern, width, and color; see [G-3] ***line_options***.

`msize(`*markersizestyle*`)` specifies the width of the cap. Option `msize()` is in fact `twoway scatter`'s *marker_option* that sets the size of the marker symbol, but here `msymbol()` is borrowed to set the cap width. See [G-4] ***markersizestyle*** for a list of size choices.

axis_choice_options associate the plot with a particular y or x axis on the graph; see [G-3] ***axis_choice_options***.

twoway_options are a set of common options supported by all `twoway` graphs. These options allow you to title graphs, name graphs, control axes and legends, add lines and text, set aspect ratios, create graphs over `by()` groups, and change some advanced settings. See [G-3] ***twoway_options***.

Remarks

Remarks are presented under the following headings:

Typical use

We have daily data recording the values for the S&P 500 in 2001:

```
. use http://www.stata-press.com/data/r12/sp500
(S&P 500)
. list date high low close in 1/5
```

	date	high	low	close
1.	02jan2001	1320.28	1276.05	1283.27
2.	03jan2001	1347.76	1274.62	1347.56
3.	04jan2001	1350.24	1329.14	1333.34
4.	05jan2001	1334.77	1294.95	1298.35
5.	08jan2001	1298.35	1276.29	1295.86

We will use the first 37 observations from these data:

```
. twoway rcap high low date in 1/37
```

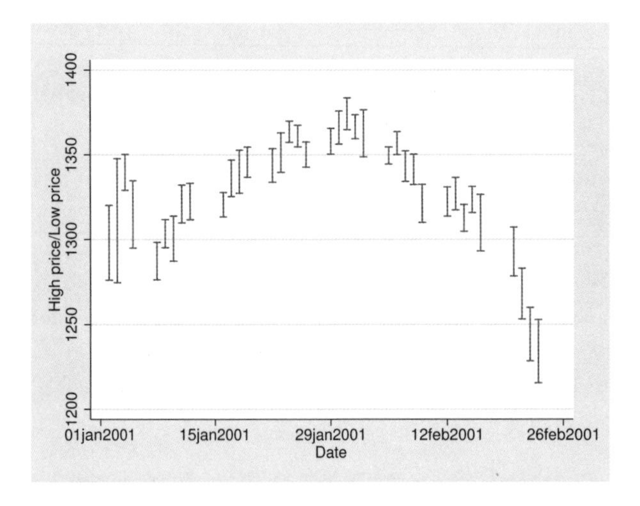

Advanced use

`twoway rcap` works well when combined with a horizontal line representing a base value:

```
. use http://www.stata-press.com/data/r12/sp500, clear
(S&P 500)
. generate month = month(date)
. sort month
. by month: egen lo = min(volume)
. by month: egen hi = max(volume)
. format lo hi %10.0gc
. summarize volume

    Variable |       Obs        Mean    Std. Dev.       Min        Max
-------------+--------------------------------------------------------
      volume |       248    12320.68    2585.929       4103    23308.3
. by month: keep if _n==_N
(236 observations deleted)
. twoway rcap lo hi month,
    xlabel(1 "J"  2 "F"  3 "M"  4 "A"  5 "M"  6 "J"
           7 "J"  8 "A"  9 "S" 10 "O" 11 "N" 12 "D")
    xtitle("Month of 2001")
    ytitle("High and Low Volume")
    yaxis(1 2) ylabel(12321 "12,321 (mean)", axis(2) angle(0))
    ytitle("", axis(2))
    yline(12321, lstyle(foreground))
    msize(*2)
    title("Volume of the S&P 500", margin(b+2.5))
    note("Source:  Yahoo!Finance and Commodity Systems Inc.")
```

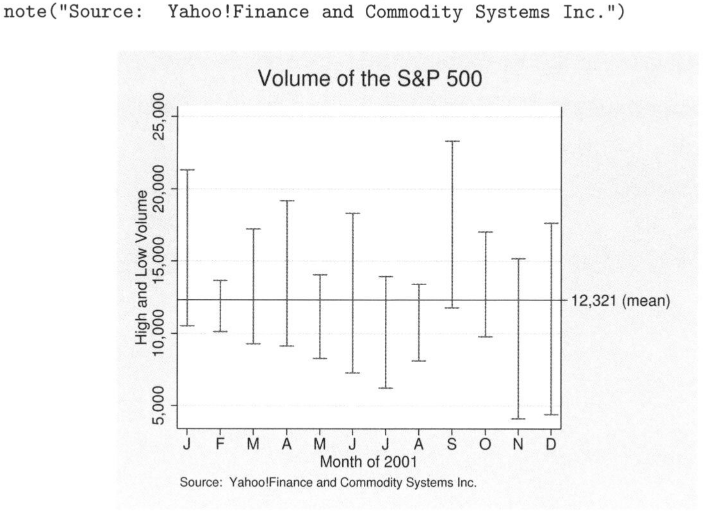

Advanced use 2

`twoway rcap` also works well when combined with a scatterplot to produce hi-lo-middle graphs. Returning to the first 37 observations of the S&P 500 used in the first example, we add a scatterplot of the closing value:

```
. use http://www.stata-press.com/data/r12/sp500, clear
(S&P 500)
. twoway rcap high low date || scatter close date
```

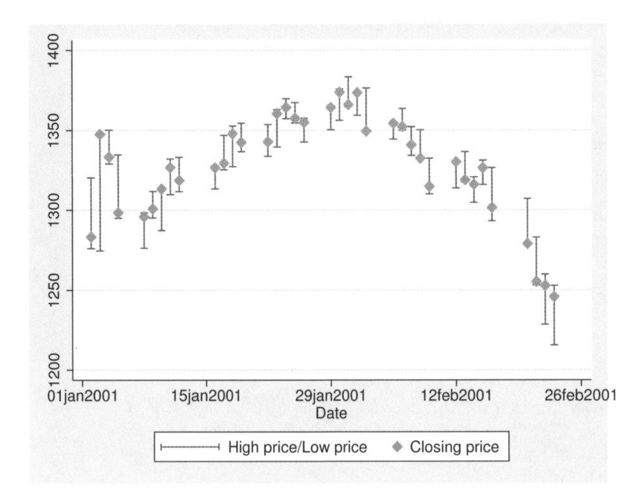

Also see

[G-2] **graph twoway rarea** — Range plot with area shading

[G-2] **graph twoway rbar** — Range plot with bars

[G-2] **graph twoway rspike** — Range plot with spikes

[G-2] **graph twoway rcapsym** — Range plot with spikes capped with marker symbols

[G-2] **graph twoway rline** — Range plot with lines

[G-2] **graph twoway rconnected** — Range plot with connected lines

[G-2] **graph twoway rscatter** — Range plot with markers

Title

[G-2] graph twoway rcapsym — Range plot with spikes capped with marker symbols

Syntax

`twoway rcapsym` *y1var y2var xvar* [*if*] [*in*] [, *options*]

options	Description
`vertical`	vertical spikes; the default
`horizontal`	horizontal spikes
line_options	change look of spike lines
marker_options	change look of markers (color, size, etc.)
marker_label_options	add marker labels; change look or position
axis_choice_options	associate plot with alternative axis
twoway_options	titles, legends, axes, added lines and text, by, regions, name, aspect ratio, etc.

See [G-3] ***line_options***, [G-3] ***marker_options***, [G-3] ***marker_label_options***, [G-3] ***axis_choice_options***, and [G-3] ***twoway_options***.

All explicit options are *rightmost*, except `vertical` and `horizontal`, which are *unique*; see [G-4] **concept: repeated options**.

Menu

Graphics > Twoway graph (scatter, line, etc.)

Description

A range plot has two y variables, such as high and low daily stock prices or upper and lower 95% confidence limits.

`twoway rcapsym` plots a range, using spikes capped with marker symbols.

Options

`vertical` and `horizontal` specify whether the high and low y values are to be presented vertically (the default) or horizontally.

In the default `vertical` case, *y1var* and *y2var* record the minimum and maximum (or maximum and minimum) y values to be graphed against each *xvar* value.

If `horizontal` is specified, the values recorded in *y1var* and *y2var* are plotted in the x direction and *xvar* is treated as the y value.

line_options specify the look of the lines used to draw the spikes, including pattern, width, and color; see [G-3] ***line_options***.

marker_options specify how the markers look, including shape, size, color, and outline; see [G-3] ***marker_options***. The same marker is used on both ends of the spikes.

marker_label_options specify if and how the markers are to be labeled. Because the same marker label would be used to label both ends of the spike, these options are of limited use here. See [G-3] ***marker_label_options***.

axis_choice_options associate the plot with a particular y or x axis on the graph; see [G-3] ***axis_choice_options***.

twoway_options are a set of common options supported by all `twoway` graphs. These options allow you to title graphs, name graphs, control axes and legends, add lines and text, set aspect ratios, create graphs over `by()` groups, and change some advanced settings. See [G-3] ***twoway_options***.

Remarks

We have daily data recording the values for the S&P 500 in 2001:

```
. use http://www.stata-press.com/data/r12/sp500
(S&P 500)

. list date high low close in 1/5
```

	date	high	low	close
1.	02jan2001	1320.28	1276.05	1283.27
2.	03jan2001	1347.76	1274.62	1347.56
3.	04jan2001	1350.24	1329.14	1333.34
4.	05jan2001	1334.77	1294.95	1298.35
5.	08jan2001	1298.35	1276.29	1295.86

We will use the first 37 observations from these data:

```
. twoway rcapsym high low date in 1/37
```

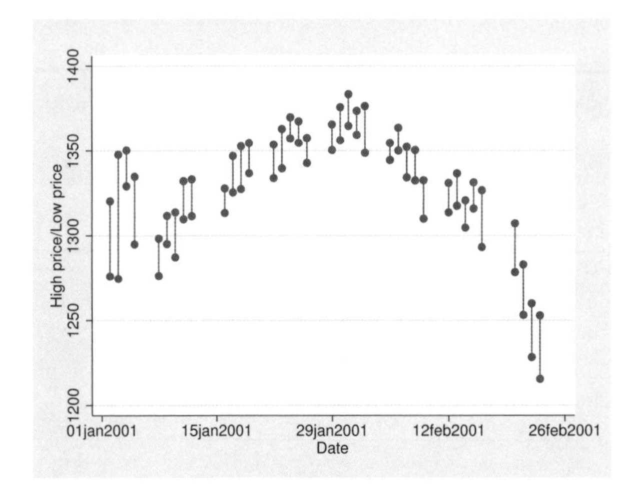

Also see

[G-2] **graph twoway rarea** — Range plot with area shading

[G-2] **graph twoway rbar** — Range plot with bars

[G-2] **graph twoway rspike** — Range plot with spikes

[G-2] **graph twoway rcap** — Range plot with capped spikes

[G-2] **graph twoway rline** — Range plot with lines

[G-2] **graph twoway rconnected** — Range plot with connected lines

[G-2] **graph twoway rscatter** — Range plot with markers

Title

[G-2] graph twoway rconnected — Range plot with connected lines

Syntax

`twoway rconnected` *y1var y2var xvar* [*if*] [*in*] [, *options*]

options	Description
`vertical`	vertical plot; the default
`horizontal`	horizontal plot
connect_options	change rendition of lines connecting points
marker_options	change look of markers (color, size, etc.)
marker_label_options	add marker labels; change look or position
axis_choice_options	associate plot with alternative axis
twoway_options	titles, legends, axes, added lines and text, by, regions, name, aspect ratio, etc.

See [G-3] ***connect_options***, [G-3] ***marker_options***, [G-3] ***marker_label_options***, [G-3] ***axis_choice_options***, and [G-3] ***twoway_options***.

All explicit options are *rightmost*, except `vertical` and `horizontal`, which are *unique*; see [G-4] **concept: repeated options**.

Menu

Graphics > Twoway graph (scatter, line, etc.)

Description

A range plot has two y variables, such as high and low daily stock prices or upper and lower 95% confidence limits.

`twoway rconnected` plots the upper and lower ranges by using connected lines.

Options

`vertical` and `horizontal` specify whether the high and low y values are to be presented vertically (the default) or horizontally.

In the default `vertical` case, *y1var* and *y2var* record the minimum and maximum (or maximum and minimum) y values to be graphed against each *xvar* value.

If `horizontal` is specified, the values recorded in *y1var* and *y2var* are plotted in the x direction and *xvar* is treated as the y value.

connect_options change the rendition of the lines connecting the plotted points, including sorting, handling missing observations, and the look of the line—line thickness, pattern, and color. For details, see [G-3] ***connect_options***.

marker_options specify how the markers look, including shape, size, color, and outline; see [G-3] ***marker_options***. The same symbol is used for both lines.

marker_label_options specify if and how the markers are to be labeled. Because the same marker label would be used to label both lines, these options are of limited use here. See [G-3] ***marker_label_options***.

axis_choice_options associate the plot with a particular y or x axis on the graph; see [G-3] ***axis_choice_options***.

twoway_options are a set of common options supported by all `twoway` graphs. These options allow you to title graphs, name graphs, control axes and legends, add lines and text, set aspect ratios, create graphs over `by()` groups, and change some advanced settings. See [G-3] ***twoway_options***.

Remarks

Visually, there is no difference between

```
. twoway rconnected y1var y2var  xvar
```

and

```
. twoway connected y1var xvar || connected y2var xvar, pstyle(p1)
```

The two connected lines are presented in the same overall style, meaning symbol selection and color and line color, thickness, and pattern.

Also see

[G-2] **graph twoway rarea** — Range plot with area shading

[G-2] **graph twoway rbar** — Range plot with bars

[G-2] **graph twoway rspike** — Range plot with spikes

[G-2] **graph twoway rcap** — Range plot with capped spikes

[G-2] **graph twoway rcapsym** — Range plot with spikes capped with marker symbols

[G-2] **graph twoway rline** — Range plot with lines

[G-2] **graph twoway rscatter** — Range plot with markers

Title

[G-2] graph twoway rline — Range plot with lines

Syntax

`twoway rline` *y1var y2var xvar* [*if*] [*in*] [, *options*]

options	Description
`vertical`	vertical plot; the default
`horizontal`	horizontal plot
connect_options	change rendition of lines connecting points
axis_choice_options	associate plot with alternative axis
twoway_options	titles, legends, axes, added lines and text, by, regions, name, aspect ratio, etc.

See [G-3] ***connect_options***, [G-3] ***axis_choice_options***, and [G-3] ***twoway_options***.

All explicit options are *rightmost*, except `vertical` and `horizontal`, which are *unique*; see [G-4] **concept: repeated options**.

Menu

Graphics > Twoway graph (scatter, line, etc.)

Description

A range plot has two y variables, such as high and low daily stock prices or upper and lower 95% confidence limits.

`twoway rline` plots the upper and lower ranges by using lines.

Options

`vertical` and `horizontal` specify whether the high and low y values are to be presented vertically (the default) or horizontally.

In the default `vertical` case, *y1var* and *y2var* record the minimum and maximum (or maximum and minimum) y values to be graphed against each *xvar* value.

If `horizontal` is specified, the values recorded in *y1var* and *y2var* are plotted in the x direction and *xvar* is treated as the y value.

connect_options change the rendition of the lines connecting the points, including sorting, handling missing observations, and the look of the line—line thickness, pattern, and color. For details, see [G-3] ***connect_options***.

axis_choice_options associate the plot with a particular y or x axis on the graph; see [G-3] ***axis_choice_options***.

twoway_options are a set of common options supported by all `twoway` graphs. These options allow you to title graphs, name graphs, control axes and legends, add lines and text, set aspect ratios, create graphs over `by()` groups, and change some advanced settings. See [G-3] ***twoway_options***.

Remarks

Visually, there is no difference between

```
. twoway rline y1var y2var xvar
```

and

```
. twoway line y1var xvar || line y2var xvar, pstyle(p1)
```

The two lines are presented in the same overall style, meaning color, thickness, and pattern.

Also see

[G-2] **graph twoway rarea** — Range plot with area shading

[G-2] **graph twoway rbar** — Range plot with bars

[G-2] **graph twoway rspike** — Range plot with spikes

[G-2] **graph twoway rcap** — Range plot with capped spikes

[G-2] **graph twoway rcapsym** — Range plot with spikes capped with marker symbols

[G-2] **graph twoway rconnected** — Range plot with connected lines

[G-2] **graph twoway rscatter** — Range plot with markers

Title

[G-2] graph twoway rscatter — Range plot with markers

Syntax

`twoway rscatter` *y1var y2var xvar* [*if*] [*in*] [, *options*]

options	Description
`vertical`	vertical plot; the default
`horizontal`	horizontal plot
marker_options	change look of marker (color, size, etc.)
marker_label_options	add marker labels; change look or position
axis_choice_options	associate plot with alternative axis
twoway_options	titles, legends, axes, added lines and text, by, regions, name, aspect ratio, etc.

See [G-3] ***marker_options***, [G-3] ***marker_label_options***, [G-3] ***axis_choice_options***, and [G-3] ***twoway_options***.

All explicit options are *rightmost*; see [G-4] **concept: repeated options**.

Menu

Graphics > Twoway graph (scatter, line, etc.)

Description

A range plot has two y variables, such as high and low daily stock prices or upper and lower 95% confidence limits.

`twoway rscatter` plots the upper and lower ranges as scatters.

Options

`vertical` and `horizontal` specify whether the high and low y values are to be presented vertically (the default) or horizontally.

In the default `vertical` case, *y1var* and *y2var* record the minimum and maximum (or maximum and minimum) y values to be graphed against each *xvar* value.

If `horizontal` is specified, the values recorded in *y1var* and *y2var* are plotted in the x direction and *xvar* is treated as the y value.

marker_options specify how the markers look, including shape, size, color, and outline; see [G-3] ***marker_options***. The same marker is used for both points.

marker_label_options specify if and how the markers are to be labeled. Because the same marker label would be used to label both points, these options are of limited use in this case. See [G-3] ***marker_label_options***.

axis_choice_options associate the plot with a particular y or x axis on the graph; see [G-3] ***axis_choice_options***.

twoway_options are a set of common options supported by all `twoway` graphs. These options allow you to title graphs, name graphs, control axes and legends, add lines and text, set aspect ratios, create graphs over `by()` groups, and change some advanced settings. See [G-3] ***twoway_options***.

Remarks

Visually, there is no difference between

```
. twoway rscatter y1var y2var xvar
```

and

```
. twoway scatter y1var xvar || scatter y2var xvar, pstyle(p1)
```

The two scatters are presented in the same overall style, meaning that the markers (symbol shape and color) are the same.

Also see

[G-2] **graph twoway rarea** — Range plot with area shading

[G-2] **graph twoway rbar** — Range plot with bars

[G-2] **graph twoway rspike** — Range plot with spikes

[G-2] **graph twoway rcap** — Range plot with capped spikes

[G-2] **graph twoway rcapsym** — Range plot with spikes capped with marker symbols

[G-2] **graph twoway rline** — Range plot with lines

[G-2] **graph twoway rconnected** — Range plot with connected lines

Title

[G-2] graph twoway rspike — Range plot with spikes

Syntax

`twoway rspike` *y1var y2var xvar* [*if*] [*in*] [, *options*]

options	Description
`vertical`	vertical spikes; the default
`horizontal`	horizontal spikes
line_options	change look of spike lines
axis_choice_options	associate plot with alternative axis
twoway_options	titles, legends, axes, added lines and text, by, regions, name, aspect ratio, etc.

See [G-3] ***line_options***, [G-3] ***axis_choice_options***, and [G-3] ***twoway_options***.

All explicit options are *rightmost*, except `vertical` and `horizontal`, which are *unique*; see [G-4] **concept: repeated options**.

Menu

Graphics > Twoway graph (scatter, line, etc.)

Description

A range plot has two y variables, such as high and low daily stock price or upper and lower 95% confidence limits.

`twoway rspike` plots a range, using spikes to connect the high and low values.

Also see [G-2] **graph twoway spike** for another style of spike chart.

Options

`vertical` and `horizontal` specify whether the high and low y values are to be presented vertically (the default) or horizontally.

In the default `vertical` case, *y1var* and *y2var* record the minimum and maximum (or maximum and minimum) y values to be graphed against each *xvar* value.

If `horizontal` is specified, the values recorded in *y1var* and *y2var* are plotted in the x direction and *xvar* is treated as the y value.

line_options specify the look of the lines used to draw the spikes, including pattern, width, and color; see [G-3] ***line_options***.

axis_choice_options associate the plot with a particular y or x axis on the graph; see [G-3] ***axis_choice_options***.

twoway_options are a set of common options supported by all `twoway` graphs. These options allow you to title graphs, name graphs, control axes and legends, add lines and text, set aspect ratios, create graphs over `by()` groups, and change some advanced settings. See [G-3] ***twoway_options***.

Remarks

Remarks are presented under the following headings:

Typical use
Advanced use
Advanced use 2

Typical use

We have daily data recording the values for the S&P 500 in 2001:

```
. use http://www.stata-press.com/data/r12/sp500
(S&P 500)
. list date high low close in 1/5
```

	date	high	low	close
1.	02jan2001	1320.28	1276.05	1283.27
2.	03jan2001	1347.76	1274.62	1347.56
3.	04jan2001	1350.24	1329.14	1333.34
4.	05jan2001	1334.77	1294.95	1298.35
5.	08jan2001	1298.35	1276.29	1295.86

We will use the first 57 observations from these data:

```
. twoway rspike high low date in 1/57
```

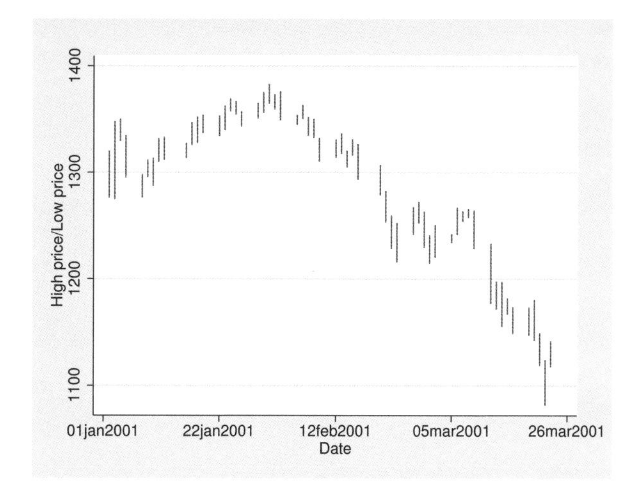

Advanced use

twoway rspike can be usefully combined with other twoway plottypes (see [G-2] **graph twoway**):

```
. twoway rspike high low date, lcolor(gs11) ||
         line close date || in 1/57
```

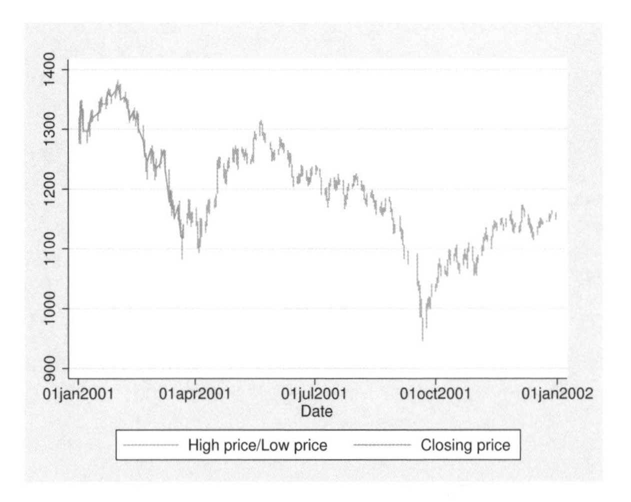

We specified lcolor(gs11) to tone down the spikes and give the line plot more prominence.

Advanced use 2

A popular financial graph is

```
. use http://www.stata-press.com/data/r12/sp500, clear
(S&P 500)
. replace volume = volume/1000
. twoway
        rspike hi low date ||
        line   close  date ||
        bar    volume date, barw(.25) yaxis(2) ||
  in 1/57
  , ysca(axis(1) r(900 1400))
    ysca(axis(2) r(  9   45))
    ytitle("                          Price -- High, Low, Close")
    ytitle(" Volume (millions)", axis(2) astext just(left))
    legend(off)
    subtitle("S&P 500", margin(b+2.5))
    note("Source:  Yahoo!Finance and Commodity Systems, Inc.")
```

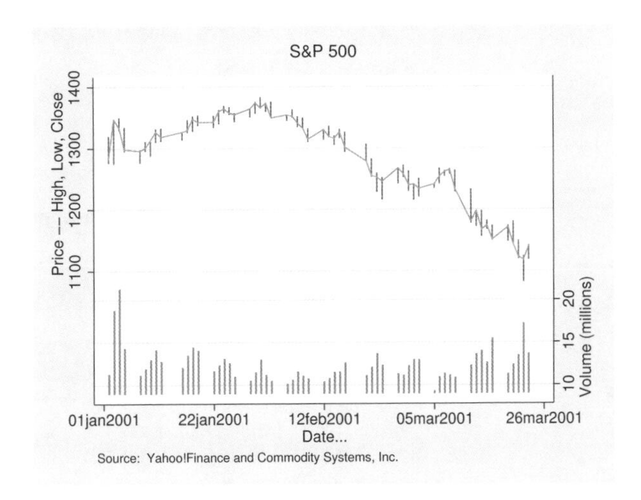

Also see

[G-2] **graph twoway spike** — Twoway spike plots

[G-2] **graph twoway rarea** — Range plot with area shading

[G-2] **graph twoway rbar** — Range plot with bars

[G-2] **graph twoway rcap** — Range plot with capped spikes

[G-2] **graph twoway rcapsym** — Range plot with spikes capped with marker symbols

[G-2] **graph twoway rline** — Range plot with lines

[G-2] **graph twoway rconnected** — Range plot with connected lines

[G-2] **graph twoway rscatter** — Range plot with markers

Title

[G-2] graph twoway scatter — Twoway scatterplots

Syntax

[twoway] scatter *varlist* [*if*] [*in*] [*weight*] [, *options*]

where *varlist* is

y_1 [y_2[...]] x

options	Description
marker_options	change look of markers (color, size, etc.)
marker_label_options	add marker labels; change look or position
connect_options	change look of lines or connecting method
composite_style_option	overall style of the plot
jitter_options	jitter marker positions using random noise
axis_choice_options	associate plot with alternative axis
twoway_options	titles, legends, axes, added lines and text, by, regions, name, aspect ratio, etc.

Each is defined below.

marker_options	Description
msymbol(*symbolstylelist*)	shape of marker
mcolor(*colorstylelist*)	color of marker, inside and out
msize(*markersizestylelist*)	size of marker
mfcolor(*colorstylelist*)	inside or "fill" color
mlcolor(*colorstylelist*)	color of outline
mlwidth(*linewidthstylelist*)	thickness of outline
mlstyle(*linestylelist*)	overall style of outline
mstyle(*markerstylelist*)	overall style of marker

See [G-3] ***marker_options***.

marker_label_options	Description
mlabel(*varlist*)	specify marker variables
mlabposition(*clockposlist*)	where to locate label
mlabvposition(*varname*)	where to locate label 2
mlabgap(*relativesizelist*)	gap between marker and label
mlabangle(*anglestylelist*)	angle of label
mlabsize(*textsizestylelist*)	size of label
mlabcolor(*colorstylelist*)	color of label
mlabtextstyle(*textstylelist*)	overall style of text
mlabstyle(*markerlabelstylelist*)	overall style of label

See [G-3] ***marker_label_options***.

connect_options	Description
connect(*connectstylelist*)	how to connect points
sort[(*varlist*)]	how to order data before connecting
cmissing({ y | n } ...)	missing values are ignored
lpattern(*linepatternstylelist*)	line pattern (solid, dashed, etc.)
lwidth(*linewidthstylelist*)	thickness of line
lcolor(*colorstylelist*)	color of line
lstyle(*linestylelist*)	overall style of line

See [G-3] ***connect_options***.

composite_style_option	Description
pstyle(*pstylelist*)	all the ...style() options above

See *Appendix: Styles and composite styles* under *Remarks* below.

jitter_options	Description
jitter(*relativesizelist*)	perturb location of point
jitterseed(*#*)	random-number seed for jitter()

See *Jittered markers* under *Remarks* below.

axis_choice_options	Description
yaxis(*#* [*#* ...])	which y axis to use
xaxis(*#* [*#* ...])	which x axis to use

See [G-3] ***axis_choice_options***.

twoway_options	Description
added_line_options	draw lines at specified y or x values
added_text_options	display text at specified (y,x value)
axis_options	labels, ticks, grids, log scales
title_options	titles, subtitles, notes, captions
legend_options	legend explaining what means what
`scale(`*#*`)`	resize text and markers
region_options	outlining, shading, aspect ratio
aspect_option	constrain aspect ratio of plot region
`scheme(`*schemename*`)`	overall look
`play(`*recordingname*`)`	play edits from *recordingname*
`by(`*varlist*`, ...)`	repeat for subgroups
`nodraw`	suppress display of graph
`name(`*name*`, ...)`	specify name for graph
`saving(`*filename*`, ...)`	save graph in file
advanced_options	difficult to explain

See [G-3] ***twoway_options***.

`aweights`, `fweights`, and `pweights` are allowed; see [U] **11.1.6 weight**.

Menu

Graphics > Twoway graph (scatter, line, etc.)

Description

`scatter` draws scatterplots and is the mother of all the `twoway` plottypes, such as `line` and `lfit` (see [G-2] **graph twoway line** and [G-2] **graph twoway lfit**).

`scatter` is both a command and a *plottype* as defined in [G-2] **graph twoway**. Thus the syntax for `scatter` is

```
. graph twoway scatter ...
. twoway scatter ...
. scatter ...
```

Being a plottype, `scatter` may be combined with other plottypes in the `twoway` family (see [G-2] **graph twoway**), as in,

```
. twoway (scatter ...) (line ...) (lfit ...) ...
```

which can equivalently be written

```
. scatter ... || line ... || lfit ... || ...
```

Options

marker_options specify how the points on the graph are to be designated. Markers are the ink used to mark where points are on a plot. Markers have shape, color, and size, and other characteristics. See [G-3] ***marker_options*** for a description of markers and the options that specify them.

`msymbol(O D S T + X o d s t smplus x)` is the default. `msymbol(i)` will suppress the appearance of the marker altogether.

marker_label_options specify labels to appear next to or in place of the markers. For instance, if you were plotting country data, marker labels would allow you to have "Argentina", "Bolivia", ..., appear next to each point and, with a few data, that might be desirable. See [G-3] ***marker_label_options*** for a description of marker labels and the options that control them.

By default, no marker labels are displayed. If you wish to display marker labels in place of the markers, specify `mlabposition(0)` and `msymbol(i)`.

connect_options specify how the points are to be connected. The default is not to connect the points.

`connect()` specifies whether points are to be connected and, if so, how the line connecting them is to be shaped. The line between each pair of points can connect them directly or in stairstep fashion.

`sort` specifies that the data be sorted by the x variable before the points are connected. Unless you are after a special effect or your data are already sorted, do not forget to specify this option. If you are after a special effect, and if the data are not already sorted, you can specify `sort(`*varlist*`)` to specify exactly how the data should be sorted. Understand that specifying `sort` or `sort(`*varlist*`)` when it is not necessary will slow Stata down a little. You must specify `sort` if you wish to connect points, and you must specify the *twoway_option* `by()` with `total`.

`cmissing(y)` and `cmissing(n)` specify whether missing values are ignored when points are connected; whether the line should have a break in it. The default is `cmissing(y)`, meaning that there will be no breaks.

`lpattern()` specifies how the style of the line is to be drawn: solid, dashed, etc.

`lwidth()` specifies the width of the line.

`lcolor()` specifies the color of the line.

`lstyle()` specifies the overall style of the line.

See [G-3] ***connect_options*** for more information on these and related options. See [G-4] **concept: lines** for an overview of lines.

`pstyle(`*pstyle*`)` specifies the overall style of the plot and is a composite of `mstyle()`, `mlabstyle()`, `lstyle()`, `connect()`, and `cmissing()`. The default is `pstyle(p1)` for the first plot, `pstyle(p2)` for the second, and so on. See *Appendix: Styles and composite styles* under *Remarks*.

`jitter(`*relativesize*`)` adds spherical random noise to the data before plotting. This is useful when plotting data which otherwise would result in points plotted on top of each other. See *Jittered markers* under *Remarks*.

Commonly specified are `jitter(5)` or `jitter(6)`; `jitter(0)` is the default. See [G-4] ***relativesize*** for a description of relative sizes.

`jitterseed(`*#*`)` specifies the seed for the random noise added by the `jitter()` option. # should be specified as a positive integer. Use this option to reproduce the same plotted points when the `jitter()` option is specified.

axis_choice_options are for use when you have multiple x or y axes.
See [G-3] ***axis_choice_options*** for more information.

twoway_options include

added_line_options, which specify that horizontal or vertical lines be drawn on the graph; see [G-3] ***added_line_options***. If your interest is in drawing grid lines through the plot region, see *axis_options* below.

added_text_options, which specify text to be displayed on the graph (inside the plot region); see [G-3] ***added_text_options***.

axis_options, which allow you to specify labels, ticks, and grids. These options also allow you to obtain logarithmic scales; see [G-3] ***axis_options***.

title_options allow you to specify titles, subtitles, notes, and captions to be placed on the graph; see [G-3] ***title_options***.

legend_options, which allows specifying the legend explaining the symbols and line styles used; see [G-3] ***legend_options***.

`scale(`*#*`)`, which makes all the text and markers on a graph larger or smaller (`scale(1)` means no change); see [G-3] ***scale_option***.

region_options, which allow you to control the aspect ratio and to specify that the graph be outlined, or given a background shading; see [G-3] ***region_options***.

`scheme(`*schemename*`)`, which specifies the overall look of the graph; see [G-3] ***scheme_option***.

`play(`*recordingname*`)` applies the edits from *recordingname* to the graph, where *recordingname* is the name under which edits previously made in the Graph Editor have been recorded and stored. See *Graph Recorder* in [G-1] **graph editor**.

`by(`*varlist*`, ...)`, which allows drawing multiple graphs for each subgroup of the data; see [G-3] ***by_option***.

`nodraw`, which prevents the graph from being displayed; see [G-3] ***nodraw_option***.

`name(`*name*`)`, which allows you to save the graph in memory under a name different from `Graph`; see [G-3] ***name_option***.

`saving(`*filename*[`, asis replace`]`)`, which allows you to save the graph to disk; see [G-3] ***saving_option***.

other options that allow you to suppress the display of the graph, to name the graph, etc.

See [G-3] ***twoway_options***.

Remarks

Remarks are presented under the following headings:

Typical use

The scatter plottype by default individually marks the location of each point:

```
. use http://www.stata-press.com/data/r12/uslifeexp2
(U.S. life expectancy, 1900-1940)
. scatter le year
```

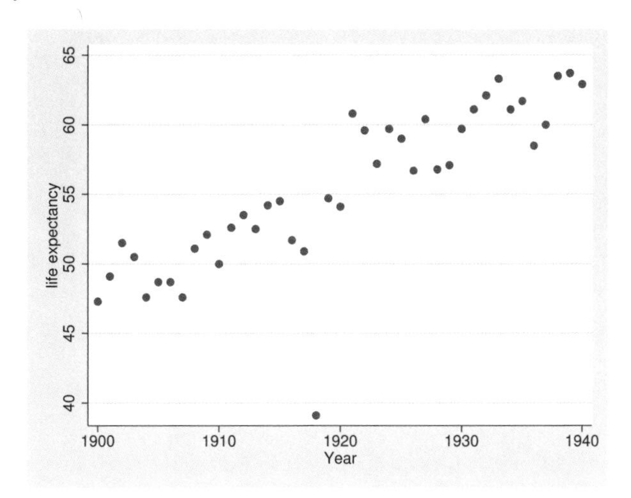

With the specification of options, you can produce the same effect as `twoway connected` (see [G-2] **graph twoway connected**),

```
. scatter le year, connect(l)
```

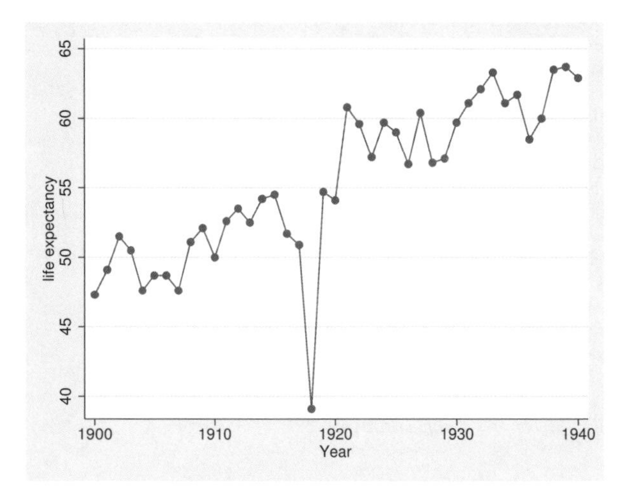

or `twoway line`:

```
. scatter le year, connect(l) msymbol(i)
```

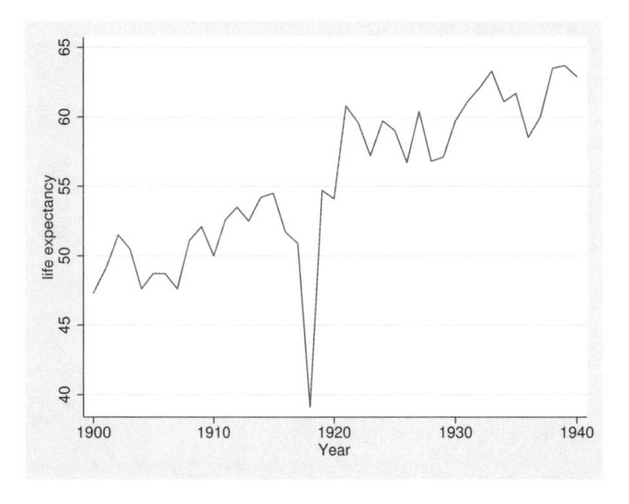

In fact, all the other twoway plottypes eventually work their way back to executing `scatter`. `scatter` literally is the mother of all twoway graphs in Stata.

Scatter syntax

See [G-2] **graph twoway** for an overview of `graph twoway` syntax. Especially for `graph twoway scatter`, the only thing to know is that if more than two variables are specified, all but the last are given the interpretation of being y variables. For example,

. `scatter` *y1var y2var xvar*

would plot *y1var* versus *xvar* and overlay that with a plot of *y2var* versus *xvar*, so it is the same as typing

```
. scatter y1var xvar || scatter y2var xvar
```

If, using the multiple-variable syntax, you specify `scatter`-level options (that is, all options except *twoway_options* as defined in the syntax diagram), you specify arguments for *y1var*, *y2var*, ..., separated by spaces. That is, you might type

```
. scatter y1var y2var xvar, ms(O i) c(. l)
```

`ms()` and `c()` are abbreviations for `msymbol()` and `connect()`; see [G-3] ***marker_options*** and [G-3] ***connect_options***. In any case, the results from the above are the same as if you typed

```
. scatter y1var xvar, ms(O) c(.) || scatter y2var xvar, ms(i) c(l)
```

There need not be a one-to-one correspondence between options and y variables when you use the multiple-variable syntax. If you typed

```
. scatter y1var y2var xvar, ms(O) c(l)
```

then options `ms()` and `c()` will have default values for the second scatter, and if you typed

```
. scatter y1var y2var xvar, ms(O S i) c(l l l)
```

the extra options for the nonexistent third variable would be ignored.

If you wish to specify the default for one of the *y* variables, you may specify period (`.`):

```
. scatter y1var y2var xvar, ms(. O) c(. l)
```

There are other shorthands available to make specifying multiple arguments easier; see [G-4] ***stylelists***.

Because multiple variables are interpreted as multiple y variables, to produce graphs containing multiple x variables, you must chain together separate `scatter` commands:

```
. scatter yvar x1var, ... || . scatter yvar x2var, ...
```

The overall look for the graph

The overall look of the graph is mightily affected by the scheme, and there is a `scheme()` option that will allow you to specify which scheme to use. We showed earlier the results of `scatter le year`. Here is the same graph repeated using the `economist` scheme:

```
. use http://www.stata-press.com/data/r12/uslifeexp2, clear
(U.S. life expectancy, 1900-1940)

. scatter le year,
        title("Scatterplot")
        subtitle("Life expectancy at birth, U.S.")
        note("1")
        caption("Source: National Vital Statistics Report,
         Vol. 50 No. 6")
        scheme(economist)
```

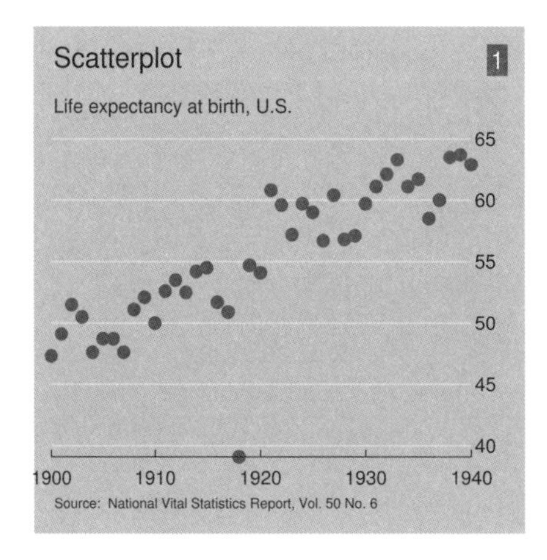

See [G-4] **schemes intro**.

The size and aspect ratio of the graph

The size and aspect ratio of the graph are controlled by the *region_options* `ysize(#)` and `xsize(#)`, which specify the height and width in inches of the graph. For instance,

```
. scatter yvar xvar, xsize(4) ysize(4)
```

would produce a 4×4 inch square graph. See [G-3] ***region_options***.

Titles

By default, no titles appear on the graph, but the *title_options* `title()`, `subtitle()`, `note()`, `caption()`, and `legend()` allow you to specify the titles that you wish to appear, as well as to control their position and size. For instance,

```
. scatter yvar xvar, title("My title")
```

would draw the graph and include the title "My title" (without the quotes) at the top. Multiple-line titles are allowed. Typing

```
. scatter yvar xvar, title("My title" "Second line")
```

would create a two-line title. The above, however, would probably look better as a title followed by a subtitle:

```
. scatter yvar xvar, title("My title") subtitle("Second line")
```

In any case, see [G-3] ***title_options***.

Axis titles

Titles do, by default, appear on the y and x axes. The axes are titled with the variable names being plotted or, if the variables have variable labels, with their variable labels. The *axis_title_options* `ytitle()` and `xtitle()` allow you to override that. If you specify

```
. scatter yvar xvar, ytitle("")
```

the title on the y axis would disappear. If you specify

```
. scatter yvar xvar, ytitle("Rate of change")
```

the y-axis title would become "Rate of change". As with all titles, multiple-line titles are allowed:

```
. scatter yvar xvar, ytitle("Time to event" "Rate of change")
```

See [G-3] ***axis_title_options***.

Axis labels and ticking

By default, approximately five major ticks and labels are placed on each axis. The *axis_label_options* `ylabel()` and `xlabel()` allow you to control that. Typing

```
. scatter yvar xvar, ylabel(#10)
```

would put approximately 10 labels and ticks on the y axis. Typing

```
. scatter yvar xvar, ylabel(0(1)9)
```

would put exactly 10 labels at the values 0, 1, ..., 9.

`ylabel()` and `xlabel()` have other features, and options are also provided for minor labels and minor ticks; see [G-3] ***axis_label_options***.

Grid lines

If you use a member of the `s2` family of schemes—see [G-4] **scheme s2**—grid lines are included in y but not x, by default. You can specify option `xlabel(,grid)` to add x grid lines, and you can specify `ylabel(,nogrid)` to suppress y grid lines.

Grid lines are considered an extension of ticks and are specified as suboptions inside the *axis_label_options* `ylabel()` and `xlabel()`. For instance,

```
. use http://www.stata-press.com/data/r12/auto, clear
(1978 Automobile Data)
. scatter mpg weight, xlabel(,grid)
```

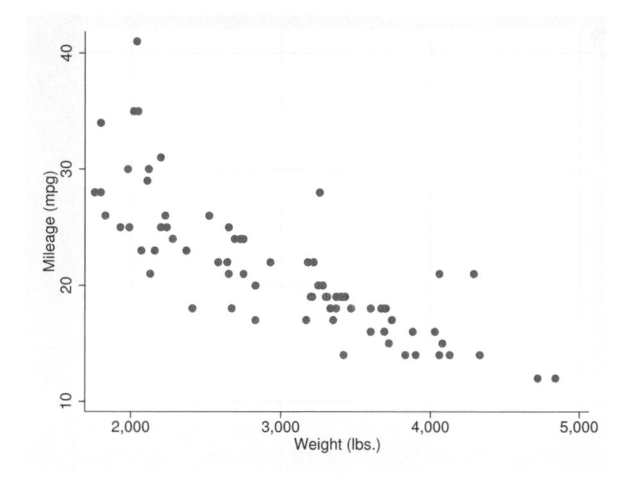

In the above example, the grid lines are placed at the same values as the default ticks and labels, but you can control that, too. See [G-3] ***axis_label_options***.

Added lines

Lines may be added to the graph for emphasis by using the *added_line_options* `yline()` and `xline()`; see [G-3] ***added_line_options***.

Axis range

The extent or range of an axis is set according to all the things that appear on it—the data being plotted and the values on the axis being labeled or ticked. In the graph that just appeared above,

```
. use http://www.stata-press.com/data/r12/auto, clear
(1978 Automobile Data)
. scatter mpg weight
```

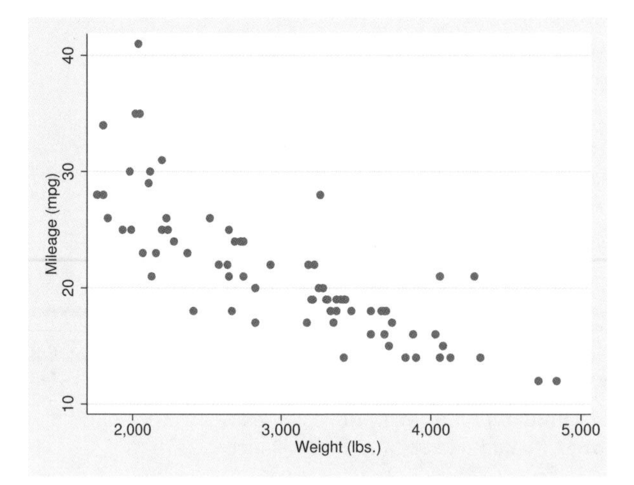

variable `mpg` varies between 12 and 41 and yet the y axis extends from 10 to 41. The axis was extended to include $10 < 12$ because the value 10 was labeled. Variable `weight` varies between 1,760 and 4,840; the x axis extends from 1,760 to 5,000. This axis was extended to include $5{,}000 > 4{,}840$ because the value 5,000 was labeled.

You can prevent axes from being extended by specifying the `ylabel(minmax)` and `xlabel(minmax)` options. `minmax` specifies that only the minimum and maximum are to be labeled:

```
. scatter mpg weight, ylabel(minmax) xlabel(minmax)
```

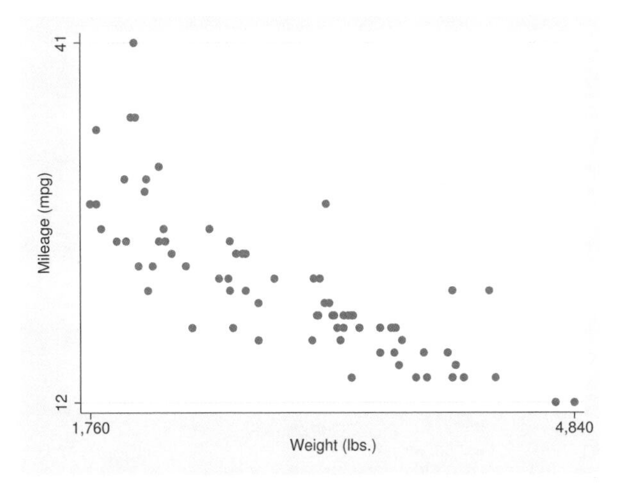

In other cases, you may wish to widen the range of an axis. This you can do by specifying the `range()` descriptor of the *axis_scale_options* `yscale()` or `xscale()`. For instance,

```
. scatter mpg weight, xscale(range(1000 5000))
```

would widen the x axis to include 1,000–5,000. We typed out the name of the option, but most people would type

```
. scatter mpg weight, xscale(r(1000 5000))
```

`range()` can widen, but never narrow, the extent of an axis. Typing

```
. scatter mpg weight, xscale(r(1000 4000))
```

would not omit cars with `weight`> 4000 from the plot. If that is your desire, type

```
. scatter mpg weight if weight<=4000
```

See [G-3] ***axis_scale_options*** for more information on `range()`, `yscale()`, and `xscale()`; see [G-3] ***axis_label_options*** for more information on `ylabel(minmax)` and `xlabel(minmax)`.

Log scales

By default, arithmetic scales for the axes are used. Log scales can be obtained by specifying the `log` suboption of `yscale()` and `xscale()`. For instance,

```
. use http://www.stata-press.com/data/r12/lifeexp, clear
(Life expectancy, 1998)
. scatter lexp gnppc, xscale(log) xlab(,g)
```

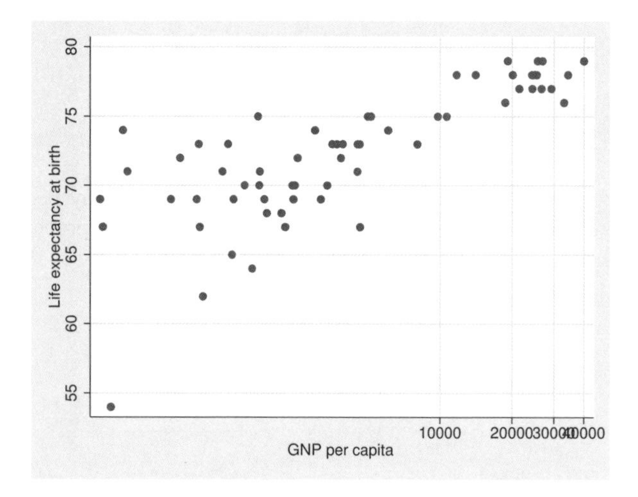

The important option above is `xscale(log)`, which caused `gnppc` to be presented on a log scale.

We included `xlab(,g)` (abbreviated form of `xlabel(,grid)`) to obtain x grid lines. The values 30,000 and 40,000 are overprinted. We could improve the graph by typing

```
. generate gnp000 = gnppc/1000
. label var gnp000 "GNP per capita, thousands of dollars"
. scatter lexp gnp000, xsca(log) xlab(.5 2.5 10(10)40, grid)
```

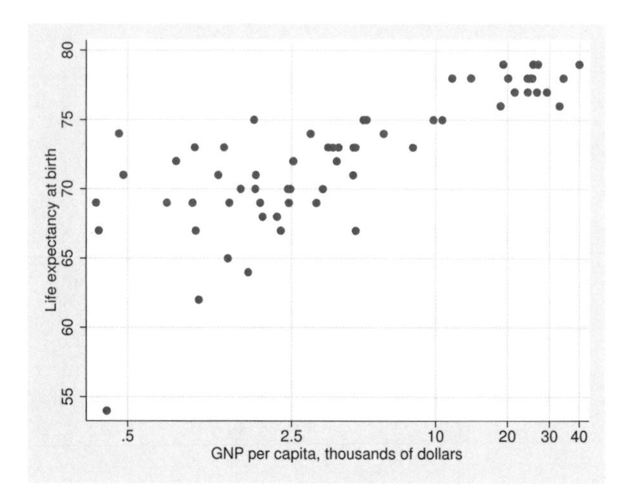

See [G-3] ***axis_options***.

Multiple axes

Graphs may have more than one y axis and more than one x axis. There are two reasons to do this: you might include an extra axis so that you have an extra place to label special values or so that you may plot multiple variables on different scales. In either case, specify the `yaxis()` or `xaxis()` option. See [G-3] ***axis_choice_options***.

Markers

Markers are the ink used to mark where points are on the plot. Many people think of markers in terms of their shape (circles, diamonds, etc.), but they have other properties, including, most importantly, their color and size. The shape of the marker is specified by the `msymbol()` option, its color by the `mcolor()` option, and its size by the `msize()` option.

By default, solid circles are used for the first y variable, solid diamonds for the second, solid squares for the third, and so on; see *marker_options* under *Options* for the remaining details, if you care. In any case, when you type

```
. scatter yvar xvar
```

results are as if you typed

```
. scatter yvar xvar, msymbol(O)
```

You can vary the symbol used by specifying other `msymbol()` arguments. Similarly, you can vary the color and size of the symbol by specifying the `mcolor()` and `msize()` options. See [G-3] ***marker_options***.

In addition to the markers themselves, you can request that the individual points be labeled. These marker labels are numbers or text that appear beside the marker symbol—or in place of it—to identify the points. See [G-3] ***marker_label_options***.

Weighted markers

If weights are specified—see [U] **11.1.6 weight**—the size of the marker is scaled according to the size of the weights. `aweights`, `fweights`, and `pweights` are allowed and all are treated the same; `iweights` are not allowed because `scatter` would not know what to do with negative values. Weights affect the size of the marker and nothing else about the plot.

Below we use U.S. state–averaged data to graph the divorce rate in a state versus the state's median age. We scale the symbols to be proportional to the population size:

```
. use http://www.stata-press.com/data/r12/census, clear
(1980 Census data by state)
. generate drate = divorce / pop18p
. label var drate "Divorce rate"
. scatter drate medage [w=pop18p] if state!="Nevada", msymbol(Oh)
      note("Stata data excluding Nevada"
           "Area of symbol proportional to state's population aged 18+")
```

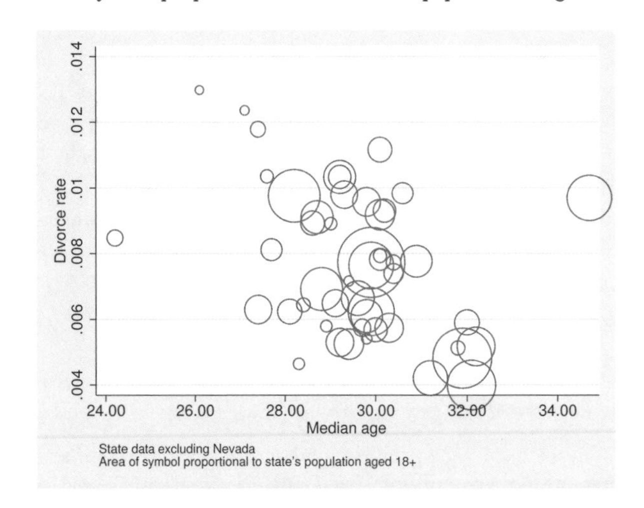

Note the use of the `msymbol(Oh)` option. Hollow scaled markers look much better than solid ones.

`scatter` scales the symbols so that the sizes are a fair representation when the weights represent population weights. If all the weights except one are 1,000 and the exception is 999, the symbols will all be of almost equal size. The weight 999 observation will not be a dot and the weight 1,000 observation giant circles as would be the result if the exception had weight 1.

When weights are specified, option `msize()` (which also affects the size of the marker), if specified, is ignored. See [G-3] ***marker_options***.

Weights are ignored when the `mlabel()` option is specified. See [G-3] ***marker_label_options***.

Jittered markers

`scatter` will add spherical random noise to your data before plotting if you specify `jitter(#)`, where # represents the size of the noise as a percentage of the graphical area. This can be useful for creating graphs of categorical data when, were the data not jittered, many of the points would be on top of each other, making it impossible to tell whether the plotted point represented one or 1,000 observations.

For instance, in a variation on `auto.dta` used below, `mpg` is recorded in units of 5 mpg, and `weight` is recorded in units of 500 pounds. A standard scatter has considerable overprinting:

```
. use http://www.stata-press.com/data/r12/autornd, clear
(1978 Automobile Data)
. scatter mpg weight
```

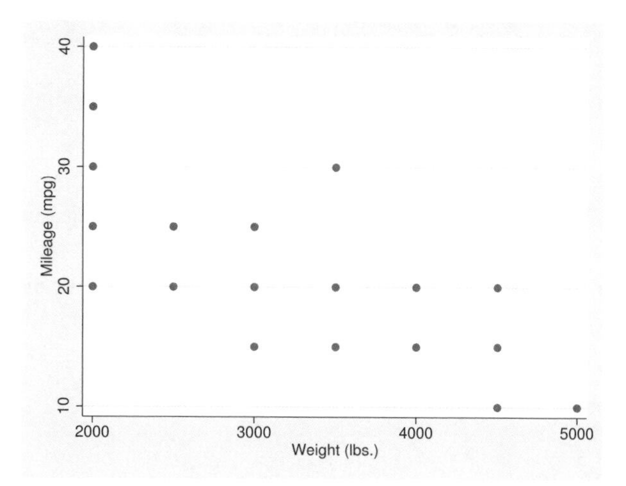

There are 74 points in the graph, even though it appears because of overprinting as if there are only 19. Jittering solves that problem:

```
. scatter mpg weight, jitter(7)
```

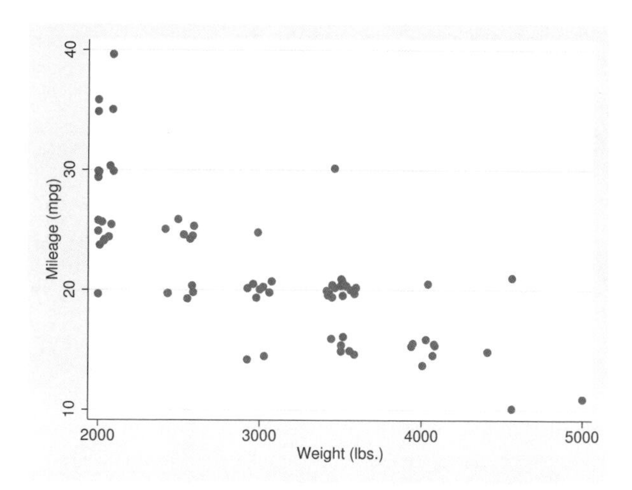

Connected lines

The `connect()` option allows you to connect the points of a graph. The default is not to connect the points.

If you want connected points, you probably want to specify `connect(l)`, which is usually abbreviated `c(l)`. The `l` means that the points are to be connected with straight lines. Points can be connected in other ways (such as a stairstep fashion), but usually `c(l)` is the right choice. The command

```
. scatter yvar xvar, c(l)
```

will plot *yvar* versus *xvar*, marking the points in the usual way, and drawing straight lines between the points. It is common also to specify the `sort` option,

```
. scatter yvar xvar, c(l) sort
```

because otherwise points are connected in the order of the data. If the data are already in the order of *xvar*, the `sort` is unnecessary. You can also omit the `sort` when creating special effects.

`connect()` is often specified with the `msymbol(i)` option to suppress the display of the individual points:

```
. scatter yvar xvar, c(l) sort m(i)
```

See [G-3] ***connect_options***.

Graphs by groups

Option `by()` specifies that graphs are to be drawn separately for each of the different groups and the results arrayed into one display. Below we use country data and group the results by region of the world:

```
. use http://www.stata-press.com/data/r12/lifeexp, clear
(Life expectancy, 1998)
. scatter lexp gnppc, by(region)
```

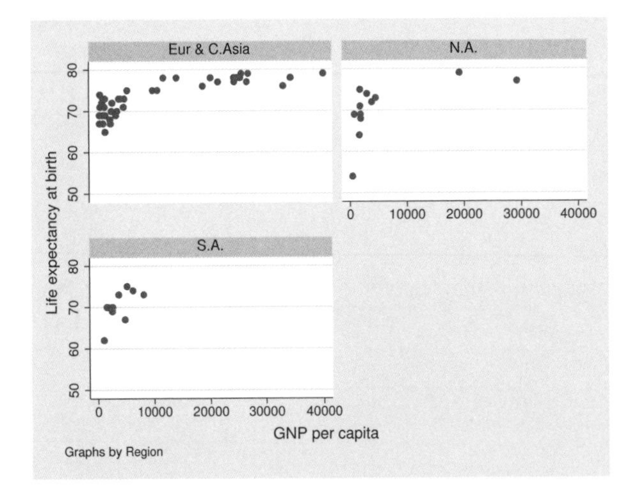

Variable `region` is a numeric variable taking on values 1, 2, and 3. Separate graphs were drawn for each value of region. The graphs were titled "Eur & C. Asia", "N.A.", and "S.A." because numeric variable `region` had been assigned a value label, but results would have been the same had variable `region` been a string directly containing "Eur & C. Asia", "N.A.", and "S.A.".

See [G-3] ***by_option*** for more information on this useful option.

Saving graphs

To save a graph to disk for later printing or reviewing, include the saving() option,

```
. scatter ..., ... saving(filename)
```

or use the graph save command afterward:

```
. scatter ...
. graph save filename
```

See [G-3] ***saving_option*** and [G-2] **graph save**. Also see [G-4] **concept: gph files** for information on how files such as *filename*.gph can be put to subsequent use.

Appendix: Styles and composite styles

Many options end in the word style, including mstyle(), mlabstyle(), and lstyle(). Option mstyle(), for instance, is described as setting the "overall look" of a marker. What does that mean?

How something looks—a marker, a marker label, a line—is specified by many detail options. For markers, option msymbol() specifies its shape, mcolor() specifies its color, msize() specifies its size, and so on.

A *style* specifies a composite of related option settings. If you typed option mstyle(p1), you would be specifying a whole set of values for msymbol(), mcolor(), msize(), and all the other m*() options. p1 is called the name of a style, and p1 contains the settings.

Concerning mstyle() and all the other options ending in the word style, throughout this manual you will read statements such as

> Option *whatever*style() specifies the overall look of *whatever*, such as its *(insert list here)*. The other options allow you to change the attributes of a *whatever*, but *whatever*style() is the starting point.
>
> You need not specify *whatever*style() just because there is something you want to change about the look of a *whatever*, and in fact, most people seldom specify the *whatever*style() option. You specify *whatever*style() when another style exists that is exactly what you desire or when another style would allow you to specify fewer changes to obtain what you want.

Styles actually come in two forms called *composite styles* and *detail styles*, and the above statement applies only to composite styles and appears only in manual entries concerning composite styles. Composite styles are specified in options that end in the word style. The following are examples of composite styles:

mstyle(*symbolstyle*)
mlstyle(*linestyle*)
mlabstyle(*markerlabelstyle*)
lstyle(*linestyle*)
pstyle(*pstyle*)

The following are examples of detail styles:

mcolor(*colorstyle*)
mlwidth(*linewidthstyle*)
mlabsize(*textsizestyle*)
lpattern(*linepatternstyle*)

In the above examples, distinguish carefully between option names such as `mcolor()` and option arguments such as *colorstyle*. *colorstyle* is an example of a detail style because it appears in the option `mcolor()`, and the option name does not end in the word style.

Detail styles specify precisely how an attribute of something looks, and composite styles specify an "overall look" in terms of detail-style values.

Composite styles sometimes contain other composite styles as members. For instance, when you specify the `mstyle()` option—which specifies the overall look of markers—you are also specifying an `mlstyle()`—which specifies the overall look of the lines that outline the shape of the markers. That does not mean you cannot specify the `mlstyle()` option, too. It just means that specifying `mstyle()` implies an `mlstyle()`. The order in which you specify the options does not matter. You can type

```
. scatter ..., ... mstyle(...) ... mlstyle(...) ...
```

or

```
. scatter ..., ... mlstyle(...) ... mstyle(...) ...
```

and, either way, `mstyle()` will be set as you specify, and then `mlstyle()` will be reset as you wish. The same applies for mixing composite-style and detail-style options. Option `mstyle()` implies an `mcolor()` value. Even so, you may type

```
. scatter ..., ... mstyle(...) ... mcolor(...) ...
```

or

```
. scatter ..., ... mcolor(...) ... mstyle(...) ...
```

and the outcome will be the same.

The grandest composite style of them all is `pstyle(`*pstyle*`)`. It contains all the other composite styles and `scatter` (`twoway`, in fact) makes great use of this grand style. When you type

```
. scatter y1var y2var xvar, ...
```

results are as if you typed

```
. scatter y1var y2var xvar, pstyle(p1 p2) ...
```

That is, *y1var* versus *xvar* is plotted using `pstyle(p1)`, and *y2var* versus *xvar* is plotted using `pstyle(p2)`. It is the `pstyle(p1)` that sets all the defaults—which marker symbols are used, what color they are, etc.

The same applies if you type

```
. scatter y1var xvar, ... || scatter y2var xvar, ...
```

y1var versus *xvar* is plotted using `pstyle(p1)`, and *y2var* versus *xvar* is plotted using `pstyle(p2)`, just as if you had typed

```
. scatter y1var xvar, pstyle(p1) ... || scatter y2var xvar, pstyle(p2) ...
```

The same applies if you mix scatter with other plottypes:

```
. scatter y1var xvar, ... || line y2var xvar, ...
```

is equivalent to

```
. scatter y1var xvar, pstyle(p1) ... || line y2var xvar, pstyle(p2) ...
```

and

```
. twoway (..., ...) (..., ...), ...
```

is equivalent to

```
. twoway (..., pstyle(p1) ...) (..., pstyle(p2) ...), ...
```

which is why we said that it is twoway, and not just scatter, that exploits scheme().

You can put this to use. Pretend that you have a dataset on husbands and wives and it contains the variables

hinc	husband's income
winc	wife's income
hed	husband's education
wed	wife's education

You wish to draw a graph of income versus education, drawing no distinctions between husbands and wives. You type

```
. scatter hinc hed || scatter winc wed
```

You intend to treat husbands and wives the same in the graph, but in the above example, they are treated differently because msymbol(O) will be used to mark the points of hinc versus hed and msymbol(D) will be used to designate winc versus wed. The color of the symbols will be different, too.

You could address that problem in many different ways. You could specify the msymbol() and mcolor() options (see [G-3] ***marker_options***), along with whatever other detail options are necessary to make the two scatters appear the same. Being knowledgeable, you realize you do not have to do that. There is, you know, a composite style that specifies this. So you get out your manuals, flip through, and discover that the relevant composite style for the marker symbols is mstyle().

Easiest of all, however, would be to remember that pstyle() contains all the other styles. Rather than resetting mstyle(), just reset pstyle(), and whatever needs to be set to make the two plots the same will be set. Type

```
. scatter hinc hed || scatter winc wed, pstyle(p1)
```

or, if you prefer,

```
. scatter hinc hed, pstyle(p1) || scatter winc wed, pstyle(p1)
```

You do not need to specify pstyle(p1) for the first plot, however, because that is the default.

As another example, you have a dataset containing

mpg	Mileage ratings of cars
weight	Each car's weight
prediction	A predicted mileage rating based on weight

You wish to draw the graph

```
. scatter mpg weight || line prediction weight
```

but you wish the appearance of the line to "match" that of the markers used to plot `mpg` versus `weight`. You could go digging to find out which option controlled the line style and color and then dig some more to figure out which line style and color goes with the markers used in the first plot, but much easier is simply to type

```
. scatter mpg weight || line prediction weight, pstyle(p1)
```

References

Cox, N. J. 2005a. Stata tip 24: Axis labels on two or more levels. *Stata Journal* 5: 469.

——. 2005b. Stata tip 27: Classifying data points on scatter plots. *Stata Journal* 5: 604–606.

Friendly, M., and D. Denis. 2005. The early origins and development of the scatterplot. *Journal of the History of the Behavioral Sciences* 41: 103–130.

Royston, P., and N. J. Cox. 2005. A multivariable scatterplot smoother. *Stata Journal* 5: 405–412.

Winter, N. J. G. 2005. Stata tip 23: Regaining control over axis ranges. *Stata Journal* 5: 467–468.

Also see

[G-2] **graph twoway** — Twoway graphs

[G-3] ***marker_options*** — Options for specifying markers

[G-3] ***marker_label_options*** — Options for specifying marker labels

[G-3] ***connect_options*** — Options for connecting points with lines

[G-3] ***axis_choice_options*** — Options for specifying the axes on which a plot appears

[G-3] ***twoway_options*** — Options for twoway graphs

Title

[G-2] graph twoway scatteri — Scatter with immediate arguments

Syntax

`twoway scatteri` *immediate_values* [, *options*]

where *immediate_values* is one or more of

$\#_y$ $\#_x$ [($\#_{\text{clockposstyle}}$)] ["*text for label*"]

See [G-4] ***clockposstyle*** for a description of $\#_{\text{clockposstyle}}$.

Menu

Graphics > Twoway graph (scatter, line, etc.)

Description

`scatteri` is an immediate version of `twoway scatter`; see [U] **19 Immediate commands** and [G-2] **graph twoway scatter**. `scatteri` is intended for programmer use but can be useful interactively.

Options

options are as defined in [G-2] **graph twoway scatter**, with the following modifications:

If "*text for label*" is specified among any of the immediate arguments, option `mlabel()` is assumed.

If ($\#_{\text{clockposstyle}}$) is specified among any of the immediate arguments, option `mlabvposition()` is assumed.

Remarks

Immediate commands are commands that obtain data from numbers typed as arguments. Typing

```
. twoway scatteri 1 1  2 2, any_options
```

produces the same graph as typing

```
. clear
. input y x
                y           x
  1. 1 1
  2. 2 2
  3. end
. twoway scatter y x, any_options
```

`twoway scatteri` does not modify the data in memory.

`scatteri` is intended for programmer use but can be used interactively. In [G-3] ***added_text_options***, we demonstrated the use of option `text()` to add text to a graph:

```
. twoway qfitci  mpg weight, stdf ||
         scatter mpg weight, ms(O)
                 text(41 2040 "VW Diesel", place(e))
                 text(28 3260 "Plymouth Arrow", place(e))
                 text(35 2050 "Datsun 210 and Subaru", place(e))
```

Below we use `scatteri` to obtain similar results:

```
. twoway qfitci  mpg weight, stdf ||
         scatter mpg weight, ms(O) ||
         scatteri 41 2040 (3) "VW Diesel"
                  28 3260 (3) "Plymouth Arrow"
                  35 2050 (3) "Datsun 210 and Subaru"
                  , msymbol(i)
```

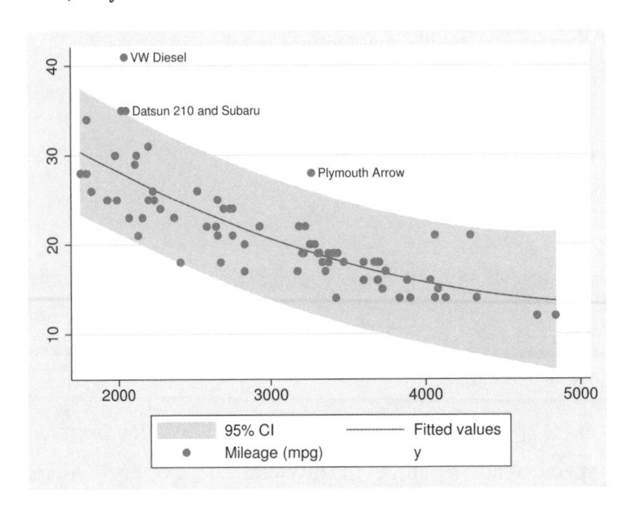

We translated `text(..., place(e))` to `(3)`, 3 o'clock being the *clockposstyle* notation for the east *compassdirstyle*. Because labels are by default positioned at 3 o'clock, we could omit `(3)` altogether:

```
. twoway qfitci  mpg weight, stdf ||
         scatter mpg weight, ms(O) ||
         scatteri 41 2040 "VW Diesel"
                  28 3260 "Plymouth Arrow"
                  35 2050 "Datsun 210 and Subaru"
                  , msymbol(i)
```

We specified `msymbol(i)` option to suppress displaying the marker symbol.

❑ Technical note

Programmers: Note carefully `scatter`'s *advanced_option* `recast()`; see [G-3] ***advanced_options***. It can be used to good effect, such as using `scatteri` to add areas, bars, spikes, and dropped lines.
❑

Also see

[G-2] **graph twoway scatter** — Twoway scatterplots

Title

[G-2] graph twoway spike — Twoway spike plots

Syntax

`twoway spike` *yvar xvar* [*if*] [*in*] [, *options*]

options	Description
`vertical`	vertical spike plot; the default
`horizontal`	horizontal spike plot
`base(#)`	value to drop to; default is 0
line_options	change look of spike lines
axis_choice_options	associate plot with alternative axis
twoway_options	titles, legends, axes, added lines and text, by, regions, name, aspect ratio, etc.

See [G-3] ***line_options***, [G-3] ***axis_choice_options***, and [G-3] ***twoway_options***.

All explicit options are *rightmost*, except `vertical` and `horizontal`, which are *unique*; see [G-4] **concept: repeated options**.

Menu

Graphics > Twoway graph (scatter, line, etc.)

Description

`twoway spike` displays numerical (y,x) data as spikes. `twoway spike` is useful for drawing spike plots of time-series data or other equally spaced data and is useful as a programming tool. For sparse data, also see [G-2] **graph bar**.

Options

`vertical` and `horizontal` specify either a vertical or a horizontal spike plot. `vertical` is the default. If `horizontal` is specified, the values recorded in *yvar* are treated as x values, and the values recorded in *xvar* are treated as y values. That is, to make horizontal plots, do not switch the order of the two variables specified.

In the `vertical` case, spikes are drawn at the specified *xvar* values and extend up or down from 0 according to the corresponding *yvar* values. If 0 is not in the range of the y axis, spikes extend up or down to the x axis.

In the `horizontal` case, spikes are drawn at the specified *xvar* values and extend left or right from 0 according to the corresponding *yvar* values. If 0 is not in the range of the x axis, spikes extend left or right to the y axis.

base(#) specifies the value from which the spike should extend. The default is base(0); in the above description of options vertical and horizontal, this default was assumed.

line_options specify the look of the lines used to draw the spikes, including pattern, width, and color; see [G-3] ***line_options***.

axis_choice_options associate the plot with a particular y or x axis on the graph; see [G-3] ***axis_choice_options***.

twoway_options are a set of common options supported by all twoway graphs. These options allow you to title graphs, name graphs, control axes and legends, add lines and text, set aspect ratios, create graphs over by() groups, and change some advanced settings. See [G-3] ***twoway_options***.

Remarks

Remarks are presented under the following headings:

Typical use
Advanced use
Cautions

Typical use

We have daily data recording the values for the S&P 500 in 2001:

```
. use http://www.stata-press.com/data/r12/sp500
(S&P 500)
. list date close change in 1/5
```

	date	close	change
1.	02jan2001	1283.27	.
2.	03jan2001	1347.56	64.29004
3.	04jan2001	1333.34	-14.22009
4.	05jan2001	1298.35	-34.98999
5.	08jan2001	1295.86	-2.48999

The example in [G-2] **graph twoway bar** graphed the first 57 observations of these data by using bars. Here is the same graph presented as spikes:

```
. twoway spike change date in 1/57
```

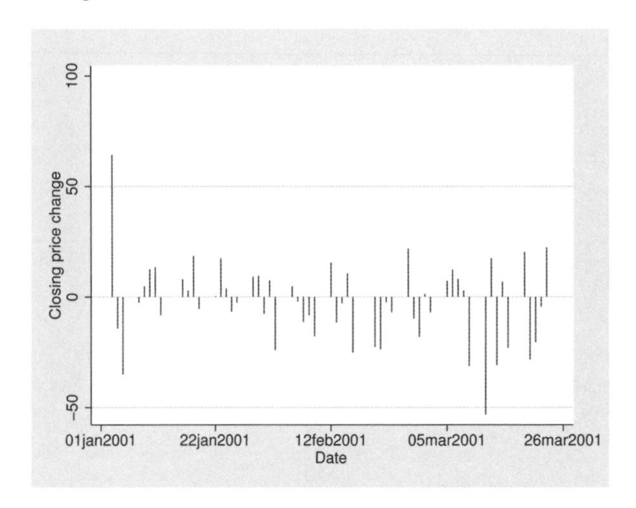

Spikes are especially useful when there are a lot of data. The graph below shows the data for the entire year:

```
. twoway spike change date
```

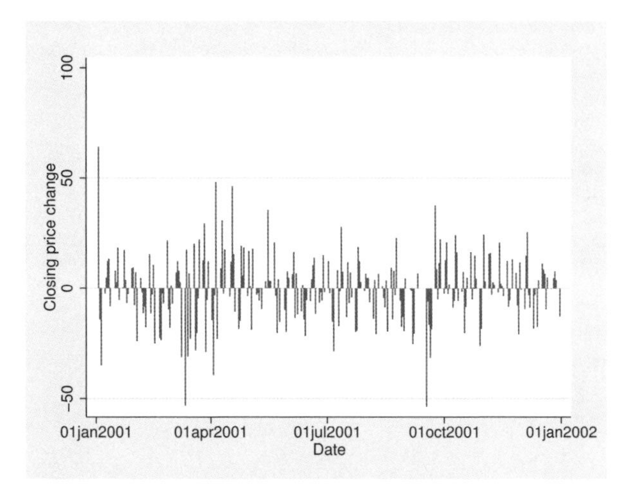

Advanced use

The useful thing about `twoway spike` is that it can be combined with other `twoway` plottypes (see [G-2] **graph twoway**):

```
. twoway line close date || spike change date
```

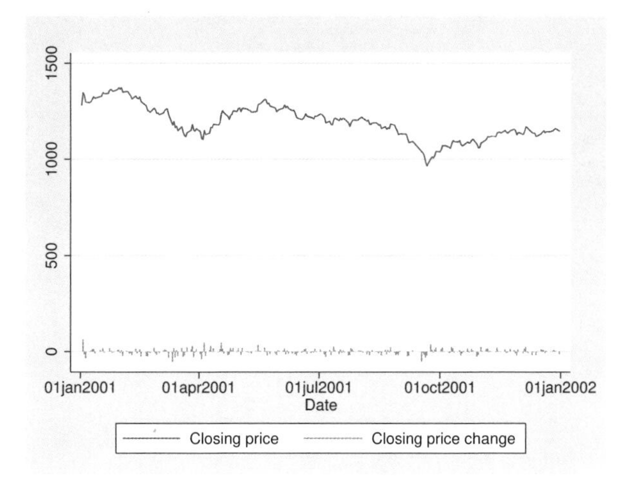

We can improve this graph by typing

```
. twoway
        line close date, yaxis(1)
    ||
         spike change date, yaxis(2)
    ||,
         ysca(axis(1) r(700  1400)) ylab(1000(100)1400, axis(1))
         ysca(axis(2) r(-50 300)) ylab(-50 0 50, axis(2))
                ytick(-50(25)50, axis(2) grid)
         legend(off)
         xtitle("Date")
         title("S&P 500")
         subtitle("January - December 2001")
         note("Source:  Yahoo!Finance and Commodity Systems, Inc.")
         yline(950, axis(1) lstyle(foreground))
```

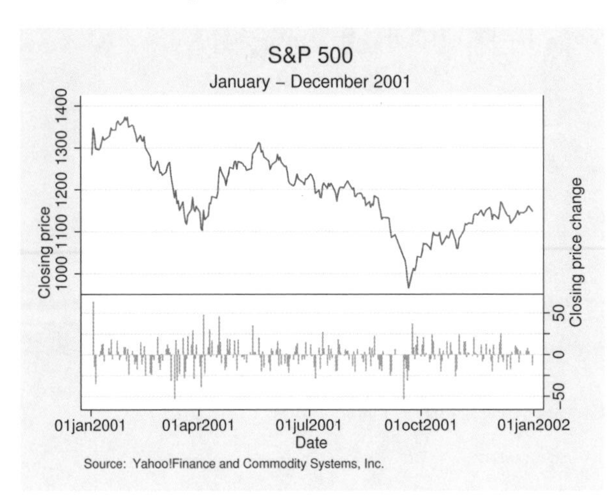

Concerning our use of

```
yline(950, axis(1) lstyle(foreground))
```

see *Advanced use: Overlaying* in [G-2] **graph twoway bar**.

Cautions

See *Cautions* in [G-2] **graph twoway bar**, which applies equally to `twoway spike`.

Also see

[G-2] **graph twoway scatter** — Twoway scatterplots

[G-2] **graph twoway bar** — Twoway bar plots

[G-2] **graph twoway dot** — Twoway dot plots

[G-2] **graph twoway dropline** — Twoway dropped-line plots

Title

[G-2] graph twoway tsline — Twoway line plots

Syntax

Time-series line plot

[twoway] tsline *varlist* [*if*] [*in*] [, *scatter_options twoway_options*]

Time-series range plot with lines

[twoway] tsrline y_1 y_2 [*if*] [*in*] [, *rline_options twoway_options*]

where the time variable is assumed set by `tsset`, *varlist* has the interpretation $y_1 [y_2 \dots y_k]$.

Menu

Graphics > Twoway graph (scatter, line, etc.)

Description

`tsline` draws line plots for time-series data.

`tsrline` draws a range plot with lines for time-series data.

For complete documentation of `tsline` and `tsrline`; see [TS] **tsline**.

Also see

[G-2] **graph twoway** — Twoway graphs

[XT] **xtline** — Panel-data line plots

Title

[G-2] graph use — Display graph stored on disk

Syntax

`graph use` *filename* [, *options*]

options	Description
`nodraw`	do not draw the graph
`name(`*name* [`, replace`]`)`	specify new name for graph
`scheme(`*schemename*`)`	overall look
`play(`*recordingname*`)`	play edits from *recordingname*

Description

`graph use` displays (draws) the graph previously saved in a `.gph` file and, if the graph was stored in live format, loads it.

If *filename* is specified without an extension, `.gph` is assumed.

Options

`nodraw` specifies that the graph not be displayed. If the graph was stored in live format, it is still loaded; otherwise, `graph use` does nothing. See [G-3] ***nodraw_option***.

`name(`*name*[`, replace`]`)` specifies the name under which the graph is to be stored in memory, assuming that the graph was saved in live format. *filename* is the default name, where any path component in *filename* is excluded. For example,

```
. graph use mydir\mygraph.gph
```

will draw a graph with the name `mygraph`.

If the default name already exists `graph#` is used instead, where # is chosen to create a unique name.

If the graph is not stored in live format, the graph can only be displayed, not loaded, and the `name()` is irrelevant.

`scheme(`*schemename*`)` specifies the scheme controlling the overall look of the graph to be used; see [G-3] ***scheme_option***. If `scheme()` is not specified, the default is the *schemename* recorded in the graph being loaded.

`play(`*recordingname*`)` applies the edits from *recordingname* to the graph, where *recordingname* is the name under which edits previously made in the Graph Editor have been recorded and stored. See *Graph Recorder* in [G-1] **graph editor**.

Remarks

Graphs can be saved at the time you draw them either by specifying the `saving()` option or by subsequently using the `graph save` command; see [G-3] ***saving_option*** and [G-2] **graph save**. Modern graphs are saved in live format or as-is format; see [G-4] **concept: gph files**. Regardless of how the graph was saved or the format in which it was saved, `graph use` can redisplay the graph; simply type

```
. graph use filename
```

In a prior session, you drew a graph by typing

```
. twoway qfitci  mpg weight, stdf ||
         scatter mpg weight       ||
  , by(foreign, total row(1)) saving(cigraph)
```

The result of this was to create file `cigraph.gph`. At a later date, you can see the contents of the file by typing

```
. graph use cigraph
```

You might now edit the graph (see [G-1] **graph editor**), or print a copy of the graph.

Also see

[G-2] **graph combine** — Combine multiple graphs

[G-2] **graph save** — Save graph to disk

[G-3] ***saving_option*** — Option for saving graph to disk

[G-4] **concept: gph files** — Using gph files

[G-3] ***name_option*** — Option for naming graph in memory

Title

[G-2] palette — Display palettes of available selections

Syntax

`palette color` *colorstyle* [*colorstyle*] [, `scheme(`*schemename*`)` `cmyk`]

`palette linepalette` [, `scheme(`*schemename*`)`]

`palette symbolpalette` [, `scheme(`*schemename*`)`]

Description

`palette` produces graphs showing various selections available.

`palette color` shows how a particular color looks and allows you to compare two colors; see [G-4] ***colorstyle***.

`palette linepalette` shows you the different *linepatternstyles*; see [G-4] ***linepatternstyle***.

`palette symbolpalette` shows you the different *symbolstyles*; see [G-4] ***symbolstyle***.

Options

`scheme(`*schemename*`)` specifies the scheme to be used to draw the graph. With this command, `scheme()` is rarely specified. We recommend specifying `scheme(color)` if you plan to print the graph on a color printer; see [G-3] ***scheme_option***.

`cmyk` specifies that the color value be reported in CMYK rather than in RGB; see [G-4] ***colorstyle***.

Remarks

The `palette` command is more a part of the documentation of `graph` than a useful command in its own right.

Also see

[G-2] **graph** — The graph command

[G-2] **graph query** — List available schemes and styles

Title

[G-2] set graphics — Set whether graphs are displayed

Syntax

`query graphics`

`set graphics { on | off }`

Description

`query graphics` shows the graphics settings.

`set graphics` allows you to change whether graphs are displayed.

Remarks

If you type

```
. set graphics off
```

when you type a `graph` command, such as

```
. scatter yvar xvar, saving(mygraph)
```

the graph will be "drawn" and saved in file `mygraph.gph`, but it will not be displayed. If you type

```
. set graphics on
```

graphs will be displayed once again.

Drawing graphs without displaying them is sometimes useful in programming contexts, although in such contexts, it is better to specify the `nodraw` option; see [G-3] ***nodraw_option***. Typing

```
. scatter yvar xvar, saving(mygraph) nodraw
```

has the same effect as typing

```
. set graphics off
. scatter yvar xvar, saving(mygraph)
. set graphics on
```

The advantage of the former is not only does it require less typing, but if the user should press **Break**, `set graphics` will not be left `off`.

Also see

[G-3] ***nodraw_option*** — Option for suppressing display of graph

Title

[G-2] set printcolor — Set how colors are treated when graphs are printed

Syntax

query graphics

set printcolor { automatic | asis | gs1 | gs2 | gs3 } [, permanently]

set copycolor { automatic | asis | gs1 | gs2 | gs3 } [, permanently]

Description

query graphics shows the graphics settings.

set printcolor determines how colors are handled when graphs are printed.

set copycolor (Mac and Windows only) determines how colors are handled when graphs are copied to the clipboard.

Option

permanently specifies that, in addition to making the change right now, the setting be remembered and become the default setting when you invoke Stata.

Remarks

printcolor and copycolor can be set one of five ways: automatic, asis, and gs1, gs2, or gs3. Four of the settings—asis and gs1, gs2, and gs3—specify how colors should be rendered when graphs are printed or copied. The remaining setting—automatic—specifies that Stata determine by context whether asis or gs1 is used.

In the remarks below, copycolor can be used interchangeably with printcolor, the only difference being the ultimate destination of the graph.

Remarks are presented under the following headings:

What set printcolor affects
The problem set printcolor solves
set printcolor automatic
set printcolor asis
set printcolor gs1, gs2, and gs3
The scheme matters, not the background color you set

What set printcolor affects

set printcolor affects how graphs are printed when you select **File > Print graph** or when you use the graph print command; see [G-2] **graph print**.

set printcolor also affects the behavior of the graph export command when you use it to translate .gph files into another format, such as PostScript; see [G-2] **graph export**.

We will refer to all of the above in what follows as "printing graphs" or, equivalently, as "rendering graphs".

The problem set printcolor solves

If you should choose a scheme with a black background—see [G-4] **schemes intro**—and if you were then to print that graph, do you really want black ink poured onto the page so that what you get is exactly what you saw? Probably not. The purpose of `set printcolor` is to avoid such results.

set printcolor automatic

`set printcolor`'s default setting—`automatic`—looks at the graph to be printed and determines whether it should be rendered exactly as you see it on the screen or if instead the colors should be reversed and the graph printed in a monochrome gray scale.

`set printcolor automatic` bases its decision on the background color used by the scheme. If it is white (or light), the graph is printed `asis`. If it is black (or dark), the graph is printed `grayscale`.

set printcolor asis

If you specify `set printcolor asis`, all graphs will be rendered just as you see them on the screen, regardless of the background color of the scheme.

set printcolor gs1, gs2, and gs3

If you specify `set printcolor gs1`, `gs2`, or `gs3`, all graphs will be rendered according to a gray scale. If the scheme sets a black or dark background, the gray scale will be reversed (black becomes white and white becomes black).

`gs1`, `gs2`, and `gs3` vary how colors are mapped to grays. `gs1` bases its mapping on the average RGB value, `gs2` on "true grayscale", and `gs3` on the maximum RGB value. In theory, true grayscale should work best, but we have found that average generally works better with Stata graphs.

The scheme matters, not the background color you set

In all of the above, the background color you set using the *region_options* `graphregion(fcolor())` and `plotregion(fcolor())` plays no role in the decision that is made. Decisions are made based exclusively on whether the scheme naturally has a light or dark background. See [G-3] ***region_options***.

You may set background colors but remember to start with the appropriate scheme. Set light background colors with light-background schemes and dark background colors with dark-background schemes.

Also see

[G-2] **graph export** — Export current graph

[G-2] **graph print** — Print a graph

Title

[G-2] set scheme — Set default scheme

Syntax

query graphics

set scheme *schemename* [, permanently]

For a list of available *schemenames*, see [G-4] **schemes intro**.

Description

query graphics shows the graphics settings, which includes the graphics scheme.

set scheme allows you to set the graphics scheme to be used. The default setting is s2color.

Option

permanently specifies that in addition to making the change right now, the scheme setting be remembered and become the default setting when you invoke Stata.

Remarks

The graphics scheme specifies the overall look for the graph. You can specify the scheme to be used for an individual graph by specifying the scheme() option on the graph command, or you can specify the scheme once and for all by using set scheme.

See [G-4] **schemes intro** for a description of schemes and a list of available *schemenames*.

One of the available *schemenames* is economist, which roughly corresponds to the style used by *The Economist* magazine. If you wanted to make the economist scheme the default for the rest of this session, you could type

```
. set scheme economist
```

and if you wanted to make economist your default, even in subsequent sessions, you could type

```
. set scheme economist, permanently
```

Also see

[G-4] **schemes intro** — Introduction to schemes

[G-3] ***scheme_option*** — Option for specifying scheme

[G-3] Options

Title

[G-3] ***added_line_options*** — Options for adding lines to twoway graphs

Syntax

added_line_options	Description
`yline(`*linearg*`)`	add horizontal lines at specified y values
`xline(`*linearg*`)`	add vertical lines at specified x values
`tline(`*time_linearg*`)`	add vertical lines at specified t values

`yline()`, `xline()`, and `tline()` are *merged-implicit*; see [G-4] **concept: repeated options** and see *Interpretation of repeated options* below.

where *linearg* is

numlist [, *suboptions*]

For a description of *numlist*, see [U] **11.1.8 numlist**.

and where *time_linearg* is

datelist [, *suboptions*]

For a description of *datelist*, see [U] **11.1.9 datelist**.

suboptions	Description
`axis(`*#*`)`	which axis to use, $1 \leq \# \leq 9$
`style(`*addedlinestyle*`)`	overall style of added line
[`no`]`extend`	extend line through plot region's margins
`lstyle(`*linestyle*`)`	overall style of line
`lpattern(`*linepatternstyle*`)`	line patter (solid, dashed, etc.)
`lwidth(`*linewidthstyle*`)`	thickness of line
`lcolor(`*colorstyle*`)`	color of line

See [G-4] ***addedlinestyle***, [G-4] ***linestyle***, [G-4] ***linepatternstyle***, [G-4] ***linewidthstyle***, and [G-4] ***colorstyle***.

Description

`yline()`, `xline()`, and `tline()` are used with `twoway` to add lines to the plot region. `tline()` is an extension to `xline()`; see [TS] **tsline** for examples using `tline()`.

Options

`yline(`*linearg*`)`, `xline(`*linearg*`)`, and `tline(`*time_linearg*`)` specify the y, x, and t (time) values where lines should be added to the plot.

Suboptions

`axis(`*#*`)` is for use only when multiple y, x, or t axes are being used (see [G-3] ***axis_choice_options***). `axis()` specifies to which axis the `yline()`, `xline()`, or `tline()` is to be applied.

`style(`*addedlinestyle*`)` specifies the overall style of the added line, which includes [`no`]`extend` and `lstyle(`*linestyle*`)` documented below. See [G-4] ***addedlinestyle***. The [`no`]`extend` and `lstyle()` options allow you to change the added line's attributes individually, but `style()` is the starting point.

You need not specify `style()` just because there is something that you want to change, and in fact, most people seldom specify the `style()` option. You specify `style()` when another style exists that is exactly what you desire or when another style would allow you to specify fewer changes to obtain what you want.

`extend` and `noextend` specify whether the line should extend through the plot region's margin and touch the axis; see [G-3] ***region_options***. Usually `noextend` is the default, and `extend` is the option, but that is determined by the overall `style()` and, of course, the scheme; see [G-4] **schemes intro**.

`lstyle(`*linestyle*`)`, `lpattern(`*linepatternstyle*`)`, `lwidth(`*linewidthstyle*`)`, and `lcolor(`*colorstyle*`)` specify the look of the line; see [G-2] **graph twoway line**. `lstyle()` can be of particular use:

To create a line with the same look as the lines used to draw axes, specify `lstyle(foreground)`.

To create a line with the same look as the lines used to draw grid lines, specify `lstyle(grid)`.

Remarks

`yline()` and `xline()` add lines where specified. If, however, your interest is in obtaining grid lines, see the `grid` option in [G-3] ***axis_label_options***.

Remarks are presented under the following headings:

Typical use
Interpretation of repeated options

Typical use

`yline()` or `xline()` are typically used to add reference values:

```
. scatter yvar xvar, yline(10)
. scatter yvar year, xline(1944 1989)
```

To give the line in the first example the same look as used to draw an axis, we could specify

```
. scatter yvar xvar, yline(10, lstyle(foreground))
```

If we wanted to give the lines used in the second example the same look as used to draw grids, we could specify

```
. scatter yvar year, xline(1944 1989, lstyle(grid))
```

Interpretation of repeated options

Options `yline()` and `xline()` may be repeated, and each is executed separately. Thus different styles can be used for different lines on the same graph:

```
. scatter yvar year, xline(1944) xline(1989, lwidth(3))
```

Reference

Cox, N. J. 2009. Stata tip 82: Grounds for grids on graphs. *Stata Journal* 9: 648–651.

Also see

[G-4] ***addedlinestyle*** — Choices for overall look of added lines

[G-4] ***colorstyle*** — Choices for color

[G-4] ***linestyle*** — Choices for overall look of lines

[G-4] ***linepatternstyle*** — Choices for whether lines are solid, dashed, etc.

[G-4] ***linewidthstyle*** — Choices for thickness of lines

Title

[G-3] ***added_text_options*** — Options for adding text to twoway graphs

Syntax

added_text_options	Description
`text(`*text_arg*`)`	add text at specified y x
`ttext(`*text_arg*`)`	add text at specified y t

The above options are *merged-implicit*; see [G-4] **concept: repeated options**.

where *text_arg* is

loc_and_text [*loc_and_text* ...] [, *textoptions*]

and where *loc_and_text* is

$\#_y$ $\#_x$ "*text*" ["*text*" ...]

textoptions	Description
`yaxis(`#`)`	how to interpret $\#_y$
`xaxis(`#`)`	how to interpret $\#_x$
`placement(`*compassdirstyle*`)`	where to locate relative to $\#_y$ $\#_x$
textbox_options	look of text

See [G-4] ***compassdirstyle*** and [G-3] ***textbox_options***. `placement()` is also a textbox option, but ignore the description of `placement()` found there in favor of the one below.

Description

`text()` adds the specified text to the specified location in the plot region.

`ttext()` is an extension to `text()`, accepting a date in place of $\#_x$ when the time axis has a time format; see [U] **11.1.9 datelist**.

Options

`text(`*text_arg*`)` and `ttext(`*text_arg*`)` specify the location and text to be displayed.

Suboptions

`yaxis(`#`)` and `xaxis(`#`)` specify how $\#_y$ and $\#_x$ are to be interpreted when there are multiple y, x, or t axis scales; see [G-3] ***axis_choice_options***.

In the usual case, there is one y axis and one x axis, so options `yaxis()` and `xaxis()` are not specified. $\#_y$ is specified in units of the y scale and $\#_x$ in units of the x scale.

In the multiple-axis case, specify `yaxis(#)` and/or `xaxis(#)` to specify which units you wish to use. `yaxis(1)` and `xaxis(1)` are the defaults.

`placement(`*compassdirstyle*`)` specifies where the textbox is to be displayed relative to $\#_y$ $\#_x$. The default is usually `placement(center)`. The default is controlled both by the scheme and by the *textbox_option* `tstyle(`*textboxstyle*`)`; see [G-4] **schemes intro** and [G-3] ***textbox_options***. The available choices are

compassdirstyle	Location of text
`c`	centered on the point, vertically and horizontally
`n`	above the point, centered
`ne`	above and to the right of the point
`e`	right of the point, vertically centered
`se`	below and to the right of the point
`s`	below point, centered
`sw`	below and to the left of the point
`w`	left of the point, vertically centered
`nw`	above and to the left of the point

north	*northwest northeast*
west X *east*	X
south	*southwest southeast*

You can see [G-4] ***compassdirstyle***, but that will just give you synonyms for `c`, `n`, `ne`, ..., `nw`.

textbox_options specifies the look of the text; see [G-3] ***textbox_options***.

Remarks

Remarks are presented under the following headings:

Typical use
Advanced use
Use of the textbox option width( )

Typical use

`text()` is used for placing annotations on graphs. One example is the labeling of outliers. For instance, type

```
. use http://www.stata-press.com/data/r12/auto
(1978 Automobile Data)
. twoway qfitci mpg weight, stdf || scatter mpg weight
(graph omitted)
```

There are four outliers. First, we find the outliers by typing

```
. quietly regress mpg weight
. predict hat
. predict s, stdf
. generate upper = hat + 1.96*s
```

```
. list make mpg weight if mpg>upper
```

	make	mpg	weight
13.	Cad. Seville	21	4,290
42.	Plym. Arrow	28	3,260
57.	Datsun 210	35	2,020
66.	Subaru	35	2,050
71.	VW Diesel	41	2,040

Now we can remake the graph and label the outliers:

```
. twoway qfitci  mpg weight, stdf ||
        scatter mpg weight, ms(O)
                text(41 2040 "VW Diesel", place(e))
                text(28 3260 "Plymouth Arrow", place(e))
                text(35 2050 "Datsun 210 and Subaru", place(e))
```

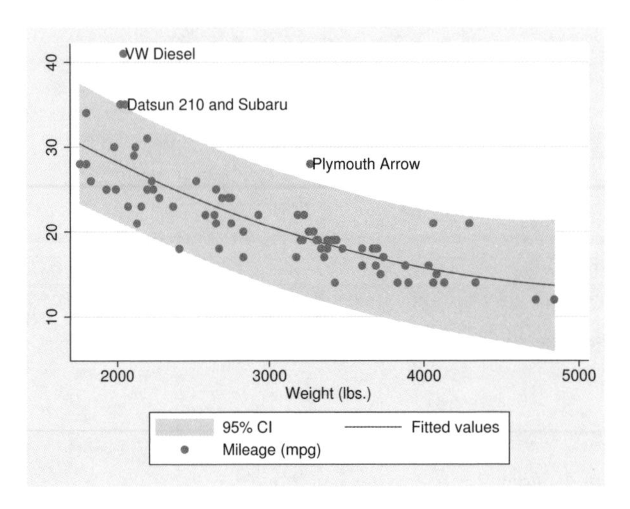

Advanced use

Another common use of *text* is to add an explanatory box of text inside the graph:

```
. use http://www.stata-press.com/data/r12/uslifeexp, clear
(U.S. life expectancy, 1900-1999)
. twoway line  le year ||
        fpfit le year ||
  , ytitle("Life Expectancy, years")
    xlabel(1900 1918 1940(20)2000)
    title("Life Expectancy at Birth")
    subtitle("U.S., 1900-1999")
    note("Source:  National Vital Statistics Report, Vol. 50 No. 6")
    legend(off)
    text( 48.5 1923
         "The 1918 Influenza Pandemic was the worst epidemic"
         "known in the U.S."
         "More citizens died than in all combat deaths of the"
         "20th century."
         , place(se) box just(left) margin(l+4 t+1 b+1) width(85) )
```

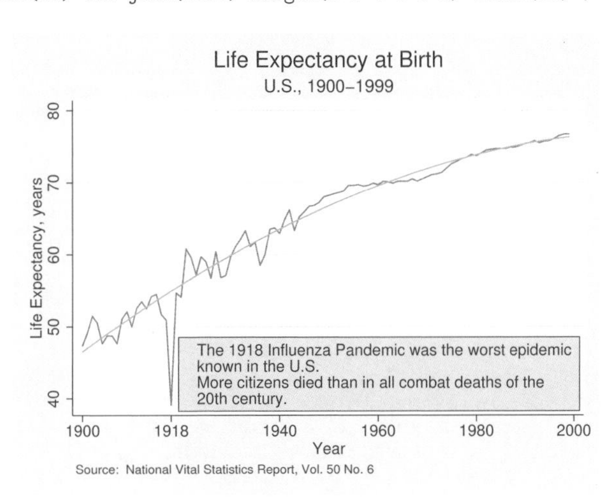

The only thing to note in the above command is the `text()` option:

```
text( 48.5 1923
         "The 1918 Influenza Pandemic was the worst epidemic"
         "known in the U.S."
         "More citizens died than in all combat deaths of the"
         "20th century."
         , place(se) box just(left) margin(l+4 t+1 b+1) width(85) )
```

and, in particular, we want to draw your eye to the location of the text and the suboptions:

```
text( 48.5 1923
         ...
         , place(se) box just(left) margin(l+4 t+1 b+1) width(85) )
```

We placed the text at $y = 48.5$, $x = 1923$, `place(se)`, meaning the box will be placed below and to the right of $y = 48.5$, $x = 1923$.

The other suboptions, `box just(left) margin(l+4 t+1 b+1) width(85)`, are *textbox_options*. We specified `box` to draw a border around the textbox, and we specified `just(left)`—an abbreviation for `justification(left)`—so that the text was left-justified inside the box. `margin(l+4 t+1 b+1)` made the text in the box look better. On the left we added 4%, and on the top and bottom we added 1%; see [G-3] ***textbox_options*** and [G-4] ***relativesize***. `width(85)` was specified to solve the problem described below.

Use of the textbox option width()

Let us look at the results of the above command, omitting the `width()` suboption. What you would see on your screen—or in a printout—might look virtually identical to the version we just drew, or it might look like this

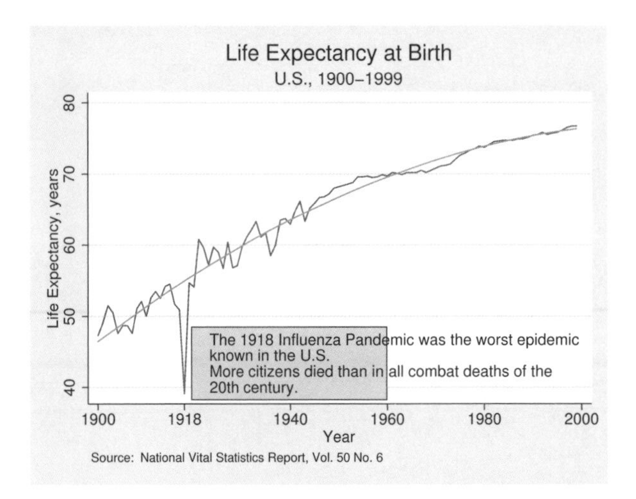

or like this:

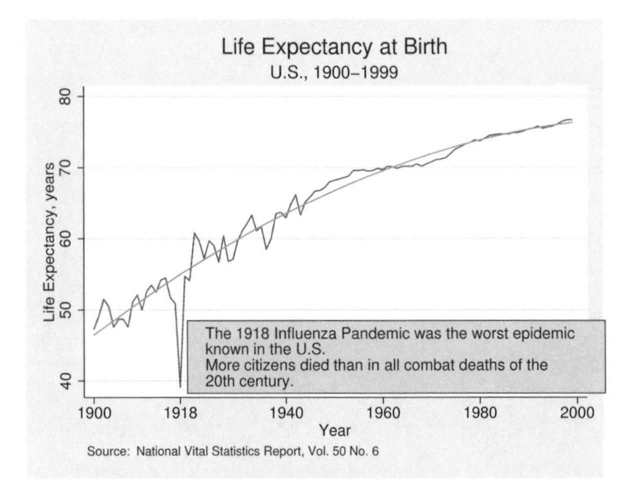

That is, Stata might make the textbox too narrow or too wide. In the above illustrations, we have exaggerated the extent of the problem, but it is common for the box to run a little narrow or a little wide. Moreover, with respect to this one problem, how the graph appears on your screen is no guarantee of how it will appear when printed.

This problem arises because Stata uses an approximation formula to determine the width of the text. This approximation is good for some fonts and poorer for others.

When the problem arises, use the *textbox_option* `width(`*relativesize*`)` to work around it. `width()` overrides Stata's calculation. In fact, we drew the two examples above by purposely misstating the `width()`. In the first case, we specified `width(40)`, and in the second, `width(95)`.

Getting the `width()` right is a matter of trial and error. The correct width will nearly always be between 0 and 100.

Corresponding to `width(`*relativesize*`)`, there is also the *textbox_option* `height(`*relativesize*`)`, but Stata never gets the height incorrect.

Also see

[G-4] ***compassdirstyle*** — Choices for location

[G-3] ***textbox_options*** — Options for textboxes and concept definition

[U] **11.1.9 datelist**

Title

[G-3] *addplot_option* — Option for adding additional twoway plots to command

Syntax

command ... [, ... addplot(*plot* ... [|| *plot* ... [...]] [, below]) ...]

where *plot* may be any subcommand of graph twoway (see [G-2] **graph twoway**), such as scatter, line, or histogram.

Description

Some commands that draw graphs (but do not start with the word graph) are documented in the other reference manuals. Many of those commands allow the addplot() option. This option allows them to overlay their results on top of graph twoway plots; see [G-2] **graph twoway**.

Option

addplot(*plots* [, below]) specifies the rest of the graph twoway subcommands to be added to the graph twoway command issued by *command*.

below is a suboption of the addplot() option and specifies that the added plots be drawn before the plots drawn by the command. Thus the added plots will appear below the plots drawn by *command*. The default is to draw the added plots after the command's plots so that they appear above the command's plots. below affects only the added plots that are drawn on the same x and y axes as the command's plots.

Remarks

Remarks are presented under the following headings:

Commands that allow the addplot( ) option
Advantage of graph twoway commands
Advantages of graphic commands implemented outside graph twoway
Use of the addplot( ) option

Commands that allow the addplot( ) option

graph commands never allow the addplot() option. The addplot() option is allowed by commands outside graph that are implemented in terms of graph twoway.

For instance, the histogram command—see [R] **histogram**—allows addplot(). graph twoway histogram—see [G-2] **graph twoway histogram**—does not.

Advantage of graph twoway commands

The advantage of `graph twoway` commands is that they can be overlaid, one on top of the other. For instance, you can type

```
. graph twoway scatter yvar xvar || lfit yvar xvar
```

and the separate graphs produced, `scatter` and `lfit`, are combined. The variables to which each refers need not even be the same:

```
. graph twoway scatter yvar xvar || lfit y2var x2var
```

Advantages of graphic commands implemented outside graph twoway

Graphic commands implemented outside `graph twoway` can have simpler syntax. For instance, the `histogram` command has an option, `normal`, that will overlay a normal curve on top of the histogram:

```
. histogram myvar, normal
```

That is easier than typing

```
. summarize myvar
. graph twoway histogram myvar ||
    function normalden(x,`r(mean)',`r(sd)'), range(myvar)
```

which is the `graph twoway` way of producing the same thing.

Thus the trade-off between `graph` and non`graph` commands is one of greater flexibility versus easier use.

Use of the addplot() option

The `addplot()` option attempts to give back flexibility to non`graph` graphic commands. Such commands are, in fact, implemented in terms of `graph twoway`. For instance, when you type

```
. histogram ...
```

or you type

```
. sts graph ...
```

the result is that those commands construct a complicated `graph twoway` command

```
→  graph twoway something_complicated
```

and then run that for you. When you specify the `addplot()` option, such as in

```
. histogram ..., addplot(your_contribution)
```

or

```
. sts graph, addplot(your_contribution)
```

the result is that the commands construct

```
→ graph twoway something_complicated || your_contribution
```

Let us assume that you have survival data and wish to visually compare the Kaplan–Meier (that is, the empirical survivor function) with the function that would be predicted if the survival times were assumed to be exponentially distributed. Simply typing

```
. use http://www.stata-press.com/data/r12/cancer, clear
(Patient Survival in Drug Trial)
. quietly stset studytime, fail(died) noshow
```

```
. sts graph
```

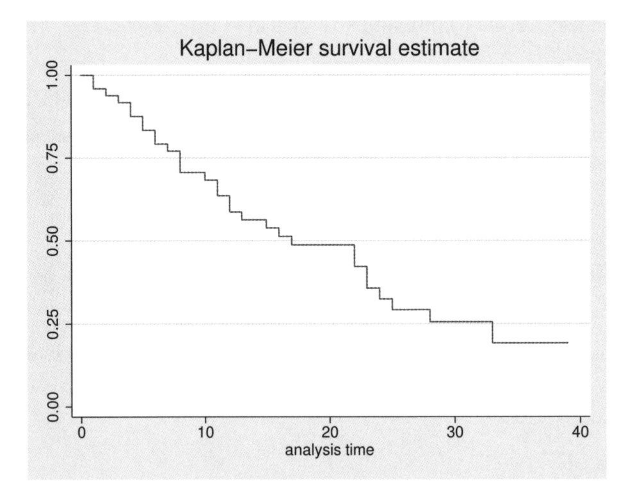

will obtain a graph of the empirical estimate. To obtain the exponential estimate, you might type

```
. quietly streg, distribution(exponential)
. predict S, surv
. graph twoway line S _t, sort
```

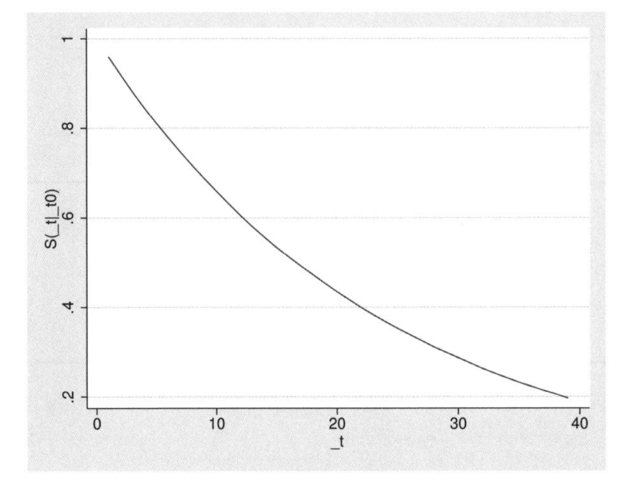

To put these two graphs together, you can type

```
. sts graph, addplot(line S _t, sort)
```

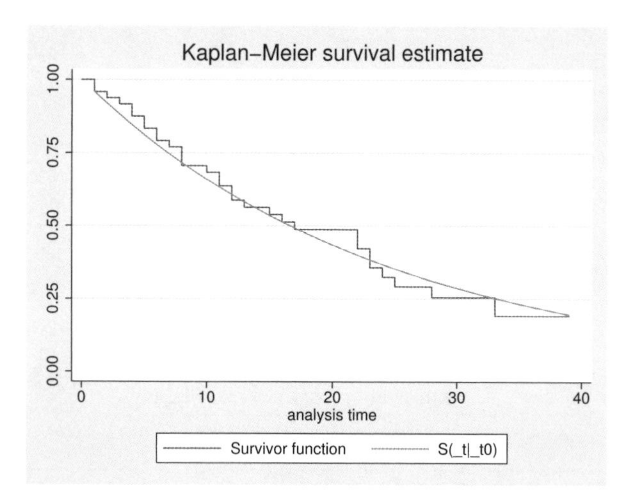

The result is just as if you typed

```
. sts graph || line S _t, sort
```

if only that were allowed.

Also see

[G-2] **graph twoway** — Twoway graphs

Title

[G-3] *advanced_options* — Rarely specified options for use with graph twoway

Syntax

title_options	Description
`pcycle(#)`	plots before pstyles recycle
`yvarlabel(quoted_strings)`	respecify y-variable labels
`xvarlabel(quoted_string)`	respecify x-variable label
`yvarformat(%fmt [...])`	respecify y-variable formats
`xvarformat(%fmt)`	respecify x-variable format
`yoverhangs`	adjust margins for y-axis labels
`xoverhangs`	adjust margins for x-axis labels
`recast(newplottype)`	treat plot as *newplottype*

The above options are *rightmost*; see [G-4] **concept: repeated options.**

where *quoted_string* is one quoted string and *quoted_strings* are one or more quoted strings, such as

"*plot 1 label*"

"*plot 1 label*" "*plot 2 label*"

newplottype	Description
`scatter`	treat as `graph twoway scatter`
`line`	treat as `graph twoway line`
`connected`	treat as `graph twoway connected`
`bar`	treat as `graph twoway bar`
`area`	treat as `graph twoway area`
`spike`	treat as `graph twoway spike`
`dropline`	treat as `graph twoway dropline`
`dot`	treat as `graph twoway dot`
`rarea`	treat as `graph twoway rarea`
`rbar`	treat as `graph twoway rbar`
`rspike`	treat as `graph twoway rspike`
`rcap`	treat as `graph twoway rcap`
`rcapsym`	treat as `graph twoway rcapsym`
`rline`	treat as `graph twoway rline`
`rconnected`	treat as `graph twoway rconnected`
`rscatter`	treat as `graph twoway rscatter`

pcspike	treat as graph twoway pcspike
pccapsym	treat as graph twoway pccapsym
pcarrow	treat as graph twoway pcarrow
pcbarrow	treat as graph twoway pcbarrow
pcscatter	treat as graph twoway pcscatter

newplottypes in each grouping (scatter through dot, rarea though rscatter, and pcspike through pcscatter) should be recast only among themselves.

Description

The *advanced_options* are not so much advanced as they are difficult to explain and are rarely used. They are also invaluable when you need them.

Options

pcycle(*#*) specifies how many plots are drawn before the pstyle (see [G-4] ***pstyle***) of the next plot begins again at p1, with the plot after the next plot using p2, and so on. The default # for most schemes is pcycle(15).

yvarlabel(*quoted_strings*) and xvarlabel(*quoted_string*) specify strings that are to be treated as if they were the variable labels of the first, second, ..., y variables and of the x variable.

yvarformat(*%fmt*) and xvarformat(*%fmt*) specify display formats that are to be treated as if they were the display formats of the first, second, ..., y variables and of the x variable.

yoverhangs and xoverhangs attempt to adjust the graph region margins to prevent long labels on the y or x axis from extending off the edges of the graph. Only the labels for the smallest and largest tick values on the axes are considered when making the adjustment. yoverhangs and xoverhangs are ignored if by() is specified; see [G-3] ***by_option***.

recast(*newplottype*) specifies the new plottype to which the original graph twoway *plottype* command is to be recast; see [G-2] **graph twoway** to see the available *plottypes*.

Remarks

Remarks are presented under the following headings:

Use of yvarlabel( ) and xvarlabel( )

When you type, for instance,

```
. scatter mpg weight
```

the axes are titled using the variable labels of mpg and weight or, if the variables have no variable labels, using the names of the variables themselves. Options yvarlabel() and xvarlabel() allow you to specify strings that will be used in preference to both the variable label and the name.

```
. scatter mpg weight, yvarl("Miles per gallon")
```

would label the y axis "Miles per gallon" (omitting the quotes), regardless of how variable mpg was labeled. Similarly,

```
. scatter mpg weight, xvarl("Weight in pounds")
```

would label the x axis "Weight in pounds", regardless of how variable weight was labeled.

Obviously, you could specify both options.

In neither case will the actual variable label be changed. Options yvarlabel() and xvarlabel() treat the specified strings as if they were the variable labels. yvarlabel() and xvarlabel() are literal in this treatment. If you specified xvarlabel(""), for instance, the variable label would be treated as if it were nonexistent, and thus the variable name would be used to title the x axis.

What makes these two options "advanced" is not only that they affect the way axes are titled but also that they substitute the specified strings for the variable labels wherever the variable label might be used. Variable labels are also used, for instance, in the construction of legends (see [G-3] ***legend_options***).

Use of yvarformat() and xvarformat()

Options yvarformat() and xvarformat() work much like yvarlabel() and xvarlabel(), except that, rather than overriding the variable labels, they override the variable formats. If you type

```
. scatter mpg weight, yvarformat(%9.2f)
```

the values on the y axis will be labeled 10.00, 20.00, 30.00, and 40.00 rather than 10, 20, 30, and 40.

Use of recast()

scatter, line, histogram, ... —the word that appears directly after graph twoway—is called a *plottype*. Plottypes come in two forms: *base plottypes* and *derived plottypes*.

Base plottypes plot the data as given according to some style. scatter and line are examples of base plottypes.

Derived plottypes do not plot the data as given but instead derive something from the data and then plot that according to one of the base plottypes. histogram is an example of a derived plottype. It derives from the data the values for the frequencies at certain x ranges, and then it plots that derived data using the base plottype graph twoway bar. lfit is another example of a derived plottype. It takes the data, fits a linear regression, and then passes that result along to graph twoway line.

recast() is useful when using derived plottypes. It specifies that the data are to be derived just as they would be ordinarily, but rather than passing the derived data to the default base plottype for plotting, they are passed to the specified base plottype.

For instance, if we typed

```
. twoway lfit mpg weight, pred(resid)
```

we would obtain a graph of the residuals as a line plot because the lfit plottype produces line plots. If we typed

```
. twoway lfit mpg weight, pred(resid) recast(scatter)
```

we would obtain a scatterplot of the residuals. graph twoway lfit would use graph twoway scatter rather than graph twoway line to plot the data it derives.

recast(*newplottype*) may be used with both derived and base plottypes, although it is most useful when combined with derived plots.

❑ Technical note

The syntax diagram shown for `scatter` in [G-2] **graph twoway scatter**, although extensive, is incomplete, and so are all the other plottype syntax diagrams shown in this manual.

Consider what would happen if you specified

```
. scatter ... , ... recast(bar)
```

You would be specifying that `scatter` be treated as a `bar`. Results would be the same as if you typed

```
. twoway bar ... , ...
```

but let's ignore that and pretend that you typed the `recast()` version. What if you wanted to specify the look of the bars? You could type

```
. scatter ... , ... bar_options recast(bar)
```

That is, `scatter` allows `graph twoway bar`'s options, even though they do not appear in `scatter`'s syntax diagram. Similarly, `graph twoway bar` allows all of `scatter`'s options; you might type

```
. twoway bar ..., ... scatter_options recast(scatter)
```

The same is true for all other pairs of base plottypes, with the result that all base plottypes allow all base plottype options. The emphasis here is on base: the derived plottypes do not allow this sharing.

If you use a base plottype without `recast()` and if you specify irrelevant options from other base types, that is not an error, but the irrelevant options are ignored. In the syntax diagrams for the base plottypes, we have listed only the options that matter under the assumption that you do not specify `recast`.

❑

Also see

[G-2] **graph twoway** — Twoway graphs

Title

[G-3] *area_options* — Options for specifying the look of special areas

Syntax

area_options	Description
`color(`*colorstyle*`)`	outline and fill color
`fcolor(`*colorstyle*`)`	fill color
`fintensity(`*intensitystyle*`)`	fill intensity
`lcolor(`*colorstyle*`)`	outline color
`lwidth(`*linewidthstyle*`)`	thickness of outline
`lpattern(`*linepatternstyle*`)`	outline pattern (solid, dashed, etc.)
`lstyle(`*linestyle*`)`	overall look of outline
`astyle(`*areastyle*`)`	overall look of area, all settings above
`pstyle(`*pstyle*`)`	overall plot style, including areastyle
`recast(`*newplottype*`)`	advanced; treat plot as *newplottype*

See [G-4] ***colorstyle***, [G-4] ***intensitystyle***, [G-4] ***linewidthstyle***, [G-4] ***linepatternstyle***, [G-4] ***linestyle***, [G-4] ***areastyle***, [G-4] ***pstyle***, and [G-3] ***advanced_options***.

All options are *merged-implicit*; see [G-4] **concept: repeated options**.

Description

The *area_options* determine the look of, for instance, the areas created by `twoway area` (see [G-2] **graph twoway area**) or the "rectangles" used by `graph dot` (see [G-2] **graph twoway dot**). The *area_options* and the *barlook_options* (see [G-3] ***barlook_options***) are synonymous when used on `graph twoway` (see [G-2] **graph twoway**) and may be used interchangeably.

Options

`color(`*colorstyle*`)` specifies one color to be used both to outline the shape of the area and to fill its interior. See [G-4] ***colorstyle*** for a list of color choices.

`fcolor(`*colorstyle*`)` specifies the color to be used to fill the interior of the area. See [G-4] ***colorstyle*** for a list of color choices.

`fintensity(`*intensitystyle*`)` specifies the intensity of the color used to fill the interior of the area. See [G-4] ***intensitystyle*** for a list of intensity choices.

`lcolor(`*colorstyle*`)` specifies the color to be used to outline the area. See [G-4] ***colorstyle*** for a list of color choices.

`lwidth(`*linewidthstyle*`)` specifies the thickness of the line to be used to outline the area. See [G-4] ***linewidthstyle*** for a list of choices.

`lpattern(`*linepatternstyle*`)` specifies whether the line used to outline the area is solid, dashed, etc. See [G-4] ***linepatternstyle*** for a list of pattern choices.

`lstyle(`*linestyle*`)` specifies the overall style of the line used to outline the area, including its pattern (solid, dashed, etc.), thickness, and color. The three options listed above allow you to change the line's attributes, but `lstyle()` is the starting point. See [G-4] ***linestyle*** for a list of choices.

`astyle(`*areastyle*`)` specifies the overall look of the area. The options listed above allow you to change each attribute, but `style()` provides a starting point.

You need not specify `style()` just because there is something you want to change. You specify `style()` when another style exists that is exactly what you desire or when another style would allow you to specify fewer changes to obtain what you want.

See [G-4] ***areastyle*** for a list of available area styles.

`pstyle(`*pstyle*`)` specifies the overall style of the plot, including not only the *areastyle*, but also all other settings for the look of the plot. Only the *areastyle* affects the look of areas. See [G-4] ***pstyle*** for a list of available plot styles.

`recast(`*newplottype*`)` is an advanced option allowing the plot to be recast from one type to another, for example, from an area plot to a line plot; see [G-3] ***advanced_options***. Most, but not all, plots allow `recast()`.

Remarks

Remarks are presented under the following headings:

Use with twoway
Use with graph dot

Use with twoway

area_options are allowed as options with any `graph twoway` plottype that creates shaded areas, for example, `graph twoway area` and `graph twoway rarea`, as in

```
. graph twoway area yvar xvar, color(blue)
```

The above would set the area enclosed by *yvar* and the x axis to be blue; see [G-2] **graph twoway area** and [G-2] **graph twoway rarea**.

The `lcolor()`, `lwidth()`, `lpattern()`, and `lstyle()` options are also used to specify how plotted lines and spikes look for all of `graph twoway`'s range plots, paired-coordinate plots, and for area plots, bar plots, spike plots, and dropline plots. For example,

```
. graph twoway rspike y1var y2var xvar, lcolor(red)
```

will set the color of the horizontal spikes between values of *y1var* and *y2var* to red.

Use with graph dot

If you specify `graph dot`'s `linetype(rectangle)` option, the dot chart will be drawn with rectangles substituted for the dots. Then the *area_options* determine the look of the rectangle. The *area_options* are specified inside `graph dot`'s `rectangles()` option:

```
. graph dot ..., ... linetype(rectangle) rectangles(area_options) ...
```

If, for instance, you wanted to make the rectangles green, you could specify

```
. graph dot ..., ... linetype(rectangle) rectangles(fcolor(green)) ...
```

See [G-2] **graph dot**.

Also see

[G-2] **graph dot** — Dot charts (summary statistics)

Title

[G-3] *aspect_option* — Option for controlling the aspect ratio of the plot region

Syntax

aspect_option	Description
`aspectratio(#` [, *pos_option*] `)`	set plot region aspect ratio to #

pos_option	Description
`placement(`*compassdirstyle*`)`	placement of plot region

See [G-4] ***compassdirstyle***.

Description

The `aspectratio()` option controls the relationship between the height and width of a graph's plot region. For example, when # = 1, the height and width will be equal (their ratio is 1), and the plot region will be square.

Option

`aspectratio(#` [, *pos_option*] `)` specifies the aspect ratio and, optionally, the placement of the plot region.

Suboption

`placement(`*compassdirstyle*`)` specifies where the plot region is to be placed to take up the area left over by restricting the aspect ratio. See [G-4] ***compassdirstyle***.

Remarks

The `aspectratio(#)` option constrains the ratio of the plot region to #. So, if # is 1, the plot region is square; if it is 2, the plot region is twice as tall as it is wide; and, if it is .25, the plot region is one-fourth as tall as it is wide. The most common use is `aspectratio(1)`, which produces a square plot region.

The overall size of the graph is not changed by the `aspectratio()` option. Thus constraining the aspect ratio will generally leave some additional space around the plot region in either the horizontal or vertical dimension. By default, the plot region will be centered in this space, but you can use the `placement()` option to control where the plot region is located. `placement(right)` will place the plot region all the way to the right in the extra space, leaving all the blank space to the left; `placement(top)` will place the plot region at the top of the extra space, leaving all the blank space at the bottom; `placement(left)` and `placement(right)` work similarly.

Specifying an aspect ratio larger than the default for a graph causes the width of the plot region to become narrower. Conversely, specifying a small aspect ratio causes the plot region to become shorter. Because titles and legends can be wider than the plot region, and because most schemes do not allow titles and legends to span beyond the width of the plot region, this can sometimes lead to surprising spacing of some graph elements; for example, axes may be forced away from their plot region. If this occurs, the spacing can be improved by adding the `span` suboption to the `title()`, `subtitle()`, `legend()`, or other options. The `span` option must be added to each element that is wider than the plot region. See *Spanning* in [G-3] ***title_options*** for a diagram.

Reference

Cox, N. J. 2004. Stata tip 12: Tuning the plot region aspect ratio. *Stata Journal* 4: 357–358.

Also see

[G-2] **graph twoway** — Twoway graphs

[G-2] **graph bar** — Bar charts

[G-2] **graph box** — Box plots

[G-2] **graph dot** — Dot charts (summary statistics)

Title

[G-3] *axis_choice_options* — Options for specifying the axes on which a plot appears

Syntax

axis_choice_options	Description
yaxis(# [# ...])	which y axis to use, $1 \le \# \le 9$
xaxis(# [# ...])	which x axis to use, $1 \le \# \le 9$

yaxis() and xaxis() are *unique*; see [G-4] **concept: repeated options**.

These options are allowed with any of the *plottypes* (scatter, line, etc.) allowed by graph twoway; see [G-2] **graph twoway**.

Description

The *axis_choice_options* determine the y and x axis (or axes) on which the plot is to appear.

Options

yaxis(# [# ...]) and xaxis(# [# ...]) specify the y or x axis to be used. The default is yaxis(1) and xaxis(1).

Typically, yaxis() and xaxis() are treated as if their syntax is yaxis(#) and xaxis(#)—that is, just one number is specified. In fact, however, more than one number may be specified, and specifying a second is sometimes useful with yaxis(). The first y axis appears on the left, and the second (if there is a second) appears on the right. Specifying yaxis(1 2) allows you to force there to be two identical y axes. You could use the one on the left in the usual way and the one on the right to label special values.

Remarks

Options yaxis() and xaxis() are used when you wish to create one graph with multiple axes. These options are specified with twoway's scatter, line, etc., to specify which axis is to be used for each individual plot.

Remarks are presented under the following headings:

Usual case: one set of axes

Normally, when you construct a `twoway` graph with more than one plot, as in

```
. scatter y1 y2 x
```

or equivalently,

```
. twoway (scatter y1 x) (scatter y2 x)
```

the two plots share common axes for y and for x.

Special case: multiple axes due to multiple scales

Sometimes you want the two y plots graphed on separate scales. Then you type

```
. twoway (scatter gnp year, c(l) yaxis(1))
         (scatter r   year, c(l) yaxis(2))
```

`yaxis(1)` specified on the first `scatter` says, "This scatter is to appear on the first y axis." `yaxis(2)` specified on the second `scatter` says, "This scatter is to appear on the second y axis."

The result is that two y axes will be constructed. The one on the left will correspond to `gnp` and the one on the right to `r`. If we had two x axes instead, one would appear on the bottom and one on the top:

```
. twoway (scatter year gnp, c(l) xaxis(1))
         (scatter year r,   c(l) xaxis(2))
```

You are not limited to having just two y axes or two x axes. You could have two of each:

```
. twoway (scatter y1var x1var, c(l) yaxis(1) xaxis(1))
         (scatter y2var x2var, c(l) yaxis(2) xaxis(2))
```

You may have up to nine y axes and nine x axes, although graphs become pretty well unreadable by that point. When there are three or more y axes (or x axes), the axes are stacked up on the left (on the bottom). In any case, you specify `yaxis(#)` and `xaxis(#)` to specify which axis applies to which plot.

Also, you may reuse axes:

```
. twoway (scatter gnp year, c(l) yaxis(1))
         (scatter nnp year, c(l) yaxis(1))
         (scatter r   year, c(l) yaxis(2))
         (scatter r2  year, c(l) yaxis(2))
```

The above graph has two y axes, one on the left and one on the right. The left axis is used for `gnp` and `nnp`; the right axis is used for `r` and `r2`.

The order in which we type the plots is not significant; the following would result in the same graph,

```
. twoway (scatter gnp year, c(l) yaxis(1))
         (scatter r   year, c(l) yaxis(2))
         (scatter nnp year, c(l) yaxis(1))
         (scatter r2  year, c(l) yaxis(2))
```

except that the symbols, colors, and *linestyles* associated with each plot would change.

yaxis(1) and xaxis(1) are the defaults

In the first multiple-axis example,

```
. twoway (scatter gnp year, c(l) yaxis(1))
         (scatter r   year, c(l) yaxis(2))
```

`xaxis(1)` is assumed because we did not specify otherwise. The command is interpreted as if we had typed

```
. twoway (scatter gnp year, c(l) yaxis(1) xaxis(1))
         (scatter r   year, c(l) yaxis(2) xaxis(1))
```

Because `yaxis(1)` is the default, you need not bother to type it. Similarly, because `xaxis(1)` is the default, you could omit typing it, too:

```
. twoway (scatter gnp year, c(l))
         (scatter r   year, c(l) yaxis(2))
```

Notation style is irrelevant

Whether you use the `()`-binding notation or the `||`-separator notation never matters. You could just as well type

```
. scatter gnp year, c(l) || scatter r year, c(l) yaxis(2)
```

yaxis( ) and xaxis( ) are plot options

Unlike all the other axis options, `yaxis()` and `xaxis()` are options of the individual plots and not of `twoway` itself. You may not type

```
. scatter gnp year, c(l) || scatter r year, c(l) ||, yaxis(2)
```

because `twoway` would have no way of knowing whether you wanted `yaxis(2)` to apply to the first or to the second `scatter`. Although it is true that how the axes appear is a property of `twoway`—see [G-3] ***axis_options***—which axes are used for which plots is a property of the plots themselves.

For instance, options `ylabel()` and `xlabel()` are options that specify the major ticking and labeling of an axis (see [G-3] ***axis_label_options***). If you want the x axis to have 10 ticks with labels, you can type

```
. scatter gnp year, c(l) ||
  scatter r   year, c(l) yaxis(2) ||, xlabel(#10)
```

and indeed you are "supposed" to type it that way to illustrate your deep understanding that `xlabel()` is a `twoway` option. Nonetheless, you may type

```
. scatter gnp year, c(l) ||
  scatter r   year, c(l) yaxis(2) xlabel(#10)
```

or

```
. scatter gnp year, c(l) xlabel(#10) ||
  scatter r   year, c(l) yaxis(2)
```

because `twoway` can reach inside the individual plots and pull out options intended for it. What `twoway` cannot do is redistribute options specified explicitly as `twoway` back to the individual plots.

Specifying the other axes options with multiple axes

Continuing with our example,

```
. scatter gnp year, c(l) ||
  scatter r   year, c(l) yaxis(2) ||
  , xlabel(#10)
```

say that you also wanted 10 ticks with labels on the first y axis and eight on the second. You type

```
. scatter gnp year, c(l) ||
  scatter r   year, c(l) yaxis(2) ||
  , xlabel(#10)  ylabel(#10, axis(1))  xlabel(#8, axis(2))
```

Each of the other axis options (see [G-3] ***axis_options***) has an axis(#) option that specifies to which axis the option applies. When you do not specify that suboption, axis(1) is assumed.

As always, even though the other axis options are options of twoway, you can let them run together with the options of individual plots:

```
. scatter gnp year, c(l) ||
  scatter r   year, c(l) yaxis(2) xlabel(#10) ylabel(#10, axis(1))
                    ylabel(#8, axis(2))
```

Each plot may have at most one x scale and one y scale

Each scatter, line, connected, etc.—that is, each plot—may have only one y scale and one x scale, so you may not type the shorthand

```
. scatter gnp r year, c(l l) yaxis(1 2)
```

to put gnp on one axis and r on another. In fact, yaxis(1 2) is not an error—we will get to that in the next section—but it will not put gnp on one axis and r on another. To do that, you must type

```
. twoway (scatter gnp year, c(l) yaxis(1))
         (scatter r   year, c(l) yaxis(2))
```

which, of course, you may type as

```
. scatter gnp year, c(l) yaxis(1) || scatter r year, c(l) yaxis(2)
```

The overall graph may have multiple scales, but the individual plots that appear in it may not.

Special case: Multiple axes with a shared scale

It is sometimes useful to have multiple axes just so that you have extra places to label special values. Consider graphing blood pressure versus concentration of some drug:

```
. scatter bp concentration
```

Perhaps you would like to add a line at bp = 120 and label that value specially. One thing you might do is

```
. scatter bp concentration, yaxis(1 2) ylabel(120, axis(2))
```

The ylabel(120, axis(2)) part is explained in [G-3] ***axis_label_options***; it caused the second axis to have the value 120 labeled. The option yaxis(1 2) caused there to be a second axis, which you could label. When you specify yaxis() (or xaxis()) with more than one number, you are specifying that the axes be created sharing the same scale.

To better understand what `yaxis(1 2)` does, compare the results of

```
. scatter bp concentration
```

with

```
. scatter bp concentration, yaxis(1 2)
```

In the first graph, there is one y axis on the left. In the second graph, there are two y axes, one on the left and one on the right, and they are labeled identically.

Now compare

```
. scatter bp concentration
```

with

```
. scatter bp concentration, xaxis(1 2)
```

In the first graph, there is one x axis on the bottom. In the second graph, there are two x axes, one on the bottom and one on the top, and they are labeled identically.

Finally, try

```
. scatter bp concentration, yaxis(1 2) xaxis(1 2)
```

In this graph, there are two y axes and two x axes: left and right, and top and bottom.

Reference

Wiggins, V. 2010. Stata tip 93: Handling multiple y axes on twoway graphs. *Stata Journal* 10: 689–690.

Also see

[G-3] ***axis_label_options*** — Options for specifying axis labels

[G-3] ***axis_options*** — Options for specifying numeric axes

[G-3] ***axis_scale_options*** — Options for specifying axis scale, range, and look

[G-3] ***axis_title_options*** — Options for specifying axis titles

Title

[G-3] ***axis_label_options*** — Options for specifying axis labels

Syntax

axis_label_options are a subset of *axis_options*; see [G-3] ***axis_options***. *axis_label_options* control the placement and the look of ticks and labels on an axis.

axis_label_options	Description
{ y | x | t | z }label(*rule_or_values*)	major ticks plus labels
{ y | x | t | z }tick(*rule_or_values*)	major ticks only
{ y | x | t | z }mlabel(*rule_or_values*)	minor ticks plus labels
{ y | x | t | z }mtick(*rule_or_values*)	minor ticks only

The above options are *merged-explicit*; see [G-4] **concept: repeated options**.

where *rule_or_values* is defined as

[*rule*] [*numlist* ["*label*" [*numlist* ["*label*" [...]]]]] [, *suboptions*]

Either *rule* or *numlist* must be specified, and both may be specified.

rule	Example	Description
`##`	`#6`	approximately 6 nice values
`###`	`##10`	10 − 1 = 9 values between major ticks; allowed with `mlabel()` and `mtick()` only
`#(#)#`	`-4(.5)3`	specified range: −4 to 3 in steps of .5
`minmax`	`minmax`	minimum and maximum values
`none`	`none`	label no values
`.`	`.`	skip the rule

where *numlist* is as described in [U] **11.1.8 numlist**.

`tlabel()`, `ttick()`, `tmlabel()`, and `tmtick()` also accept a *datelist* and an extra type of *rule*

rule	Example	Description
date(#)*date*	`1999m1(1)1999m12`	specified date range: each month assuming the axis has the `%tm` format

where *date* and *datelist* may contain dates, provided that the *t* (time) axis has a date format; see [U] **11.1.9 datelist**.

suboptions	Description
axis(#)	which axis, $1 \leq \# \leq 9$
add	combine options
[no]ticks	suppress ticks
[no]labels	suppress labels
valuelabel	label values using first variable's value label
format(%*fmt*)	format values per %*fmt*
angle(*anglestyle*)	angle the labels
alternate	offset adjacent labels
norescale	do not rescale the axis
tstyle(*tickstyle*)	labels and ticks: overall style
labgap(*relativesize*)	labels: margin between tick and label
labstyle(*textstyle*)	labels: overall style
labsize(*textsizestyle*)	labels: size of text
labcolor(*colorstyle*)	labels: color of text
tlength(*relativesize*)	ticks: length
tposition(outside | crossing | inside)	ticks: position/direction
tlstyle(*linestyle*)	ticks: linestyle of
tlwidth(*linewidthstyle*)	ticks: thickness of line
tlcolor(*colorstyle*)	ticks: color of line
custom	tick- and label-rendition options apply only to these labels
[no]grid	grid: include
[no]gmin	grid: grid line at minimum
[no]gmax	grid: grid line at maximum
gstyle(*gridstyle*)	grid: overall style
[no]gextend	grid: extend into plot region margin
glstyle(*linestyle*)	grid: linestyle of
glwidth(*linewidthstyle*)	grid: thickness of line
glcolor(*colorstyle*)	grid: color of line
glpattern(*linepatternstyle*)	grid: line pattern of line

See [G-4] ***anglestyle***, [G-4] ***tickstyle***, [G-4] ***relativesize***, [G-4] ***textstyle***, [G-4] ***textsizestyle***, [G-4] ***colorstyle***, [G-4] ***linestyle***, [G-4] ***linewidthstyle***, [G-4] ***linepatternstyle***, and [G-4] ***gridstyle***.

Description

axis_label_options control the placement and the look of ticks and labels on an axis.

Options

`ylabel(`*rule_or_values*`)`, `xlabel(`*rule_or_values*`)`, `tlabel(`*rule_or_values*`)`, and
`zlabel(`*rule_or_values*`)` specify the major values to be labeled and ticked along the axis. For instance, to label the values 0, 5, 10, ..., 25 along the x axis, specify `xlabel(0(5)25)`. If the t axis has the `%tm` format, `tlabel(1999m1(1)1999m12)` will label all the months in 1999.

`ytick(`*rule_or_values*`)`, `xtick(`*rule_or_values*`)`, `ttick(`*rule_or_values*`)`, and
`ztick(`*rule_or_values*`)` specify the major values to be ticked but not labeled along the axis. For instance, to tick the values 0, 5, 10, ..., 25 along the x axis, specify `xtick(0(5)25)`. Specify `ttick(1999m1(1)1999m12)` to place ticks for each month in the year 1999.

`ymlabel(`*rule_or_values*`)`, `xmlabel(`*rule_or_values*`)`, `tmlabel(`*rule_or_values*`)`, and
`zmlabel(`*rule_or_values*`)` specify minor values to be labeled and ticked along the axis.

`ymtick(`*rule_or_values*`)`, `xmtick(`*rule_or_values*`)`, `tmtick(`*rule_or_values*`)`, and
`zmtick(`*rule_or_values*`)` specify minor values to be ticked along the axis.

`zlabel(`*rule_or_values*`)`, `ztick(`*rule_or_values*`)`, `zmlabel(`*rule_or_values*`)`, and
`zmtick(`*rule_or_values*`)`; see *Contour axes—zlabel(), etc.* below.

Suboptions

`axis(`*#*`)` specifies to which scale this axis belongs and is specified when dealing with multiple x (t) or y axes; see [G-3] ***axis_choice_options***.

`add` specifies what is to be added to any `xlabel()`, `ylabel()`, `xtick()`, ..., or `ymtick()` option previously specified. Labels or ticks are added to any default labels or ticks or to any labels or ticks specified in previous `xlabel()`, `ylabel()`, `xtick()`, ..., or `ymtick()` options. Only value specifications are added; rule specifications always replace any existing rule. See *Interpretation of repeated options* below.

`noticks` and `ticks` suppress/force the drawing of ticks. `ticks` is the usual default, so `noticks` makes { `y` | `x` }`label()` and { `y` | `x` }`mlabel()` display the labels only.

`nolabels` and `labels` suppress/force the display of the labels. `labels` is the usual default, so `nolabels` turns { `y` | `x` }`label()` into { `y` | `x` }`tick()` and { `y` | `x` }`mlabel()` into { `y` | `x` }`mtick()`. Why anyone would want to do this is difficult to imagine.

`valuelabel` specifies that values should be mapped through the first y variable's value label (`y*()` options) or the x variable's value label (`x*()` options). Consider the command `scatter yvar xvar` and assume that `xvar` has been previously given a value label:

```
. label define cat 1 "Low" 2 "Med" 3 "Hi"
. label values xvar cat
```

Then

```
. scatter yvar xvar, xlabel(1 2 3, valuelabel)
```

would, rather than putting the numbers 1, 2, and 3, put the words Low, Med, and Hi on the x axis. It would have the same effect as

```
. scatter yvar xvar, xlabel(1 "Low" 2 "Med" 3 "Hi")
```

`format(%`*fmt*`)` specifies how numeric values on the axes should be formatted. The default `format()` is obtained from the variables specified with the `graph` command, which for `ylabel()`, `ytick()`, `ymlabel()`, and `ymtick()` usually means the first y variable, and for `xlabel()`, ..., `xmtick()`, means the x variable. For instance, in

```
. scatter y1var y2var xvar
```

the default format for the y axis would be `y1var`'s format, and the default for the x axis would be `xvar`'s format.

You may specify the `format()` suboption (or any suboption) without specifying values if you want the default labeling presented differently. For instance,

```
. scatter y1var y2var xvar, ylabel(,format(%9.2fc))
```

would present default labeling of the y axis, but the numbers would be formatted with the `%9.2fc` format. Note carefully the comma in front of `format`. Inside the `ylabel()` option, we are specifying suboptions only.

`angle(`*anglestyle*`)` causes the labels to be presented at an angle. See [G-4] ***anglestyle***.

`alternate` causes adjacent labels to be offset from one another and is useful when many values are being labeled. For instance, rather than obtaining

```
1.0   1.1   1.2   1.3   1.4   1.5   1.6
```

with `alternate`, you would obtain

```
1.0         1.2         1.4         1.6
      1.1         1.3         1.5
```

`norescale` specifies that the ticks or labels in the option be placed directly on the graph without rescaling the axis or associated plot region for the new values. By default, label options automatically rescale the axis and plot region to include the range of values in the new labels or ticks. `norescale` allows you to plot ticks or labels outside the normal bounds of an axis.

`tstyle(`*tickstyle*`)` specifies the overall look of ticks and labels; see [G-4] ***tickstyle***. The options documented below will allow you to change each attribute of a tick and its label, but the *tickstyle* specifies the starting point.

You need not specify `tstyle()` just because there is something you want to change about the look of ticks. You specify `tstyle()` when another style exists that is exactly what you desire or when another style would allow you to specify fewer changes to obtain what you want.

`labgap(`*relativesize*`)`, `labstyle(`*textstyle*`)`, `labsize(`*textsizestyle*`)`, and `labcolor(`*colorstyle*`)` specify details about how the labels are presented. See [G-4] ***relativesize***, [G-4] ***textstyle***, [G-4] ***textsizestyle***, and [G-4] ***colorstyle***.

`tlength(`*relativesize*`)` specifies the overall length of the ticks; see [G-4] ***relativesize***.

`tposition(outside | crossing | inside)` specifies whether the ticks are to extend `outside` (from the axis out, the usual default), `crossing` (crossing the axis line, extending in and out), or `inside` (from the axis into the plot region).

`tlstyle(`*linestyle*`)`, `tlwidth(`*linewidthstyle*`)`, and `tlcolor(`*colorstyle*`)` specify other details about the look of the ticks. See [G-4] ***linestyle***, [G-4] ***linewidthstyle***, and [G-4] ***colorstyle***. Ticks are just lines. See [G-4] **concept: lines** for more information.

`custom` specifies that the label-rendition suboptions, the tick-rendition options, and the `angle()` option apply only to the labels added on the current { `y` | `x` | `t` } [`m`] `label()` option, rather than being applied to all major or minor labels on the axis. Customizable suboptions are `tstyle()`, `labgap()`, `labstyle()`, `labsize()`, `labcolor()`, `tlength()`, `tposition()`, `tlstyle()`, `tlwidth()`, and `tlcolor()`.

`custom` is usually combined with suboption `add` to emphasize points on the axis by extending the length of the tick, changing the color or size of the label, or otherwise changing the look of the custom labels or ticks.

`grid` and `nogrid` specify whether grid lines are to be drawn across the plot region in addition to whatever else is specified in the {`y`|`x`}[`m`]`label()` or {`y`|`x`}[`m`]`tick()` option in which `grid` or `nogrid` appears. Typically, `nogrid` is the default, and `grid` is the option for all except `ylabel()`, where things are reversed and `grid` is the default and `nogrid` is the option. (Which is the default and which is the option is controlled by the scheme; see [G-4] **schemes intro**.)

For instance, specifying option

`ylabel(, nogrid)`

would suppress the grid lines in the y direction and specifying

`xlabel(, grid)`

would add them in the x. Specifying

`xlabel(0(1)10, grid)`

would place major labels, major ticks, and grid lines at $x = 0, 1, 2, \dots, 10$.

[`no`]`gmin` and [`no`]`gmax` are relevant only if `grid` is in effect (because `grid` is the default and `nogrid` was not specified or because `grid` was specified). [`no`]`gmin` and [`no`]`gmax` specify whether grid lines are to be drawn at the minimum and maximum values. Consider

```
. scatter yvar xvar, xlabel(0(1)10, grid)
```

Clearly the values 0, 1, ..., 10 are to be ticked and labeled, and clearly, grid lines should be drawn at 1, 2, ..., 9; but should grid lines be drawn at 0 and 10? If 0 and 10 are at the edge of the plot region, you probably do not want grid lines there. They will be too close to the axis and border of the graph.

What you want will differ from graph to graph, so the `graph` command tries to be smart, meaning that neither `gmin` nor `nogmin` (and neither `gmax` nor `nogmax`) is the default: The default is for `graph` to decide which looks best; the options force the decision one way or the other.

If `graph` decided to suppress the grids at the extremes and you wanted them, you could type

```
. scatter yvar xvar, xlabel(0(1)10, grid gmin gmax)
```

`gstyle(`*gridstyle*`)` specifies the overall style of the grid lines, including whether the lines extend beyond the plot region and into the plot region's margins, along with the style, color, width, and pattern of the lines themselves. The options that follow allow you to change each attribute, but the *gridstyle* provides the starting point. See [G-4] ***gridstyle***.

You need not specify `gstyle()` just because there is something you want to change. You specify `gstyle()` when another style exists that is exactly what you desire or when another style would allow you to specify fewer changes to obtain what you want.

`gextend` and `nogextend` specify whether the grid lines should extend beyond the plot region and pass through the plot region's margins; see [G-3] ***region_options***. The default is determined by the `gstyle()` and scheme, but usually, `nogextend` is the default and `gextend` is the option.

`glstyle(`*linestyle*`)`, `glwidth(`*linewidthstyle*`)`, `glcolor(`*colorstyle*`)`, and `glpattern(`*linepatternstyle*`)` specify other details about the look of the grid. See [G-4] ***linestyle***, [G-4] ***linewidthstyle***, [G-4] ***colorstyle***, and [G-4] ***linepatternstyle***. Grids are just lines. See [G-4] **concept: lines** for more information. Of these options, `glpattern()` is of particular interest because, with it, you can make the grid lines dashed.

Remarks

axis_label_options are a subset of *axis_options*; see [G-3] ***axis_options*** for an overview. The other appearance options are

axis_scale_options	(see [G-3] ***axis_scale_options***)
axis_title_options	(see [G-3] ***axis_title_options***)

Remarks are presented under the following headings:

Default labeling and ticking
Controlling the labeling and ticking
Adding extra ticks
Adding minor labels and ticks
Adding grid lines
Suppressing grid lines
Substituting text for labels
Contour axes—zlabel(), etc.
Appendix: Details of syntax
Suboptions without rules, numlists, or labels
Rules
Rules and numlists
Rules and numlists and labels
Interpretation of repeated options

Default labeling and ticking

By default, approximately five values are labeled and ticked on each axis. For example, in

```
. use http://www.stata-press.com/data/r12/auto
(1978 Automobile Data)
. scatter mpg weight
```

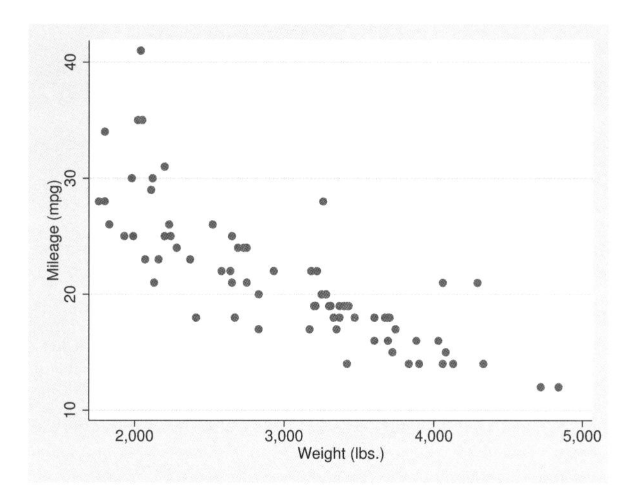

four values are labeled on each axis because choosing five would have required widening the scale too much.

Controlling the labeling and ticking

We would obtain the same results as we did in the above example if we typed

```
. scatter mpg weight, ylabel(#5) xlabel(#5)
```

Options `ylabel()` and `xlabel()` specify the values to be labeled and ticked, and `#5` specifies that Stata choose approximately five values for us. If we wanted many values labeled, we might type

```
. scatter mpg weight, ylabel(#10) xlabel(#10)
```

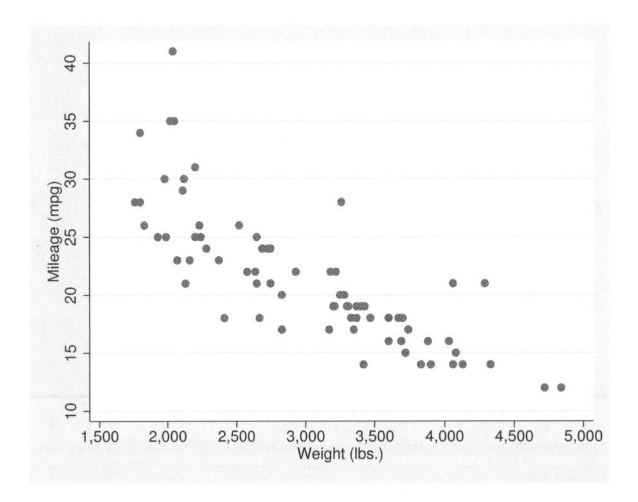

As with `#5`, `#10` was not taken too seriously; we obtained seven labels on the y axis and eight on the x axis.

We can also specify precisely the values we want labeled by specifying *#*(*#*)*#* or by specifying a list of numbers:

```
. scatter mpg weight, ylabel(10(5)45)
                      xlabel(1500 2000 3000 4000 4500 5000)
```

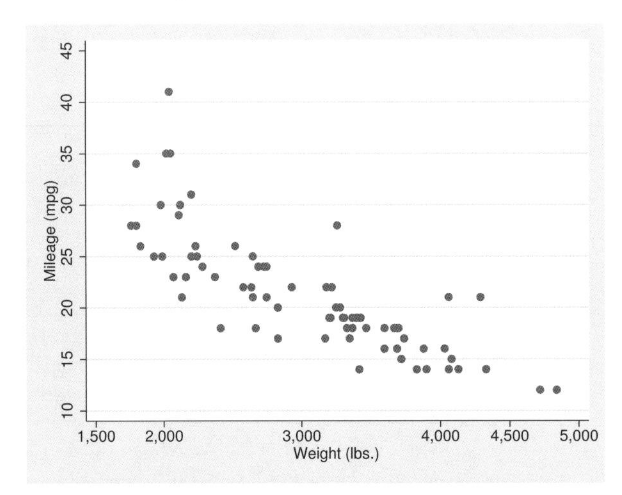

In option `ylabel()`, we specified the rule `10(5)45`, which means to label from 10 to 45 in steps of 5. In option `xlabel()`, we typed out the values to be labeled.

Adding extra ticks

Options ylabel() and xlabel() draw ticks plus labels. Options ytick() and xtick() draw ticks only, so you can do things such as

```
. scatter mpg weight, ytick(#10) xtick(#15)
```

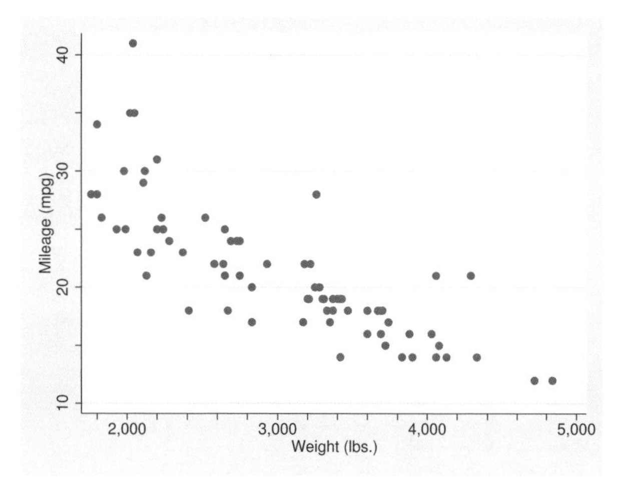

Of course, as with ylabel() and xlabel(), you can specify the exact values you want ticked.

Adding minor labels and ticks

Minor ticks and minor labels are smaller than regular ticks and regular labels. Options ymlabel() and xmlabel() allow you to place minor ticks with labels, and ymtick() and xmtick() allow you to place minor ticks without labels. When using minor ticks and labels, in addition to the usual syntax of #5 to mean approximately 5 values, 10(5)45 to mean 10 to 45 in steps of 5, and a list of numbers, there is an additional syntax: ##5. ##5 means that each major interval is divided into 5 minor intervals.

The graph below is intended more for demonstration than as an example of a good-looking graph:

```
. scatter mpg weight, ymlabel(##5) xmtick(##10)
```

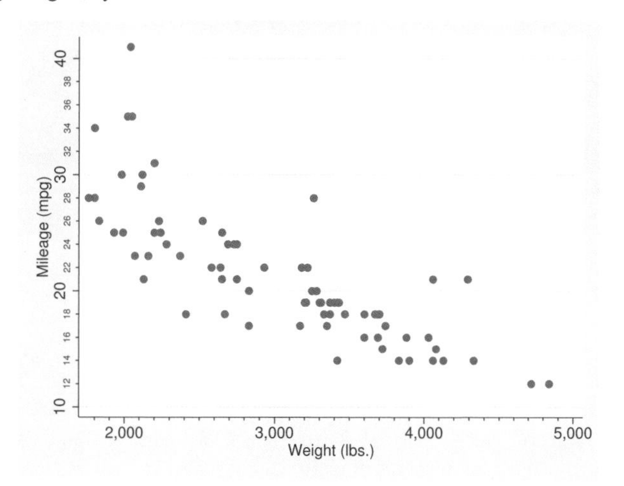

##5 means four ticks, and ##10 means nine ticks because most people think in reciprocals they say to themselves, "I want to tick the fourths so I want 4 ticks between," or, "I want to tick the tenths so I want 10 ticks between". They think incorrectly. They should think that if they want fourths, they want $4-1=3$ ticks between, or if they want tenths, they want $10-1=9$ ticks between. Stata subtracts one so that they can think—and correctly—when they want fourths that they want ##4 ticks between and that when they want tenths they want ##10 ticks between.

For ##*#* rules to work, the major ticks must be evenly spaced. This format is guaranteed only when the major ticks or labels are specified using the *#(#)#* rule. The ##*#* rule also works in almost all cases, the exception being daily data where the date variable is specified in the %td format. Here "nice" daily labels often do not have a consistent number of days between the ticks and thus the space between each major tick cannot be evenly divided. If the major ticks are not evenly spaced, the ##*#* rule does not produce any minor ticks.

Adding grid lines

To obtain grid lines, specify the grid suboption of ylabel(), xlabel(), ymlabel(), or xmlabel(). grid specifies that, in addition to whatever else the option would normally do, grid lines be drawn at the same values. In the example below,

```
. use http://www.stata-press.com/data/r12/uslifeexp, clear
(U.S. life expectancy, 1900-1999)
. line le year, xlabel(,grid)
```

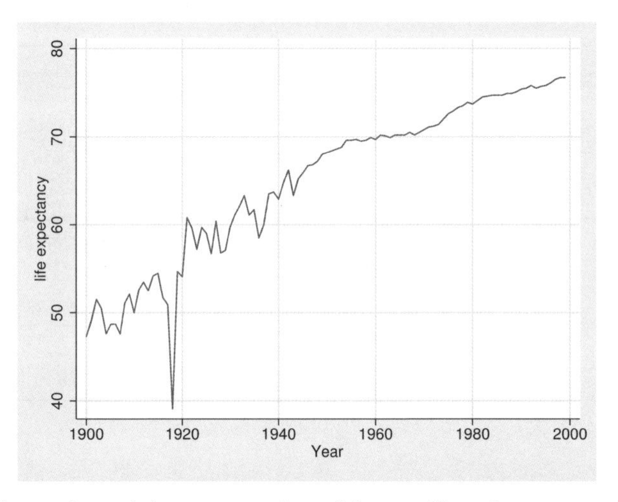

we specify xlabel(,grid), omitting any mention of the specific values to use. Thus xlabel() did what it does ordinarily (labeled approximately five nice values), and it drew grid lines at those same values.

Of course, we could have specified the values to be labeled and gridded:

```
. line le year, xlabel(#10, grid)
. line le year, xlabel(1900(10)2000, grid)
. line le year, xlabel(1900 1918 1940(20)2000, grid)
```

The grid suboption is usually specified with xlabel() (and with ylabel() if, given the scheme, grid is not the default), but it may be specified with any of the *axis_label_options*. In the example below, we "borrow" ymtick() and xmtick(), specify grid to make them draw grids, and specify style(none) to make the ticks themselves invisible:

```
. use http://www.stata-press.com/data/r12/auto, clear
(1978 Automobile Data)
. scatter mpg weight, ymtick(#20, grid tstyle(none))
                      xmtick(#20, grid tstyle(none))
```

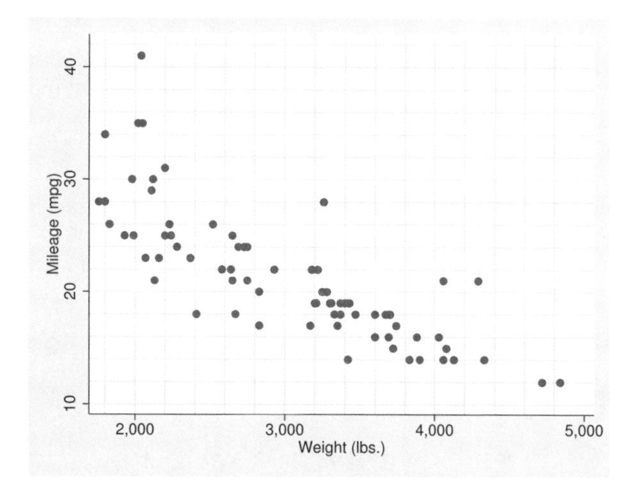

If you look carefully at the graph above, you will find that no grid line was drawn at $x = 5{,}000$. Stata suppresses grid lines when they get too close to the axes or borders of the graph. If you want to force Stata to draw them anyway, you can specify the `gmin` and `gmax` options:

```
. scatter mpg weight, ymtick(#20, grid tstyle(none))
                      xmtick(#20, grid tstyle(none) gmax)
```

Suppressing grid lines

Some commands, and option `ylabel()`, usually draw grid lines by default. For instance, in the following, results are the same as if you specified `ylabel(,grid)`:

```
. use http://www.stata-press.com/data/r12/auto, clear
(1978 Automobile Data)
. scatter mpg weight, by(foreign)
```

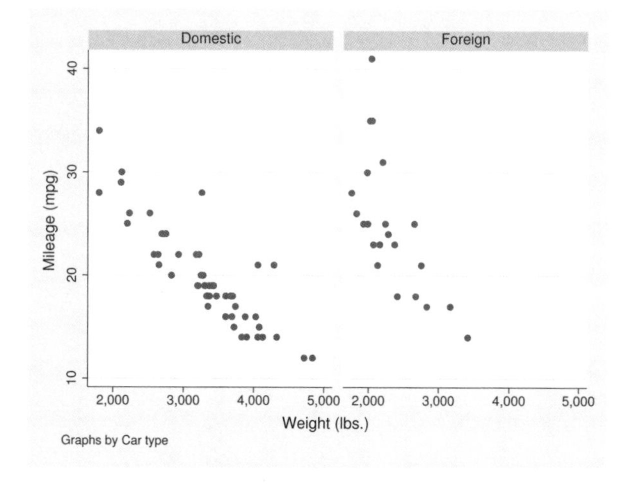

To suppress the grid lines, specify `ylabel(,nogrid)`:

```
. scatter mpg weight, by(foreign) ylabel(,nogrid)
```

Substituting text for labels

In addition to specifying explicitly the values to be labeled by specifying things such as `ylabel(10(10)50)` or `ylabel(10 20 30 40 50)`, you can specify text to be substituted for the label. If you type

```
. graph ... , ... ylabel(10 20 30 "mean" 40 50)
```

The values 10, 20, ..., 50 will be labeled, just as you would expect, but for the middle value, rather than the text "30" appearing, the text "mean" (without the quotes) would appear.

In the advanced example below, we specify

```
xlabel(1 "J" 2 "F" 3 "M" 4 "A" 5 "M" 6 "J" 7 "J" 8 "A" 9 "S" 10 "O" 11 "N" 12 "D")
```

so that rather than seeing the numbers 1, 2, ..., 12 (which are month numbers), we see J, F, ..., D; and we specify

```
ylabel(12321 "12,321 (mean)", axis(2) angle(0))
```

so that we label 12321 but, rather than seeing 12321, we see "12,321 (mean)". The `axis(2)` option puts the label on the second y axis (see [G-3] ***axis_choice_options***) and `angle(0)` makes the text appear horizontally rather than vertically (see *Options* above):

```
. use http://www.stata-press.com/data/r12/sp500, clear
(S&P 500)
. generate month = month(date)
. sort month
. by month: egen lo = min(volume)
. by month: egen hi = max(volume)
. format lo hi %10.0gc
. summarize volume

    Variable |       Obs        Mean    Std. Dev.       Min        Max
-------------+--------------------------------------------------------
      volume |       248    12320.68    2585.929       4103    23308.3

. by month: keep if _n==_N
(236 observations deleted)
```

```
. twoway rcap lo hi month,
          xlabel(1 "J"  2 "F"  3 "M"  4 "A"  5 "M"  6 "J"
                 7 "J"  8 "A"  9 "S" 10 "O" 11 "N" 12 "D")
          xtitle("Month of 2001")
          ytitle("High and Low Volume")
          yaxis(1 2) ylabel(12321 "12,321 (mean)", axis(2) angle(0))
          ytitle("", axis(2))
          yline(12321, lstyle(foreground))
          msize(*2)
          title("Volume of the S&P 500", margin(b+2.5))
          note("Source:  Yahoo!Finance and Commodity Systems Inc.")
```

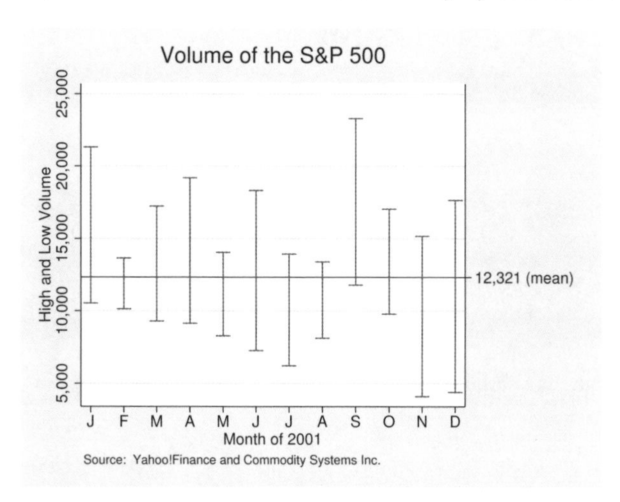

Contour axes—zlabel(), etc.

The `zlabel()`, `ztick()`, `zmlabel()`, and `zmtick()` options are unusual in that they apply not to axes on the plot region, but to the axis that shows the scale of a contour legend. They have effect only when the graph includes a `twoway contour` plot; see [G-2] **graph twoway contour**. In all other respects, they act like the `x*`, `y*`, and `t*` options.

For an example using `zlabel()`, see *Controlling the number of contours and their values* in [G-2] **graph twoway contour**.

The options associated with grids have no effect when specified on contour axes.

Appendix: Details of syntax

Suboptions without rules, numlists, or labels

What may appear in each of the options { `y` | `x` }{ `label` | `tick` | `mlabel` | `mtick` }`()` is a rule or numlist followed by suboptions:

[*rule*] [*numlist* ["*label*" [*numlist* ["*label*" [...]]]]] [, *suboptions*]

rule, *numlist*, and *label* are optional. If you remove those, you are left with

, *suboptions*

That is, the options { y | x }{ label | tick | mlabel | mtick }() may be specified with just suboptions and, in fact, they are often specified that way. If you want default labeling of the y axis and x axis, but you want grid lines in the x direction as well as the y, specify

```
. scatter yvar xvar , xlabel(,grid)
```

When you do not specify the first part—the *rule*, *numlist*, and *label*—you are saying that you do not want that part to change. You are saying that you merely wish to change how the *rule*, *numlist*, and *label* are displayed.

Of course, you may specify more than one suboption. You might type

```
. scatter yvar xvar, xlabel(,grid format(%9.2f))
```

if, in addition to grid lines, you wanted the numbers presented on the x axis to be presented in a `%9.2f` format.

Rules

What may appear in each of the axis-label options is a rule or numlist

[*rule*] [*numlist* ["*label*" [*numlist* ["*label*" [...]]]]] [, *suboptions*]

where either *rule* or *numlist* must be specified and both may be specified. Let us ignore the "*label*" part right now. Then the above simplifies to

[*rule*] [*numlist*] [, *suboptions*]

where *rule* or *numlist* must be specified, both may be specified, and most often you will simply specify the *rule*, which may be any of the following:

rule	Example	Description
`##`	`#6`	6 nice values
`###`	`##10`	$10 - 1 = 9$ values between major ticks; allowed with `mlabel()` and `mtick()` only
`#(#)#`	`-4(.5)3`	specified range: -4 to 3 in steps of .5
`minmax`	`minmax`	minimum and maximum values
`none`	`none`	label no values
`.`	`.`	skip the rule

The most commonly specified rules are `##` and `###`.

Specifying `##` says to choose # nice values. Specifying `#5` says to choose five nice values, `#6` means to choose six, and so on. If you specify `ylabel(#5)`, then five values will be labeled (on the y axis). If you also specify `ymtick(#10)`, then 10 minor ticks will also be placed on the axis. Actually, `ylabel(#5)` and `ymtick(#10)` will result in approximately five labels and 10 minor ticks because the choose-a-nice-number routine will change your choice a little if, in its opinion, that would yield a nicer overall result. You may not agree with the routine about what is nice, and then the `#(#)#` rule will let you specify exactly what you want, assuming that you want evenly spaced labels and numbers.

`###` is allowed only with the { y | x }`mlabel()` and { y | x }`mtick()` options—the options that result in minor ticks. `###` says to put #−1 minor ticks between the major ticks. `##5` would put four, and `##10` would put nine. Here # is taken seriously, at least after subtraction, and you are given exactly what you request.

#(#)# can be used with major or minor labels and ticks. This rule says to label the first number specified, increment by the second number, and keep labeling, as long as the result is less than or equal to the last number specified. `ylabel(1(1)10)` would label (and tick) the values 1, 2, ..., 10. `ymtick(1(.5)10)` would put minor ticks at 1, 1.5, 2, 2.5, ..., 10. It would be perfectly okay to specify both of those options. When specifying rules, minor ticks and labels will check what is specified for major ticks and labels and remove the intersection so as not to overprint. The results will be the same as if you specified `ymtick(1.5(1)9.5)`.

The rule `minmax` specifies that you want the minimum and maximum. `ylabel(minmax)` would label only the minimum and maximum.

Rule `none` means precisely that: the rule that results in no labels and no ticks.

Rule `.` makes sense only when `add` is specified, although it is allowed at other times, and then `.` means the same as `none`.

Rules and numlists

After the *rule*—or instead of it—you can specify a *numlist*. A numlist is a list of numbers, for instance, "`1 2 5 7`" (without the quotes) or "`3/9`" (without the quotes). Other shorthands are allowed (see [U] **11.1.8 numlist**), and in fact, one of *numlist*'s syntaxes looks just like a *rule*: *#(#)#*. It has the same meaning, too.

There is, however, a subtle distinction between, for example,

`ylabel(1(1)10)` (a *rule*) and `ylabel(none 1(1)10)` (a *numlist*)

Rules are more efficient. Visually, however, there is no difference.

Use numlists when the values you wish to label or to tick are unequally spaced,

```
ylabel(none 1 2 5 7)
```

or when there is one or more extra values you want to label or to tick:

```
ylabel(1(1)10 3.5 7.5)
```

Rules and numlists and labels

Numlists serve an additional purpose—you can specify text that is to be substituted for the value to be labeled. For instance,

```
ylabel(1(1)10 3.5 "Low" 7.5 "Hi")
```

says to label 1, 2, ..., 10 (that is the *rule* part) and to label the special values 3.5 and 7.5. Rather than actually printing "3.5" and "7.5" next to the ticks at 3.5 and 7.5, however, `graph` will instead print the words "Low" and "Hi".

Interpretation of repeated options

Each of the axis-label options may be specified more than once in the same command. If you do that and you do not specify suboption `add`, the rightmost of each is honored. If you specify suboption `add`, then the option just specified and the previous options are merged. `add` specifies that any new ticks or labels are added to any existing ticks or labels on the axis. All suboptions are *rightmost*; see [G-4] **concept: repeated options**.

Reference

Cox, N. J. 2007. Stata tip 55: Better axis labeling for time points and time intervals. *Stata Journal* 7: 590–592.

Also see

[G-3] ***axis_options*** — Options for specifying numeric axes

[G-3] ***axis_scale_options*** — Options for specifying axis scale, range, and look

[G-3] ***axis_title_options*** — Options for specifying axis titles

Title

[G-3] ***axis_options*** — Options for specifying numeric axes

Syntax

axis_scale_options	Description
{ y | x | t | z }scale(*axis_description*)	log scales, range, appearance

See [G-3] ***axis_scale_options***.

axis_label_options	Description
{ y | x | t | z }label(*rule_or_values*)	major ticks plus labels
{ y | x | t | z }tick(*rule_or_values*)	major ticks only
{ y | x | t | z }mlabel(*rule_or_values*)	minor ticks plus labels
{ y | x | t | z }mtick(*rule_or_values*)	minor ticks only

(also allows control of grid lines; see [G-3] ***axis_label_options***)

axis_title_options	Description
{ y | x | t | z }title(*axis_title*)	specify axis title

See [G-3] ***axis_title_options***.

Description

Axes are the graphical elements that indicate the scale.

Options

yscale(), xscale(), tscale(), and zscale() specify how the y, x, t, and z axes are scaled (arithmetic, log, reversed), the range of the axes, and the look of the lines that are the axes. See [G-3] ***axis_scale_options***. tscale() is an extension of xscale(). zscale() applies to the axis in the contour legend of a graph with a contour plot.

ylabel(), ytick(), ymlabel(), ymtick(), xlabel(), ..., xmtick(), tlabel(), ..., tmtick(), and zlabel(), ..., zmtick() specify how the axes should be labeled and ticked. These options allow you to control the placement of major and minor ticks and labels. Also, these options allow you to add or to suppress grid lines on your graphs. See [G-3] ***axis_label_options***. tlabel(), ..., tmtick() are extensions of xlabel(), ..., xmtick(), respectively.

ytitle(), xtitle(), ttitle(), and ztitle() specify the titles to appear next to the axes. See [G-3] ***axis_title_options***. ttitle() is a synonym of xtitle().

Remarks

Numeric axes are allowed with `graph twoway` (see [G-2] **graph twoway**) and `graph matrix` (see [G-2] **graph matrix**) and are allowed for one of the axes of `graph bar` (see [G-2] **graph bar**), `graph dot` (see [G-2] **graph dot**), and `graph box` (see [G-2] **graph box**). They are also allowed on the contour key of a legend on a contour plot. How the numeric axes look is affected by the *axis_options*.

Remarks are presented under the following headings:

Use of axis-appearance options with graph twoway

When you type

```
. scatter yvar xvar
```

the resulting graph will have y and x axes. How the axes look will be determined by the scheme; see [G-4] **schemes intro**. The *axis_options* allow you to modify the look of the axes in terms of whether the y axis is on the left or on the right, whether the x axis is on the bottom or on the top, the number of major and minor ticks that appear on each axis, the values that are labeled, and the titles that appear along each.

For instance, you might type

```
. scatter yvar xvar, ylabel(#6) ymtick(##10) ytitle("values of y") xlabel(#6)
        xmtick(##10) xtitle("values of x")
```

to draw a graph of `yvar` versus `xvar`, putting on each axis approximately six labels and major ticks and 10 minor ticks between major ticks, and labeling the y axis "values of y" and the x axis "values of x".

```
. scatter yvar xvar, ylabel(0(5)30) ymtick(0(1)30) ytitle("values of y")
        xlabel(0(10)100) xmtick(0(5)100) xtitle("values of x")
```

would draw the same graph, putting major ticks on the y axis at 0, 5, 10, ..., 30 and minor ticks at every integer over the same range, and putting major ticks on the x axis at 0, 10, ..., 100 and minor ticks at every five units over the same range.

The way we have illustrated it, it appears that the axis options are options of `scatter`, but that is not so. Here they are options of `twoway`, and the "right" way to write the last command is

```
. twoway (scatter yvar xvar), ylabel(0(5)30) ymtick(0(1)30) ytitle("values of y")
        xlabel(0(10)100) xmtick(0(5)100) xtitle("values of x")
```

The parentheses around `(scatter yvar xvar)` and the placing of the axis-appearance options outside the parentheses make clear that the options are aimed at `twoway` rather than at `scatter`. Whether you use the `||`-separator notation or the `()`-binding notation makes no difference, but it is important to understand that there is only one set of axes, especially when you type more complicated commands, such as

```
. twoway (scatter yvar xvar)
         (scatter y2var x2var)
               , ylabel(0(5)30) ymtick(0(1)30) ytitle("values of y")
               xlabel(0(10)100) xmtick(0(5)100) xtitle("values of x")
```

There is one set of axes in the above, and it just so happens that both `yvar` versus `xvar` and `y2var` versus `x2var` appear on it. You are free to type the above command how you please, such as

```
. scatter yvar  xvar ||
  scatter y2var x2var ||,
        ylabel(0(5)30) ymtick(0(1)30) ytitle("values of y")
        xlabel(0(10)100) xmtick(0(5)100) xtitle("values of x")
```

or

```
. scatter yvar  xvar ||
  scatter y2var x2var, ylabel(0(5)30) ymtick(0(1)30)
                       ytitle("values of y") xlabel(0(10)100)
                       xmtick(0(5)100) xtitle("values of x")
```

or

```
. scatter yvar xvar, ylabel(0(5)30) ymtick(0(1)30)
                     ytitle("values of y") xlabel(0(10)100)
                     xmtick(0(5)100) xtitle("values of x") ||
  scatter y2var x2var
```

The above all result in the same graph, even though the last makes it appear that the axis options are associated with just the first `scatter`, and the next to the last makes it appear that they are associated with just the second. However you type it, the command is really `twoway`. `twoway` draws twoway graphs with one set of axes (or one set per by-group), and all the plots that appear on the twoway graph share that set.

Multiple y and x scales

Actually, a twoway graph can have more than one set of axes. Consider the command:

```
. twoway (scatter yvar xvar) (scatter y2var x2var, yaxis(2))
```

The above graphs `yvar` versus `xvar` and `y2var` versus `x2var`, but two y scales are provided. The first (which will appear on the left) applies to `yvar`, and the second (which will appear on the right) applies to `y2var`. The `yaxis(2)` option says that the y axis of the specified scatter is to appear on the second y scale.

See [G-3] ***axis_choice_options***.

Axis on the left, axis on the right?

When there is only one y scale, whether the axis appears on the left or the right is determined by the scheme; see [G-4] **schemes intro**. The default scheme puts the y axis on the left, but the scheme that mirrors the style used by *The Economist* puts it on the right:

```
scatter yvar xvar, scheme(economist)
```

Specifying `scheme(economist)` will change other things about the appearance of the graph, too. If you just want to move the y axis to the right, you can type

```
scatter yvar xvar, yscale(alt)
```

As explained in [G-3] ***axis_scale_options***, `yscale(alt)` switches the axis from one side to the other, so if you typed

```
scatter yvar xvar, scheme(economist) yscale(alt)
```

you would get *The Economist* scheme but with the y axis on the left.

`xscale(alt)` switches the x axis from the bottom to the top or from the top to the bottom; see [G-3] ***axis_scale_options***.

Contour axes—zscale(), zlabel(), etc.

The `zscale()`, `zlabel()`, `ztitle()`, and other `z` options are unusual in that they apply not to axes on the plot region, but to the axis that shows the scale of a contour legend. They have effect only when the graph includes a `twoway contour` plot; see [G-2] **graph twoway contour**. In all other respects, they act like the `x*`, `y*`, and `t*` options.

Also see

[G-3] ***axis_label_options*** — Options for specifying axis labels

[G-3] ***axis_scale_options*** — Options for specifying axis scale, range, and look

[G-3] ***axis_title_options*** — Options for specifying axis titles

[G-3] ***axis_choice_options*** — Options for specifying the axes on which a plot appears

[G-3] ***region_options*** — Options for shading and outlining regions and controlling graph size

Title

[G-3] *axis_scale_options* — Options for specifying axis scale, range, and look

Syntax

axis_scale_options are a subset of *axis_options*; see [G-3] ***axis_options***.

axis_scale_options	Description
`yscale(`*axis_suboptions*`)`	how y axis looks
`xscale(`*axis_suboptions*`)`	how x axis looks
`tscale(`*axis_suboptions*`)`	how t (time) axis looks
`zscale(`*axis_suboptions*`)`	how contour legend axis looks

The above options are *merged-implicit*; see [G-4] **concept: repeated options**.

axis_suboptions	Description
`axis(`*#*`)`	which axis to modify; $1 \leq \# \leq 9$
[`no`]`log`	use logarithmic scale
[`no`]`reverse`	reverse scale to run from max to min
`range(`*numlist*`)`	expand range of axis
`range(`*datelist*`)`	expand range of t axis (`tscale()` only)
`off` and `on`	suppress/force display of axis
`fill`	allocate space for axis even if `off`
`alt`	move axis from left to right or from top to bottom
`fextend`	extend axis line through plot region and plot region's margin
`extend`	extend axis line through plot region
`noextend`	do not extend axis line at all
`noline`	do not draw axis line
`line`	force drawing of axis line
`titlegap(`*relativesize*`)`	margin between axis title and tick labels
`outergap(`*relativesize*`)`	margin outside axis title
`lstyle(`*linestyle*`)`	overall style of axis line
`lcolor(`*colorstyle*`)`	color of axis line
`lwidth(`*linewidthstyle*`)`	thickness of axis line
`lpattern(`*linepatternstyle*`)`	axis pattern (solid, dashed, etc.)

See [G-4] ***relativesize***, [G-4] ***linestyle***, [G-4] ***colorstyle***, [G-4] ***linewidthstyle***, and [G-4] ***linepatternstyle***.

Description

The *axis_scale_options* determine how axes are scaled (arithmetic, log, reversed), the range of the axes, and the look of the lines that are the axes.

Options

`yscale(`*axis_suboptions*`)`, `xscale(`*axis_suboptions*`)`, and `tscale(`*axis_suboptions*`)` specify the look of the y, x, and t axes. The t axis is an extension of the x axis. Inside the parentheses, you specify *axis_suboptions*.

`zscale(`*axis_suboptions*`)`; see *Contour axes—zscale()* below.

yscale(), xscale(), and tscale() suboptions

`axis(`*#*`)` specifies to which scale this axis belongs and is specified when dealing with multiple y or x axes; see [G-3] ***axis_choice_options***.

`log` and `nolog` specify whether the scale should be logarithmic or arithmetic. `nolog` is the usual default, so `log` is the option. See *Obtaining log scales* under *Remarks* below.

`reverse` and `noreverse` specify whether the scale should run from the maximum to the minimum or from the minimum to the maximum. `noreverse` is the usual default, so `reverse` is the option. See *Obtaining reversed scales* under *Remarks* below.

`range(`*numlist*`)` specifies that the axis be expanded to include the numbers specified. Missing values, if specified, are ignored. See *Specifying the range of a scale* under *Remarks* below.

`range(`*datelist*`)` (`tscale()` only) specifies that the axis be expanded to include the specified dates; see [U] **11.1.9 datelist**. Missing values, if specified, are ignored. See [TS] **tsline** for examples.

`off` and `on` suppress or force the display of the axis. `on` is the default and `off` the option. See *Suppressing the axes* under *Remarks* below.

`fill` goes with `off` and is seldom specified. If you turned an axis off but still wanted the space to be allocated for the axis, you could specify `fill`.

`alt` specifies that, if the axis is by default on the left, it be on the right; if it is by default on the bottom, it is to be on the top. The following would draw a scatterplot with the y axis on the right:

```
. scatter yvar xvar, yscale(alt)
```

`fextend`, `extend`, `noextend`, `line`, and `noline` determine how much of the line representing the axis is to be drawn. They are alternatives.

`noline` specifies that the line not be drawn at all. The axis is there, ticks and labels will appear, but the line that is the axis itself will not be drawn.

`line` is the opposite of `noline`, for use if the axis line somehow got turned off.

`noextend` specifies that the axis line not extend beyond the range of the axis. Say that the axis extends from -1 to $+20$. With `noextend`, the axis line begins at -1 and ends at $+20$.

`extend` specifies that the line be longer than that and extend all the way across the plot region. For instance, -1 and $+20$ might be the extent of the axis, but the scale might extend from -5 to $+25$, with the range $[-5,-1)$ and $(20,25]$ being unlabeled on the axis. With `extend`, the axis line begins at -5 and ends at 25.

`fextend` specifies that the line be longer than that and extend across the plot region and across the plot region's margins. For a definition of the plot region's margins, see [G-3] ***region_options***. If the plot region has no margins (which would be rare), `fextend` means the same as `extend`. If the plot region does have margins, `extend` would result in the y and x axes not meeting. With `fextend`, they touch.

`fextend` is the default with most schemes.

`titlegap(`*relativesize*`)` specifies the margin to be inserted between the axis title and the axis tick labels; see [G-4] ***relativesize***.

`outergap(`*relativesize*`)` specifies the margin to be inserted outside the axis title; see [G-4] ***relativesize***.

`lstyle(`*linestyle*`)`, `lcolor(`*colorstyle*`)`, `lwidth(`*linewidthstyle*`)`, and `lpattern(`*linepatternstyle*`)` determine the overall look of the line that is the axis; see [G-4] **concept: lines**.

Remarks

axis_scale_options are a subset of *axis_options*; see [G-3] ***axis_options*** for an overview. The other appearance options are

axis_label_options	(see [G-3] ***axis_label_options***)
axis_title_options	(see [G-3] ***axis_title_options***)

Remarks are presented under the following headings:

Use of the yscale( ) and xscale( )

`yscale()` and `xscale()` specify the look of the y and x axes. Inside the parentheses, you specify *axis_suboptions*, for example:

```
. twoway (scatter ...) ... , yscale(range(0 10) titlegap(1))
```

`yscale()` and `xscale()` may be abbreviated `ysc()` and `xsc()`, suboption `range()` may be abbreviated `r()`, and `titlegap()` may be abbreviated `titleg()`:

```
. twoway (scatter ...) ... , ysc(r(0 10) titleg(1))
```

Multiple `yscale()` and `xscale()` options may be specified on the same command, and their results will be combined. Thus the above command could also be specified

```
. twoway (scatter ...) ... , ysc(r(0 10)) ysc(titleg(1))
```

Suboptions may also be specified more than once, either within one `yscale()` or `xscale()` option, or across multiple options, and the rightmost suboption will take effect. In the following command, `titlegap()` will be 2, and `range()` 0 and 10:

```
. twoway (scatter ...) ... , ysc(r(0 10)) ysc(titleg(1)) ysc(titleg(2))
```

Specifying the range of a scale

To specify the range of a scale, specify the { `y` | `x` }`scale(range(`*numlist*`))` option. This option specifies that the axis be expanded to include the numbers specified.

Consider the graph

```
. scatter yvar xvar
```

Assume that it resulted in a graph where the y axis varied over 1–100 and assume that, given the nature of the y variable, it would be more natural if the range of the axis were expanded to go from 0 to 100. You could type

```
. scatter yvar xvar, ysc(r(0))
```

Similarly, if the range without the `yscale(range())` option went from 1 to 99 and you wanted it to go from 0 to 100, you could type

```
. scatter yvar xvar, ysc(r(0 100))
```

If the range without `yscale(range())` went from 0 to 99 and you wanted it to go from 0 to 100, you could type

```
. scatter yvar xvar, ysc(r(100))
```

Specifying missing for a value leaves the current minimum or maximum unchanged; specifying a nonmissing value changes the range, but only if the specified value is outside the value that would otherwise have been chosen. `range()` never narrows the scale of an axis or causes data to be omitted from the plot. If you wanted to graph *yvar* versus *xvar* for the subset of *xvar* values between 10 and 50, typing

```
. scatter yvar xvar, xsc(r(10 50))
```

would not suffice. You need to type

```
. scatter yvar xvar if xvar >=10 & xvar <=50
```

Obtaining log scales

To obtain log scales specify the { y | x }`scale(log)` option. Ordinarily when you draw a graph, you obtain arithmetic scales:

```
. use http://www.stata-press.com/data/r12/lifeexp, clear
(Life expectancy, 1998)
. scatter lexp gnppc
```

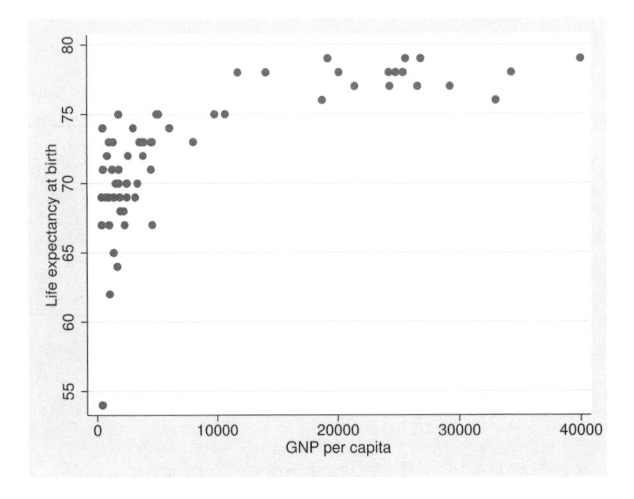

To obtain the same graph with a log x scale, we type

```
. scatter lexp gnppc, xscale(log)
```

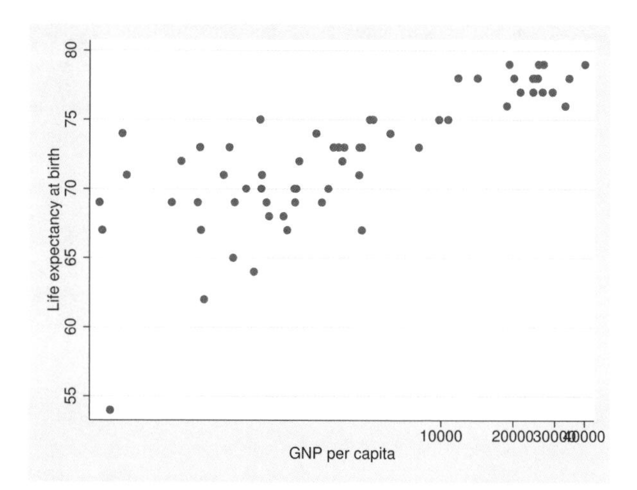

We obtain the same graph as if we typed

```
. generate log_gnppc = log(gnppc)
. scatter lexp log_gnppc
```

The difference concerns the labeling of the axis. When we specify $\{$ y | x $\}$`scale(log)`, the axis is labeled in natural units. Here the overprinting of the 30,000 and 40,000 is unfortunate, but we could fix that by dividing `gnppc` by 1,000.

Obtaining reversed scales

To obtain reversed scales—scales that run from high to low—specify the $\{$ y | x $\}$`scale(reverse)` option:

```
. use http://www.stata-press.com/data/r12/auto, clear
(1978 Automobile Data)
. scatter mpg weight, yscale(rev)
```

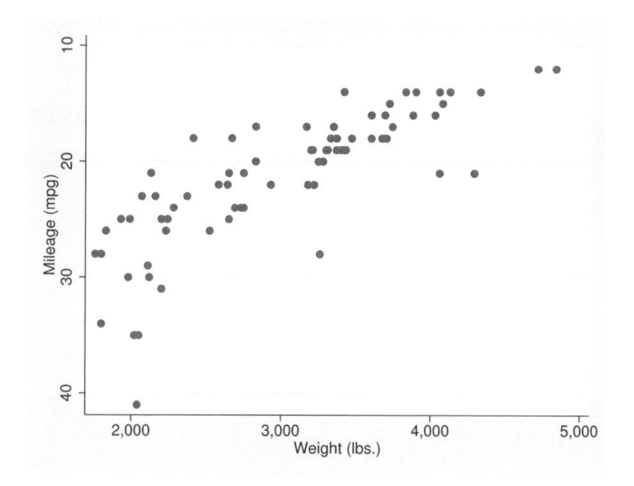

Suppressing the axes

There are two ways to suppress the axes. The first is to turn them off completely, which means that the axis line is suppressed, along with all of its ticking, labeling, and titling. The second is to simply suppress the axis line while leaving the ticking, labeling, and titling in place.

The first is done by { y | x }`scale(off)` and the second by { y | x }`scale(noline)`. Also, you will probably need to specify the `plotregion(style(none))` option; see [G-3] ***region_options***.

The axes and the border around the plot region are right on top of each other. Specifying `plotregion(style(none))` will do away with the border and reveal the axes to us:

```
. use http://www.stata-press.com/data/r12/auto, clear
(1978 Automobile Data)
. scatter mpg weight, plotregion(style(none))
```

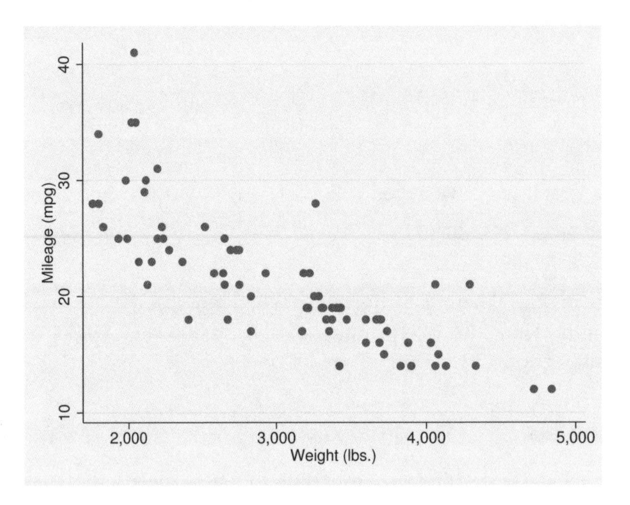

To eliminate the axes, type

```
. scatter mpg weight, plotregion(style(none))
                      yscale(off) xscale(off)
```

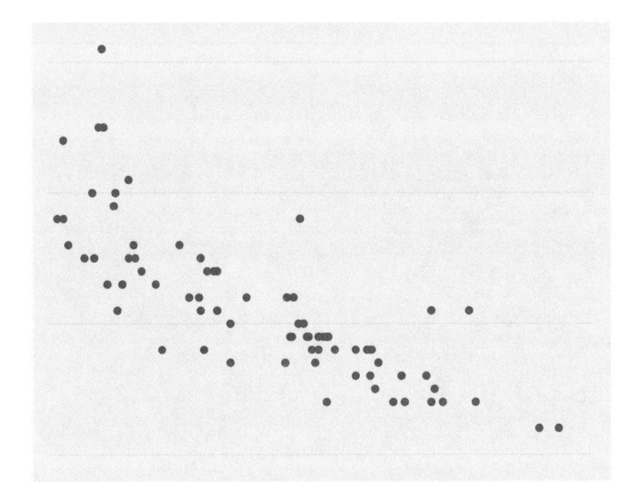

To eliminate the lines that are the axes while leaving in place the labeling, ticking, and titling, type

```
. scatter mpg weight, plotregion(style(none))
                      yscale(noline) xscale(noline)
```

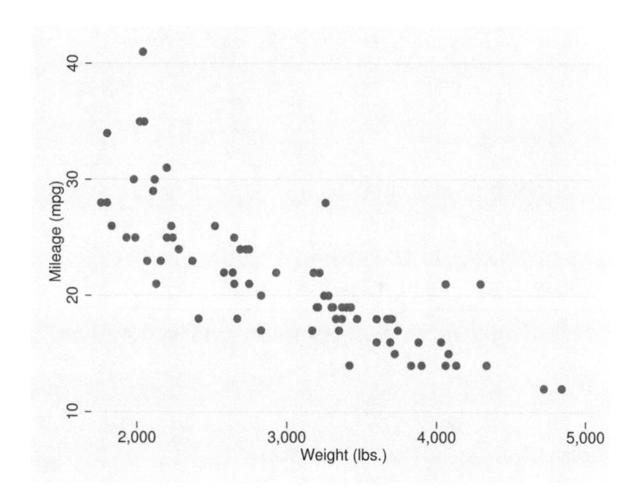

Rather than using { y | x }`scale(noline)`, you may specify { y | x }`scale(lstyle(noline))` or { y | x }`scale(lstyle(none))`. They all mean the same thing.

Contour axes—zscale()

The `zscale()` option is unusual in that it applies not to axes on the plot region, but to the axis that shows the scale of a contour legend. It has effect only when the graph includes a `twoway contour` plot; see [G-2] **graph twoway contour**. In all other respects, it acts like `xscale()`, `yscale()`, and `tscale()`.

Reference

Cox, N. J. 2008. Stata tip 59: Plotting on any transformed scale. *Stata Journal* 8: 142–145.

Also see

[G-3] ***axis_options*** — Options for specifying numeric axes

[G-3] ***axis_label_options*** — Options for specifying axis labels

[G-3] ***axis_title_options*** — Options for specifying axis titles

[G-3] ***region_options*** — Options for shading and outlining regions and controlling graph size

[U] **11.1.9 datelist**

[TS] **tsline** — Plot time-series data

Title

[G-3] ***axis_title_options*** — Options for specifying axis titles

Syntax

axis_title_options are a subset of *axis_options*; see [G-3] ***axis_options***. *axis_title_options* control the titling of an axis.

axis_title_options	Description
`ytitle(`*axis_title*`)`	specify y axis title
`xtitle(`*axis_title*`)`	specify x axis title
`ttitle(`*axis_title*`)`	specify t (time) axis title
`ztitle(`*axis_title*`)`	specify contour legend axis title

The above options are *merged-explicit*; see [G-4] **concept: repeated options**.

where *axis_title* is

"*string*" ["*string*" [...]] [, *suboptions*]

suboptions	Description
`axis(`*#*`)`	which axis, $1 \leq \# \leq 9$
`prefix`	combine options
`suffix`	combine options
textbox_options	control details of text appearance; see [G-3] ***textbox_options***

Description

axis_title_options specify the titles to appear on axes.

Options

`ytitle(`*axis_title*`)`, `xtitle(`*axis_title*`)`, and `ttitle(`*axis_title*`)` specify the titles to appear on the y, x, and t axes. `ttitle()` is a synonym for `xtitle()`.

`ztitle(`*axis_title*`)`; see *Contour axes—ztitle()* below.

Suboptions

`axis(`*#*`)` specifies to which axis this title belongs and is specified when dealing with multiple y axes or multiple x axes; see [G-3] ***axis_choice_options***.

`prefix` and `suffix` specify that what is specified in this option is to be added to any previous `xtitle()` or `ytitle()` options previously specified. See *Interpretation of repeated options* below.

textbox_options specifies the look of the text. See [G-3] ***textbox_options***.

Remarks

axis_title_options are a subset of *axis_options*; see [G-3] ***axis_options*** for an overview. The other appearance options are

axis_scale_options	(see [G-3] ***axis_scale_options***)
axis_label_options	(see [G-3] ***axis_label_options***)

Remarks are presented under the following headings:

Default axis titles
Overriding default titles
Specifying multiline titles
Suppressing axis titles
Interpretation of repeated options
Titles with multiple y axes or multiple x axes
Contour axes—ztitle()

Default axis titles

Even if you do not specify the `ytitle()` or `xtitle()` options, axes will usually be titled. In those cases, { y | x }`title()` changes the title. If an axis is not titled, specifying { y | x }`title()` adds a title.

Default titles are obtained using the corresponding variable's variable label or, if it does not have a label, using its name. For instance, in

```
. twoway scatter yvar xvar
```

the default title for the y axis will be obtained from variable `yvar`, and the default title for the x axis will be obtained from `xvar`. Sometimes the plottype substitutes a different title; for instance,

```
. twoway lfit yvar xvar
```

labels the y axis "Fitted values" regardless of the name or variable label associated with variable `yvar`.

If multiple variables are associated with the same axis, their individual titles (variable label, variable name, or as substituted) are joined, with a slash (/) in between. For instance, in

```
. twoway scatter y1var xvar || line y2var xvar || lfit y1var xvar
```

the y axis will be titled

y1var_title/*y2var_title*/`Fitted values`

When many plots are overlaid, this often results in titles that run off the end of the graph.

Overriding default titles

You may specify the title to appear on the y axis using `ytitle()` and the title to appear on the x axis using `xtitle()`. You specify the text—surrounded by double quotes—inside the option:

`ytitle("My y title")`

`xtitle("My x title")`

For `scatter`, the command might read

```
. scatter yvar xvar, ytitle("Price") xtitle("Quantity")
```

Specifying multiline titles

Titles may include more than one line. Lines are specified one after the other, each enclosed in double quotes:

```
ytitle("First line" "Second line")
xtitle("First line" "Second line" "Third line")
```

Suppressing axis titles

To eliminate an axis title, specify { y | x }title("").

To eliminate the title on a second, third, ..., axis, specify { y | x }title("", axis(#)). See *Titles with multiple y axes or multiple x axes* below.

Interpretation of repeated options

xtitle() and ytitle() may be specified more than once in the same command. When you do that, the rightmost one takes effect.

See *Interpretation of repeated options* in [G-3] ***axis_label_options***. Multiple ytitle() and xtitle() options work the same way. The twist for the title options is that you specify whether the extra information is to be prefixed or suffixed onto what came before.

For instance, pretend that sts graph produced the x axis title "analysis time". If you typed

```
. sts graph, xtitle("My new title")
```

the title you specified would replace that. If you typed

```
. sts graph, xtitle("in days", suffix)
```

the x axis title would be (first line) "analysis time" (second line) "in days". If you typed

```
. sts graph, xtitle("Time to failure", prefix)
```

the x axis title would be (first line) "Time to failure" (second line) "analysis time".

Titles with multiple y axes or multiple x axes

When you have more than one y or x axis (see [G-3] ***axis_choice_options***), remember to specify the axis(#) suboption to indicate to which axis you are referring.

Contour axes—ztitle()

The ztitle() option is unusual in that it applies not to axes on the plot region, but to the axis that shows the scale of a contour legend. It has effect only when the graph includes a twoway contour plot; see [G-2] **graph twoway contour**. In all other respects, it acts like xtitle(), ytitle(), and ttitle().

Also see

[G-3] ***axis_options*** — Options for specifying numeric axes

[G-3] ***axis_label_options*** — Options for specifying axis labels

[G-3] ***axis_scale_options*** — Options for specifying axis scale, range, and look

[G-4] ***text*** — Text in graphs

Title

[G-3] *barlook_options* — Options for setting the look of bars

Syntax

barlook_options	Description
color(*colorstyle*)	outline and fill color
fcolor(*colorstyle*)	fill color
fintensity(*intensitystyle*)	fill intensity
lcolor(*colorstyle*)	outline color
lwidth(*linewidthstyle*)	thickness of outline
lpattern(*linepatternstyle*)	outline pattern (solid, dashed, etc.)
lstyle(*linestyle*)	overall look of outline
bstyle(*areastyle*)	overall look of bars, all settings above
pstyle(*pstyle*)	overall plot style, including areastyle

See [G-4] ***colorstyle***, [G-4] ***intensitystyle***, [G-4] ***linewidthstyle***, [G-4] ***linepatternstyle***, [G-4] ***linestyle***, [G-4] ***areastyle***, and [G-4] ***pstyle***.

All options are *merged-implicit*; see [G-4] **concept: repeated options**.

Description

The *barlook_options* determine the look of bars produced by `graph bar` (see [G-2] **graph bar**), `graph hbar` (see [G-2] **graph bar**), `graph twoway bar` (see [G-2] **graph twoway bar**), and several other commands that render bars. The *barlook_options* and the *area_options* (see [G-3] ***area_options***) are synonyms, and the options may be used interchangeably.

Options

`color(`*colorstyle*`)` specifies one color to be used both to outline the shape of the bar and to fill its interior. See [G-4] ***colorstyle*** for a list of color choices.

`fcolor(`*colorstyle*`)` specifies the color to be used to fill the interior of the bar. See [G-4] ***colorstyle*** for a list of color choices.

`fintensity(`*intensitystyle*`)` specifies the intensity of the color used to fill the interior of the bar. See [G-4] ***intensitystyle*** for a list of intensity choices.

`lcolor(`*colorstyle*`)` specifies the color to be used to outline the bar. See [G-4] ***colorstyle*** for a list of color choices.

`lwidth(`*linewidthstyle*`)` specifies the thickness of the line to be used to outline the bar. See [G-4] ***linewidthstyle*** for a list of choices.

`lpattern(`*linepatternstyle*`)` specifies whether the line used to outline the bar is solid, dashed, etc. See [G-4] ***linepatternstyle*** for a list of pattern choices.

`lstyle(`*linestyle*`)` specifies the overall style of the line used to outline the bar, including its pattern (solid, dashed, etc.), thickness, and color. The three options listed above allow you to change the line's attributes, but `lstyle()` is the starting point. See [G-4] ***linestyle*** for a list of choices.

`bstyle(`*areastyle*`)` specifies the look of the bar. The options listed below allow you to change each attribute, but `bstyle()` provides the starting point.

You need not specify `bstyle()` just because there is something you want to change. You specify `bstyle()` when another style exists that is exactly what you desire or when another style would allow you to specify fewer changes to obtain what you want.

See [G-4] ***areastyle*** for a list of available area styles.

`pstyle(`*pstyle*`)` specifies the overall style of the plot, including not only the *areastyle*, but also all other settings for the look of the plot. Only the *areastyle* affects the look of areas. See [G-4] ***pstyle*** for a list of available plot styles.

Remarks

The *barlook_options* are allowed inside **graph bar**'s and **graph hbar**'s option `bar(#,` *barlook_options*`)`, as in

```
. graph bar yvar1 yvar2, bar(1, color(green)) bar(2, color(red))
```

The command above would set the bar associated with *yvar1* to be green and the bar associated with *yvar2* to red; see [G-2] **graph bar**.

barlook_options are also allowed as options with `graph twoway bar` and `graph twoway rbar`, as in

```
. graph twoway bar yvar xvar, color(green)
```

The above would set all the bars (which are located at *xvar* and extend to *yvar*) to be green; see [G-2] **graph twoway bar** and [G-2] **graph twoway rbar**.

The `lcolor()`, `lwidth()`, `lpattern()`, and `lstyle()` options are also used to specify how plotted lines and spikes look for all of `graph twoway`'s range plots, paired-coordinate plots, and for area plots, bar plots, spike plots, and dropline plots. For example,

```
. graph twoway rspike y1var y2var xvar, lcolor(red)
```

will set the color of the horizontal spikes between values of *y1var* and *y2var* to red.

Also see

[G-4] ***areastyle*** — Choices for look of regions

[G-4] ***colorstyle*** — Choices for color

[G-4] ***linestyle*** — Choices for overall look of lines

[G-4] ***linepatternstyle*** — Choices for whether lines are solid, dashed, etc.

[G-4] ***linewidthstyle*** — Choices for thickness of lines

[G-2] **graph bar** — Bar charts

[G-2] **graph twoway bar** — Twoway bar plots

[G-2] **graph twoway rbar** — Range plot with bars

Title

[G-3] *blabel_option* — Option for labeling bars

Syntax

`graph {bar|hbar} ..., ... blabel(`*what* [`,` *where_and_how*]`) ...`

what	Description
`none`	no label; the default
`bar`	label is bar height
`total`	label is cumulative bar height
`name`	label is name of *yvar*
`group`	label is first `over()` group

where_and_how	Description
`position(outside)`	place label just above the bar (`bar`) or just to its right (`hbar`)
`position(inside)`	place label inside the bar at the top (`bar`) or at rightmost extent (`hbar`)
`position(base)`	place label inside the bar at the bar's base
`position(center)`	place label inside the bar at the bar's center
`gap(`*relativesize*`)`	distance from position
`format(`*%fmt*`)`	format if `bar` or `total`
textbox_options	look of label

See [G-4] ***relativesize*** and [G-3] ***textbox_options***.

Description

Option `blabel()` is for use with `graph bar` and `graph hbar`; see [G-2] **graph bar**. It adds a label on top of or inside each bar.

Options

`blabel(`*what*`,` *where_and_how*`)` specifies the label and where it is to be located relative to the bar. *where_and_how* is optional and is documented under *Suboptions for use with blabel()* below. *what* specifies the contents of the label.

`blabel(bar)` specifies that the label be the height of the bar. In

```
. graph bar (mean) empcost, over(division) blabel(bar)
```

the labels would be the mean employee cost.

`blabel(total)` specifies that the label be the cumulative height of the bar. `blabel(total)` is for use with `graph bar`'s `stack` option. In

```
. graph bar (sum) cost1 cost2, stack over(group) blabel(total)
```

the labels would be the total height of the stacked bar—the sum of costs. Also, the `cost1` part of the stack bar would be labeled with its height.

`blabel(name)` specifies that the label be the name of the *yvar*. In

```
. graph bar (mean) y1 y2 y3 y4, blabel(name)
```

The labels would be "`mean of y1`", "`mean of y2`", ..., "`mean of y4`". Usually, you would also want to suppress the legend here and so would type

```
. graph bar (mean) y1 y2 y3 y4, blabel(name) legend(off)
```

`blabel(group)` specifies that the label be the name of the first `over()` group. In

```
. graph bar cost, over(division) over(year) blabel(group)
```

the labels would be the name of the divisions. Usually, you would also want to suppress the appearance of the division labels on the axis:

```
. graph bar cost, over(division, axis(off)) over(year) blabel(group)
```

Suboptions for use with blabel()

`position()` specifies where the label is to appear.

`position(outside)` is the default. The label appears just above the bar (`graph bar`) or just to its right (`graph hbar`).

`position(inside)` specifies that the label appear inside the bar, at the top (`graph bar`) or at its rightmost extent (`graph hbar`).

`position(base)` specifies that the label appear inside the bar, at the bar's base; at the bottom of the bar (`graph bar`); or at the left of the bar (`graph hbar`).

`position(center)` specifies that the label appear inside the bar, at its center.

`gap(`*relativesize*`)` specifies a distance by which the label is to be offset from its location (`outside`, `inside`, `base`, or `center`). The default is usually `gap(1.7)`. The `gap()` may be positive or negative and you can specify, for instance, `gap(*1.2)` and `gap(*.8)` to increase or decrease the gap by 20%; see [G-4] ***relativesize***.

`format(%`*fmt*`)` is for use with `blabel(bar)` and `blabel(total)`; it specifies the display format to be used to format the height value. See [D] **format**.

textbox_options are any of the options allowed with a textbox. Important options include `size()`, which determines the size of the text; `box`, which draws a box around the text; and `color()`, which determines the color of the text. See [G-3] ***textbox_options***.

Remarks

`blabel()` serves two purposes: to increase the information content of the chart (`blabel(bar)` and `blabel(total)`) or to change how bars are labeled (`blabel(name)` and `blabel(group)`).

Remarks are presented under the following headings:

Increasing the information content
Changing how bars are labeled

Increasing the information content

Under the heading *Multiple bars (overlapping the bars)* in [G-2] **graph bar**, the following graph was drawn:

```
. graph bar (mean) tempjuly tempjan, over(region)
        bargap(-30)
        legend(label(1 "July") label(2 "January"))
        ytitle("Degrees Fahrenheit")
        title("Average July and January temperatures")
        subtitle("by regions of the United States")
        note("Source:  U.S. Census Bureau, U.S. Dept. of Commerce")
```

To the above, we now add

```
blabel(bar, position(inside) format(%9.1f))
```

which will add the average temperature to the bar, position the average inside the bar, at the top, and format its value by using `%9.1f`:

```
. graph bar (mean) tempjuly tempjan, over(region)
        bargap(-30)
        legend(label(1 "July") label(2 "January"))
        ytitle("Degrees Fahrenheit")
        title("Average July and January temperatures")
        subtitle("by regions of the United States")
        note("Source:  U.S. Census Bureau, U.S. Dept. of Commerce")
        blabel(bar, position(inside) format(%9.1f) color(white))      ← new
```

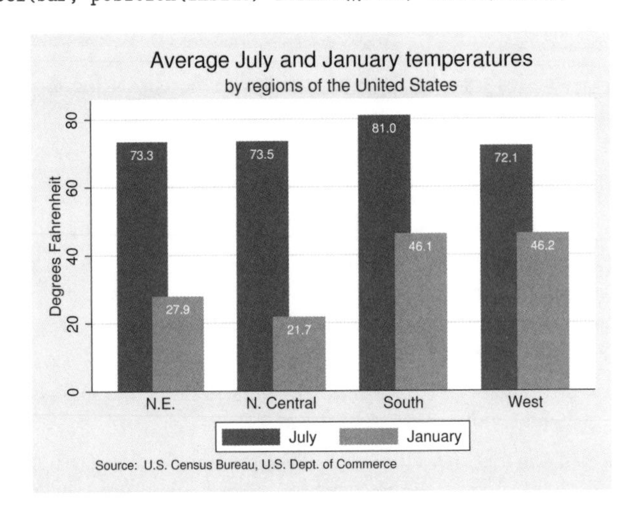

We also specified the *textbox_option* `color(white)` to change the color of the text; see [G-3] ***textbox_options***. Dark text on a dark bar would have blended too much.

Changing how bars are labeled

Placing the labels on the bars works especially well with horizontal bar charts:

```
. use http://www.stata-press.com/data/r12/nlsw88, clear
(NLSW, 1988 extract)
```

```
. graph hbar (mean) wage,
        over(occ, axis(off) sort(1))
        blabel(group, position(base) color(bg))
        ytitle("")
        by(union,
            title("Average Hourly Wage, 1988, Women Aged 34-46")
            note("Source:  1988 data from NLS, U.S. Dept. of Labor,
                  Bureau of Labor Statistics")
        )
```

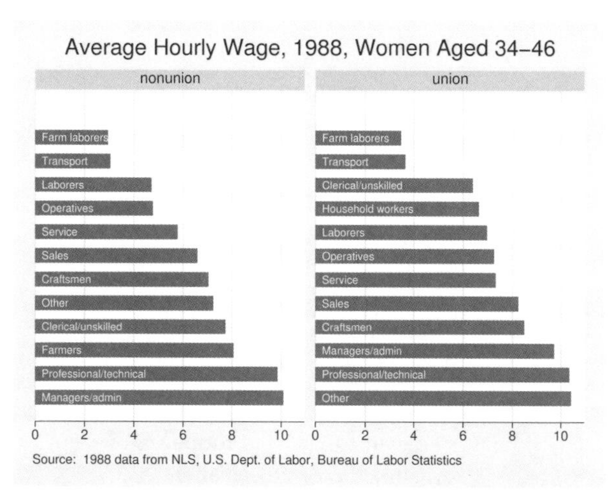

What makes moving the labels from the axis to the bars work so well here is that it saves so much horizontal space.

In the above command, note the first two option lines:

```
      over(occ, axis(off) sort(1))
   blabel(group, position(base) color(bg))
```

`blabel(group)` puts the occupation labels on top of the bars, and suboption `position(base)` located the labels at the base of each bar. We specified `over(,axis(off))` to prevent the labels from appearing on the axis. Let us run though all the options:

`over(occ, axis(off) sort(1))`
: Specified that the chart be done over occupation, that the occupation labels not be shown on the axis, and that the bars be sorted by the first (and only) *yvar*, namely, `(mean) wage`.

`ytitle("")`
: Specified that the title on the numerical y axis (the horizontal axis in this horizontal case) be suppressed.

`by(union, title(...) note(...))`
: Specified that the entire graph be repeated by values of variable `union`, and specified that the title and note be added to the overall graph. Had we specified the `title()` and `note()` options outside the `by()`, they would have been placed on each graph.

Also see

[G-2] **graph bar** — Bar charts

Title

[G-3] *by_option* — Option for repeating graph command

Syntax

by_option	Description
`by(`*varlist* [`,` *byopts*] `)`	repeat for by-groups

`by()` is *merged-implicit*; see [G-4] **concept: repeated options**.

byopts	Description
`total`	add total group
`missing`	add missing groups
`colfirst`	display down columns
`rows(`*#*`)` | `cols(`*#*`)`	display in # rows or # columns
`holes(`*numlist*`)`	positions to leave blank
`iscale(`[`*`]*#*`)`	size of text and markers
`compact`	synonym for `style(compact)`
`style(`*bystyle*`)`	overall style of presentation
[`no`]`edgelabel`	label x axes of edges
[`no`]`rescale`	separate y and x scales for each group
[`no`]`yrescale`	separate y scale for each group
[`no`]`xrescale`	separate x scale for each group
[`no`]`iyaxes`	show individual y-axes
[`no`]`ixaxes`	show individual x-axes
[`no`]`iytick`	show individual y-axes ticks
[`no`]`ixtick`	show individual x-axes ticks
[`no`]`iylabel`	show individual y-axes labels
[`no`]`ixlabel`	show individual x-axes labels
[`no`]`iytitle`	show individual y-axes titles
[`no`]`ixtitle`	show individual x-axes titles
`imargin(`*marginstyle*`)`	margin between graphs
legend_options	show legend and placement of legend
title_options	overall titles
region_options	overall outlining, shading, and aspect

See [G-4] ***bystyle***, [G-4] ***marginstyle***, [G-3] ***title_options***, and [G-3] ***region_options***.

The *title_options* and *region_options* on the command on which `by()` is appended will become the titles and regions for the individual by-groups.

Description

Option `by()` draws separate plots within one graph. *varlist* may be a numeric or a string variable.

Option

`by(`*varlist* [`,` *byopts*]`)` specifies that the `graph` command be repeated for each unique set of values of *varlist* and that the resulting individual graphs be arrayed into one graph.

byopts

`total` specifies that, in addition to the graphs for each by-group, a graph be added for all by-groups combined.

`missing` specifies that, in addition to the graphs for each by-group, graphs be added for missing values of *varlist*. Missing is defined as `.`, `.a`, `...`, `.z` for numeric variables and `""` for string variables.

`colfirst` specifies that the individual graphs be arrayed down the columns rather than across the rows. That is, if there were four groups, the graphs would be displayed

default	`colfirst`
1 2	1 3
3 4	2 4

`rows(`*#*`)` and `cols(`*#*`)` are alternatives. They specify that the resulting graphs be arrayed as # rows and however many columns are necessary, or as # columns and however many rows are necessary. The default is

`cols(`*c*`)`, c = ceil(sqrt(G))

where G is the total number of graphs to be presented and `ceil()` is the function that rounds nonintegers up to the next integer. For instance, if four graphs are to be displayed, the result will be presented in a 2×2 array. If five graphs are to be displayed, the result will be presented as a 2×3 array because `ceil(sqrt(5))==3`.

`cols(`*#*`)` may be specified as larger or smaller than c; r will be the number of rows implied by c. Similarly, `rows(`*#*`)` may be specified as larger or smaller than r.

`holes(`*numlist*`)` specifies which positions in the array are to be left unfilled. Consider drawing a graph with three groups and assume that the three graphs are being displayed in a 2×2 array. By default, the first group will appear in the graph at (1,1), the second in the graph at (1,2), and the third in the graph at (2,1). Nothing will be displayed in the (2,2) position.

Specifying `holes(3)` would cause position (2,1) to be left blank, so the third group would appear in (2,2).

The numbers that you specify in `holes()` correspond to the position number,

```
1  2     1  2  3     1  2  3  4      1  2  3  4  5
3  4     4  5  6     5  6  7  8      6  7  8  9 10
         7  8  9     9 10 11 12     11 12 13 14 15
                    12 14 15 16     16 17 18 19 20
                                    21 22 23 24 25      etc.
```

The above is the numbering when `colfirst` is not specified. If `colfirst` is specified, the positions are transposed:

```
1 3      1 4 7      1 5  9 13      1  6 11 16 21
2 4      2 5 8      2 6 10 14      2  7 12 17 22
         3 6 9      3 7 11 15      3  8 13 18 23
                    4 8 12 16      4  9 14 19 24
                                   5 10 15 20 25      etc.
```

`iscale(#)` and `iscale(*#)` specify a size adjustment (multiplier) to be used to scale the text and markers.

By default, `iscale()` gets smaller and smaller the larger is G, the number of by-groups and hence the number of graphs presented. The default is parameterized as a multiplier f(G)—$0 < \mathrm{f}(G) < 1$, $\mathrm{f}'(G) < 0$—that is used to multiply `msize()`, $\{$ `y` | `x` $\}$ `label(,labsize())`, and the like. The size of everything except the overall titles, subtitles, captions, and notes is affected by `iscale()`.

If you specify `iscale(#)`, the number you specify is substituted for f(G). `iscale(1)` means text and markers should appear at the same size as they would were each graph drawn separately. `iscale(.5)` displays text and markers at half that size. We recommend you specify a number between 0 and 1, but you are free to specify numbers larger than 1.

If you specify `iscale(*#)`, the number you specify is multiplied by f(G) and that product is used to scale text and markers. `iscale(*1)` is the default. `iscale(*1.2)` means text and markers should appear 20% larger than `graph, by()` would usually choose. `iscale(*.8)` would make them 20% smaller.

`compact` is a synonym for `style(compact)`. It makes no difference which you type. See the description of the `style()` option below, and see *By-styles* under *Remarks*.

`style(`*bystyle*`)` specifies the overall look of the by-graphs. The style determines whether individual graphs have their own axes and labels or if instead the axes and labels are shared across graphs arrayed in the same row or in the same column, how close the graphs are to be placed to each other, etc. The other options documented below will allow you to change the way the results are displayed, but the *bystyle* specifies the starting point.

You need not specify `style()` just because there is something you want to change. You specify `style()` when another style exists that is exactly what you desire or when another style would allow you to specify fewer changes to obtain what you want.

See [G-4] ***bystyle*** for a list of by-style choices. The *byopts* listed below modify the by-style:

`edgelabel` and `noedgelabel` specify whether the last graphs of a column that do not appear in the last row are to have their x axes labeled. See *Labeling the edges* under *Remarks* below.

`rescale`, `yrescale`, and `xrescale` (and `norescale`, `noyrescale`, and `noxrescale`) specify that the scales of each graph be allowed to differ (or forced to be the same). Whether *X* or `no`*X* is the default is determined by `style()`.

Usually, `no`*X* is the default and `rescale`, `yrescale`, and `xrescale` are the options. By default, all the graphs will share the same scaling for their axes. Specifying `yrescale` will allow the y scales to differ across the graphs, specifying `xrescale` will allow the x scales to differ, and specifying `rescale` is equivalent to specifying `yrescale` and `xrescale`.

`iyaxes` and `ixaxes` (and `noiyaxes` and `noixaxes`) specify whether the y axes and x axes are to be displayed with each graph. The default with most styles and schemes is to place y axes on the leftmost graph of each row and to place x axes on the bottommost graph of each column. The y and x axes include the default ticks and labels but exclude the axes titles.

`iytick` and `ixtick` (and `noiytick` and `noixtick`) are seldom specified. If you specified `iyaxis` and then wanted to suppress the ticks, you could also specify `noiytick`. In the rare event where specifying `iyaxis` did not result in the ticks being displayed (because of how the style or scheme works), specifying `iytick` would cause the ticks to be displayed.

`iylabel` and `ixlabel` (and `noiylabel` and `noixlabel`) are seldom specified. If you specified `iyaxis` and then wanted to suppress the axes labels, you could also specify `noiylabel`. In the rare event where specifying `iyaxis` did not result in the labels being displayed (because of how the style or scheme works), specifying `iylabel` would cause the labels to be displayed.

`iytitle` and `ixtitle` (and `noiytitle` and `noixtitle`) are seldom specified. If you specified `iyaxis` and then wanted to add the y-axes titles (which would make the graph appear busy), you could also specify `iytitle`. In the rare event where specifying `iyaxis` resulted in the titles being displayed (because of how the style or scheme works), specifying `noiytitle` would suppress displaying the title.

`imargin(`*marginstyle*`)` specifies the margins between the individual graphs.

legend_options used within `by()` sets whether the legend is drawn and the legend's placement; see *Use of legends with by()* below. The `legend()` option is normally *merged-implicit*, but when used inside `by()`, it is *unique*; see [G-4] **concept: repeated options**

Remarks

Remarks are presented under the following headings:

Typical use
Placement of graphs
Treatment of titles
by() uses subtitle() with graph
Placement of the subtitle()
by() uses the overall note()
Use of legends with by()
By-styles
Labeling the edges
Specifying separate scales for the separate plots
History

Typical use

One often has data that divide into different groups—person data where the persons are male or female or in different age categories (or both), country data where the countries can be categorized into different regions of the world, or, as below, automobile data where the cars are foreign or domestic. If you type

```
. scatter mpg weight
```

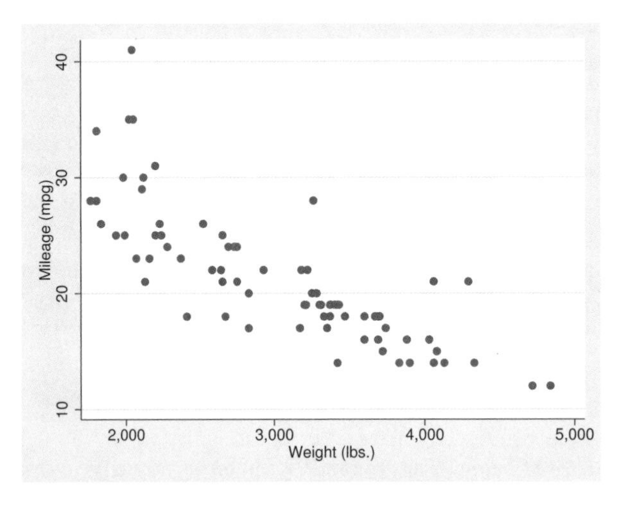

you obtain a scatterplot of `mpg` versus `weight`. If you add `by(foreign)` as an option, you obtain two graphs, one for each value of `foreign`:

```
. scatter mpg weight, by(foreign)
```

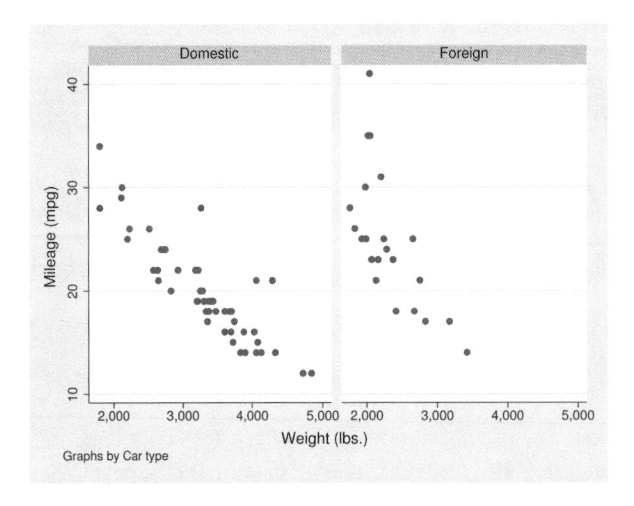

If you add `total`, another graph will be added representing the overall total:

```
. scatter mpg weight, by(foreign, total)
```

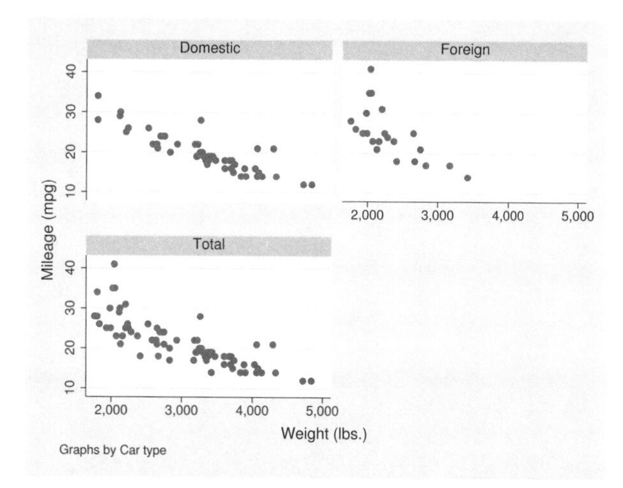

Here there were three graphs to be presented and `by()` chose to display them in a 2×2 array, leaving the last position empty.

Placement of graphs

By default, by() places the graphs in a rectangular $R \times C$ array and leaves empty the positions at the end:

Number of graphs	Array dimension	Positions left empty			
1	1×1				
2	1×2				
3	2×2	4=(2,2)			
4	2×2				
5	2×3	6=(3,3)			
6	2×3				
7	3×3	8=(3,2)	9=(3,3)		
8	3×3	9=(3,3)			
9	3×3				
10	3×4	11=(3,3)	12=(3,4)		
11	3×4	12=(3,4)			
12	3×4				
13	4×4	14=(4,2)	15=(4,3)	16=(4,4)	
14	4×4	15=(4,3)	16=(4,4)		
15	4×4	16=(4,4)			
16	4×4				
17	4×5	18=(4,3)	19=(4,4)	20=(4,5)	
18	4×5	19=(4,4)	20=(4,5)		
19	4×5	20=(4,5)			
20	4×5				
21	5×5	22=(5,2)	23=(5,3)	24=(5,4)	25=(5,5)
22	5×5	23=(5,3)	24=(5,4)	25=(5,5)	
23	5×5	24=(5,4)	25=(5,5)		
24	5×5	25=(5,5)			
25	5×5				
etc.					

Options rows(), cols(), and holes() allow you to control this behavior.

You may specify either rows() or cols(), but not both. In the previous section, we drew

```
. scatter mpg weight, by(foreign, total)
```

and had three graphs displayed in a 2×2 array with a hole at 4. We could draw the graph in a 1×3 array by specifying either rows(1) or cols(3),

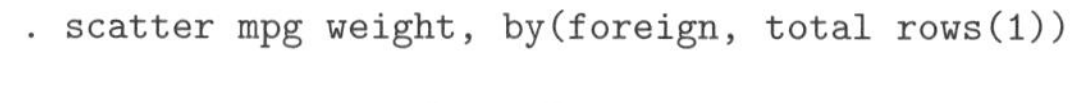

```
. scatter mpg weight, by(foreign, total rows(1))
```

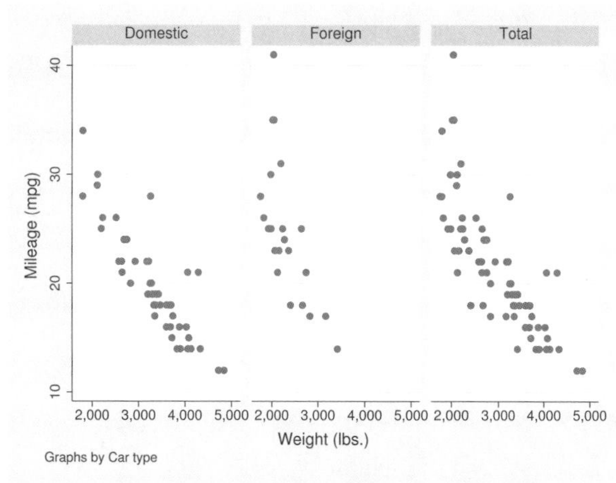

or we could stay with the 2×2 array and move the hole to 3,

```
. scatter mpg weight, by(foreign, total holes(3))
```

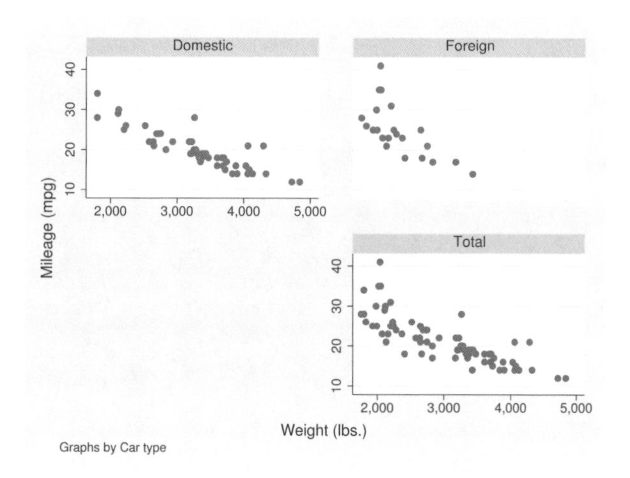

Treatment of titles

Were you to type

```
. scatter yvar xvar, title("My title") by(catvar)
```

"My title" will be repeated above each graph. `by()` repeats the entire `graph` command and then arrays the results.

To specify titles for the entire graph, specify the *title_options*—see [G-3] ***title_options***—inside the `by()` option:

```
. scatter yvar xvar, by(catvar, title("My title"))
```

by() uses subtitle() with graph

by() labels each graph by using the subtitle() *title_option*. For instance, in

```
. scatter mpg weight, by(foreign, total)
```

by() labeled the graphs "Domestic", "Foreign", and "Total". The subtitle "Total" is what by() uses when the total option is specified. The other two subtitles by() obtained from the by-variable foreign.

by() may be used with numeric or string variables. Here foreign is numeric but happens to have a value label associated with it. by() obtained the subtitles "Domestic" and "Foreign" from the value label. If foreign had no value label, the first two graphs would have been subtitled "0" and "1", the numeric values of variable foreign. If foreign had been a string variable, the subtitles would have been the string contents of foreign.

If you wish to suppress the subtitle, type

```
. scatter mpg weight, subtitle("") by(foreign, total)
```

If you wish to add "*Extra info*" to the subtitle, type

```
. scatter mpg weight, subtitle("Extra info", suffix) by(foreign, total)
```

Be aware, however, that "*Extra info*" will appear above each graph.

Placement of the subtitle()

You can use subtitle()'s suboptions to control the placement of the identifying label. For instance,

```
. scatter mpg weight,
        subtitle(, ring(0) pos(1) nobexpand) by(foreign, total)
```

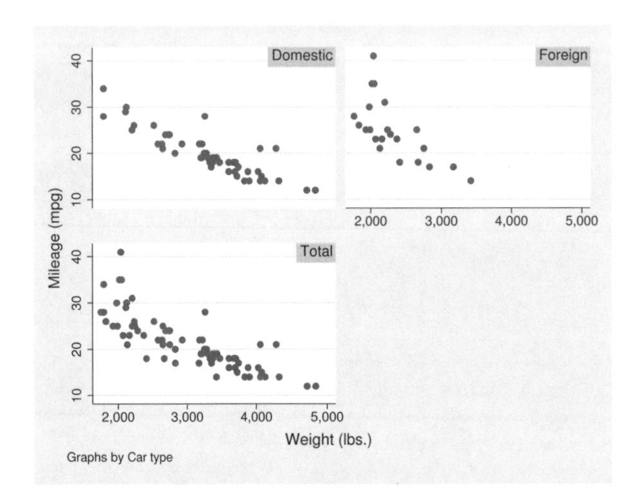

The result will be to move the identifying label inside the individual graphs, displaying it in the northeast corner of each. Type

```
. scatter mpg weight,
        subtitle(, ring(0) pos(11) nobexpand) by(foreign, total)
```

and the identifying label will be moved to the northwest corner.

`ring(0)` moves the subtitle inside the graph's plot region and `position()` defines the location, indicated as clock positions. `nobexpand` is rather strange, but just remember to specify it. By default, `by()` sets subtitles to expand to the size of the box that contains them, which is unusual but makes the default-style subtitles look good with shading.

See [G-3] ***title_options***.

by() uses the overall note()

By default, `by()` adds an overall note saying "Graphs by ...". When you type

```
. scatter yvar xvar, by(catvar)
```

results are the same as if you typed

```
. scatter yvar xvar, by(catvar, note("Graphs by ..."))
```

If you want to suppress the note, type

```
. scatter yvar xvar, by(catvar, note(""))
```

If you want to change the overall note to read "My note", type

```
. scatter yvar xvar, by(catvar, note("My note"))
```

If you want to add your note after the default note, type

```
. scatter yvar xvar, by(catvar, note("My note", suffix))
```

Use of legends with by()

If you wish to modify or suppress the default legend, you must do that differently when `by()` is specified. For instance, `legend(off)`—see [G-3] ***legend_options***—will suppress the legend, yet typing

```
. line y1 y2 x, by(group) legend(off)
```

will not have the intended effect. The `legend(off)` will seemingly be ignored. You must instead type

```
. line y1 y2 x, by(group, legend(off))
```

We moved `legend(off)` inside the `by()`.

Remember that `by()` repeats the `graph` command. If you think carefully, you will realize that the legend never was displayed at the bottom of the individual plots. It is instructive to type

```
. line y1 y2 x, legend(on) by(group)
```

This graph will have many legends: one underneath each of the plots in addition to the overall legend at the bottom of the graph! `by()` works exactly as advertised: it repeats the entire graph command for each value of `group`.

In any case, it is the overall `legend()` that we want to suppress, and that is why we must specify `legend(off)` inside the `by()` option; this is the same issue as the one discussed under *Treatment of titles* above.

The issue becomes a little more complicated when, rather than suppressing the legend, we wish to modify the legend's contents or position. Then the `legend()` option to modify the contents is specified outside the `by()` and the `legend()` option to modify the location is specified inside. See *Use of legends with by()* in [G-3] ***legend_options***.

By-styles

Option `style(`*bystyle*`)` specifies the overall look of by-graphs; see [G-4] ***bystyle*** for a list of *bystyle* choices. One *bystyle* worth noting is `compact`. Specifying `style(compact)` causes the graph to be displayed in a more compact format. Compare

```
. use http://www.stata-press.com/data/r12/lifeexp, clear
(Life expectancy, 1998)
. scatter lexp gnppc, by(region, total)
```

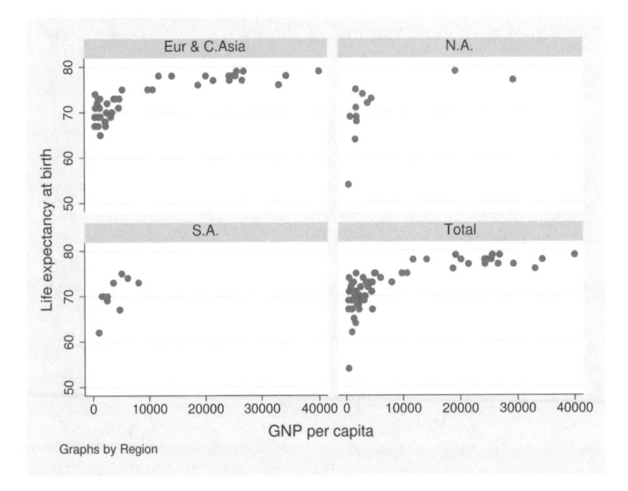

with

```
. scatter lexp gnppc, by(region, total style(compact))
```

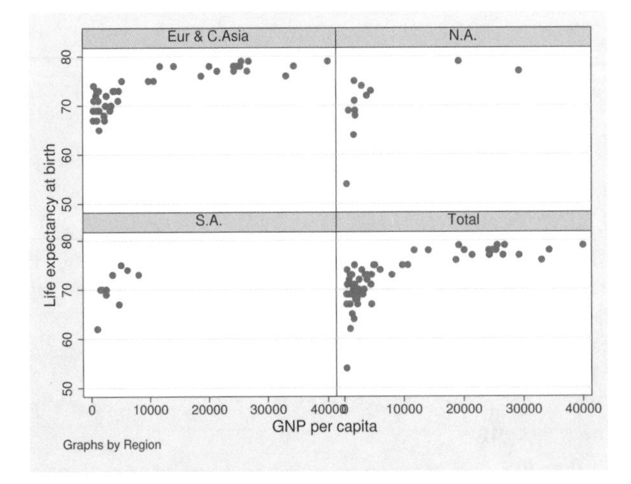

`style(compact)` pushes the graphs together horizontally and vertically, leaving more room for the individual graphs. The disadvantage is that, pushed together, the values on the axes labels sometimes run into each other, as occurred above with the 40,000 of the S.A. graph running into the 0 of the Total graph. That problem could be solved by dividing `gnppc` by 1,000.

Rather than typing out `style(compact)`, you may specify `compact`, and you may further abbreviate that as `com`.

Labeling the edges

Consider the graph

```
. sysuse auto
(1978 Automobile Data)
. scatter mpg weight, by(foreign, total)
```

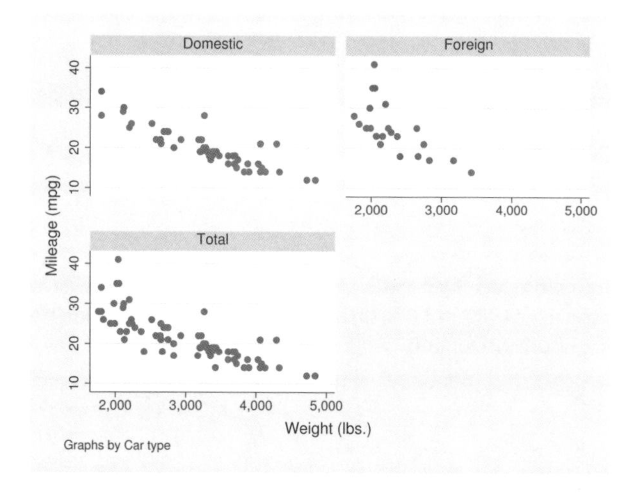

The x axis is labeled in the graph in the (1,2) position. When the last graph of a column does not appear in the last row, its x axis is referred to as an edge. In `style(default)`, the default is to label the edges, but you could type

```
. scatter mpg weight, by(foreign, total noedgelabel)
```

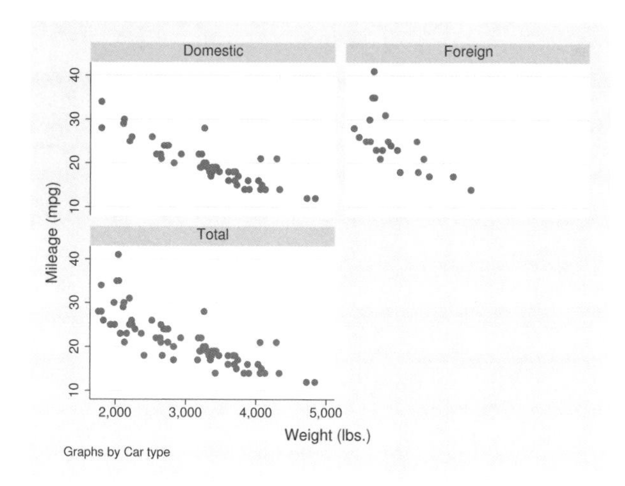

to suppress that. This results in the rows of graphs being closer to each other.

Were you to type

```
. scatter mpg weight, by(foreign, total style(compact))
```

you would discover that the x axis of the (1,2) graph is not labeled. With `style(compact)`, the default is `noedgelabel`, but you could specify `edgelabel` to override that.

Specifying separate scales for the separate plots

If you type

```
. scatter yvar xvar, by(catvar, yrescale)
```

each graph will be given a separately scaled y axis; if you type

```
. scatter yvar xvar, by(catvar, xrescale)
```

each graph will be given a separately scaled x axis; and if you type

```
. scatter yvar xvar, by(catvar, yrescale xrescale)
```

both scales will be separately set.

History

The twoway scatterplots produced by the `by()` option are similar to what are known as *casement displays* (see Chambers et al. [1983, 141–145]). A traditional casement display, however, aligns all the graphs either vertically or horizontally.

References

Buis, M. L., and M. Weiss. 2009. Stata tip 81: A table of graphs. *Stata Journal* 9: 643–647.

Chambers, J. M., W. S. Cleveland, B. Kleiner, and P. A. Tukey. 1983. *Graphical Methods for Data Analysis*. Belmont, CA: Wadsworth.

Cox, N. J. 2010. Speaking Stata: Graphing subsets. *Stata Journal* 10: 670–681.

Also see

[G-3] ***region_options*** — Options for shading and outlining regions and controlling graph size

[G-3] ***title_options*** — Options for specifying titles

Title

[G-3] ***cat_axis_label_options*** — Options for specifying look of categorical axis labels

Syntax

cat_axis_label_options	Description
`nolabels`	suppress axis labels
`ticks`	display axis ticks
`angle(`*anglestyle*`)`	angle of axis labels
`alternate`	offset adjacent labels
`tstyle(`*tickstyle*`)`	labels and ticks: overall style
`labgap(`*relativesize*`)`	labels: margin between tick and label
`labstyle(`*textstyle*`)`	labels: overall style
`labsize(`*textsizestyle*`)`	labels: size of text
`labcolor(`*colorstyle*`)`	labels: color of text
`tlength(`*relativesize*`)`	ticks: length
`tposition(outside` | `crossing` | `inside)`	ticks: position/direction
`tlstyle(`*linestyle*`)`	ticks: linestyle of
`tlwidth(`*linewidthstyle*`)`	ticks: thickness of line
`tlcolor(`*colorstyle*`)`	ticks: color of line

See [G-4] ***anglestyle***, [G-4] ***tickstyle***, [G-4] ***relativesize***, [G-4] ***textstyle***, [G-4] ***textsizestyle***, [G-4] ***colorstyle***, [G-4] ***linestyle***, and [G-4] ***linewidthstyle***.

Description

The *cat_axis_label_options* determine the look of the labels that appear on a categorical x axis produced by `graph bar`, `graph hbar`, `graph dot`, and `graph box`; see [G-2] **graph bar**, [G-2] **graph dot**, and [G-2] **graph box**. These options are specified inside `label()` of `over()`:

```
. graph ..., over(varname, ... label(cat_axis_label_options) ...)
```

The most useful *cat_axis_label_options* are `angle()`, `alternate`, `labcolor()`, and `labsize()`.

Options

`nolabels` suppresses display of category labels on the axis. For `graph bar` and `graph hbar`, the `nolabels` option is useful when combined with the `blabel()` option used to place the labels on the bars themselves; see [G-3] ***blabel_option***.

`ticks` specifies that ticks appear on the categorical x axis. By default, ticks are not presented on categorical axes, and it is unlikely that you would want them to be.

`angle(`*anglestyle*`)` specifies the angle at which the labels on the axis appear. The default is `angle(0)`, meaning horizontal. With vertical bar charts and other vertically oriented charts, it is sometimes useful to specify `angle(90)` (vertical text reading bottom to top), `angle(-90)` (vertical text reading top to bottom), or `angle(-45)` (angled text reading top left to bottom right); see [G-4] ***anglestyle***.

Unix users: if you specify `angle(-45)`, results will appear on your screen as if you specified `angle(-90)`; results will appear correctly when you print.

`alternate` causes adjacent labels to be offset from one another and is useful when there are many labels or when labels are long. For instance, rather than obtaining an axis labeled,

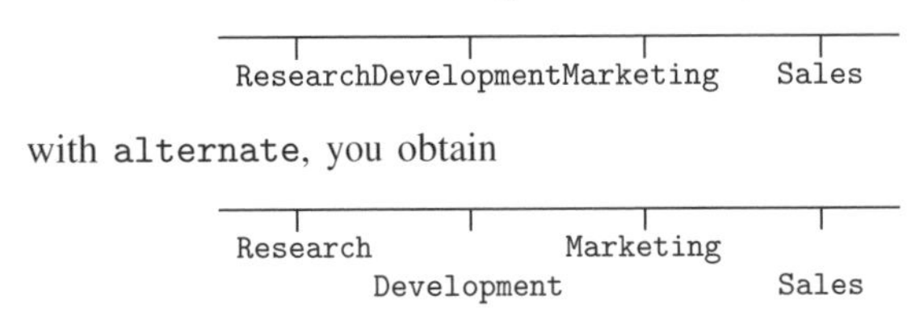

with `alternate`, you obtain

`tstyle(`*tickstyle*`)` specifies the overall look of labels and ticks; see [G-4] ***tickstyle***. Here the emphasis is on labels because ticks are usually suppressed on a categorical axis. The options documented below will allow you to change each attribute of the label and tick, but the *tickstyle* specifies the starting point.

You need not specify `tstyle()` just because there is something you want to change about the look of labels and ticks. You specify `tstyle()` when another style exists that is exactly what you desire or when another style would allow you to specify fewer changes to obtain what you want.

`labgap(`*relativesize*`)`, `labstyle(`*textstyle*`)`, `labsize(`*textsizestyle*`)`, and `labcolor(`*colorstyle*`)` specify details about how the labels are presented. Of particular interest are `labsize(`*textsizestyle*`)`, which specifies the size of the labels, and `labcolor(`*colorstyle*`)`, which specifies the color of the labels; see [G-4] ***textsizestyle*** and [G-4] ***colorstyle*** for a list of text sizes and color choices. Also see [G-4] ***relativesize*** and [G-4] ***textstyle***.

`tlength(`*relativesize*`)` specifies the overall length of the ticks; see [G-4] ***relativesize***.

`tposition(outside | crossing | inside)` specifies whether the ticks are to extend `outside` (from the axis out, the usual default), `crossing` (crossing the axis line, extending in and out), or `inside` (from the axis into the plot region).

`tlstyle(`*linestyle*`)`, `tlwidth(`*linewidthstyle*`)`, and `tlcolor(`*colorstyle*`)` specify other details about the look of the ticks. Ticks are just lines. See [G-4] **concept: lines** for more information.

Remarks

You draw a bar, dot, or box plot of `empcost` by `division`:

```
. graph ... empcost, over(division)
```

Seeing the result, you wish to make the text labeling the divisions 20% larger. You type:

```
. graph ... empcost, over(division, label(labsize(*1.2)))
```

Also see

[G-2] **graph bar** — Bar charts

[G-2] **graph box** — Box plots

[G-2] **graph dot** — Dot charts (summary statistics)

Title

[G-3] ***cat_axis_line_options*** — Options for specifying look of categorical axis line

Syntax

cat_axis_line_options	Description
`off` and `on`	suppress/force display of axis
`fill`	allocate space for axis even if `off`
`fextend`	extend axis line through plot region and plot region's margin
`extend`	extend axis line through plot region
`noextend`	do not extend axis line at all
`noline`	do not even draw axis line
`line`	force drawing of axis line
`titlegap(`*relativesize*`)`	margin between axis title and tick labels
`outergap(`*relativesize*`)`	margin outside of axis title
`lstyle(`*linestyle*`)`	overall style of axis line
`lcolor(`*colorstyle*`)`	color of axis line
`lwidth(`*linewidthstyle*`)`	thickness of axis line
`lpattern(`*linepatternstyle*`)`	whether axis solid, dashed, etc.

See [G-4] ***relativesize***, [G-4] ***linestyle***, [G-4] ***colorstyle***, [G-4] ***linewidthstyle***, and [G-4] ***linepatternstyle***.

Description

The *cat_axis_line_options* determine the look of the categorical x axis in `graph bar`, `graph hbar`, `graph dot`, and `graph box`; see [G-2] **graph bar**, [G-2] **graph dot**, [G-2] **graph box**. These options are rarely specified but when specified, they are specified inside `axis()` of `over()`:

```
. graph ..., over(varname, ... axis(cat_axis_line_options) ...)
```

Options

`off` and `on` suppress or force the display of the axis.

`fill` goes with `off` and is seldom specified. If you turned an axis off but still wanted the space to be allocated for the axis, you could specify `fill`.

`fextend`, `extend`, `noextend`, `line`, and `noline` determine how much of the line representing the axis is to be drawn. They are alternatives.

`noline` specifies that the line not be drawn at all. The axis is there, ticks and labels will appear, but the axis line itself will not be drawn.

`line` is the opposite of `noline`, for use if the axis line somehow got turned off.

`noextend` specifies that the axis line not extend beyond the range of the axis, defined by the first and last categories.

`extend` specifies that the line be longer than that and extend all the way across the plot region.

`fextend` specifies that the line be longer than that and extend across the plot region and across the plot region's margins. For a definition of the plot region's margins, see [G-3] ***region_options***. If the plot region has no margins (which would be rare), then `fextend` means the same as `extend`. If the plot region does have margins, `extend` would result in the y and x axes not meeting. With `fextend`, they touch.

`fextend` is the default with most schemes.

`titlegap(`*relativesize*`)` specifies the margin to be inserted between the axis title and the axis' tick labels; see [G-4] ***relativesize***.

`outergap(`*relativesize*`)` specifies the margin to be inserted outside the axis title; see [G-4] ***relativesize***.

`lstyle(`*linestyle*`)`, `lcolor(`*colorstyle*`)`, `lwidth(`*linewidthstyle*`)`, and `lpattern(`*linepatternstyle*`)` determine the overall look of the line that is the axis; see [G-4] **concept: lines**.

Remarks

The *cat_axis_label_options* are rarely specified.

Also see

[G-2] **graph bar** — Bar charts

[G-2] **graph box** — Box plots

[G-2] **graph dot** — Dot charts (summary statistics)

Title

[G-3] *clegend_option* — Option for controlling the contour-plot legend

Syntax

clegend_option	Description
clegend([*suboptions*])	contour-legend contents, appearance, and location

clegend() is *merged-implicit*; see [G-4] **concept: repeated options**.

suboptions	Description
Contour legend appearance	
width(*relativesize*)	width of contour key
height(*relativesize*)	height of contour key
altaxis	move the contour key's axis to the other side of the contour key
bmargin(*marginstyle*)	outer margin around legend
title_options	titles, subtitles, notes, captions
region(*roptions*)	borders and background shading
Contour legend location	
off or on	suppress or force display of legend
position(*clockposstyle*)	where legend appears
ring(*ringposstyle*)	where legend appears (detail)
bplacement(*compassdirstyle*)	placement of legend when positioned in the plotregion
at(#)	allowed with by() only

See [G-4] ***relativesize***, [G-4] ***marginstyle***, [G-3] ***title_options***.

See *Where contour legends appear* under *Remarks* below, and see *Positioning of titles* in [G-3] ***title_options*** for definitions of *clockposstyle* and *ringposstyle*.

roptions	Description
style(*areastyle*)	overall style of region
color(*colorstyle*)	line and fill color of region
fcolor(*colorstyle*)	fill color of region
lstyle(*linestyle*)	overall style of border
lcolor(*colorstyle*)	color of border
lwidth(*linewidthstyle*)	thickness of border
lpattern(*linepatternstyle*)	border pattern (solid, dashed, etc.)
margin(*marginstyle*)	margin between border and contents of legend

See [G-4] ***areastyle***, [G-4] ***colorstyle***, [G-4] ***linestyle***, [G-4] ***linewidthstyle***, [G-4] ***linepatternstyle***, and [G-4] ***marginstyle***.

Description

The `clegend()` option allows you to control the contents, appearance, and placement of the contour-plot legend.

Contour-plot legends have a single key that displays all the colors used to fill the contour areas. They also have a *c* axis that provides a scale for the key and associated contour plot. That axis is controlled using the *c*-axis option described in [G-3] ***axis_options***.

Option

`clegend(`*suboptions*`)` specifies the appearance of a contour-plot legend, along with how it is to look, and whether and where it is to be displayed.

Content and appearance suboptions for use with clegend()

`width(`*relativesize*`)` specifies the width of the contour key. See [G-4] ***relativesize***.

`height(`*relativesize*`)` specifies the height of the contour key. See [G-4] ***relativesize***.

`altaxis` specifies that the contour key's axis be placed on the alternate side of the contour key from the default side. For most schemes, this means that the axis is moved from the right side of the contour key to the left side.

`bmargin(`*marginstyle*`)` specifies the outer margin around the legend. That is, it specifies how close other things appearing near the legend can get. Also see suboption `margin()` under *Suboptions for use with clegend(region())* below for specifying the inner margin between the border and contents. See [G-4] ***marginstyle*** for a list of margin choices.

title_options allow placing titles, subtitles, notes, and captions on contour-plot legends. See [G-3] ***title_options***.

`region(`*roptions*`)` specifies the border and shading of the legend. You could give the legend a gray background tint by specifying `clegend(region(fcolor(gs9)))`. See *Suboptions for use with clegend(region())* below.

Suboptions for use with clegend(region())

`style(`*areastyle*`)` specifies the overall style of the region in which the legend appears. The other suboptions allow you to change the region's attributes individually, but `style()` provides the starting point. See [G-4] ***areastyle*** for a list of choices.

`color(`*colorstyle*`)` specifies the color of the background of the legend and the line used to outline it. See [G-4] ***colorstyle*** for a list of color choices.

`fcolor(`*colorstyle*`)` specifies the background (fill) color for the legend. See [G-4] ***colorstyle*** for a list of color choices.

`lstyle(`*linestyle*`)` specifies the overall style of the line used to outline the legend, which includes its pattern (solid, dashed, etc.), its thickness, and its color. The other suboptions listed below allow you to change the line's attributes individually, but `lstyle()` is the starting point. See [G-4] ***linestyle*** for a list of choices.

`lcolor(`*colorstyle*`)` specifies the color of the line used to outline the legend. See [G-4] ***colorstyle*** for a list of color choices.

`lwidth(`*linewidthstyle*`)` specifies the thickness of the line used to outline the legend. See [G-4] ***linewidthstyle*** for a list of choices.

`lpattern(`*linepatternstyle*`)` specifies whether the line used to outline the legend is solid, dashed, etc. See [G-4] ***linepatternstyle*** for a list of choices.

`margin(`*marginstyle*`)` specifies the inner margin between the border and the contents of the legend. Also see `bmargin()` under *Content and appearance suboptions for use with clegend()* above for specifying the outer margin around the legend. See [G-4] ***marginstyle*** for a list of margin choices.

Location suboptions for use with clegend()

`off` and `on` determine whether the legend appears. The default is `on` when a `twoway contour` plot appears in the graph. In those cases, `clegend(off)` will suppress the display of the legend.

`position(`*clockposstyle*`)`, `ring(`*ringposstyle*`)`, and `bplacement(`*compassdirstyle*`)` override the default location of the legend, which is usually to the right of the plot region. `position()` specifies a direction [*sic*] according to the hours on the dial of a 12-hour clock, and `ring()` specifies the distance from the plot region.

`ring(0)` is defined as being inside the plot region itself and allows you to place the legend inside the plot. `ring(`*k*`)`, $k > 0$, specifies positions outside the plot region; the larger the `ring()` value, the farther away the legend is from the plot region. `ring()` values may be integers or nonintegers and are treated ordinarily.

When `ring(0)` is specified, `bplacement()` further specifies where in the plot region the legend is placed. `bplacement(seast)` places the legend in the southeast (lower-right) corner of the plot region.

`position(12)` puts the legend directly above the plot region (assuming `ring()` > 0), `position(9)` directly to the left of the plot region, and so on.

See *Where contour legends appear* under *Remarks* below and *Positioning of titles* in [G-3] ***title_options*** for more information on the `position()` and `ring()` suboptions.

`at(#)` is for use only when the *twoway_option* `by()` is also specified. It specifies that the legend appear in the #th position of the $R \times C$ array of plots, using the same coding as `by(..., holes())`. See *Use of legends with by()* under *Remarks* below, and see [G-3] ***by_option***.

Remarks

Remarks are presented under the following headings:

When contour legends appear

Contour legends appear on the graph whenever the graph contains a `twoway contour` plot.

```
. use http://www.stata-press.com/data/r12/sandstone
(Subsea elevation of Lamont sandstone in an area of Ohio)
. twoway contour depth northing easting, levels(10)
```

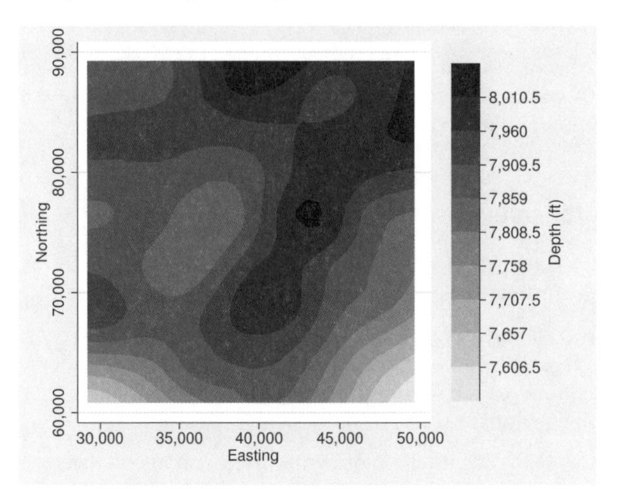

You can suppress the contour legend by specifying `clegend(off)`,

```
. twoway contour depth northing easting, levels(10) clegend(off)
```

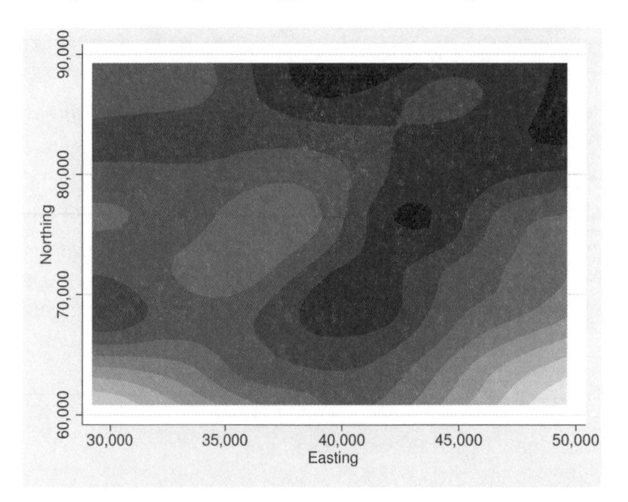

Where contour legends appear

By default, legends appear to the right of the plot region at what is technically referred to as `position(3) ring(3)`. Suboptions `position()` and `ring()` specify the location of the legend. `position()` specifies on which side of the plot region the legend appears—`position(3)` means 3 o'clock—and `ring()` specifies the distance from the plot region—`ring(3)` means farther out than the *title_option* `b2title()` but inside the *title_option* `note()`; see [G-3] ***title_options***.

If we specify `clegend(position(9))`, the legend will be moved to the 9 o'clock position:

```
. twoway contour depth northing easting, levels(10) clegend(pos(9))
```

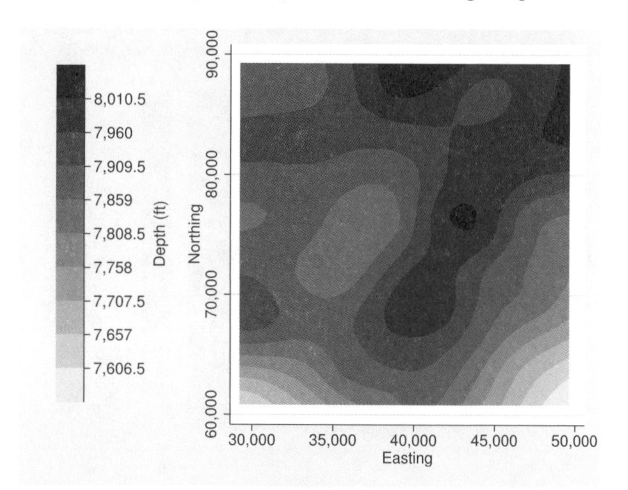

`ring()`—the suboption that specifies the distance from the plot region—is seldom specified, but, when it is specified, `ring(0)` is the most useful. `ring(0)` specifies that the legend be moved inside the plot region:

```
. twoway contour depth northing easting, levels(10) clegend(pos(5) ring(0))
```

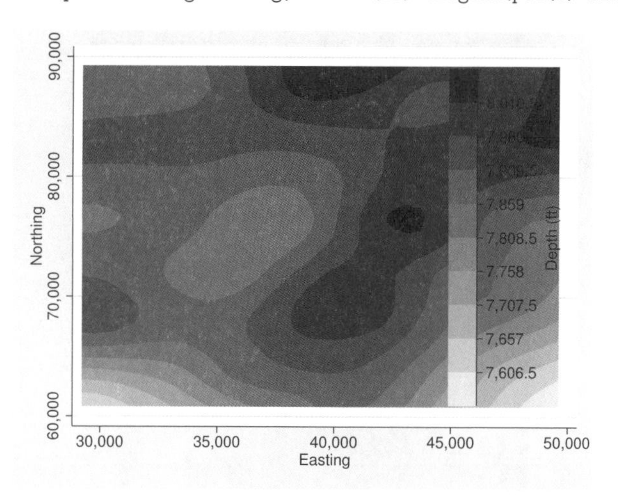

Our use of `position(5) ring(0)` put the legend inside the plot region, at 5 o'clock, meaning in the bottom right corner. Had we specified `position(2) ring(0)`, the legend would have appeared in the top right corner.

We might now add a background color to the legend:

```
. twoway contour depth northing easting, levels(10) clegend(pos(2)
> ring(0) region(fcolor(gs15)))
```

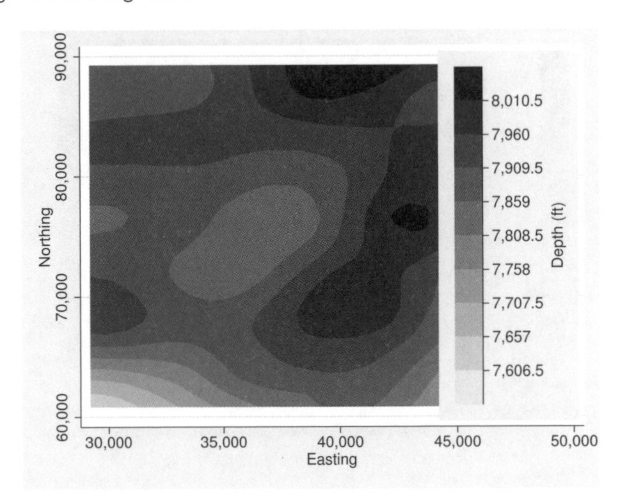

Putting titles on contour legends

By default, the z axis of a contour legend displays the z variable label or variable name as a title. You can suppress this axis title. You can also add an overall title for the legend. We do that for the previous graph by adding the `ztitle("")` and `clegend(title("Depth"))` options:

```
. twoway contour depth northing easting, levels(10) ztitle("")
> clegend(title("Depth") region(fcolor(gs15)))
```

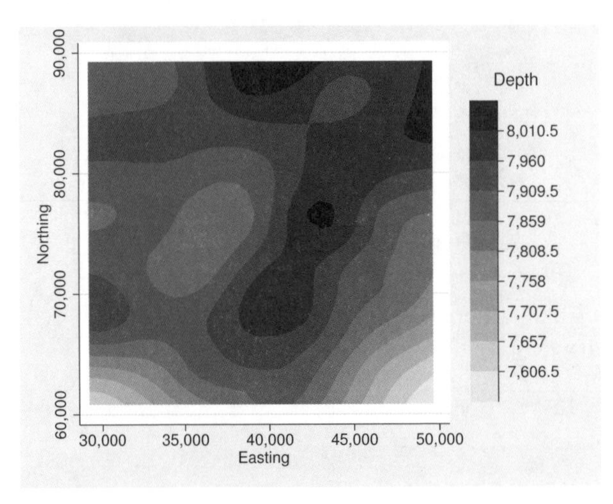

Legends may also contain `subtitles()`, `notes()`, and `captions()`, though these are rarely used; see [G-3] ***title_options***.

Controlling the axis in contour legends

Contour-plot legends contain a z axis. You control this axis just as you would the x or y axis of a graph. Here we specify cutpoints for the contours and custom tick labels using the zlabel() option,

```
. twoway contour depth northing easting, levels(10)
> zlabel(7600 "low" 7800 "medium" 8000 "high") region(fcolor(gs15)))
```

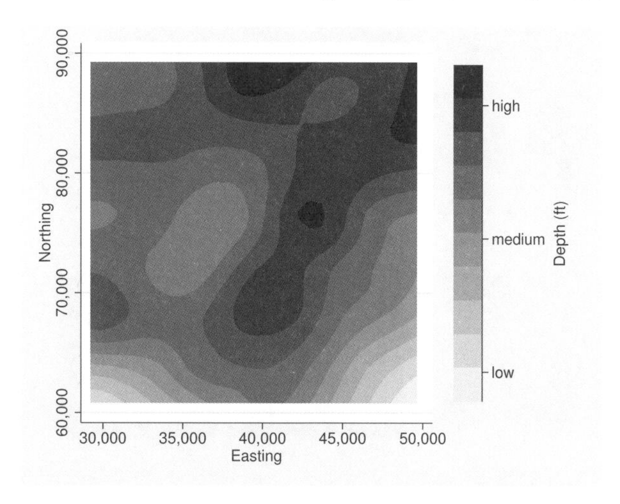

Minor ticks, axis scale (logged, reversed, etc.), and all other aspects of the z axis can be controlled using the zlabel(), zmlabel(), ztick(), zmtick(), zscale(), and ztitle() options; see [G-3] ***axis_options***.

Use of legends with by()

Legends are omitted by default when by() is specified. You can turn legends on by specifying clegend(on) within by(). It will show in the default location.

```
. use http://www.stata-press.com/data/r12/surface
(NOAA Sea Surface Temperature)
. twoway contour temperature longitude latitude, level(10)
> xlabel(,format(%9.0f)) by(date, clegend(on))
```

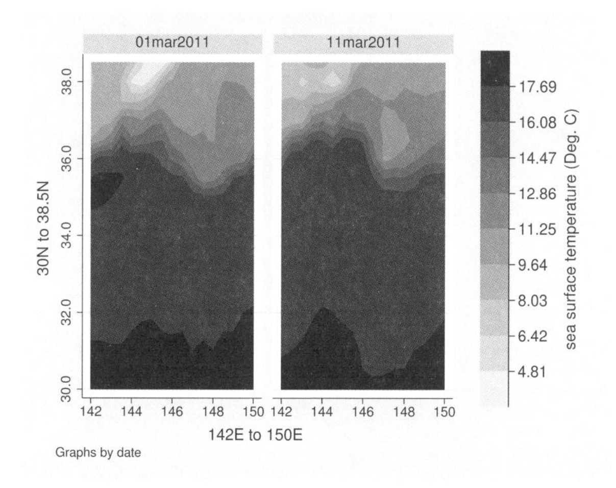

If you want to move the legend, consider the different options and their placement on the command line. *Location suboptions for use with clegend( )* should be specified within the by() option, whereas *Content and appearance suboptions for use with clegend( )* should be specified outside the by() option. For example, the position() option changes where the legend appears, so it would be specified within the by() option:

```
. twoway contour temperature longitude latitude, level(10)
> xlabel(,format(%9.0f)) by(date, clegend(on pos(9)))
```

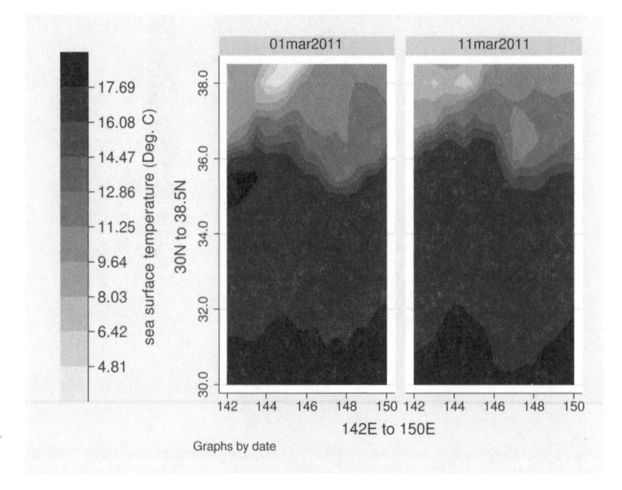

If you want to also change the appearance of the legend, specify an additional clegend() option outside the by() option:

```
. twoway contour temperature longitude latitude, level(10)
> xlabel(,format(%9.0f)) clegend(on width(15)) by(date, clegend(on pos(9)))
```

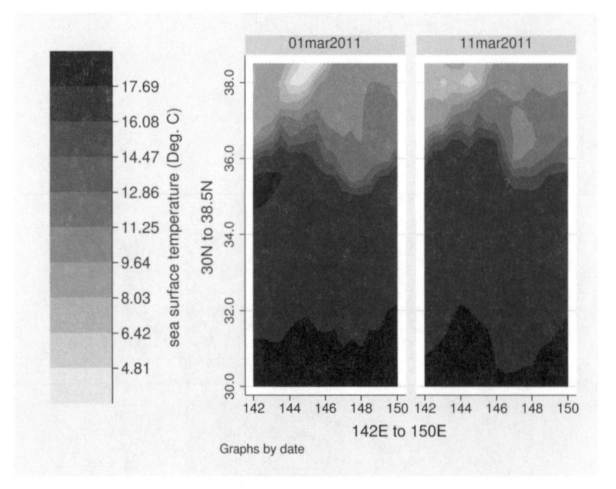

If you specify the location suboptions outside the by() option, the location suboptions will be ignored.

Also see

[G-2] **graph twoway contour** — Twoway contour plot with area shading

[G-2] **graph twoway contourline** — Twoway contour-line plot

[G-3] ***title_options*** — Options for specifying titles

Title

[G-3] *cline_options* — Options for connecting points with lines (subset of connect options)

Syntax

cline_options	Description
`connect(`*connectstyle*`)`	how to connect points
`lpattern(`*linepatternstyle*`)`	line pattern (solid, dashed, etc.)
`lwidth(`*linewidthstyle*`)`	thickness of line
`lcolor(`*colorstyle*`)`	color of line
`lstyle(`*linestyle*`)`	overall style of line
`pstyle(`*pstyle*`)`	overall plot style, including linestyle
`recast(`*newplottype*`)`	advanced; treat plot as *newplottype*

See [G-4] ***connectstyle***, [G-4] ***linepatternstyle***, [G-4] ***linewidthstyle***, [G-4] ***colorstyle***, [G-4] ***linestyle***, [G-4] ***pstyle***, and [G-3] ***advanced_options***.

All options are *rightmost*; see [G-4] **concept: repeated options**.

Some plots do not allow `recast()`.

Description

The *cline_options* specify how points on a graph are to be connected.

In certain contexts (for example, `scatter`; see [G-2] **graph twoway scatter**), the `lpattern()`, `lwidth()`, `lcolor()`, and `lstyle()` options may be specified with a list of elements, with the first element applying to the first variable, the second element to the second variable, and so on. For information on specifying lists, see [G-4] ***stylelists***.

Options

`connect(connectstyle)` specifies whether points are to be connected and, if so, how the line connecting them is to be shaped; see [G-4] ***connectstyle***. The line between each pair of points can connect them directly or in stairstep fashion.

`lpattern(`*linepatternstyle*`)`, `lwidth(`*linewidthstyle*`)`, `lcolor(`*colorstyle*`)`, and `lstyle(`*linestyle*`)` determine the look of the line used to connect the points; see [G-4] **concept: lines**. Note the `lpattern()` option, which allows you to specify whether the line is solid, dashed, etc.; see [G-4] ***linepatternstyle*** for a list of line-pattern choices.

`pstyle(`*pstyle*`)` specifies the overall style of the plot, including not only the *linestyle*, but also all other settings for the look of the plot. Only the *linestyle* affects the look of line plots. See [G-4] ***pstyle*** for a list of available plot styles.

`recast(`*newplottype*`)` is an advanced option allowing the plot to be recast from one type to another, for example, from a line plot to a scatterplot; see [G-3] ***advanced_options***. Most, but not all, plots allow `recast()`.

Remarks

An important option among all the above is `connect()`, which determines whether and how the points are connected. The points need not be connected at all (`connect(i)`), which is `scatter`'s default. Or the points might be connected by straight lines (`connect(l)`), which is `line`'s default (and is available in `scatter`). `connect(i)` and `connect(l)` are commonly specified, but there are other possibilities such as `connect(J)`, which connects in stairstep fashion and is appropriate for empirical distributions. See [G-4] ***connectstyle*** for a full list of your choices.

The remaining connect options specify how the line is to look: Is it solid or dashed? Is it red or green? How thick is it? Option `lpattern()` can be of great importance, especially when printing to a monochrome printer. For a general discussion of lines (which occur in many contexts other than connecting points), see [G-4] **concept: lines**.

Also see

[G-4] **concept: lines** — Using lines

[G-4] ***colorstyle*** — Choices for color

[G-4] ***connectstyle*** — Choices for how points are connected

[G-4] ***linestyle*** — Choices for overall look of lines

[G-4] ***linepatternstyle*** — Choices for whether lines are solid, dashed, etc.

[G-4] ***linewidthstyle*** — Choices for thickness of lines

Title

[G-3] *connect_options* — Options for connecting points with lines

Syntax

connect_options	Description
connect(*connectstyle*)	how to connect points
sort[(*varlist*)]	how to sort before connecting
cmissing({ y | n } ...)	missing values are ignored
lpattern(*linepatternstyle*)	line pattern (solid, dashed, etc.)
lwidth(*linewidthstyle*)	thickness of line
lcolor(*colorstyle*)	color of line
lstyle(*linestyle*)	overall style of line
pstyle(*pstyle*)	overall plot style, including linestyle
recast(*newplottype*)	advanced; treat plot as *newplottype*

See [G-4] ***connectstyle***, [G-4] ***linepatternstyle***, [G-4] ***linewidthstyle***, [G-4] ***colorstyle***, [G-4] ***linestyle***, [G-4] ***pstyle***, and [G-3] ***advanced_options***.

All options are *rightmost*; see [G-4] **concept: repeated options**. If both sort and sort(*varlist*) are specified, sort is ignored and sort(*varlist*) is honored.

Description

The *connect_options* specify how points on a graph are to be connected.

In certain contexts (for example, scatter; see [G-2] **graph twoway scatter**), the lstyle(), lpattern(), lwidth(), and lcolor() options may be specified with a list of elements, with the first element applying to the first variable, the second element to the second variable, and so on. For information about specifying lists, see [G-4] ***stylelists***.

Options

connect(*connectstyle*) specifies whether points are to be connected and, if so, how the line connecting them is to be shaped; see [G-4] ***connectstyle***. The line between each pair of points can connect them directly or in stairstep fashion.

sort and sort(*varlist*) specify how the data be sorted before the points are connected.

sort specifies that the data should be sorted by the x variable.

sort(*varlist*) specifies that the data be sorted by the specified variables.

sort is the option usually specified. Unless you are after a special effect or your data are already sorted, do not forget to specify this option. If you are after a special effect, and if the data are not already sorted, you can specify sort(*varlist*) to specify exactly how the data should be sorted.

Specifying sort or sort(*varlist*) when it is not necessary will slow graph down a little. It is usually necessary to specify sort if you specify the twoway option by(), and especially if you include the suboption total.

Options sort and sort(*varlist*) may not be repeated within the same plot.

`cmissing({ y | n } ...)` specifies whether missing values are to be ignored. The default is `cmissing(y ...)`, meaning that they are ignored. Consider the following data:

	rval	x
1.	.923	1
2.	3.046	2
3.	5.169	3
4.	.	.
5.	9.415	5
6.	11.538	6

Say that you graph these data by using "`line rval x`" or equivalently "`scatter rval x, c(l)`". Do you want a break in the line between 3 and 5? If so, you code

```
. line rval x, cmissing(n)
```

or equivalently

```
. scatter rval x, c(l) cmissing(n)
```

If you omit the option (or code `cmissing(y)`), the data are treated as if they contained

	rval	x
1.	.923	1
2.	3.046	2
3.	5.169	3
4.	9.415	5
5.	11.538	6

meaning that a line will be drawn between (3, 5.169) and (5, 9.415).

If you are plotting more than one variable, you may specify a sequence of `y`/`n` answers.

`lpattern(`*linepatternstyle*`)`, `lwidth(`*linewidthstyle*`)`, `lcolor(`*colorstyle*`)`, and `lstyle(`*linestyle*`)` determine the look of the line used to connect the points; see [G-4] **concept: lines**. Note the `lpattern()` option, which allows you to specify whether the line is solid, dashed, etc.; see [G-4] ***linepatternstyle*** for a list of line-pattern choices.

`pstyle(`*pstyle*`)` specifies the overall style of the plot, including not only the *linestyle*, but also all other settings for the look of the plot. Only the *linestyle* affects the look of line plots. See [G-4] ***pstyle*** for a list of available plot styles.

`recast(`*newplottype*`)` is an advanced option allowing the plot to be recast from one type to another, for example, from a line plot to a scatterplot; see [G-3] ***advanced_options***. Most, but not all, plots allow `recast()`.

Remarks

An important option among all the above is `connect()`, which determines whether and how the points are connected. The points need not be connected at all (`connect(i)`), which is `scatter`'s default. Or the points might be connected by straight lines (`connect(l)`), which is `line`'s default (and is available in `scatter`). `connect(i)` and `connect(l)` are commonly specified, but there are other possibilities such as `connect(J)`, which connects in stairstep fashion and is appropriate for empirical distributions. See [G-4] ***connectstyle*** for a full list of your choices.

Equally as important as `connect()` is `sort`. If you do not specify this, the points will be connected in the order in which they are encountered. That can be useful when you are creating special effects, but, in general, you want the points sorted into ascending order of their x variable. That is what `sort` does.

The remaining connect options specify how the line is to look: Is it solid or dashed? Is it red or green? How thick is it? Option `lpattern()` can be of great importance, especially when printing to a monochrome printer. For a general discussion of lines (which occur in many contexts other than connecting points), see [G-4] **concept: lines**.

Also see

[G-4] **concept: lines** — Using lines

[G-4] ***colorstyle*** — Choices for color

[G-4] ***connectstyle*** — Choices for how points are connected

[G-4] ***linestyle*** — Choices for overall look of lines

[G-4] ***linepatternstyle*** — Choices for whether lines are solid, dashed, etc.

[G-4] ***linewidthstyle*** — Choices for thickness of lines

Title

[G-3] *eps_options* — Options for exporting to Encapsulated PostScript

Syntax

eps_options	Description
`logo(on|off)`	whether to include Stata logo
`cmyk(on|off)`	whether to use CMYK rather than RGB colors
`preview(on|off)`	whether to include TIFF preview
`mag(#)`	magnification/shrinkage factor; default is 100
`fontface(`*fontname*`)`	default font to use
`fontfacesans(`*fontname*`)`	font to use for text in `{stSans}` "font"
`fontfaceserif(`*fontname*`)`	font to use for text in `{stSerif}` "font"
`fontfacemono(`*fontname*`)`	font to use for text in `{stMono}` "font"
`fontfacesymbol(`*fontname*`)`	font to use for text in `{stSymbol}` "font"
`fontdir(`*directory*`)`	(Unix only) directory in which TrueType fonts are stored
`orientation(portrait|landscape)`	whether vertical or horizontal

where *fontname* may be a valid font name or `default` to restore the default setting and *directory* may be a valid directory or `default` to restore the default setting.

Current default values may be listed by typing

```
. graph set eps
```

and default values may be set by typing

```
. graph set eps name value
```

where *name* is the name of an *eps_option*, omitting the parentheses.

Description

These *eps_options* are used with `graph export` when creating an Encapsulated PostScript file; see [G-2] **graph export**.

Options

`logo(on)` and `logo(off)` specify whether the Stata logo should be included at the bottom of the graph.

`cmyk(on)` and `cmyk(off)` specify whether colors in the output file should be specified as CMYK values rather than RGB values.

`preview(on)` and `preview(off)` specify whether a TIFF preview of the graph should be included in the Encapsulated PostScript file. This option allows word processors that cannot interpret PostScript to display a preview of the file. The preview is often substituted for the Encapsulated PostScript file when printing to a non-PostScript printer. This option is not available in Stata console and requires the Graph window to be visible.

`mag(#)` specifies that the graph be drawn smaller or larger than the default. `mag(100)` is the default, meaning ordinary size. `mag(110)` would make the graph 10% larger than usual, and `mag(90)` would make the graph 10% smaller than usual. # must be an integer.

`fontface(`*fontname*`)` specifies the name of the PostScript font to be used to render text for which no other font has been specified. The default is `Helvetica`, which may be restored by specifying *fontname* as `default`. If *fontname* contains spaces, it must be enclosed in double quotes.

`fontfacesans(`*fontname*`)` specifies the name of the PostScript font to be used to render text for which the `{stSans}` "font" has been specified. The default is `Helvetica`, which may be restored by specifying *fontname* as `default`. If *fontname* contains spaces, it must be enclosed in double quotes.

`fontfaceserif(`*fontname*`)` specifies the name of the PostScript font to be used to render text for which the `{stSerif}` "font" has been specified. The default is `Times`, which may be restored by specifying *fontname* as `default`. If *fontname* contains spaces, it must be enclosed in double quotes.

`fontfacemono(`*fontname*`)` specifies the name of the PostScript font to be used to render text for which the `{stMono}` "font" has been specified. The default is `Courier`, which may be restored by specifying *fontname* as `default`. If *fontname* contains spaces, it must be enclosed in double quotes.

`fontfacesymbol(`*fontname*`)` specifies the name of the PostScript font to be used to render text for which the `{stSymbol}` "font" has been specified. The default is `Symbol`, which may be restored by specifying *fontname* as `default`. If *fontname* contains spaces, it must be enclosed in double quotes.

`fontdir(`*directory*`)` specifies the directory that Stata for Unix uses to find TrueType fonts (if you specified any) for conversion to PostScript fonts when you export a graph to Encapsulated PostScript. You may specify *directory* as `default` to restore the default setting. If *directory* contains spaces, it must be enclosed in double quotes.

`orientation(portrait)` and `orientation(landscape)` specify whether the graph is to be presented vertically or horizontally.

Remarks

Remarks are presented under the following headings:

Using the eps_options
Setting defaults
Note about PostScript fonts

Using the eps_options

You have drawn a graph and wish to create an Encapsulated PostScript file for including the file in a document. You wish, however, to change text for which no other font has been specified from the default of Helvetica to Roman, which is "Times" in PostScript jargon:

```
. graph ...                                    (draw a graph)
. graph export myfile.eps, fontface(Times)
```

Setting defaults

If you always wanted `graph export` (see [G-2] **graph export**) to use Times when exporting to Encapsulated PostScript files, you could type

```
. graph set eps fontface Times
```

Later, you could type

```
. graph set eps fontface Helvetica
```

to change it back. You can list the current *eps_option* settings for Encapsulated PostScript by typing

```
. graph set eps
```

Note about PostScript fonts

Graphs exported to Encapsulated PostScript format by Stata conform to what is known as PostScript Level 2. There are 10 built-in font faces, known as the Core Font Set, some of which are available in modified forms, for example, bold or italic (a listing of the original font faces in the Core Font Set is shown at http://en.wikipedia.org/wiki/Type_1_and_Type_3_fonts#Core_Font_Set). If you change any of the `fontface*()` settings, we recommend that you use one of those 10 font faces. We do not recommend changing `fontfacesymbol()`, as doing so can lead to incorrect characters being printed.

If you specify a font face other than one that is part of the Core Font Set, Stata will first attempt to map it to the closest matching font in the Core Font Set. For example, if you specify `fontfaceserif("Times New Roman")`, Stata will map it to `fontfaceserif("Times")`.

If Stata is unable to map the font face to the Core Font Set, Stata will look in the `fontdir()` directory for a TrueType font on your system matching the font you specified. If it finds one, it will attempt to convert it to a PostScript font and, if successful, will embed the converted font in the exported Encapsulated PostScript graph. Because of the wide variety of TrueType fonts available on different systems, this conversion can be problematic, which is why we recommend that you use fonts found in the Core Font Set.

Also see

[G-2] **graph export** — Export current graph

[G-2] **graph set** — Set graphics options

[G-3] ***ps_options*** — Options for exporting or printing to PostScript

Title

[G-3] ***fcline_options*** — Options for determining the look of fitted connecting lines

Syntax

fcline_options	Description
`clpattern(`*linepatternstyle*`)`	whether line solid, dashed, etc.
`clwidth(`*linewidthstyle*`)`	thickness of line
`clcolor(`*colorstyle*`)`	color of line
`clstyle(`*linestyle*`)`	overall style of line
`pstyle(`*pstyle*`)`	overall plot style, including linestyle

See [G-4] ***linepatternstyle***, [G-4] ***linewidthstyle*** [G-4] ***colorstyle***, [G-4] ***linestyle***, and [G-4] ***pstyle***.

All options are *rightmost*; see [G-4] **concept: repeated options**.

Description

The *fcline_options* determine the look of a fitted connecting line in most contexts.

Options

`clpattern(`*linepatternstyle*`)` specifies whether the line is solid, dashed, etc. See [G-4] ***linepatternstyle*** for a list of available patterns.

`clwidth(`*linewidthstyle*`)` specifies the thickness of the line. See [G-4] ***linewidthstyle*** for a list of available thicknesses.

`clcolor(`*colorstyle*`)` specifies the color of the line. See [G-4] ***colorstyle*** for a list of available colors.

`clstyle(`*linestyle*`)` specifies the overall style of the line: its pattern, thickness, and color.

You need not specify `clstyle()` just because there is something you want to change about the look of the line. The other *fcline_options* will allow you to make changes. You specify `clstyle()` when another style exists that is exactly what you desire or when another style would allow you to specify fewer changes.

See [G-4] ***linestyle*** for a list of available line styles.

`pstyle(`*pstyle*`)` specifies the overall style of the plot, including not only the *linestyle*, but also all other settings for the look of the plot. Only the *linestyle* affects the look of lines. See [G-4] ***pstyle*** for a list of available plot styles.

Remarks

Lines occur in many contexts and, in almost all of those contexts, the above options are used to determine the look of the fitted connecting line. For instance, the `clcolor()` option in

```
. twoway lfitci y x, clcolor(red)
```

causes the line through the (y, x) points to be drawn in red.

The same option in

```
. twoway lfitci y x, title("My line", box clcolor(red))
```

causes the outline drawn around the title's box to be drawn in red. In the second command, the option `clcolor(red)` was a suboption to the `title()` option.

Title

[G-3] *fitarea_options* — Options for specifying the look of confidence interval areas

Syntax

fitarea_options	Description
`acolor(`*colorstyle*`)`	outline and fill color
`fcolor(`*colorstyle*`)`	fill color
`fintensity(`*intensitystyle*`)`	fill intensity
`alcolor(`*colorstyle*`)`	outline color
`alwidth(`*linewidthstyle*`)`	thickness of outline
`alpattern(`*linepatternstyle*`)`	outline pattern (solid, dashed, etc.)
`alstyle(`*linestyle*`)`	overall look of outline
`astyle(`*areastyle*`)`	overall look of area, all settings above
`pstyle(`*pstyle*`)`	overall plot style, including areastyle

See [G-4] ***colorstyle***, [G-4] ***intensitystyle***, [G-4] ***linewidthstyle***, [G-4] ***linepatternstyle***, [G-4] ***linestyle***, [G-4] ***areastyle***, [G-4] ***pstyle***, and [G-3] ***advanced_options***.

All options are *merged-implicit*; see [G-4] **concept: repeated options**.

Description

The *fitarea_options* determine the look of, for instance, the confidence interval areas created by `twoway fpfitci`, `twoway lfitci`, `twoway lpolyci`, and `twoway qfitci`; see [G-2] **graph twoway fpfitci**, [G-2] **graph twoway lfitci**, [G-2] **graph twoway lpolyci**, and [G-2] **graph twoway lfitci**.

Options

`acolor(`*colorstyle*`)` specifies one color to be used both to outline the shape of the area and to fill its interior. See [G-4] ***colorstyle*** for a list of color choices.

`fcolor(`*colorstyle*`)` specifies the color to be used to fill the interior of the area. See [G-4] ***colorstyle*** for a list of color choices.

`fintensity(`*intensitystyle*`)` specifies the intensity of the color used to fill the interior of the area. See [G-4] ***intensitystyle*** for a list of intensity choices.

`alcolor(`*colorstyle*`)` specifies the color to be used to outline the area. See [G-4] ***colorstyle*** for a list of color choices.

`alwidth(`*linewidthstyle*`)` specifies the thickness of the line to be used to outline the area. See [G-4] ***linewidthstyle*** for a list of choices.

`alpattern(`*linepatternstyle*`)` specifies whether the line used to outline the area is solid, dashed, etc. See [G-4] ***linepatternstyle*** for a list of pattern choices.

`alstyle(`*linestyle*`)` specifies the overall style of the line used to outline the area, including its pattern (solid, dashed, etc.), thickness, and color. The three options listed above allow you to change the line's attributes, but `lstyle()` is the starting point. See [G-4] ***linestyle*** for a list of choices.

`astyle(`*areastyle*`)` specifies the overall look of the area. The options listed above allow you to change each attribute, but `style()` provides a starting point.

You need not specify `style()` just because there is something you want to change. You specify `style()` when another style exists that is exactly what you desire or when another style would allow you to specify fewer changes to obtain what you want.

See [G-4] ***areastyle*** for a list of available area styles.

`pstyle(`*pstyle*`)` specifies the overall style of the plot, including not only the *areastyle*, but all other settings for the look of the plot. Only the *areastyle* affects the look of areas. See [G-4] ***pstyle*** for a list of available plot styles.

Remarks

fitarea_options are allowed as options with any `graph twoway` plottype that creates shaded confidence interval areas, for example, `graph twoway lfitci`, as in

```
. graph twoway lfitci yvar xvar, acolor(blue)
```

The above would set the area enclosed by *yvar* and the x axis to be blue; see [G-2] **graph twoway area** and [G-2] **graph twoway rarea**.

Title

[G-3] *legend_options* — Options for specifying legends

Syntax

legend_options	Description
legend([*contents*] [*location*])	standard legend contents and location
plegend([*contents*] [*location*])	contourline legend, contents, and location
clegend([*suboptions*])	contour plot legend; see [G-3] ***clegend_option***

legend(), plegend(), and clegend() are *merged-implicit*; see [G-4] **concept: repeated options**.

where *contents* and *location* specify the contents and the location of the legends.

contents	Description
order(*orderinfo*)	which keys appear and their order
label(*labelinfo*)	override text for a key
holes(*numlist*)	positions in legend to leave blank
all	generate keys for all symbols
style(*legendstyle*)	overall style of legend
cols(*#*)	# of keys per line
rows(*#*)	or # of rows
[no]colfirst	"1, 2, 3" in row 1 or in column 1?
[no]textfirst	symbol-text or text-symbol?
stack	symbol/text vertically stacked
rowgap(*relativesize*)	gap between lines
colgap(*relativesize*)	gap between columns
symplacement(*compassdirstyle*)	alignment/justification of key's symbol
keygap(*relativesize*)	gap between symbol-text
symysize(*relativesize*)	height for key's symbol
symxsize(*relativesize*)	width for key's symbol
textwidth(*relativesize*)	width for key's descriptive text
forcesize	always respect symysize(), symxsize(), and textwidth()
bmargin(*marginstyle*)	outer margin around legend
textbox_options	other text characteristics
title_options	titles, subtitles, notes, captions
region(*roptions*)	borders and background shading

order(), labels(), holes(), and all have no effect on plegend().
See [G-4] ***legendstyle***, [G-4] ***relativesize***, [G-4] ***compassdirstyle***, [G-4] ***marginstyle***, [G-3] ***textbox_options***, and [G-3] ***title_options***.

location	Description
`off` or `on`	suppress or force display of legend
`position(`*clockposstyle*`)`	where legend appears
`ring(`*ringposstyle*`)`	where legend appears (detail)
`bplacement(`*compassdirstyle*`)`	placement of legend when positioned in the plotregion
`span`	"centering" of legend
`at(`*#*`)`	allowed with `by()` only

See *Where legends appear* under *Remarks* below, and see *Positioning of titles* in [G-3] ***title_options*** for definitions of *clockposstyle* and *ringposstyle*.

orderinfo, the argument allowed by `legend(order())`, is defined as

$\{ \# | - \} \; [\text{"}\textit{text}\text{"} \; [\text{"}\textit{text}\text{"} \; \ldots]]$

labelinfo, the argument allowed by `legend(label())`, is defined as

$\# \; \text{"}\textit{text}\text{"} \; [\text{"}\textit{text}\text{"} \; \ldots]$

roptions, the arguments allowed by `legend(region())`, include

roptions	Description
`style(`*areastyle*`)`	overall style of region
`color(`*colorstyle*`)`	line + fill color of region
`fcolor(`*colorstyle*`)`	fill color of region
`lstyle(`*linestyle*`)`	overall style of border
`lcolor(`*colorstyle*`)`	color of border
`lwidth(`*linewidthstyle*`)`	thickness of border
`lpattern(`*linepatternstyle*`)`	border pattern (solid, dashed, etc.)
`margin(`*marginstyle*`)`	margin between border and contents of legend

See [G-4] ***areastyle***, [G-4] ***colorstyle***, [G-4] ***linestyle***, [G-4] ***linewidthstyle***, [G-4] ***linepatternstyle***, and [G-4] ***marginstyle***.

Description

The `legend()` option allows you to control the look, contents, and placement of the legend. A sample legend is

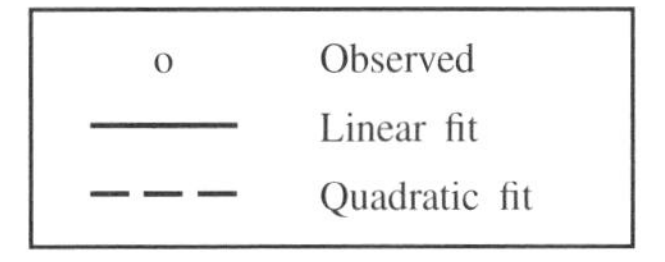

The above legend has three *keys*. Each key is composed of a *symbol* and *descriptive text* describing the symbol (whatever the symbol might be, be it a marker, a line, or a color swatch).

`contourline` and `contour` plots have their own legends and do not place keys in the standard legend—`legend()`; see [G-2] **graph twoway contourline** and [G-2] **graph twoway contour**. `contourline` plots place their keys in the `plegend()` and contour plots place their keys in the `clegend()`. The `plegend()` is similar to the `legend()` and is documented here. The `clegend()` is documented in [G-3] ***clegend_option***.

The legend options (more correctly suboptions) are discussed using the `legend()` option, but most apply equally to the `plegend()` option.

Options

`legend(`*contents*`,` *location*`)` defines the contents of the standard legend, along with how it is to look, and whether and where it is to be displayed.

`plegend(`*contents*`,` *location*`)` defines the contents of the `contourline` plot legend, along with how it is to look, and whether and where it is to be displayed.

Content suboptions for use with legend( ) and plegend( )

`order(`*orderinfo*`)` specifies which keys are to appear in the legend and the order in which they are to appear.

`order(`*# #* ...`)` is the usual syntax. `order(1 2 3)` would specify that key 1 is to appear first in the legend, followed by key 2, followed by key 3. `order(1 2 3)` is the default if there are three keys. If there were four keys, `order(1 2 3 4)` would be the default, and so on. If there were four keys and you specified `order(1 2 3)`, the fourth key would not appear in the legend. If you specified `order(2 1 3)`, first key 2 would appear, followed by key 1, followed by key 3.

A dash specifies that text be inserted into the legend. For instance, `order(1 2 - "`*text*`" 3)` specifies key 1 appear first, followed by key 2, followed by the text *text*, followed by key 3. Imagine that the default key were

o	Observed
———	Linear
— — —	Quadratic

Specifying `order(1 - "Predicted:" 2 3)` would produce

o	Observed
	Predicted:
———	Linear
— — —	Quadratic

and specifying `order(1 - " " "Predicted:" 2 3)` would produce

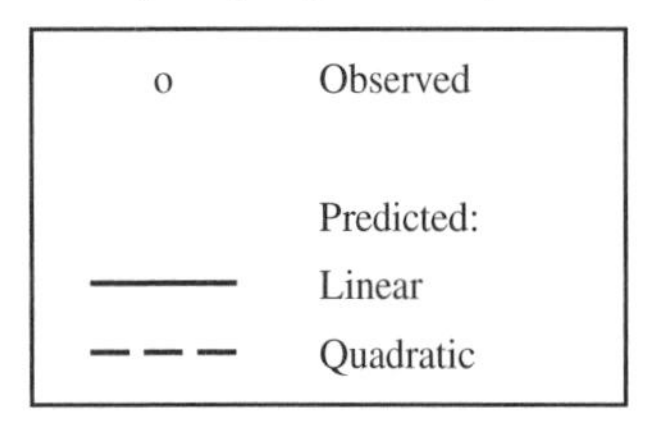

Note carefully the specification of a blank for the first line of the text insertion; we typed `" "` and not `""`. Typing `""` would insert nothing.

You may also specify quoted text after # to override the descriptive text associated with a symbol. Specifying `order(1 "Observed 1992" - " " "Predicted" 2 3)` would change "Observed" in the above to "Observed 1992". It is considered better style, however, to use the `label()` suboption to relabel symbols.

`order()` has no effect on `plegend()`.

`label(`*#* `"`*text*`"` [`"`*text*`"` ...]`)` specifies the descriptive text to be displayed next to the #th key. Multiline text is allowed. Specifying `label(1 "Observed 1992")` would change the descriptive text associated with the first key to be "Observed 1992". Specifying `label(1 "Observed" "1992-1993")` would change the descriptive text to contain two lines, "Observed" followed by "1992–1993".

The descriptive text of only one key may be changed per `label()` suboption. Specify multiple `label()` suboptions when you wish to change the text of multiple keys.

`label()` has no effect on `plegend()`.

`holes(`*numlist*`)` specifies where gaps appear in the presentation of the keys. `holes()` has an effect only if the keys are being presented in more than one row and more than one column.

Consider a case in which the default key is

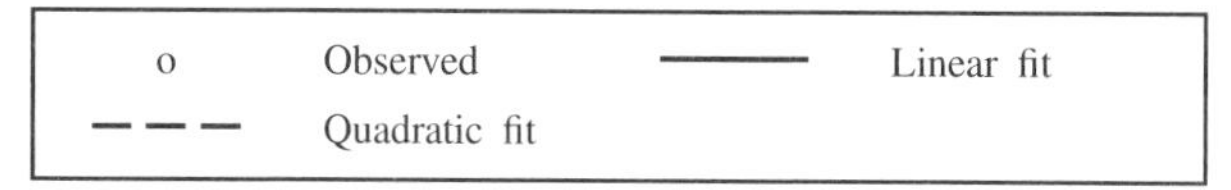

Specifying `holes(2)` would result in

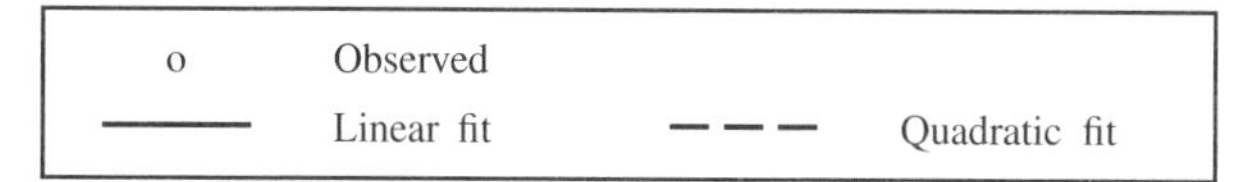

Here `holes(2)` would have the same effect as specifying `order(1 - " " 2 3)`, and as a matter of fact, there is always an `order()` command that will achieve the same result as `holes()`. `order()` has the added advantage of working in all cases.

`holes()` has no effect on `plegend()`.

`all` specifies that keys be generated for all the plots of the graph, even when the same symbol is repeated. The default is to generate keys only when the symbols are different, which is determined by the overall style. For example, in

```
. scatter ylow yhigh x, pstyle(p1 p1) || ...
```

there would be only one key generated for the variables `ylow` and `yhigh` because they share the style `p1`. That single key's descriptive text would indicate that the symbol corresponded to both variables. If, on the other hand, you typed

```
. scatter ylow yhigh x, pstyle(p1 p1) legend(all) || ...
```

then separate keys would be generated for `ylow` and `yhigh`.

In the above example, do not confuse our use of `scatter`'s option `pstyle()` with `legend()`'s suboption `legend(style())`. The `pstyle()` option sets the overall style for the rendition of the points. `legend()`'s `style()` suboption is documented directly below.

`all` has no effect on `plegend()`.

`style(`*legendstyle*`)` specifies the overall look of the legend—whether it is presented horizontally or vertically, how many keys appear across the legend if it is presented horizontally, etc. The options listed below allow you to change each attribute of the legend, but `style()` is the starting point.

You need not specify `style()` just because there is something you want to change. You specify `style()` when another style exists that is exactly what you desire or when another style would allow you to specify fewer changes to obtain what you want.

See [G-4] ***legendstyle*** for a list of available legend styles.

`cols(#)` and `rows(#)` are alternatives; they specify in how many columns or rows (lines) the keys are to be presented. The usual default is `cols(2)`, which means that legends are to take two columns:

o	Observed	———	Linear fit
— — —	Quadratic fit		

`cols(1)` would force a vertical arrangement,

o	Observed
———	Linear fit
— — —	Quadratic fit

and `rows(1)` would force a horizontal arrangement:

o	Observed	———	Linear fit	— — —	Quadratic fit

`colfirst` and `nocolfirst` determine whether, when the keys are presented in multiple columns, keys are to read down or to read across, resulting in this

o	Observed	— — —	Quadratic fit
———	Linear fit		

or this

o	Observed	———	Linear fit
— — —	Quadratic fit		

The usual default is `nocolfirst`, so `colfirst` is the option.

`textfirst` and `notextfirst` specify whether the keys are to be presented as descriptive text followed by the symbol or the symbol followed by descriptive text. The usual default is `notextfirst`, so `textfirst` is the option. `textfirst` produces keys that look like this

Observed	o	Linear fit	———
Quadratic fit	— — —		

and `textfirst cols(1)` produces

Observed	o
Linear fit	———
Quadratic fit	— — —

`stack` specifies that the symbol-text be presented vertically with the symbol on top (or with the descriptive text on top if `textfirst` is also specified). `legend(stack)` would produce

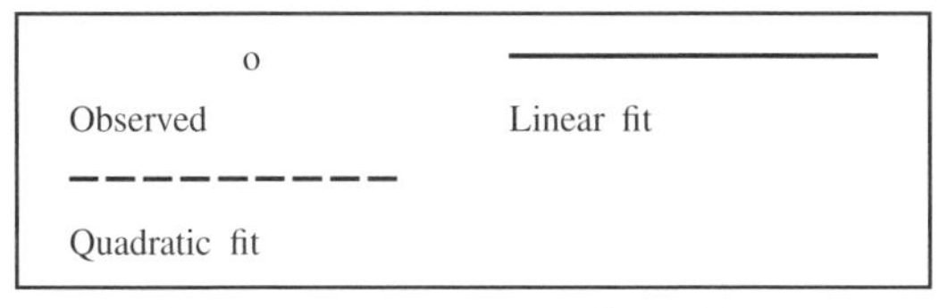

`legend(stack symplacement(left) symxsize(13) forcesize rowgap(4))` would produce

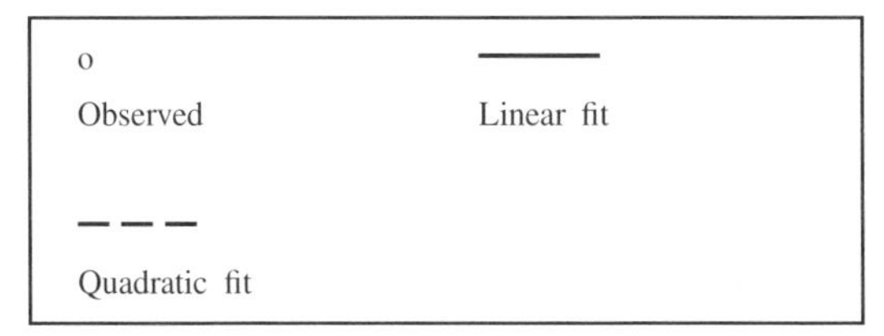

`stack` tends to be used to produce single-column keys. `legend(cols(1) stack symplacement(left) symxsize(13) forcesize rowgap(4))` produces

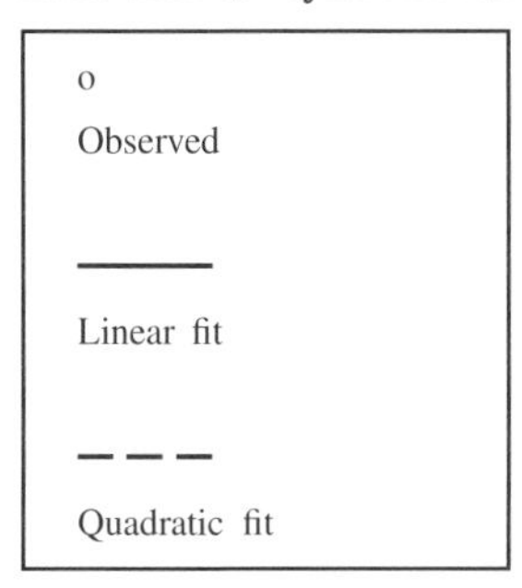

This is the real use of `stack`: to produce narrow, vertical keys.

`rowgap(`*relativesize*`)` and `colgap(`*relativesize*`)` specify the distance between lines and the distance between columns. The defaults are `rowgap(1.4)` and `colgap(4.9)`. See [G-4] ***relativesize***.

`symplacement(`*compassdirstyle*`)` specifies how symbols are justified in the key. The default is `symplacement(center)`, meaning that they are vertically and horizontally centered. The two most commonly specified alternatives are `symplacement(right)` (right alignment) and `symplacement(left)` (left alignment). See [G-4] ***compassdirstyle*** for other alignment choices.

`keygap(`*relativesize*`)`, `symysize(`*relativesize*`)`, `symxsize(`*relativesize*`)`, and `textwidth(`*relativesize*`)` specify the height and width to be allocated for the key and the key's symbols and descriptive text:

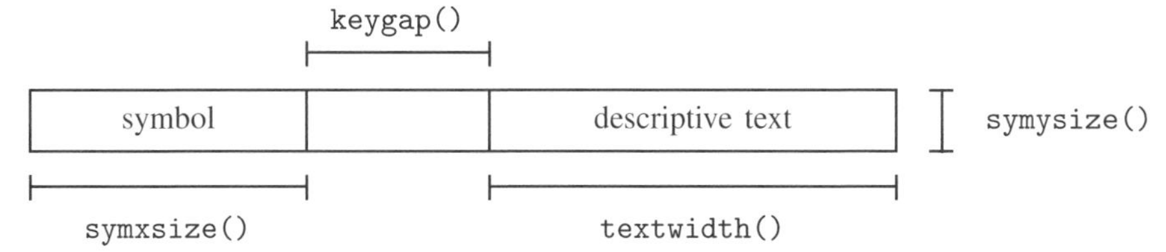

The defaults are

`symxsize()`	13
`keygap()`	2
`textwidth()`	according to longest descriptive text line
`symysize()`	according to height of font (*)

(*) The size of the font is set by the *textbox_option* `size(`*relativesize*`)`;
see *textbox_options* below.

Markers are placed in the symbol area, centered according to `symplacement()`.

Lines are placed in the symbol area vertically according to `symplacement()` and horizontally are drawn to length `symxsize()`.

Color swatches fill the `symysize()` × `symxsize()` area.

See [G-4] ***relativesize*** for information on specifying relative sizes.

`forcesize` causes the sizes specified by `symysize()` and `symxsize()` to be respected. If `forcesize` is not specified, once all the symbols have been placed for all the keys, the symbol area is compressed (or expanded) to be no larger than necessary to contain the symbols.

`bmargin(`*marginstyle*`)` specifies the outer margin around the legend. That is, it specifies how close other things appearing near to the legend can get. Also see suboption `margin()` under *Suboptions for use with legend(region())* below for specifying the inner margin between the border and contents. See [G-4] ***marginstyle*** for a list of margin choices.

textbox_options affect the rendition of the descriptive text associated with the keys. These are described in [G-3] ***textbox_options***. One of the most commonly specified *textbox_options* is `size(`*relativesize*`)`, which specifies the size of font to be used for the descriptive text.

title_options allow placing titles, subtitles, notes, and captions on legends. For instance, `legend(col(1) subtitle("Legend"))` produces

Legend

o	Observed
———	Linear fit
— — —	Quadratic fit

Note our use of `subtitle()` and not `title()`; `title()`s are nearly always too big. See [G-3] ***title_options***.

`region(`*roptions*`)` specifies the border and shading of the legend. You could remove the border around the legend by specifying `legend(region(lstyle(none)))` (thus doing away with the line) or `legend(region(lcolor(none)))` (thus making the line invisible). You could also give the legend a gray background tint by specifying `legend(region(fcolor(gs5)))`. See *Suboptions for use with legend(region())* below.

Suboptions for use with legend(region())

`style(`*areastyle*`)` specifies the overall style of the region in which the legend appears. The other suboptions allow you to change the region's attributes individually, but `style()` provides the starting point. See [G-4] ***areastyle*** for a list of choices.

`color(`*colorstyle*`)` specifies the color of the background of the legend and the line used to outline it. See [G-4] ***colorstyle*** for a list of color choices.

`fcolor(`*colorstyle*`)` specifies the background (fill) color for the legend. See [G-4] ***colorstyle*** for a list of color choices.

`lstyle(`*linestyle*`)` specifies the overall style of the line used to outline the legend, which includes its pattern (solid, dashed, etc.), its thickness, and its color. The other suboptions listed below allow you to change the line's attributes individually, but `lstyle()` is the starting point. See [G-4] ***linestyle*** for a list of choices.

`lcolor(`*colorstyle*`)` specifies the color of the line used to outline the legend. See [G-4] ***colorstyle*** for a list of color choices.

`lwidth(`*linewidthstyle*`)` specifies the thickness of the line used to outline the legend. See [G-4] ***linewidthstyle*** for a list of choices.

`lpattern(`*linepatternstyle*`)` specifies whether the line used to outline the legend is solid, dashed, etc. See [G-4] ***linepatternstyle*** for a list of choices.

`margin(`*marginstyle*`)` specifies the inner margin between the border and the contents of the legend. Also see `bmargin()` under *Content suboptions for use with legend( ) and plegend( )* above for specifying the outer margin around the legend. See [G-4] ***marginstyle*** for a list of margin choices.

Location suboptions for use with legend( )

`off` and `on` determine whether the legend appears. The default is `on` when more than one symbol (meaning marker, line style, or color swatch) appears in the legend. In those cases, `legend(off)` will suppress the display of the legend.

`position(`*clockposstyle*`)`, `ring(`*ringposstyle*`)`, and `bplacement(`*compassdirstyle*`)` override the default location of the legend, which is usually centered below the plot region. `position()` specifies a direction [*sic*] according to the hours on the dial of a 12-hour clock, and `ring()` specifies the distance from the plot region.

`ring(0)` is defined as being inside the plot region itself and allows you to place the legend inside the plot. `ring(`k`)`, $k > 0$, specifies positions outside the plot region; the larger the `ring()` value, the farther away from the plot region the legend is. `ring()` values may be integers or nonintegers and are treated ordinally.

When `ring(0)` is specified, `bplacement()` further specifies where in the plot region the legend is placed. `bplacement(seast)` places the legend in the southeast (lower-right) corner of the plot region.

`position(12)` puts the legend directly above the plot region (assuming `ring()` > 0), `position(3)` directly to the right of the plot region, and so on.

See *Where legends appear* under *Remarks* below and *Positioning of titles* in [G-3] ***title_options*** for more information on the `position()` and `ring()` suboptions.

`span` specifies that the legend be placed in an area spanning the entire width (or height) of the graph rather than an area spanning the plot region. This affects whether the legend is centered with respect to the plot region or the entire graph. See *Spanning* in [G-3] ***title_options*** for more information on `span`.

`at(#)` is for use only when the *twoway_option* `by()` is also specified. It specifies that the legend appear in the #th position of the $R \times C$ array of plots, using the same coding as `by(... , holes())`. See *Use of legends with by( )* under *Remarks* below, and see [G-3] ***by_option***.

Remarks

Remarks are presented under the following headings:

When legends appear
The contents of legends
Where legends appear
Putting titles on legends
Use of legends with by()
Problems arising with or because of legends

When legends appear

Standard legends appear on the graph whenever more than one symbol is used, where symbol is broadly defined to include markers, lines, and color swatches (such as those used to fill bars). When you draw a graph with only one symbol on it, such as

```
. use http://www.stata-press.com/data/r12/uslifeexp
(U.S. life expectancy, 1900-1999)
. line le year
```

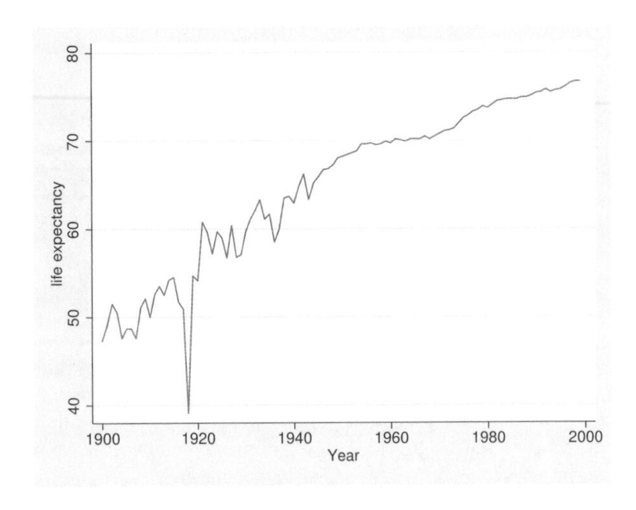

no legend appears. When there is more than one symbol, a legend is added:

```
. line le_m le_f year
```

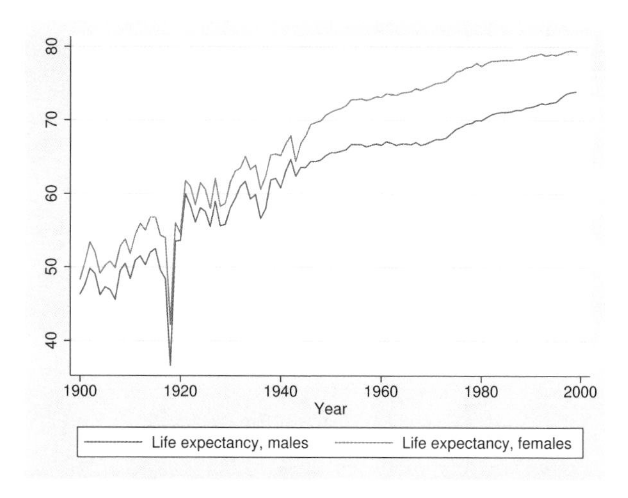

Even when there is only one symbol, a legend is constructed. It is merely not displayed. Specifying `legend(on)` forces the display of the legend:

```
. line le year, legend(on)
```

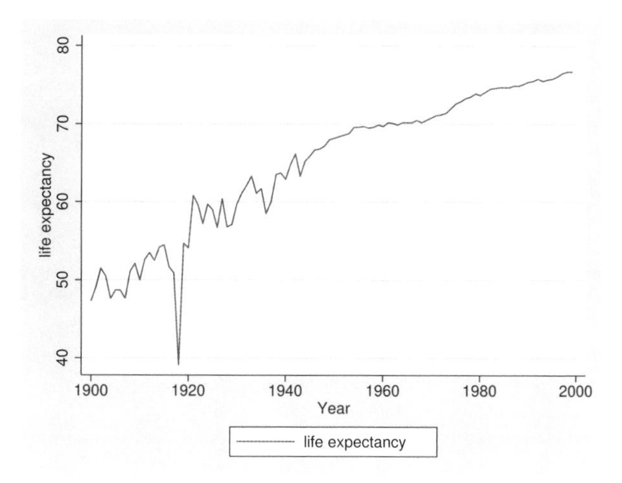

Similarly, when there is more than one symbol and you do not want the legend, you can specify `legend(off)` to suppress it:

```
. line le_m le_f year, legend(off)
```

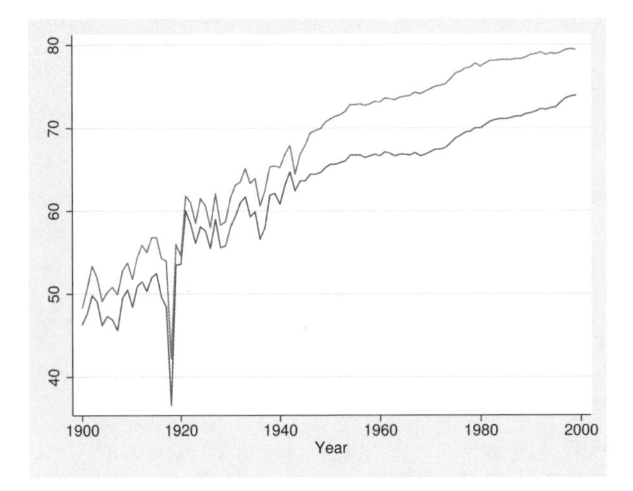

A `plegend()` appears on any graph that includes a `contourline` plot.

The contents of legends

By default, the descriptive text for legends is obtained from the variable's variable label; see [D] **label**. If the variable has no variable label, the variable's name is used. In

```
. line le_m le_f year
```

the variable `le_m` had previously been labeled "Life expectancy, males", and the variable `le_f` had been labeled "Life expectancy, females". In the legend of this graph, repeating "life expectancy" is unnecessary. The graph would be improved if we changed the labels on the variables:

```
. label var le_m "Males"
. label var le_f "Females"
. line le_m le_f year
```

We can also specify the `label()` suboption to change the descriptive text. We obtain the same visual result without relabeling our variables:

```
. line le_m le_f year, legend(label(1 "Males") label(2 "Females"))
```

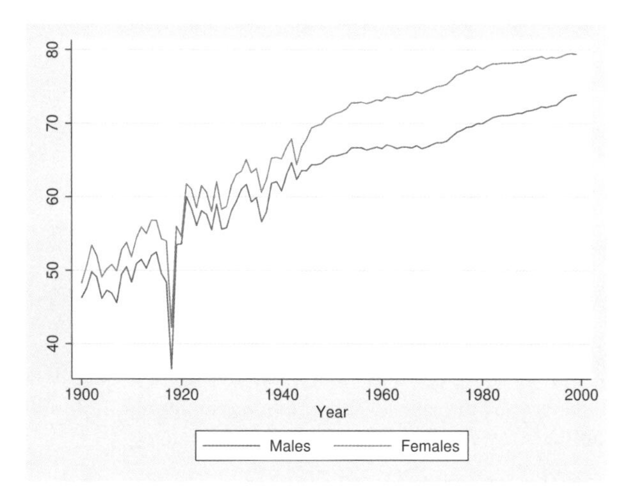

The descriptive text for `contourline` legends is the values for the contour lines of the z variable.

Where legends appear

By default, standard legends appear beneath the plot, centered, at what is technically referred to as `position(6) ring(3)`. By default, `plegends()` appear to the right of the plot region at `position(3) ring(4)`. Suboptions `position()` and `ring()` specify the location of the legend. `position()` specifies on which side of the plot region the legend appears—`position(6)` means 6 o'clock—and `ring()` specifies the distance from the plot region—`ring(3)` means farther out than the *title_option* `b2title()` but inside the *title_option* `note()`; see [G-3] ***title_options***.

If we specify `legend(position(3))`, the legend will be moved to the 3 o'clock position:

```
. line le_m le_f year, legend(pos(3))
```

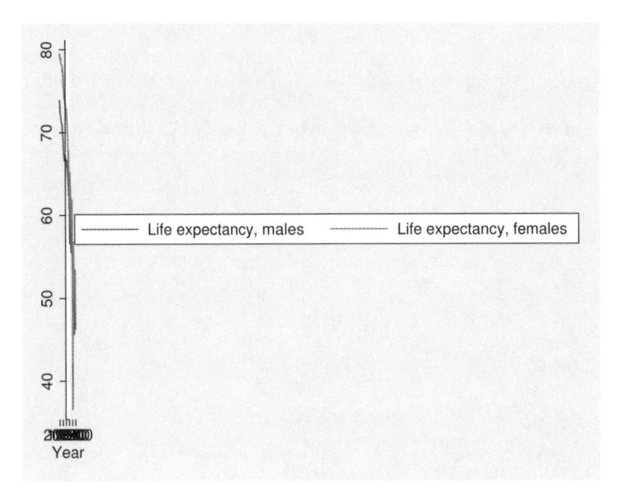

This may not be what we desired, but it is what we asked for. The legend was moved to the right of the graph and, given the size of the legend, the graph was squeezed to fit. When you move legends to the side, you invariably also want to specify the `col(1)` option:

```
. line le_m le_f year, legend(pos(3) col(1))
```

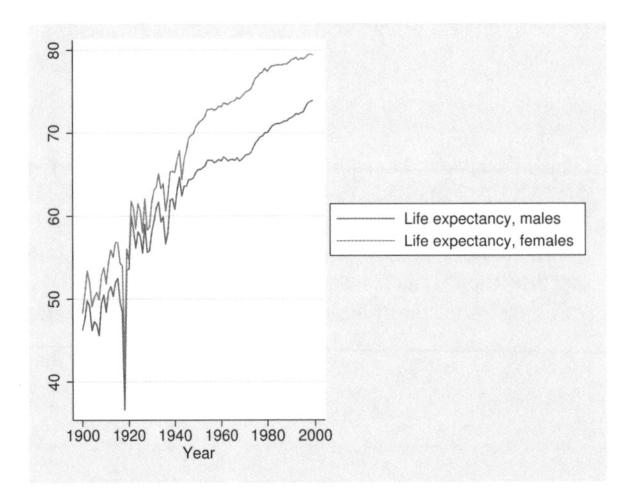

As a matter of syntax, we could have typed the above command with two `legend()` options

```
. line le_m le_f year, legend(pos(3)) legend(col(1))
```

instead of one combined: `legend(pos(3) col(1))`. We would obtain the same results either way.

If we ignore the syntax, the above graph would look better with less-descriptive text,

```
. line le_m le_f year, legend(pos(3) col(1)
                       lab(1 "Males") lab(2 "Females"))
```

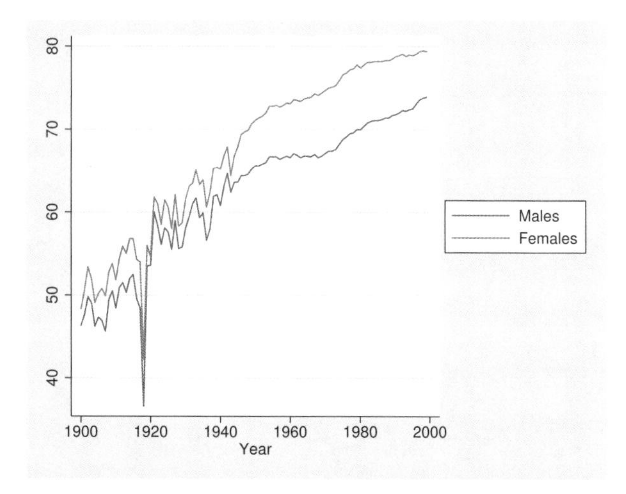

and we can further reduce the width required by the legend by specifying the `stack` suboption:

```
. line le_m le_f year, legend(pos(3) col(1)
                       lab(1 "Males") lab(2 "Females") stack)
```

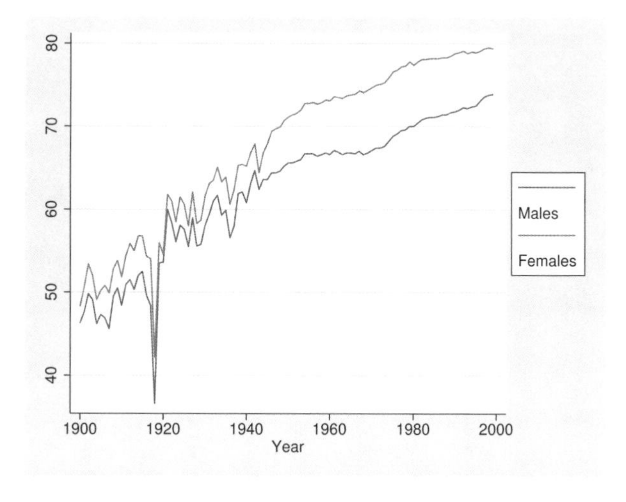

We can make this look better by placing a blank line between the first and second keys:

```
. line le_m le_f year, legend(pos(3) col(1)
                 lab(1 "Males") lab(2 "Females") stack
                           order(1 - " " 2))
```

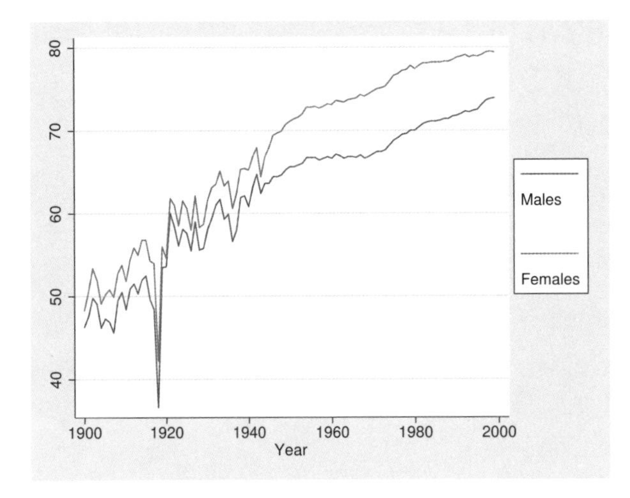

`ring()`—the suboption that specifies the distance from the plot region—is seldom specified, but, when it is specified, `ring(0)` is the most useful. `ring(0)` specifies that the legend be moved inside the plot region:

```
. line le_m le_f year, legend(pos(5) ring(0) col(1)
                          lab(1 "Males") lab(2 "Females"))
```

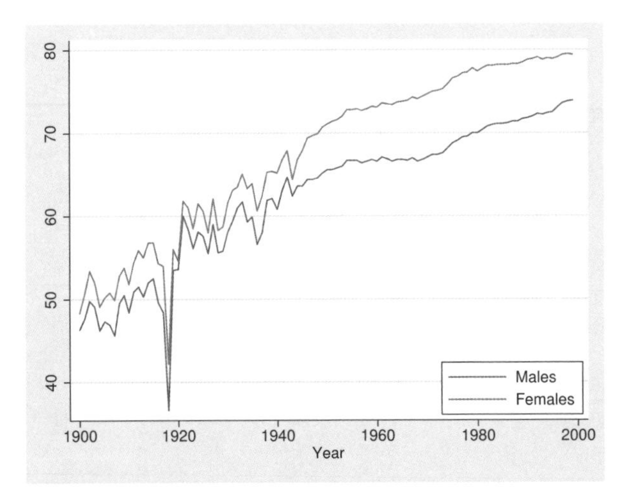

Our use of `position(5) ring(0)` put the legend inside the plot region, at 5 o'clock, meaning in the bottom right corner. Had we specified `position(2) ring(0)`, the legend would have appeared in the top left corner.

We might now add some background color to the legend:

```
. line le_m le_f year, legend(pos(5) ring(0) col(1)
                       lab(1 "Males") lab(2 "Females")
                       region(fcolor(gs15)))
```

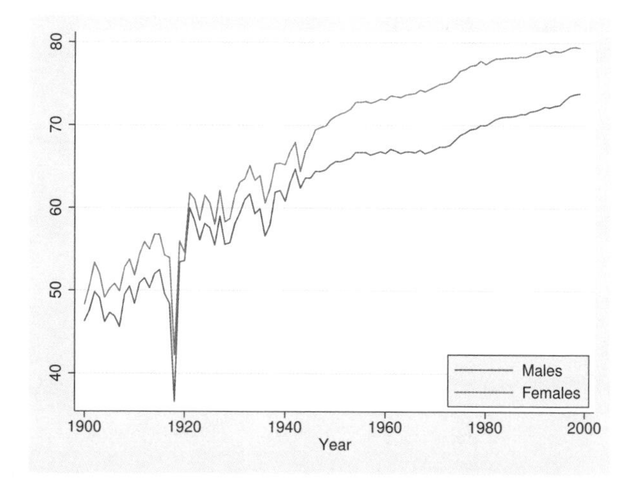

Putting titles on legends

Legends may include titles:

```
. line le_m le_f year, legend(pos(5) ring(0) col(1)
                       lab(1 "Males") lab(2 "Females")
                       region(fcolor(gs15)))
                       legend(subtitle("Legend"))
```

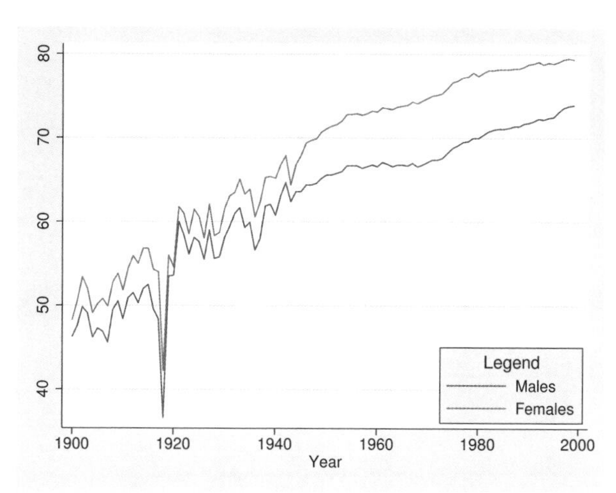

Above we specified `subtitle()` rather than `title()` because, when we tried `title()`, it seemed too big.

Legends may also contain `notes()` and `captions()`; see [G-3] ***title_options***.

Use of legends with by()

If you want the legend to be located in the default location, no special action need be taken when you use `by()`:

```
. use http://www.stata-press.com/data/r12/auto, clear
(1978 Automobile Data)
. scatter mpg weight || lfit mpg weight ||, by(foreign, total row(1))
```

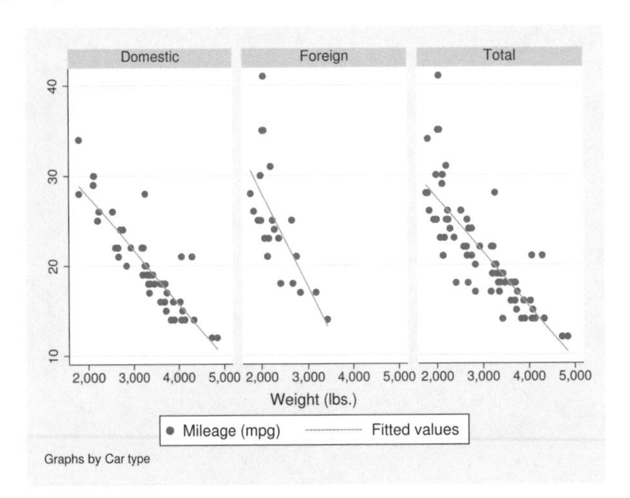

If, however, you wish to move the legend, you must distinguish between `legend(`*contents*`)` and `legend(`*location*`)`. The former must appear outside the `by()`. The latter appears inside the `by()`:

```
. scatter mpg weight || lfit mpg weight ||,
        legend(cols(1))
        by(foreign, total legend(pos(4)))
```

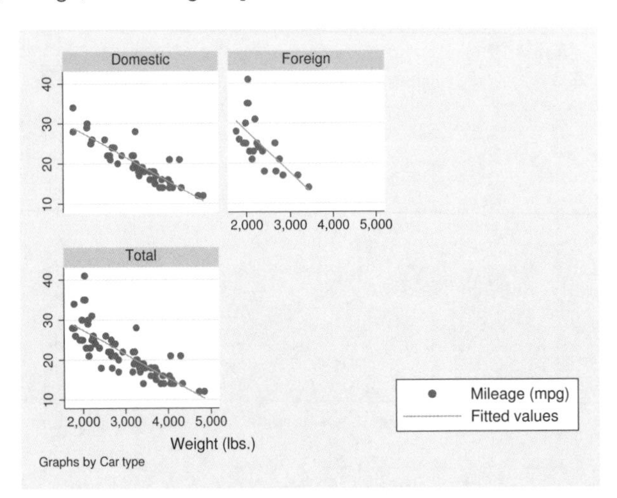

`legend(col(1))` was placed in the command just where we would place it had we not specified `by()` but that `legend(pos(4))` was moved to be inside the `by()` option. We did that because the `cols()` suboption is documented under *contents* in the syntax diagram, whereas `position()` is documented under *location*. The logic is that, at the time the individual plots are constructed, they must know what style of key they are producing. The placement of the key, however, is something that happens when the overall graph is assembled, so you must indicate to by() where the key is to be placed. Were we to forget this distinction and simply to type

```
. scatter mpg weight || lfit mpg weight ||,
          legend(cols(1) pos(4))
          by(foreign, total)
```

the cols(1) suboption would have been ignored.

Another *location* suboption is provided for use with by(): at(*#*). You specify this option to tell by() to place the legend inside the $R \times C$ array it creates:

```
. scatter mpg weight || lfit mpg weight ||,
          legend(cols(1))
          by(foreign, total legend(at(4) pos(0)))
```

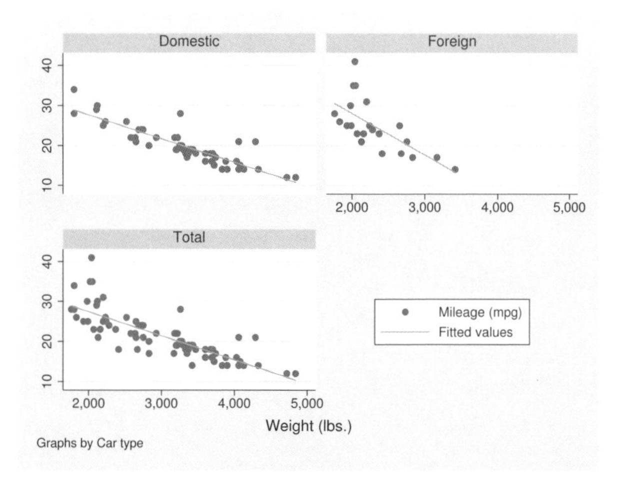

In the above, we specified at(4) to mean that the key was to appear in the fourth position of the 2×2 array, and we specified pos(0) to move the key to the middle (0 o'clock) position within the cell.

If you wish to suppress the legend, you must specify the legend(off) inside the by() option:

```
. scatter mpg weight || lfit mpg weight ||,
          by(foreign, total legend(off))
```

Problems arising with or because of legends

There are two potential problems associated with legends:

1. Text may flow outside the border of the legend box.
2. The presence of the legend may cause the title of the y axis to run into the values labeled on the axis.

The first problem arises because Stata uses an approximation to obtain the width of a text line. The solution is to specify the width(*relativesize*) *textbox_option*:

```
. graph ..., ... legend(width(#))
```

See *Use of the textbox option width()* in [G-3] ***added_text_options***.

The second problem arises when the key is in its default `position(6)` (6 o'clock) location and the descriptive text for one or more of the keys is long. In `position(6)`, the borders of the key are supposed to line up with the borders of the plot region. Usually the plot region is wider than the key, so the key is expanded to fit below it. When the key is wider than the plot region, however, it is the plot region that is widened. As the plot region expands, it will eat away at whatever is at its side, namely, the y axis labels and title. Margins will disappear. In extreme cases, the title will be printed on top of the labels, and the labels themselves may end up on top of the axis!

The solution to this problem is to shorten the descriptive text, either by using fewer words or by breaking the long description into multiple lines. Use the `legend(label(`*#* `"`*text*`"))` option to modify the longest line of the descriptive text.

Also see

[G-3] ***title_options*** — Options for specifying titles

Title

[G-3] ***line_options*** — Options for determining the look of lines

Syntax

line_options	Description
`lpattern(`*linepatternstyle*`)`	whether line solid, dashed, etc.
`lwidth(`*linewidthstyle*`)`	thickness of line
`lcolor(`*colorstyle*`)`	color of line
`lstyle(`*linestyle*`)`	overall style of line
`pstyle(`*pstyle*`)`	overall plot style, including linestyle

See [G-4] ***linepatternstyle***, [G-4] ***linewidthstyle***, [G-4] ***colorstyle***, [G-4] ***linestyle***, and [G-4] ***pstyle***.

All options are *rightmost*; see [G-4] **concept: repeated options**.

Description

The *line_options* determine the look of a line in some contexts.

Options

`lpattern(`*linepatternstyle*`)` specifies whether the line is solid, dashed, etc. See [G-4] ***linepatternstyle*** for a list of available patterns. `lpattern()` is not allowed with `graph pie`; see [G-2] **graph pie**.

`lwidth(`*linewidthstyle*`)` specifies the thickness of the line. See [G-4] ***linewidthstyle*** for a list of available thicknesses.

`lcolor(`*colorstyle*`)` specifies the color of the line. See [G-4] ***colorstyle*** for a list of available colors.

`lstyle(`*linestyle*`)` specifies the overall style of the line: its pattern, thickness, and color.

You need not specify `lstyle()` just because there is something you want to change about the look of the line. The other *line_options* will allow you to make changes. You specify `lstyle()` when another style exists that is exactly what you desire or when another style would allow you to specify fewer changes.

See [G-4] ***linestyle*** for a list of available line styles.

`pstyle(`*pstyle*`)` specifies the overall style of the plot, including not only the *linestyle*, but also all other settings for the look of the plot. Only the *linestyle* affects the look of lines. See [G-4] ***pstyle*** for a list of available plot styles.

Remarks

Lines occur in many contexts and, in some of those contexts, the above options are used to determine the look of the line. For instance, the `lcolor()` option in

```
. graph line y x, lcolor(red)
```

causes the line through the (`y`, `x`) point to be drawn in red.

The same option in the following

```
. graph line y x, title("My line", box lcolor(red))
```

causes the outline drawn around the title's box to be drawn in red. In the second command, the option `lcolor(red)` was a suboption to the `title()` option.

Also see

[G-4] **concept: lines** — Using lines

[G-2] **graph dot** — Dot charts (summary statistics)

Title

[G-3] *marker_label_options* — Options for specifying marker labels

Syntax

marker_label_options	Description
mlabel(*varname*)	specify marker variable
mlabstyle(*markerlabelstyle*)	overall style of label
mlabposition(*clockposstyle*)	where to locate the label
mlabvposition(*varname*)	where to locate the label 2
mlabgap(*relativesize*)	gap between marker and label
mlabangle(*anglestyle*)	angle of label
mlabtextstyle(*textstyle*)	overall style of text
mlabsize(*textsizestyle*)	size of label
mlabcolor(*colorstyle*)	color of label

See [G-4] ***markerlabelstyle***, [G-4] ***clockposstyle***, [G-4] ***relativesize***, [G-4] ***anglestyle***, [G-4] ***textstyle***, [G-4] ***textsizestyle***, and [G-4] ***colorstyle***.

All options are *rightmost*; see [G-4] **concept: repeated options**.

Sometimes—such as when used with scatter—lists are allowed inside the arguments. A list is a sequence of the elements separated by spaces. Shorthands are allowed to make specifying the list easier; see [G-4] ***stylelists***. When lists are allowed, option mlabel() allows a *varlist* in place of a *varname*.

Description

Marker labels are labels that appear next to (or in place of) markers. Markers are the ink used to mark where points are on a plot.

Options

mlabel(*varname*) specifies the (usually string) variable to be used that provides, observation by observation, the marker "text". For instance, you might have

```
. use http://www.stata-press.com/data/r12/auto
(1978 Automobile Data)

. list mpg weight make in 1/4
```

	mpg	weight	make
1.	22	2,930	AMC Concord
2.	17	3,350	AMC Pacer
3.	22	2,640	AMC Spirit
4.	20	3,250	Buick Century

Typing

```
. scatter mpg weight, mlabel(make)
```

would draw a scatter of `mpg` versus `weight` and label each point in the scatter according to its `make`. (We recommend that you include "`in 1/10`" on the above command. Marker labels work well only when there are few data.)

`mlabstyle(`*markerlabelstyle*`)` specifies the overall look of marker labels, including their position, their size, their text style, etc. The other options documented below allow you to change each attribute of the marker label, but `mlabstyle()` is the starting point. See [G-4] ***markerlabelstyle***.

You need not specify `mlabstyle()` just because there is something you want to change about the look of a marker and, in fact, most people seldom specify the `mlabstyle()` option. You specify `mlabstyle()` when another style exists that is exactly what you desire or when another style would allow you to specify fewer changes to obtain what you want.

`mlabposition(`*clockposstyle*`)` and `mlabvposition(`*varname*`)` specify where the label is to be located relative to the point. `mlabposition()` and `mlabvposition()` are alternatives; the first specifies a constant position for all points and the second specifies a variable that contains *clockposstyle* (a number 0–12) for each point. If both options are specified, `mlabvposition()` takes precedence.

If neither option is specified, the default is `mlabposition(3)` (3 o'clock)—meaning to the right of the point.

`mlabposition(12)` means above the point, `mlabposition(1)` means above and to the right of the point, and so on. `mlabposition(0)` means that the label is to be put directly on top of the point (in which case remember to also specify the `msymbol(i)` option so that the marker does not also display; see [G-3] ***marker_options***).

`mlabvposition(`*varname*`)` specifies a numeric variable containing values 0–12, which are used, observation by observation, to locate the labels relative to the points.

See [G-4] ***clockposstyle*** for more information on specifying *clockposstyle*.

`mlabgap(`*relativesize*`)` specifies how much space should be put between the marker and the label. See [G-4] ***relativesize***.

`mlabangle(`*anglestyle*`)` specifies the angle of text. The default is usually `mlabangle(horizontal)`. See [G-4] ***anglestyle***.

`mlabtextstyle(`*textstyle*`)` specifies the overall look of text of the marker labels, which here means their size and color. When you see [G-4] ***textstyle***, you will find that a *textstyle* defines much more, but all those other things are ignored for marker labels. In any case, the `mlabsize()` and `mlabcolor()` options documented below allow you to change the size and color, but the `mlabtextstyle` is the starting point.

As with `mlabstyle()`, you need not specify `mlabtextstyle()` just because there is something you want to change. You specify `mlabtextstyle()` when another style exists that is exactly what you desire or when another style would allow you to specify fewer changes to obtain what you want.

`mlabsize(`*textsizestyle*`)` specifies the size of the text. See [G-4] ***textsizestyle***.

`mlabcolor(`*colorstyle*`)` specifies the color of the text. See [G-4] ***colorstyle***.

Remarks

Remarks are presented under the following headings:

Typical use

Markers are the ink used to mark where points are on a plot, and marker labels optionally appear beside the markers to identify the points. For instance, if you were plotting country data, marker labels would allow you to have "Argentina", "Bolivia", ..., appear next to each point. Marker labels visually work well when there are few data.

To obtain marker labels, you specify the `mlabel(`*varname*`)` option, such as `mlabel(country)`. *varname* is the name of a variable that, observation by observation, specifies the text with which the point is to be labeled. *varname* may be a string or numeric variable, but usually it is a string. For instance, consider a subset of the life-expectancy-by-country data:

```
. use http://www.stata-press.com/data/r12/lifeexp
(Life expectancy, 1998)
. list country lexp gnppc if region==2
```

	country	lexp	gnppc
45.	Canada	79	19170
46.	Cuba	76	.
47.	Dominican Republic	71	1770
48.	El Salvador	69	1850
49.	Guatemala	64	1640
50.	Haiti	54	410
51.	Honduras	69	740
52.	Jamaica	75	1740
53.	Mexico	72	3840
54.	Nicaragua	68	1896
55.	Panama	74	2990
56.	Puerto Rico	76	.
57.	Trinidad and Tobago	73	4520
58.	United States	77	29240

We might graph these data and use labels to indicate the country by typing

```
. scatter lexp gnppc if region==2, mlabel(country)
```

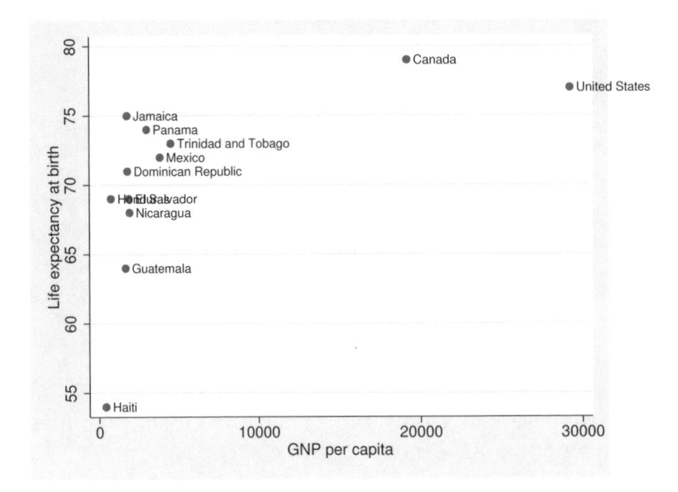

Eliminating overprinting and overruns

In the graph, the label "United States" runs off the right edge and the labels for Honduras and El Salvador are overprinted. Problems like that invariably occur when using marker labels. The `mlabposition()` allows specifying where the labels appear, and we might try

```
. scatter lexp gnppc if region==2, mlabel(country) mlabpos(9)
```

to move the labels to the 9 o'clock position, meaning to the left of the point. Here, however, that will introduce more problems than it will solve. You could try other clock positions around the point, but we could not find one that was satisfactory.

If our only problem were with "United States" running off the right, an adequate solution might be to widen the x axis so that there would be room for the label "United States" to fit:

```
. scatter lexp gnppc if region==2, mlabel(country)
                                   xscale(range(35000))
```

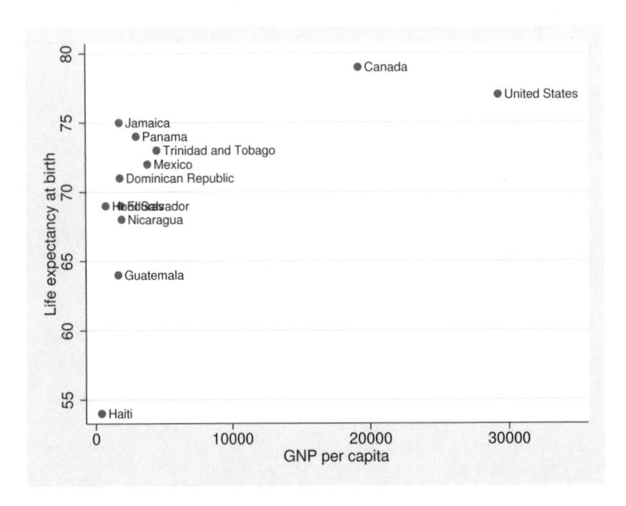

That would solve one problem but will leave us with the overprinting problem. The way to solve that problem is to move the Honduras label to being to the left of its point, and the way to do that is to specify the option `mlabvposition(`*varname*`)` rather than `mlabposition(`*clockposstyle*`)`. We will create new variable `pos` stating where we want each label:

```
. generate pos = 3
. replace pos = 9 if country=="Honduras"
. scatter lexp gnppc if region==2, mlabel(country) mlabv(pos)
                        xscale(range(35000))
```

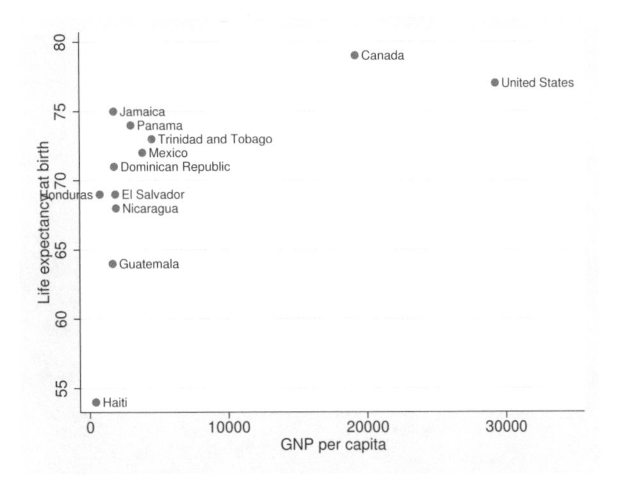

We are near a solution: Honduras is running off the left edge of the graph, but we know how to fix that. You may be tempted to solve this problem just as we solved the problem with the United States label: expand the range, say, to `range(-500 35000)`. That would be a fine solution.

Here, however, we will increase the margin between the left edge of the plot area and the y axis by adding the option `plotregion(margin(l+9))`; see [G-3] ***region_options***. `plotregion(margin(l+9))` says to increase the margin on the left by 9%, and this is really the "right" way to handle margin problems:

```
. scatter lexp gnppc if region==2, mlabel(country) mlabv(pos)
                        xscale(range(35000))
                        plotregion(margin(l+9))
```

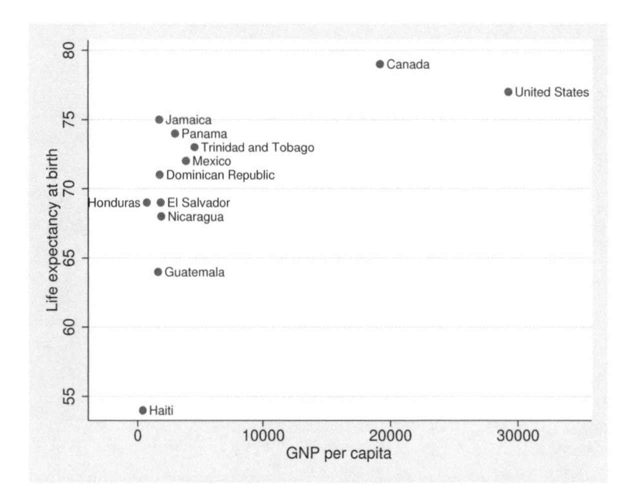

The overall result is adequate. Were we producing this graph for publication, we would move the label for United States to the left of its point, just as we did with Honduras, rather than widening the x axis.

Advanced use

Let us now consider properly graphing the life-expectancy data and graphing more of it. This time, we will include South America, as well as North and Central America, and we will graph the data on a log(GNP) scale.

```
. use http://www.stata-press.com/data/r12/lifeexp, clear
(Life expectancy, 1998)
. keep if region==2 | region==3                                      (note 1)
. replace gnppc = gnppc / 1000
. label var gnppc "GNP per capita (thousands of dollars)"            (note 2)
. generate lgnp = log(gnp)
. quietly reg lexp lgnp
. predict hat
. label var hat "Linear prediction"                                  (note 3)
. replace country = "Trinidad" if country=="Trinidad and Tobago"
. replace country = "Para" if country == "Paraguay"                  (note 4)
. generate pos = 3
. replace pos = 9 if lexp > hat                                      (note 5)
. replace pos = 3 if country == "Colombia"
. replace pos = 3 if country == "Para"
. replace pos = 3 if country == "Trinidad"
. replace pos = 9 if country == "United States"                      (note 6)
```

```
. twoway (scatter lexp gnppc, mlabel(country) mlabv(pos))
        (line hat gnppc, sort)
        , xscale(log) xlabel(.5 5 10 15 20 25 30, grid)
          legend(off)
          title("Life expectancy vs. GNP per capita")
          subtitle("North, Central, and South America")
          note("Data source:  World Bank, 1998")
          ytitle("Life expectancy at birth (years)")
```

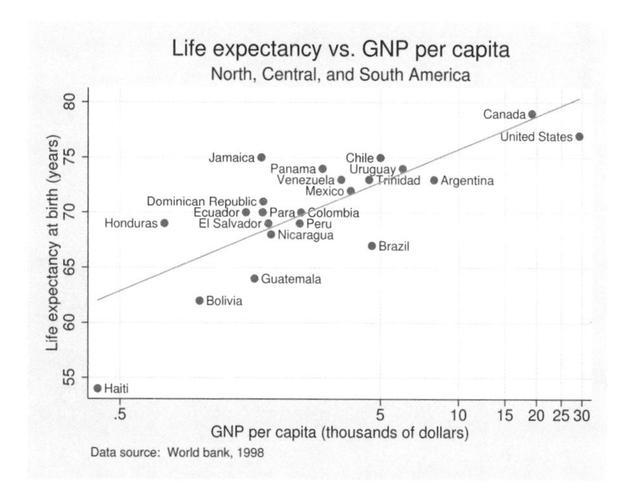

Notes:

1. In these data, region 2 is North and Central America, and region 3 is South America.
2. We divide `gnppc` by 1,000 to keep the x axis labels from running into each other.
3. We add a linear regression prediction. We cannot use `graph twoway lfit` because we want the predictions to be based on a regression of log(GNP), not GNP.
4. The first time we graphed the results, we discovered that there was no way we could make the names of these two countries fit on our graph, so we shortened them.
5. We are going to place the marker labels to the left of the marker when life expectancy is above the regression line and to the right of the marker otherwise.
6. To keep labels from overprinting, we need to override rule (5) for a few countries.

Also see [G-3] ***scale_option*** for another rendition of this graph. In that rendition, we specify one more option—`scale(1.1)`—to increase the size of the text and markers by 10%.

Using marker labels in place of markers

In addition to specifying where the marker label goes relative to the marker, you can specify that the marker label be used instead of the marker. `mlabposition(0)` means that the label is to be centered where the marker would appear. To suppress the display of the marker as well, specify option `msymbol(i)`; see [G-3] ***marker_options***.

Using the labels in place of the points tends to work well in analysis graphs where our interest is often in identifying the outliers. Below we graph the entire `lifeexp.dta` data:

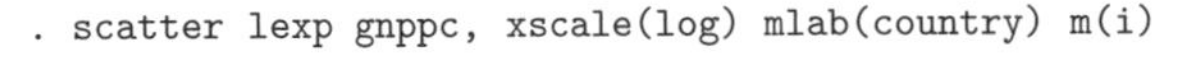

```
. scatter lexp gnppc, xscale(log) mlab(country) m(i)
```

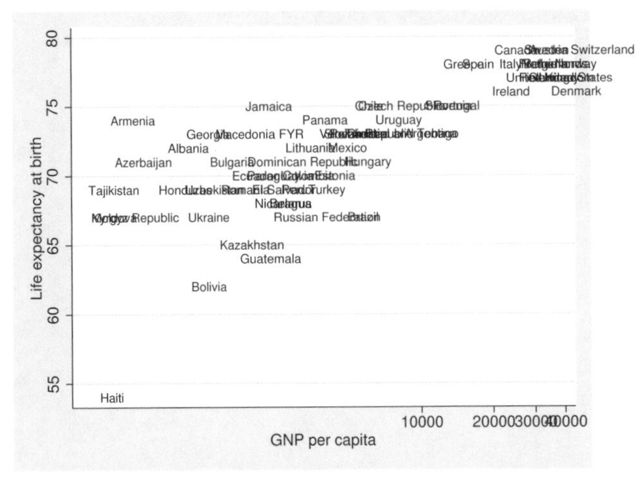

In the above graph, we also specified `xscale(log)` to convert the x axis to a log scale. A log x scale is more appropriate for these data, but had we used it earlier, the overprinting problem with Honduras and El Salvador would have disappeared, and we wanted to show how to handle the problem.

Also see

[G-2] **graph twoway scatter** — Twoway scatterplots

[G-4] ***anglestyle*** — Choices for the angle at which text is displayed

[G-4] ***clockposstyle*** — Choices for location: Direction from central point

[G-4] ***colorstyle*** — Choices for color

[G-4] ***markerlabelstyle*** — Choices for overall look of marker labels

[G-4] ***relativesize*** — Choices for sizes of objects

[G-4] ***textstyle*** — Choices for the overall look of text

[G-4] ***textsizestyle*** — Choices for the size of text

Title

[G-3] ***marker_options*** — Options for specifying markers

Syntax

marker_options	Description
`msymbol(`*symbolstyle*`)`	shape of marker
`mcolor(`*colorstyle*`)`	color of marker, inside and out
`msize(`*markersizestyle*`)`	size of marker
`mfcolor(`*colorstyle*`)`	inside or "fill" color
`mlcolor(`*colorstyle*`)`	outline color
`mlwidth(`*linewidthstyle*`)`	outline thickness
`mlstyle(`*linestyle*`)`	thickness and color, overall style of outline
`mstyle(`*markerstyle*`)`	overall style of marker; all settings above
`pstyle(`*pstyle*`)`	overall plot style, including markerstyle
`recast(`*newplottype*`)`	advanced; treat plot as *newplottype*

See [G-4] ***symbolstyle***, [G-4] ***colorstyle***, [G-4] ***markersizestyle***, [G-4] ***linewidthstyle***, [G-4] ***linestyle***, [G-4] ***markerstyle***, [G-4] ***pstyle***, and [G-3] ***advanced_options***.

All options are *rightmost*; see [G-4] **concept: repeated options**.

One example of each of the above is

```
msymbol(O)        mfcolor(red)    mlcolor(olive)     mstyle(p1)
mcolor(green)                     mlwidth(thick)     mlstyle(p1)
msize(medium)
```

Sometimes you may specify a list of elements, with the first element applying to the first variable, the second to the second, and so on. See, for instance, [G-2] **graph twoway scatter**. One example would be

```
msymbol(O o p)
mcolor(green blue black)
msize(medium medium small)

mfcolor(red red none)

mlcolor(olive olive green)
mlwidth(thick thin thick)

mstyle(p1 p2 p3)
mlstyle(p1 p2 p3)
```

For information about specifying lists, see [G-4] ***stylelists***.

Description

Markers are the ink used to mark where points are on a plot. The important options are

`msymbol(`*symbolstyle*`)`	(choice of symbol)
`mcolor(`*colorstyle*`)`	(choice of color)
`msize(`*markersizestyle*`)`	(choice of size)

Options

`msymbol(`*symbolstyle*`)` specifies the shape of the marker and is one of the more commonly specified options. See [G-4] ***symbolstyle*** for more information on this important option.

`mcolor(`*colorstyle*`)` specifies the color of the marker. This option sets both the color of the line used to outline the marker's shape and the color of the inside of the marker. Also see options `mfcolor()` and `mlcolor()` below. See [G-4] ***colorstyle*** for a list of color choices.

`msize(`*markersizestyle*`)` specifies the size of the marker. See [G-4] ***markersizestyle*** for a list of size choices.

`mfcolor(`*colorstyle*`)` specifies the color of the inside of the marker. See [G-4] ***colorstyle*** for a list of color choices.

`mlcolor(`*colorstyle*`)`, `mlwidth(`*linewidthstyle*`)`, and `mlstyle(`*linestyle*`)` specify the look of the line used to outline the shape of the marker. See [G-4] **concept: lines**, but you cannot change the line pattern of a marker.

`mstyle(`*markerstyle*`)` specifies the overall look of markers, such as their shape and their color. The other options allow you to change each attribute of the marker, but `mstyle()` is a starting point.

You need not specify `mstyle()` just because there is something you want to change about the look of the marker and, in fact, most people seldom specify the `mstyle()` option. You specify `mstyle()` when another style exists that is exactly what you desire or when another style would allow you to specify fewer changes to obtain what you want.

See [G-4] ***markerstyle*** for a list of available marker styles.

`pstyle(`*pstyle*`)` specifies the overall style of the plot, including not only the *markerstyle*, but also the *markerlabelstyle* and all other settings for the look of the plot. Only the *markerstyle* and *markerlabelstyle* affect the look of markers. See [G-4] ***pstyle*** for a list of available plot styles.

`recast(`*newplottype*`)` is an advanced option allowing the plot to be recast from one type to another, for example, from a scatterplot to a line plot; see [G-3] ***advanced_options***. Most, but not all, plots allow `recast()`.

Remarks

You will never need to specify all nine marker options, and seldom will you even need to specify more than one or two of them. Many people think that there is just one important marker option,

`msymbol(`*symbolstyle*`)`

`msymbol()` specifies the shape of the symbol; see [G-4] ***symbolstyle*** for choice of symbol. A few people would add to the important list a second option, mcolor(*colorstyle*)

mcolor() specifies the marker's color; see [G-4] ***colorstyle*** for choice of color. Finally, a few would add

msize(*markersizestyle*)

msize() specifies the marker's size; see [G-4] ***markersizestyle*** for choice of sizes.

After that, we are really into the details. One of the remaining options, however, is of interest:

mstyle(*markerstyle*)

A marker has a set of characteristics:

$\{$shape, color, size, inside details, outside details$\}$

Each of the options other than mstyle() modifies something in that set. mstyle() sets the values of the entire set. It is from there that the changes you specify are made. See [G-4] ***markerstyle***.

Also see

[G-4] ***symbolstyle*** — Choices for the shape of markers

[G-4] ***colorstyle*** — Choices for color

[G-4] ***markersizestyle*** — Choices for the size of markers

[G-4] ***linewidthstyle*** — Choices for thickness of lines

[G-4] ***linestyle*** — Choices for overall look of lines

[G-4] ***markerstyle*** — Choices for overall look of markers

Title

[G-3] *name_option* — Option for naming graph in memory

Syntax

name_option	Description
name(*name*[, replace])	specify name

name() is *unique*; see [G-4] **concept: repeated options**.

Description

Option name() specifies the name of the graph being created.

Option

name(*name*[, replace]) specifies the name of the graph. If name() is not specified, name(Graph, replace) is assumed.

In fact, name(Graph) has the same effect as name(Graph, replace) because replace is assumed when the name is Graph. For all other *names*, you must specify suboption replace if a graph under that name already exists.

Remarks

When you type, for instance,

```
. scatter yvar xvar
```

you see a graph. The graph is also stored in memory. For instance, try the following: close the Graph window, and then type

```
. graph display
```

Your graph will reappear.

Every time you draw a graph, that previously remembered graph is discarded, and the new graph replaces it.

You can have more than one graph stored in memory. When you do not specify the name under which the graph is to be remembered, it is remembered under the default name Graph. For instance, if you were now to type

```
. scatter y2var xvar, name(g2)
```

You would now have two graphs stored in memory: Graph and g2. If you typed

```
. graph display
```

or

```
. graph display Graph
```

you would see your first graph. Type

```
. graph display g2
```

and you will see your second graph.

Do not confuse Stata's storing of graphs in memory with the saving of graphs to disk. Were you now to `exit` Stata, the graphs you have saved in memory would be gone forever. If you want to save your graphs, you want to specify the `saving()` option (see [G-3] ***saving_option***) or you want to use the `graph save` command (see [G-2] **graph save**); either result in the same outcome.

You can find out what graphs you have in memory by using `graph dir`, drop them by using `graph drop`, rename them by using `graph rename`, and so on, and of course, you can redisplay them by using `graph display`. See [G-2] **graph manipulation** for the details on all those commands.

You can drop all graphs currently stored in memory by using `graph drop _all` or `discard`; see [G-2] **graph drop**.

Also see

[G-2] **graph display** — Display graph stored in memory

[G-2] **graph drop** — Drop graphs from memory

[G-3] ***saving_option*** — Option for saving graph to disk

[G-2] **graph save** — Save graph to disk

[G-2] **graph manipulation** — Graph manipulation commands

Title

[G-3] ***nodraw_option*** — Option for suppressing display of graph

Syntax

nodraw_option	Description
nodraw	suppress display of graph

nodraw is *unique*; see [G-4] **concept: repeated options**.

Description

Option nodraw prevents the graph from being displayed. Graphs drawn with nodraw may not be printed or exported, though they may be saved.

Option

nodraw specifies that the graph not be displayed.

Remarks

When you type, for instance,

```
. scatter yvar xvar, saving(mygraph)
```

a graph is displayed and is stored in file mygraph.gph. If you type

```
. scatter yvar xvar, saving(mygraph) nodraw
```

the graph will still be saved in file mygraph.gph, but it will not be displayed. The result is the same as if you typed

```
. set graphics off
. scatter yvar xvar, saving(mygraph)
. set graphics on
```

Here, however, the graph may also be printed or exported.

You need not specify saving() (see [G-3] ***saving_option***) to use nodraw. You could type

```
. scatter yvar xvar, nodraw
```

and later type (or code in an ado-file)

```
. graph display Graph
```

See [G-2] **graph display**.

Also see

[R] **set** — Overview of system parameters

Title

[G-3] ***play_option*** — Option for playing graph recordings

Syntax

play_option	Description
`play(`*recordingname*`)`	play edits from *recordingname*

`play()` is *unique*; see [G-4] **concept: repeated options**.

Description

Option `play()` replays edits that were previously recorded using the Graph Recorder.

Option

`play(`*recordingname*`)` applies the edits from *recordingname* to the graph, where *recordingname* is the name under which edits previously made in the Graph Editor have been recorded and stored. See *Graph Recorder* in [G-1] **graph editor**.

Remarks

Edits made in the Graph Editor (see [G-1] **graph editor**) can be saved as a recording and the edits subsequently played on another graph. In addition to being played from the Graph Editor, these recordings can be played when a graph is created or used from disk with the option `play()`.

If you have previously created a recording named `xyz` and you are drawing a scatterplot of `y` on `x`, you can replay the edits from that recording on your new graph by adding the option `play(xyz)` to your graph command:

```
. scatter y x, play(xyz)
```

To learn about creating recordings, see *Graph Recorder* in [G-1] **graph editor**.

Also see

[G-1] **graph editor** — Graph Editor

[G-2] **graph play** — Apply edits from a recording on current graph

Title

[G-3] *png_options* — Options for exporting to portable network graphics (PNG) format

Syntax

png_options	Description
`width(#)`	width of graph in pixels
`height(#)`	height of graph in pixels

Description

The *png_options* are used with `graph export` when creating graphs in PNG format; see [G-2] **graph export**.

Options

`width(#)` specifies the width of the graph in pixels. `width()` must contain an integer between 8 and 16,000.

`height(#)` specifies the height of the graph in pixels. `height()` must contain an integer between 8 and 16,000.

Remarks

Remarks are presented under the following headings:

Using png_options
Specifying the width or height

Using png_options

You have drawn a graph and wish to create a PNG file to include in a document. You wish, however, to set the width of the graph to 800 pixels and the height to 600 pixels:

```
. graph ...                                        (draw a graph)
. graph export myfile.png, width(800) height(600)
```

Specifying the width or height

If the width is specified but not the height, Stata determines the appropriate height from the graph's aspect ratio. If the height is specified but not the width, Stata determines the appropriate width from the graph's aspect ratio. If neither the width nor the height is specified, Stata will export the graph on the basis of the current size of the Graph window.

Also see

[G-2] **graph export** — Export current graph

[G-2] **graph set** — Set graphics options

Title

[G-3] *pr_options* — Options for use with graph print

Syntax

pr_options	Description
`tmargin(#)`	top margin, in inches, $0 \leq \# \leq 20$
`lmargin(#)`	left margin, in inches, $0 \leq \# \leq 20$
`logo(on|off)`	whether to display Stata logo

Current default values may be listed by typing

```
. graph set print
```

The defaults may be changed by typing

```
. graph set print name value
```

where *name* is the name of a *pr_option*, omitting the parentheses.

Description

The *pr_options* are used with `graph print`; see [G-2] **graph print**.

Options

`tmargin(#)` and `lmargin(#)` set the top and left page margins—the distance from the edge of the page to the start of the graph. # is specified in inches, must be between 0 and 20, and may be fractional.

`logo(on)` and `logo(off)` specify whether the Stata logo should be included at the bottom of the graph.

Remarks

Remarks are presented under the following headings:

Using the pr_options
Setting defaults
Note for Unix users

Using the pr_options

You have drawn a graph and wish to print it. You wish, however, to suppress the Stata logo (although we cannot imagine why you would want to do that):

```
. graph ...                              (draw a graph)
. graph print, logo(off)
```

Setting defaults

If you always wanted `graph print` to suppress the Stata logo, you could type

```
. graph set print logo off
```

At a future date, you could type

```
. graph set print logo on
```

to set it back. You can determine your default *pr_options* settings by typing

```
. graph set print
```

Note for Unix users

In addition to the options documented above, there are other options you may specify. Under Stata for Unix, the *pr_options* are in fact *ps_options*; see [G-3] ***ps_options***.

Also see

[G-2] **graph print** — Print a graph

Title

[G-3] *ps_options* — Options for exporting or printing to PostScript

Syntax

ps_options	Description
tmargin(*#*)	top margin in inches
lmargin(*#*)	left margin in inches
logo(on | off)	whether to include Stata logo
cmyk(on | off)	whether to use CMYK rather than RGB colors
mag(*#*)	magnification/shrinkage factor; default is 100
fontface(*fontname*)	default font to use
fontfacesans(*fontname*)	font to use for text in {stSans} "font"
fontfaceserif(*fontname*)	font to use for text in {stSerif} "font"
fontfacemono(*fontname*)	font to use for text in {stMono} "font"
fontfacesymbol(*fontname*)	font to use for text in {stSymbol} "font"
fontdir(*directory*)	(Unix only) directory in which TrueType fonts are stored
orientation(portrait | landscape)	whether vertical or horizontal
pagesize(letter | legal | executive | A4 | custom)	size of page
pageheight(*#*)	inches; relevant only if pagesize(custom)
pagewidth(*#*)	inches; relevant only if pagesize(custom)

where *fontname* may be a valid font name or default to restore the default setting and *directory* may be a valid directory or default to restore the default setting.

Current default values may be listed by typing

```
. graph set ps
```

and default values may be set by typing

```
. graph set ps name value
```

where *name* is the name of a *ps_option*, omitting the parentheses.

Description

These *ps_options* are used with graph export when creating a PostScript file; see [G-2] **graph export**.

Also, in Stata for Unix, these options are used with graph print; see [G-2] **graph print**.

Options

`tmargin(`*#*`)` and `lmargin(`*#*`)` set the top and left page margins—the distance from the edge of the page to the start of the graph. *#* is specified in inches, must be between 0 and 20, and may be fractional.

`logo(on)` and `logo(off)` specify whether the Stata logo should be included at the bottom of the graph.

`cmyk(on)` and `cmyk(off)` specify whether colors in the output file should be specified as CMYK values rather than RGB values.

`mag(`*#*`)` specifies that the graph be drawn smaller or larger than the default. `mag(100)` is the default, meaning ordinary size. `mag(110)` would make the graph 10% larger than usual and `mag(90)` would make the graph 10% smaller than usual. *#* must be an integer.

`fontface(`*fontname*`)` specifies the name of the PostScript font to be used to render text for which no other font has been specified. The default is `Helvetica`, which may be restored by specifying *fontname* as `default`. If *fontname* contains spaces, it must be enclosed in double quotes.

`fontfacesans(`*fontname*`)` specifies the name of the PostScript font to be used to render text for which the `{stSans}` "font" has been specified. The default is `Helvetica`, which may be restored by specifying *fontname* as `default`. If *fontname* contains spaces, it must be enclosed in double quotes.

`fontfaceserif(`*fontname*`)` specifies the name of the PostScript font to be used to render text for which the `{stSerif}` "font" has been specified. The default is `Times`, which may be restored by specifying *fontname* as `default`. If *fontname* contains spaces, it must be enclosed in double quotes.

`fontfacemono(`*fontname*`)` specifies the name of the PostScript font to be used to render text for which the `{stMono}` "font" has been specified. The default is `Courier`, which may be restored by specifying *fontname* as `default`. If *fontname* contains spaces, it must be enclosed in double quotes.

`fontfacesymbol(`*fontname*`)` specifies the name of the PostScript font to be used to render text for which the `{stSymbol}` "font" has been specified. The default is `Symbol`, which may be restored by specifying *fontname* as `default`. If *fontname* contains spaces, it must be enclosed in double quotes.

`fontdir(`*directory*`)` specifies the directory that Stata for Unix uses to find TrueType fonts (if you specified any) for conversion to PostScript fonts when you export a graph to Encapsulated PostScript. You may specify *directory* as `default` to restore the default setting. If *directory* contains spaces, it must be enclosed in double quotes.

`orientation(portrait)` and `orientation(landscape)` specify whether the graph is to be presented vertically or horizontally.

`pagesize()` specifies the size of the page. `pagesize(letter)`, `pagesize(legal)`, `pagesize(executive)`, and `pagesize(A4)` are prerecorded sizes. `pagesize(custom)` specifies that you wish to explicitly specify the size of the page by using the `pageheight()` and `pagewidth()` options.

`pageheight(`*#*`)` and `pagewidth(`*#*`)` are relevant only if `pagesize(custom)` is specified. They specify the height and width of the page in inches. *#* is specified in inches, must be between 0 and 20, and may be fractional.

Remarks

Remarks are presented under the following headings:

Using the ps_options
Setting defaults
Note about PostScript fonts
Note for Unix users

Using the ps_options

You have drawn a graph and wish to create a PostScript file. You wish, however, to change text for which no other font has been specified from the default of Helvetica to Roman, which is "Times" in PostScript jargon:

```
. graph ...                                    (draw a graph)
. graph export myfile.ps, fontface(Times)
```

Setting defaults

If you always wanted `graph export` (see [G-2] **graph export**) to use Times when exporting to PostScript files, you could type

```
. graph set ps fontface Times
```

Later, you could type

```
. graph set ps fontface Helvetica
```

to set it back. You can list the current *ps_option* settings for PostScript by typing

```
. graph set ps
```

Note about PostScript fonts

Graphs exported to PostScript format by Stata conform to what is known as PostScript Level 2. There are 10 built-in font faces, known as the Core Font Set, some of which are available in modified forms, for example, bold or italic (a listing of the original font faces in the Core Font Set is shown at http://en.wikipedia.org/wiki/Type_1_and_Type_3_fonts#Core_Font_Set). If you change any of the `fontface*()` settings, we recommend that you use one of those 10 font faces. We do not recommend changing `fontfacesymbol()`, because doing so can lead to incorrect characters being printed.

If you specify a font face other than one that is part of the Core Font Set, Stata will first attempt to map it to the closest matching font in the Core Font Set. For example, if you specify `fontfaceserif("Times New Roman")`, Stata will map it to `fontfaceserif("Times")`.

If Stata is unable to map the font face to the Core Font Set, Stata will look in the `fontdir()` directory for a TrueType font on your system matching the font you specified. If it finds one, it will attempt to convert it to a PostScript font and, if successful, will embed the converted font in the exported PostScript graph. Because of the wide variety of TrueType fonts available on different systems, this conversion can be problematic, which is why we recommend that you use fonts found in the Core Font Set.

Note for Unix users

The PostScript settings are used not only by `graph export` when creating a PostScript file but also by `graph print`. In [G-3] ***pr_options***, you are told that you may list and set defaults by typing

```
. graph set print ...
```

That is true, but under Unix, `print` is a synonym for `ps`, so whether you type `graph set print` or `graph set ps` makes no difference.

Also see

[G-2] **graph export** — Export current graph

[G-2] **graph set** — Set graphics options

[G-3] ***eps_options*** — Options for exporting to Encapsulated PostScript

Title

[G-3] ***rcap_options*** — Options for determining the look of range plots with capped spikes

Syntax

rcap_options	Description
line_options	change look of spike and cap lines
`msize(`*markersizestyle*`)`	width of cap
`recast(`*newplottype*`)`	advanced; treat plot as *newplottype*

See [G-3] ***line_options***, [G-4] ***markersizestyle***, and [G-3] ***advanced_options***.

All options are *rightmost*; see [G-4] **concept: repeated options**.

Description

The *rcap_options* determine the look of spikes (lines connecting two points vertically or horizontally) and their endcaps.

Options

line_options specify the look of the lines used to draw the spikes and their caps, including pattern, width, and color; see [G-3] ***line_options***.

`msize(`*markersizestyle*`)` specifies the width of the cap. Option `msize()` is in fact `twoway scatter`'s *marker_option* that sets the size of the marker symbol, but here `msymbol()` is borrowed to set the cap width. See [G-4] ***markersizestyle*** for a list of size choices.

`recast(`*newplottype*`)` is an advanced option allowing the plot to be recast from one type to another, for example, from a range-capped plot to an area plot; see [G-3] ***advanced_options***. Most, but not all, plots allow `recast()`.

Remarks

Range-capped plots are used in many contexts. They are sometimes the default for confidence intervals. For instance, the `lcolor()` suboption of `ciopts()` in

```
. tabodds died age, ciplot ciopts(lcolor(green))
```

causes the color of the horizontal lines representing the confidence intervals in the graph to be drawn in green.

Also see

[G-4] **concept: lines** — Using lines

[G-3] ***advanced_options*** — Rarely specified options for use with graph twoway

[G-4] ***markersizestyle*** — Choices for the size of markers

Title

[G-3] *region_options* — Options for shading and outlining regions and controlling graph size

Syntax

region_options	Description
`ysize(#)`	height of *available area* (in inches)
`xsize(#)`	width of *available area* (in inches)
`graphregion(`*suboptions*`)`	attributes of *graph region*
`plotregion(`*suboptions*`)`	attributes of *plot region*

Options `ysize()` and `xsize()` are *unique*; options `graphregion()` and `plotregion()` are *merged-implicit*; see [G-4] **concept: repeated options**.

suboptions	Description
`style(`*areastyle*`)`	overall style of outer region
`color(`*colorstyle*`)`	line and fill color of outer region
`fcolor(`*colorstyle*`)`	fill color of outer region
`lstyle(`*linestyle*`)`	overall style of outline
`lcolor(`*colorstyle*`)`	color of outline
`lwidth(`*linewidthstyle*`)`	thickness of outline
`lpattern(`*linepatternstyle*`)`	outline pattern (solid, dashed, etc.)
`istyle(`*areastyle*`)`	overall style of inner region
`icolor(`*colorstyle*`)`	line and fill color of inner region
`ifcolor(`*colorstyle*`)`	fill color of inner region
`ilstyle(`*linestyle*`)`	overall style of outline
`ilcolor(`*colorstyle*`)`	color of outline
`ilwidth(`*linewidthstyle*`)`	thickness of outline
`ilpattern(`*linepatternstyle*`)`	outline pattern (solid, dashed, etc.)
`margin(`*marginstyle*`)`	margin between inner and outer regions

See [G-4] ***areastyle***, [G-4] ***colorstyle***, [G-4] ***linestyle***, [G-4] ***linewidthstyle***, [G-4] ***linepatternstyle***, and [G-4] ***marginstyle***.

The *available area*, *graph region*, and *plot region* are defined

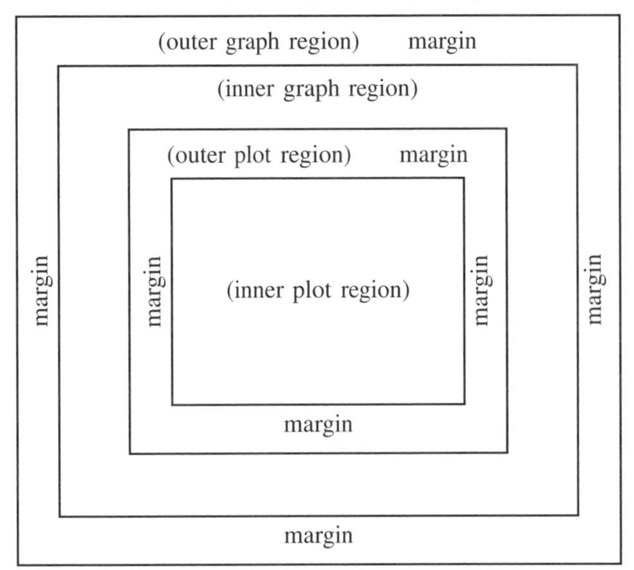

titles appear outside the borders of outer plot region

axes appear on the borders of the outer plot region

plot appears in inner plot region

Note: what are called the "graph region" and the "plot region" are sometimes the inner and sometimes the outer regions.

The available area and outer graph region are almost coincident; they differ only by the width of the border.

The borders of the outer plot or graph region are sometimes called the outer borders of the plot or graph region.

Description

The *region_options* set the size, margins, and color of the area in which the graph appears.

Options

`ysize(`*#*`)` and `xsize(`*#*`)` specify in inches the height and width of the *available area*. The defaults are usually `ysize(4)` and `xsize(5)`, but this, of course, is controlled by the scheme; see [G-4] **schemes intro**. These two options can be used to control the overall aspect ratio of a graph. See *Controlling the aspect ratio* below.

`graphregion(`*suboptions*`)` and `plotregion(`*suboptions*`)` specify attributes for the *graph region* and *plot region*.

Suboptions

`style(`*areastyle*`)` and `istyle(`*areastyle*`)` specify the overall style of the outer and inner regions. The other suboptions allow you to change the region's attributes individually, but `style()` and `istyle()` provide the starting points. See [G-4] ***areastyle*** for a list of choices.

`color(`*colorstyle*`)` and `icolor(`*colorstyle*`)` specify the color of the line used to outline the outer and inner regions; see [G-4] ***colorstyle*** for a list of choices.

`fcolor(`*colorstyle*`)` and `ifcolor(`*colorstyle*`)` specify the fill color for the outer and inner regions; see [G-4] ***colorstyle*** for a list of choices.

`lstyle(`*linestyle*`)` and `ilstyle(`*linestyle*`)` specify the overall style of the line used to outline the outer and inner regions, which includes its pattern (solid, dashed, etc.), thickness, and color. The other suboptions listed below allow you to change the line's attributes individually, but `lstyle()` and `ilstyle()` are the starting points. See [G-4] ***linestyle*** for a list of choices.

`lcolor(`*colorstyle*`)` and `ilcolor(`*colorstyle*`)` specify the color of the line used to outline the outer and inner regions; see [G-4] ***colorstyle*** for a list of choices.

`lwidth(`*linewidthstyle*`)` and `ilwidth(`*linewidthstyle*`)` specify the thickness of the line used to outline the outer and inner regions; see [G-4] ***linewidthstyle*** for a list of choices.

`lpattern(`*linepatternstyle*`)` and `ilpattern(`*linepatternstyle*`)` specify whether the line used to outline the outer and inner regions is solid, dashed, etc.; see [G-4] ***linepatternstyle*** for a list of choices.

`margin(`*marginstyle*`)` specifies the margin between the outer and inner regions; see [G-4] ***marginstyle***.

Remarks

Remarks are presented under the following headings:

Setting the offset between the axes and the plot region

By default, most schemes (see [G-4] **schemes intro**) offset the axes from the region in which the data are plotted. This offset is specified by `plotregion(margin(`*marginstyle*`))`; see [G-4] ***marginstyle***.

If you do not want the axes offset from the contents of the plot, specify `plotregion(margin(zero))`. Compare the next two graphs:

```
. use http://www.stata-press.com/data/r12/auto
(1978 Automobile Data)
. scatter price mpg
```

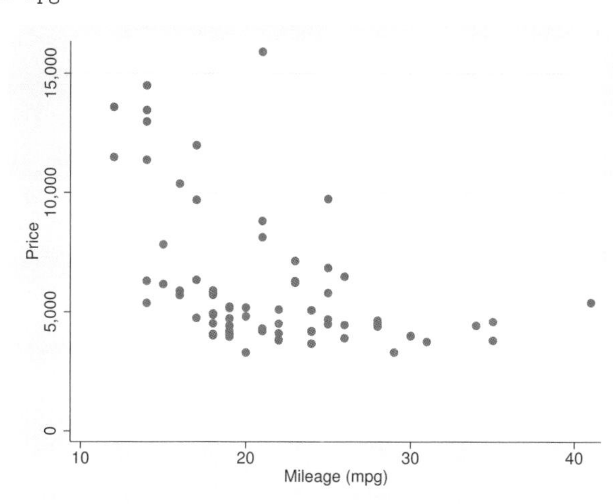

```
. scatter price mpg, plotr(m(zero))
```

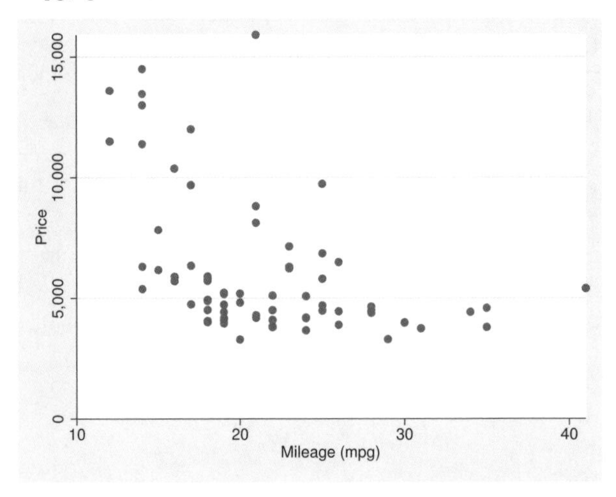

Controlling the aspect ratio

Here we discuss controlling the overall aspect ratio of a graph. To control the aspect ratio of a plot region for `twoway`, `graph bar`, `graph box`, or `graph dot`, see [G-3] ***aspect_option***.

The way to control the aspect ratio of the overall graph is by specifying the `xsize()` or `ysize()` options. For instance, you draw a graph and find that the graph is too wide given its height. To address the problem, either increase `ysize()` or decrease `xsize()`. The usual defaults (which of course are determined by the scheme; see [G-4] **schemes intro**) are `ysize(4)` and `xsize(5.5)`, so you might try

```
. graph ..., ... ysize(5)
```

or

```
. graph ..., ... xsize(4.5)
```

For instance, compare

```
. scatter mpg weight
```

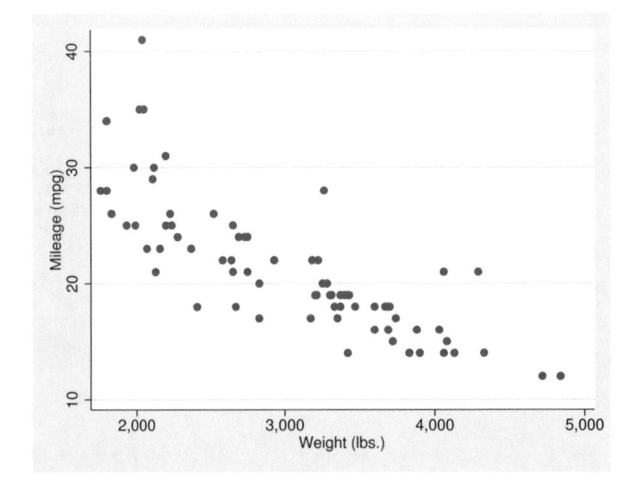

```
. scatter mpg weight, ysize(5)
```

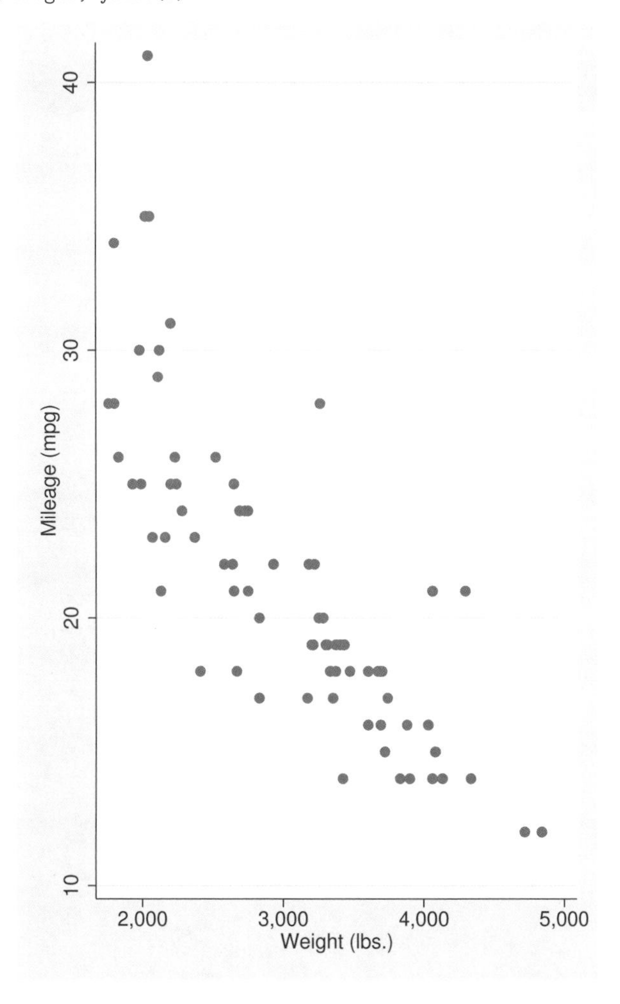

Another way to control the aspect ratio is to add to the outer margin of the *graph area*. This will keep the overall size of the graph the same while using less of the *available area*. For instance,

```
. scatter mpg weight, graphregion(margin(l+10 r+10))
```

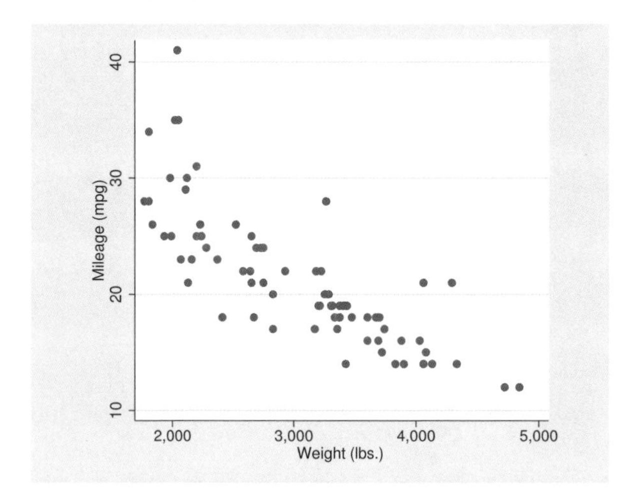

This method is especially useful when using `graph, by()`, but remember to specify the `graphregion(margin())` option inside the `by()` so that it affects the entire graph:

```
. scatter mpg weight, by(foreign, total graphr(m(l+10 r+10)))
```

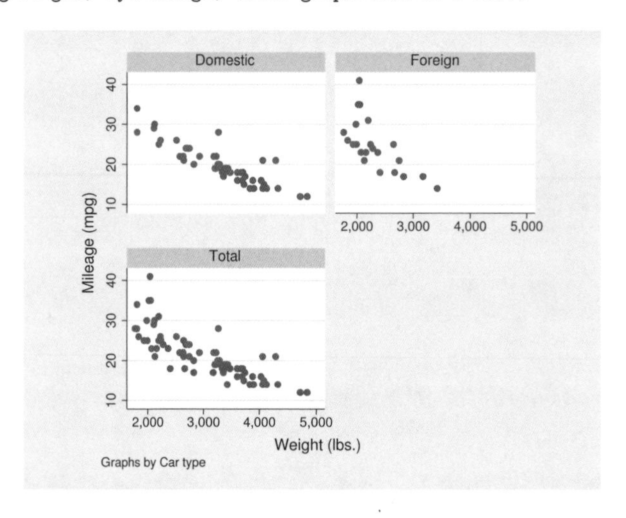

Compare the above with

```
. scatter mpg weight, by(foreign, total)
```

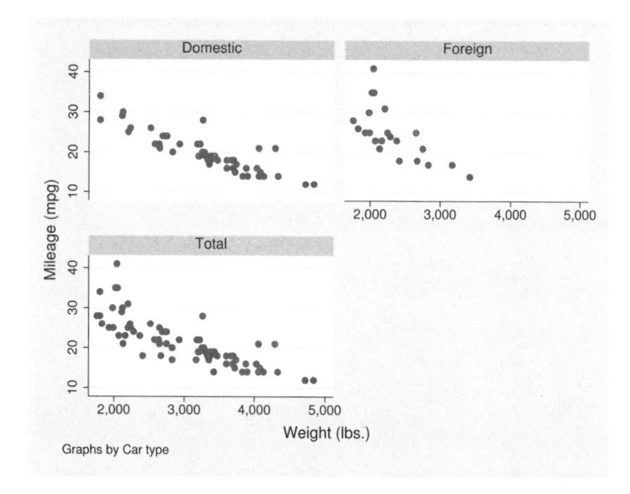

A similar, and often preferable, effect can be obtained by constraining the aspect ratio of the plot region itself; see [G-3] ***aspect_option***.

You do not have to get the aspect ratio or size right the first time you draw a graph; using `graph display`, you can change the aspect ratio of an already drawn graph—even a graph saved in a `.gph` file. See *Changing the size and aspect ratio* in [G-2] **graph display**.

Suppressing the border around the plot region

To eliminate the border around the plot region, specify `plotregion(style(none))`:

```
. use http://www.stata-press.com/data/r12/auto, clear
(1978 Automobile Data)
. scatter mpg weight, plotregion(style(none))
```

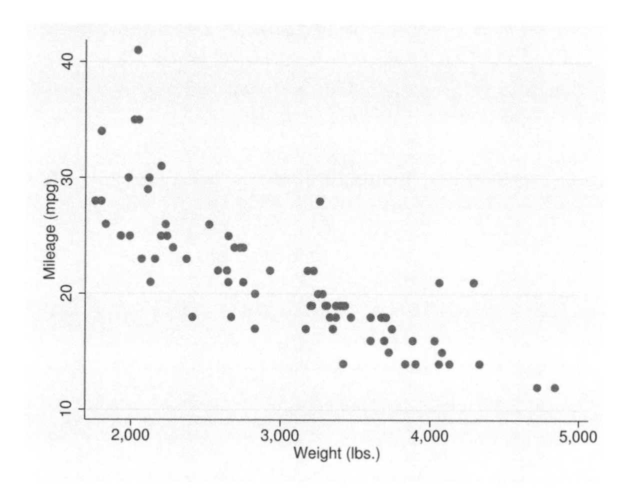

Setting background and fill colors

The background color of a graph is determined by default by the scheme you choose—see [G-4] **schemes intro**—and is usually black or white, perhaps with a tint. Option `graphregion(fcolor(`*colorstyle*`))` allows you to override the scheme's selection. When doing this, choose a light background color for schemes that are naturally white and a dark background color for schemes that are naturally black, or you will have to type many options to make your graph look good.

Below we draw a graph, using a light gray background:

```
. use http://www.stata-press.com/data/r12/auto, clear
(1978 Automobile Data)
. scatter mpg weight, graphregion(fcolor(gs13))
```

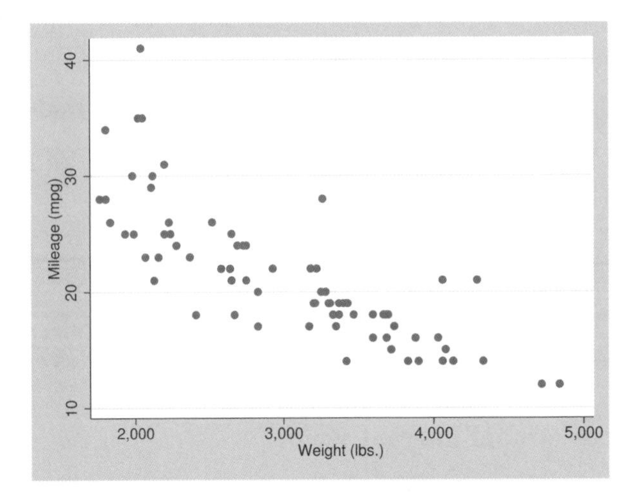

See [G-4] ***colorstyle*** for information on what you may specify inside the `graphregion(fcolor())` option.

In addition to `graphregion(fcolor())`, there are three other fill-color options:

`graphregion(ifcolor())`	fills *inner graph region*	←*of little use*
`plotregion(fcolor())`	fills *outer plot region*	←*useful*
`plotregion(ifcolor())`	fills *inner plot region*	←*could be useful*

`plotregion(fcolor())` is worth remembering. Below we make the plot region a light gray:

```
. scatter mpg weight, plotr(fcolor(gs13))
```

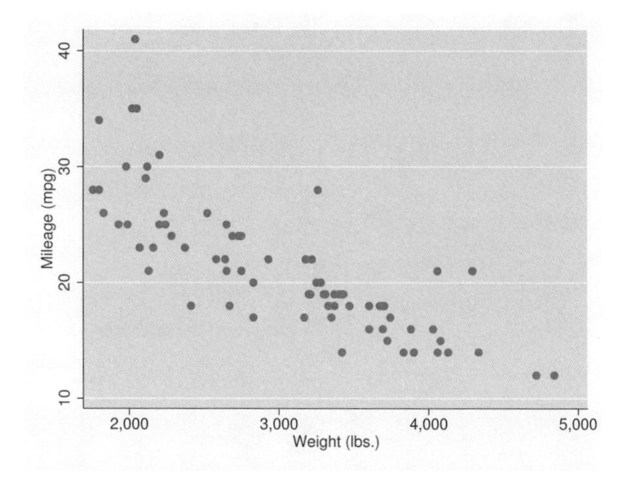

The other two options—`graphregion(ifcolor())` and `plotregion(ifcolor())`—fill the *inner graph region* and *inner plot region*. Filling the *inner graph region* serves little purpose. Filling the *inner plot region*—which is the same as the *outer plot region* except that it omits the margin between the *inner plot region* and the axes—generally makes graphs appear too busy.

How graphs are constructed

`graph` works from the outside in, with the result that the dimensions of the *plot region* are what are left over.

`graph` begins with the *available area*, the size of which is determined by the `xsize()` and `ysize()` options. `graph` indents on all four sides by `graphregion(margin())`, so it defines the outer border of the *graph region*, the interior of which is the *inner graph region*.

Overall titles (if any) are now placed on the graph, and on each of the four sides, those titles are allocated whatever space they require. Next are placed any axis titles and labels, and they too are allocated whatever space necessary. That then determines the outer border of the *plot region* (or, more properly, the border of the *outer plot region*).

The axis (if any) is placed right on top of that border. `graph` now indents on all four sides by `plotregion(margin())`, and that determines the inner border of the plot region, meaning the border of the (*inner*) *plot region*. It is inside this that the data are plotted.

An implication of the above is that, if `plotregion(margin(zero))`, the axes are not offset from the region in which the data are plotted.

Now consider the lines used to outline the regions and the fill colors used to shade their interiors.

Starting once again with the *available area*, `graph` outlines its borders by using `graphregion(lstyle())`—which is usually `graphregion(lstyle(none))`—and fills the area with the `graphregion(fcolor())`.

`graph` now moves to the inner border of the *graph region*, outlines it using `graphregion(ilstyle())`, and fills the *graph region* with `graphregion(ifcolor())`.

`graph` moves to the outer border of the *plot region*, outlines it using `plotregion(lstyle())`, and fills the *outer plot region* with `plotregion(fcolor())`.

Finally, `graph` moves to the inner border of the *plot region*, outlines it using `plotregion(ilstyle())`, and fills the (*inner*) *plot region* with `plotregion(ifcolor())`.

Also see

[G-4] ***areastyle*** — Choices for look of regions

[G-4] ***colorstyle*** — Choices for color

[G-4] ***linestyle*** — Choices for overall look of lines

[G-4] ***linewidthstyle*** — Choices for thickness of lines

[G-4] ***linepatternstyle*** — Choices for whether lines are solid, dashed, etc.

[G-4] ***marginstyle*** — Choices for size of margins

Title

[G-3] *rspike_options* — Options for determining the look of range spikes

Syntax

rspike_options	Description
`lpattern(`*linepatternstyle*`)`	whether spike line is solid, dashed, etc.
`lwidth(`*linewidthstyle*`)`	thickness of spike line
`lcolor(`*colorstyle*`)`	color of spike line
`lstyle(`*linestyle*`)`	overall style of spike line
`pstyle(`*pstyle*`)`	overall plot style, including line style
`recast(`*newplottype*`)`	advanced; treat plot as *newplottype*

See [G-4] ***linepatternstyle***, [G-4] ***linewidthstyle***, [G-4] ***colorstyle***, [G-4] ***linestyle***, [G-4] ***pstyle***, and [G-3] ***advanced_options***.

All options are *rightmost*; see [G-4] **concept: repeated options**.

Description

The *rspike_options* determine the look of spikes (lines connecting two points vertically or horizontally) in most contexts.

Options

`lpattern(`*linepatternstyle*`)` specifies whether the line for the spike is solid, dashed, etc. See [G-4] ***linepatternstyle*** for a list of available patterns.

`lwidth(`*linewidthstyle*`)` specifies the thickness of the line for the spike. See [G-4] ***linewidthstyle*** for a list of available thicknesses.

`lcolor(`*colorstyle*`)` specifies the color of the line for the spike. See [G-4] ***colorstyle*** for a list of available colors.

`lstyle(`*linestyle*`)` specifies the overall style of the line for the spike: its pattern, thickness, and color.

You need not specify `lstyle()` just because there is something you want to change about the look of the spike. The other *rspike_options* will allow you to make changes. You specify `lstyle()` when another style exists that is exactly what you want or when another style would allow you to specify fewer changes.

See [G-4] ***linestyle*** for a list of available line styles.

`pstyle(`*pstyle*`)` specifies the overall style of the plot, including not only the *linestyle*, but also all other settings for the look of the plot. Only the *linestyle* affects the look of spikes. See [G-4] ***pstyle*** for a list of available plot styles.

`recast(`*newplottype*`)` is an advanced option allowing the plot to be recast from one type to another, for example, from a range spike plot to a range area plot; see [G-3] ***advanced_options***. Most, but not all, plots allow `recast()`.

Remarks

Range spikes are used in many contexts. They are sometimes the default for confidence intervals. For instance, the `lcolor()` suboption of `ciopts()` in

```
. ltable age, graph ciopts(lcolor(red))
```

causes the color of the horizontal lines representing the confidence intervals in the life-table graph to be drawn in red.

Also see

[G-4] **concept: lines** — Using lines

[G-3] ***advanced_options*** — Rarely specified options for use with graph twoway

[G-4] ***colorstyle*** — Choices for color

[G-4] ***linepatternstyle*** — Choices for whether lines are solid, dashed, etc.

[G-4] ***linewidthstyle*** — Choices for thickness of lines

[G-4] ***pstyle*** — Choices for overall look of plot

Title

[G-3] ***saving_option*** — Option for saving graph to disk

Syntax

saving_option	Description
saving(*filename* [, *suboptions*])	save graph to disk

saving() is *unique*; see [G-4] **concept: repeated options**.

suboptions	Description
asis	freeze graph and save as is
replace	okay to replace existing *filename*

Description

Option saving() saves the graph to disk.

Option

saving(*filename* [, *suboptions*]) specifies the name of the diskfile to be created or replaced. If *filename* is specified without an extension, .gph will be assumed.

Suboptions

asis specifies that the graph be frozen and saved just as it is. The alternative—and the default if asis is not specified—is known as *live format*. In live format, the graph can continue to be edited in future sessions, and the overall look of the graph continues to be controlled by the chosen scheme (see [G-4] **schemes intro**).

Say that you type

```
. scatter yvar xvar, ... saving(mygraph)
```

That will create file mygraph.gph. Now pretend you send that file to a colleague. The way the graph appears on your colleague's computer might be different from how it appears on yours. Perhaps you display titles on the top and your colleague has set his scheme to display titles on the bottom. Or perhaps your colleague prefers the y axis on the right rather than the left. It will still be the same graph, but it might have a different look.

Or perhaps you just file away mygraph.gph for use later. If you store it in the default live format, you can come back to it later and change the way it looks by specifying a different scheme or can edit it.

If, on the other hand, you specify `asis`, the graph will look forever just as it looked the instant it was saved. You cannot edit it; you cannot change the scheme. If you send the as-is graph to colleagues, they will see it in exactly the form you see it.

Whether a graph is saved as-is or live makes no difference for printing. As-is graphs usually require fewer bytes to store, and they generally display more quickly, but that is all.

`replace` specifies that the file may be replaced if it already exists.

Remarks

To save a graph permanently, you add `saving()` to the end of the `graph` command (or any place among the options):

```
. graph ..., ... saving(myfile) ...
(file myfile.gph saved)
```

You can also achieve the same result in two steps:

```
. graph ..., ...
. graph save myfile
(file myfile.gph saved)
```

The advantage of the two-part construction is that you can edit the graph between the time you first draw it and save it. The advantage of the one-part construction is that you will not forget to save it.

Also see

[G-2] **graph save** — Save graph to disk

[G-2] **graph export** — Export current graph

[G-4] **concept: gph files** — Using gph files

[G-2] **graph manipulation** — Graph manipulation commands

Title

[G-3] *scale_option* — Option for resizing text, markers, and line widths

Syntax

scale_option	Description
`scale(#)`	specify scale; default is `scale(1)`

`scale()` is *unique*; see [G-4] **concept: repeated options**.

Description

Option `scale()` makes all the text, markers, and line widths on a graph larger or smaller.

Option

`scale(#)` specifies a multiplier that affects the size of all text, markers, and line widths on a graph. `scale(1)` is the default.

To increase the size of all text, markers, and line widths on a graph by 20%, specify `scale(1.2)`. To reduce the size of all text, markers, and line widths on a graph by 20%, specify `scale(.8)`.

Remarks

Under *Advanced use* in [G-3] ***marker_label_options***, we showed the following graph,

```
. twoway (scatter lexp gnppc, mlabel(country) mlabv(pos))
         (line hat gnppc, sort)
         , xsca(log) xlabel(.5 5 10 15 20 25 30, grid)
           legend(off)
           title("Life expectancy vs. GNP per capita")
           subtitle("North, Central, and South America")
           note("Data source:  World Bank, 1998")
           ytitle("Life expectancy at birth (years)")
```

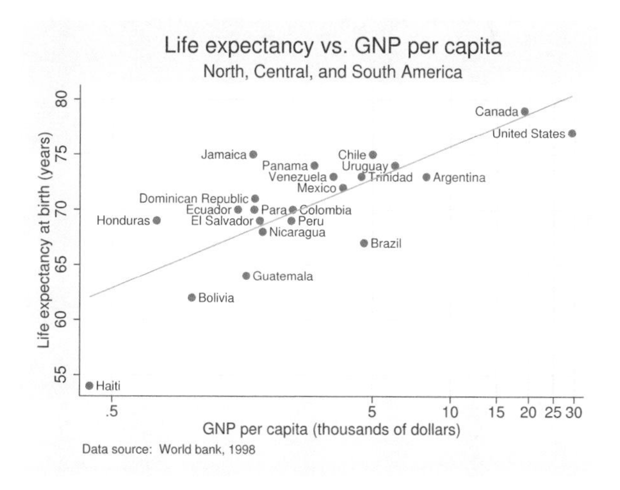

Here is the same graph with the size of all text, markers, and line widths increased by 10%:

```
. twoway (scatter lexp gnppc, mlabel(country) mlabv(pos))
         (line hat gnppc, sort)
         , xsca(log) xlabel(.5 5 10 15 20 25 30, grid)
           legend(off)
           title("Life expectancy vs. GNP per capita")
           subtitle("North, Central, and South America")
           note("Data source:  World Bank, 1998")
           ytitle("Life expectancy at birth (years)")
           scale(1.1)                                              (new)
```

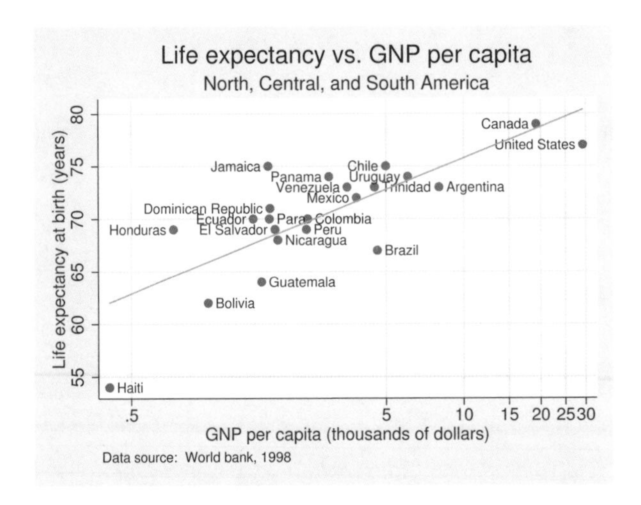

All we did was add the option `scale(1.1)` to the original command.

Also see

[G-2] **graph** — The graph command

Title

[G-3] ***scheme_option*** — Option for specifying scheme

Syntax

scheme_option	Description
`scheme(`*schemename*`)`	specify scheme to be used

`scheme()` is *unique*; see [G-4] **concept: repeated options**.

For a list of available *schemenames*, see [G-4] **schemes intro**.

Description

Option `scheme()` specifies the graphics scheme to be used. The scheme specifies the overall look of the graph.

Option

`scheme(`*schemename*`)` specifies the scheme to be used. If `scheme()` is not specified the default scheme is used; see [G-4] **schemes intro**.

Remarks

See [G-4] **schemes intro**.

Also see

[G-4] **schemes intro** — Introduction to schemes

[G-2] **set scheme** — Set default scheme

Title

[G-3] ***std_options*** — Options for use with graph construction commands

Syntax

std_options	Description
title_options	titles, subtitles, notes, captions
`scale(`*#*`)`	resize text and markers
region_options	outlining, shading, graph size
`scheme(`*schemename*`)`	overall look
`play(`*recordingname*`)`	play edits from *recordingname*
`nodraw`	suppress display of graph
`name(`*name*`, ...)`	specify name for graph
`saving(`*filename*`, ...)`	save graph in file

See [G-3] ***title_options***, [G-3] ***scale_option***, [G-3] ***region_options***, [G-3] ***nodraw_option***, [G-3] ***name_option***, and [G-3] ***saving_option***.

Description

The above options are allowed with

Command	Manual entry
`graph bar` and `graph hbar`	[G-2] **graph bar**
`graph dot`	[G-2] **graph dot**
`graph box`	[G-2] **graph box**
`graph pie`	[G-2] **graph pie**

See [G-3] ***twoway_options*** for the standard options allowed with `graph twoway`.

Options

title_options allow you to specify titles, subtitles, notes, and captions to be placed on the graph; see [G-3] ***title_options***.

`scale(`*#*`)` specifies a multiplier that affects the size of all text and markers in a graph. `scale(1)` is the default, and `scale(1.2)` would make all text and markers 20% larger.
See [G-3] ***scale_option***.

region_options allow outlining the plot region (such as placing or suppressing a border around the graph), specifying a background shading for the region, and the controlling of the graph size. See [G-3] ***region_options***.

`scheme(`*schemename*`)` specifies the overall look of the graph; see [G-3] ***scheme_option***.

`play(`*recordingname*`)` applies the edits from *recordingname* to the graph, where *recordingname* is the name under which edits previously made in the Graph Editor have been recorded and stored. See *Graph Recorder* in [G-1] **graph editor**.

`nodraw` causes the graph to be constructed but not displayed; see [G-3] ***nodraw_option***.

`name(`*name*[`, replace`]`)` specifies the name of the graph. `name(Graph, replace)` is the default. See [G-3] ***name_option***.

`saving(`*filename*[`, asis replace`]`)` specifies that the graph be saved as *filename*. If *filename* is specified without an extension, `.gph` is assumed. `asis` specifies that the graph be saved just as it is. `replace` specifies that, if the file already exists, it is okay to replace it. See [G-3] ***saving_option***.

Remarks

The above options may be used with any of the `graph` commands listed above.

Also see

[G-3] ***name_option*** — Option for naming graph in memory

[G-3] ***nodraw_option*** — Option for suppressing display of graph

[G-3] ***region_options*** — Options for shading and outlining regions and controlling graph size

[G-3] ***saving_option*** — Option for saving graph to disk

[G-3] ***scale_option*** — Option for resizing text, markers, and line widths

[G-3] ***title_options*** — Options for specifying titles

Title

[G-3] ***textbox_options*** — Options for textboxes and concept definition

Syntax

Textboxes contain one or more lines of text. The appearance of textboxes is controlled by the following options:

textbox_options	Description
`tstyle(`*textboxstyle*`)`	overall style
`orientation(`*orientationstyle*`)`	whether vertical or horizontal
`size(`*textsizestyle*`)`	size of text
`color(`*colorstyle*`)`	color of text
`justification(`*justificationstyle*`)`	text left, centered, right-justified
`alignment(`*alignmentstyle*`)`	text top, middle, bottom baseline
`margin(`*marginstyle*`)`	margin from text to border
`linegap(`*relativesize*`)`	space between lines
`width(`*relativesize*`)`	width of textbox override
`height(`*relativesize*`)`	height of textbox override
`box` or `nobox`	whether border is drawn around box
`bcolor(`*colorstyle*`)`	color of background and border
`fcolor(`*colorstyle*`)`	color of background
`lstyle(`*linestyle*`)`	overall style of border
`lpattern(`*linepatternstyle*`)`	line pattern of border
`lwidth(`*linewidthstyle*`)`	thickness of border
`lcolor(`*colorstyle*`)`	color of border
`bmargin(`*marginstyle*`)`	margin from border outwards
`bexpand`	expand box in direction of text
`placement(`*compassdirstyle*`)`	location of textbox override

See [G-4] ***textboxstyle***, [G-4] ***orientationstyle***, [G-4] ***textsizestyle***, [G-4] ***colorstyle***, [G-4] ***justificationstyle***, [G-4] ***alignmentstyle***, [G-4] ***marginstyle***, [G-4] ***relativesize***, [G-4] ***linestyle***, [G-4] ***linepatternstyle***, [G-4] ***linewidthstyle***, and [G-4] ***compassdirstyle***.

The above options invariably occur inside other options. For instance, the syntax of `title()` (see [G-3] ***title_options***) is

`title("`*string*`"` [`"`*string*`"` [...]] [, *title_options textbox_options*]`)`

so any of the options above can appear inside the `title()` option:

```
. graph ..., ... title("My title", color(green) box) ...
```

Description

A textbox contains one or more lines of text. The textbox options listed above specify how the text and textbox should appear.

Options

`tstyle(`*textboxstyle*`)` specifies the overall style of the textbox. Think of a textbox as a set of characteristics that include, in addition to the text, the size of font, the color, whether lines are drawn around the box, etc. The *textboxstyle* you choose specifies all of those things, and it is from there that the changes you make by specifying the other operations take effect.

The default is determined by the overall context of the text (such as whether it is due to `title()`, `subtitle()`, etc.), and that in turn is specified by the scheme (see [G-4] **schemes intro**). That is, identifying the name of the default style in a context is virtually impossible.

Option `tstyle()` is rarely specified. Usually, you simply let the overall style be whatever it is and specify the other textbox options to modify it. Do not, however, dismiss the idea of looking for a better overall style that more closely matches your desires.

See [G-4] ***textboxstyle***.

`orientation(`*orientationstyle*`)` specifies whether the text and box are to be oriented horizontally or vertically (text reading from bottom to top or text reading from top to bottom). See [G-4] ***orientationstyle***.

`size(`*textsizestyle*`)` specifies the size of the text that appears inside the textbox. See [G-4] ***textsizestyle***.

`color(`*colorstyle*`)` specifies the color of the text that appears inside the textbox. See [G-4] ***colorstyle***.

`justification(`*justificationstyle*`)` specifies how the text is to be "horizontally" aligned in the box. Choices include `left`, `right`, and `center`. Think of the textbox as being horizontal, even if it is vertical when specifying this option. See [G-4] ***justificationstyle***.

`alignment(`*alignmentstyle*`)` specifies how the text is to be "vertically" aligned in the box. Choices include `baseline`, `middle`, and `top`. Think of the textbox as being horizontal, even if it is vertical when specifying this option. See [G-4] ***alignmentstyle***.

`margin(`*marginstyle*`)` specifies the margin around the text (the distance from the text to the borders of the box). The text that appears in a box, plus `margin()`, determine the overall size of the box. See [G-4] ***marginstyle***.

When dealing with rotated textboxes—textboxes for which `orientation(vertical)` or `orientation(rvertical)` has been specified—the margins for the left, right, bottom, and top refer to the margins before rotation.

`linegap(`*relativesize*`)` specifies the distance between lines. See [G-4] ***relativesize*** for argument choices.

`width(`*relativesize*`)` and `height(`*relativesize*`)` override Stata's usual determination of the width and height of the textbox on the basis of its contents. See *Width and height* under *Remarks* below. See [G-4] ***relativesize*** for argument choices.

`box` and `nobox` specify whether a box is to be drawn outlining the border of the textbox. The default is determined by the `tstyle()`, which in turn is determined by context, etc. In general, the default is not to outline boxes, so the option to outline boxes is `box`. If an outline appears by default, `nobox` is the option to suppress the outlining of the border. No matter what the default, you can specify `box` or `nobox`.

`bcolor(`*colorstyle*`)` specifies the color of both the background of the box and the color of the outlined border. This option is typically not specified because it results in the border disappearing into the background of the textbox; see options `fcolor()` and `lcolor()` below for alternatives. The color matters only if `box` is also specified; otherwise, `bcolor()` is ignored. See [G-4] ***colorstyle*** for a list of color choices.

`fcolor(`*colorstyle*`)` specifies the color of the background of the box. The background of the box is filled with the `fcolor()` only if `box` is also specified; otherwise, `fcolor()` is ignored. See [G-4] ***colorstyle*** for a list of color choices.

`lstyle(`*linestyle*`)` specifies the overall style of the line used to outline the border. The style includes the line's pattern (solid, dashed, etc.), thickness, and color.

You need not specify `lstyle()` just because there is something you want to change about the look of the line. Options `lpattern`, `lwidth()`, and `lcolor()` will allow you to change the attributes individually. You specify `lstyle()` when there is a style that is exactly what you desire or when another style would allow you to specify fewer changes.

See [G-4] ***linestyle*** for a list of style choices and see [G-4] **concept: lines** for a discussion of lines in general.

`lpattern(`*linepatternstyle*`)` specifies the pattern of the line outlining the border. See [G-4] ***linepatternstyle***. Also see [G-4] **concept: lines** for a discussion of lines in general.

`lwidth(`*linewidthstyle*`)` specifies the thickness of the line outlining the border. See [G-4] ***linewidthstyle***. Also see [G-4] **concept: lines** for a discussion of lines in general.

`lcolor(`*colorstyle*`)` specifies the color of the border of the box. The border color matters only if `box` is also specified; otherwise, the `lcolor()` is ignored. See [G-4] ***colorstyle*** for a list of color choices.

`bmargin(`*marginstyle*`)` specifies the margin between the border and the containing box. See [G-4] ***marginstyle***.

`bexpand` specifies that the textbox be expanded in the direction of the text, made wider if the text is horizontal, and made longer if the text is vertical. It is expanded to the borders of its containing box. See [G-3] ***title_options*** for a demonstration of this option.

`placement(`*compassdirstyle*`)` overrides default placement; see *Appendix: Overriding default or context-specified positioning* below. See [G-4] ***compassdirstyle*** for argument choices.

Remarks

Remarks are presented under the following headings:

Definition of a textbox
Position
Justification
Position and justification combined
Margins
Width and height
Appendix: Overriding default or context-specified positioning

Definition of a textbox

A textbox is one or more lines of text

single-line textbox

1st line of multiple-line textbox
2nd line of multiple-line textbox

for which the borders may or may not be visible (controlled by the `box/nobox` option). Textboxes can be horizontal or vertical

horizontal

vertical

rvertical

in an `orientation(vertical)` *textbox, letters are rotated 90 degrees counterclockwise;* `orientation(vertical)` *reads bottom to top*

in an `orientation(rvertical)` *textbox, letters are rotated 90 degrees clockwise;* `orientation(rvertical)` *reads top to bottom*

Even in vertical textboxes, options are stated in horizontal terms of left and right. Think horizontally, and imagine the rotation as being performed at the end.

Position

Textboxes are first formed and second positioned on the graph. The *textbox_options* affect the construction of the textbox, not its positioning. The options that control its positioning are provided by the context in which the textbox is used. For instance, the syntax of the `title()` option—see [G-3] ***title_options***—is

`title("`*string*`"` ... [`, position(...) ring(...) span(...)` ... *textbox_options*] `)`

`title()`'s `position()`, `ring()`, and `span()` options determine where the title (that is, textbox) is positioned. Once the textbox is formed, its contents no longer matter; it is just a box to be positioned on the graph.

Textboxes are positioned inside other boxes. For instance, the textbox might be

title

and, because of the `position()`, `ring()`, and `span()` options specified, `title()` might position that box somewhere on the top "line":

There are many ways the smaller box could be fit into the larger box, which is the usual case, and forgive us for combining two discussions: how boxes fit into each other and the controlling of placement. If you specified `title()`'s `position(11)` option, the result would be

title

If you specified `title()`'s `position(12)` option, the result would be

title

If you specified `title()`'s `position(1)` option, the result would be

title

Justification

An implication of the above is that it is not the textbox option `justification()` that determines whether the title is centered; it is `title()`'s `position()` option.

Remember, textbox options describe the construction of textboxes, not their use. `justification(left|right|center)` determines how text is placed in multiple-line textboxes:

Example of multiple-line textbox
`justification(left)`

Example of multiple-line textbox
`justification(right)`

Example of multiple-line textbox
`justification(center)`

Textboxes are no wider than the text of their longest line. `justification()` determines how lines shorter than the longest are placed inside the box. In a one-line textbox,

single-line textbox

it does not matter how the text is justified.

Position and justification combined

With positioning options provided by the context in which the textbox is being used, and the `justification()` option, you can create many different effects in the presentation of multiple-line textboxes. For instance, considering `title()`, you could produce

First line of title
Second line (1)

or

First line of title
Second line (2)

or

First line of title
Second line (3)

or

First line of title
Second line (4)

or

First line of title
Second line (5)

or

First line of title
Second line (6)

or many others. The corresponding commands would be

```
. graph ..., title("First line of title" "Second line",       (1)
                              position(12) justification(left))
. graph ..., title("First line of title" "Second line",       (2)
                              position(12) justification(center))
. graph ..., title("First line of title" "Second line",       (3)
                              position(12) justification(right))
. graph ..., title("First line of title" "Second line",       (4)
                              position(1) justification(left))
. graph ..., title("First line of title" "Second line",       (5)
                              position(1) justification(center))
. graph ..., title("First line of title" "Second line",       (6)
                              position(1) justification(right))
```

Margins

There are two margins: `margin()` and `bmargin()`. `margin()` specifies the margin between the text and the border. `bmargin()` specifies the margin between the border and the containing box.

By default, textboxes are the smallest rectangle that will just contain the text. If you specify `margin()`, you add space between the text and the borders of the bounding rectangle:

margin(zero) textbox

textbox with ample margin on all four sides

`margin(`*marginstyle*`)` allows different amounts of padding to be specified above, below, left, and right of the text; see [G-4] ***marginstyle***. `margin()` margins make the textbox look better when the border is outlined via the `box` option and/or the box is shaded via the `bcolor()` or `fcolor()` option.

`bmargin()` margins are used to move the textbox a little or a lot when the available positioning options are inadequate. Consider specifying the `caption()` option (see [G-3] ***title_options***) so that it is inside the plot region:

```
. graph ..., caption("My caption", ring(0) position(7))
```

Seeing the result, you decide that you want to shift the box up and to the right a bit:

```
. graph ..., caption("My caption", ring(0) position(7)
             bmargin("2 0 2 0"))
```

The `bmargin()` numbers (and `margin()` numbers) are the top, bottom, left, and right amounts, and the amounts are specified as relative sizes (see [G-4] ***relativesize***). We specified a 2% bottom margin and a 2% left margin, thus pushing the caption box up and to the right.

Width and height

The width and the height of a textbox are determined by its contents (the text width and number of lines) plus the margins just discussed. The width calculation, however, is based on an approximation, with the result that the textbox that should look like this

Stata approximates the width of textboxes

can end up looking like this

Stata approximates the width of textboxes

or like this

Stata approximates the width of textboxes

You will not notice this problem unless the borders are being drawn (option `box`) because, without borders, in all three cases you would see

Stata approximates the width of textboxes

For an example of this problem and the solution, see *Use of the textbox option width()* in [G-3] ***added_text_options***. If the problem arises, use `width(`*relativesize*`)` to work around it. Getting the `width()` right is a matter of trial and error. The correct width will nearly always be between 0 and 100.

Corresponding to `width(`*relativesize*`)`, there is `height(`*relativesize*`)`. This option is less useful because Stata never gets the height incorrect.

Appendix: Overriding default or context-specified positioning

What follows is really a footnote. We said previously that where a textbox is located is determined by the context in which it is used and by the positioning options provided by that context. Sometimes you wish to override that default, or the context may not provide such control. In such cases, the option `placement()` allows you to take control.

Let us begin by correcting a misconception we introduced. We previously said that textboxes are fit inside other boxes when they are positioned. That is not exactly true. For instance, what happens when the textbox is bigger than the box into which it is being placed? Say that we have the textbox

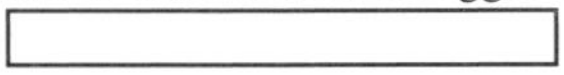

and we need to put it "in" the box

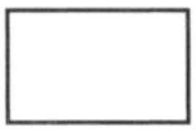

The way things work, textboxes are not put inside other boxes; they are merely positioned so that they align a certain way with the preexisting box. Those alignment rules are such that, if the preexisting box is larger than the textbox, the result will be what is commonly meant by "inside". The alignment rules are either to align one of the four corners or to align and center on one of the four edges.

In the example just given, the textbox could be positioned so that its northwest corner is coincident with the northwest corner of the preexisting box,

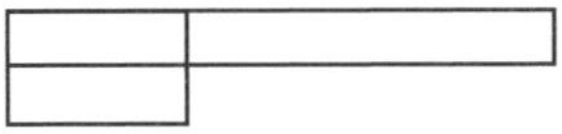

`placement(nw)`

or so that their northeast corners are coincident,

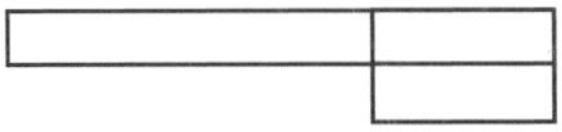

`placement(ne)`

or so that their southwest corners are coincident,

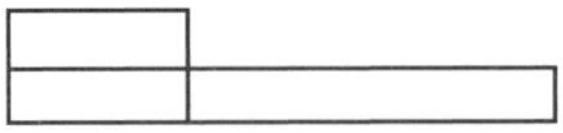

`placement(sw)`

or so that their southeast corners are coincident,

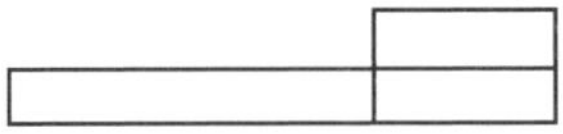

`placement(se)`

or so that the midpoint of the top edges are the same,

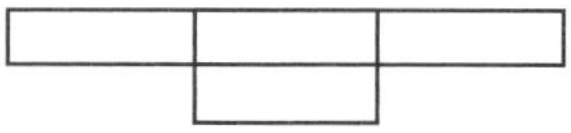

placement(n)

or so that the midpoint of the left edges are the same,

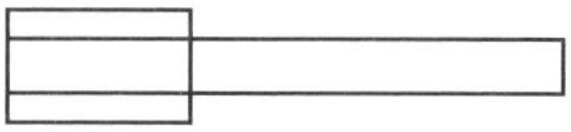

placement(w)

or so that the midpoint of the right edges are the same,

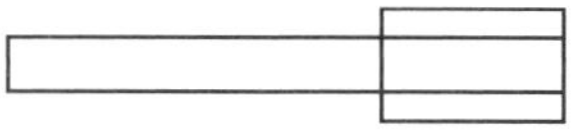

placement(e)

or so that the midpoint of the bottom edges are the same,

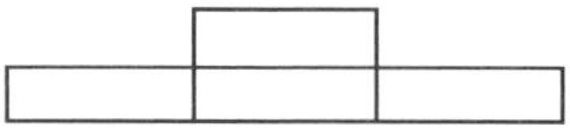

placement(s)

or so that the center point of the boxes are the same:

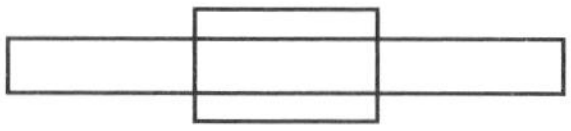

placement(c)

If you have trouble seeing any of the above, consider what you would obtain if the preexisting box were larger than the textbox. Below we show the preexisting box with eight different textboxes:

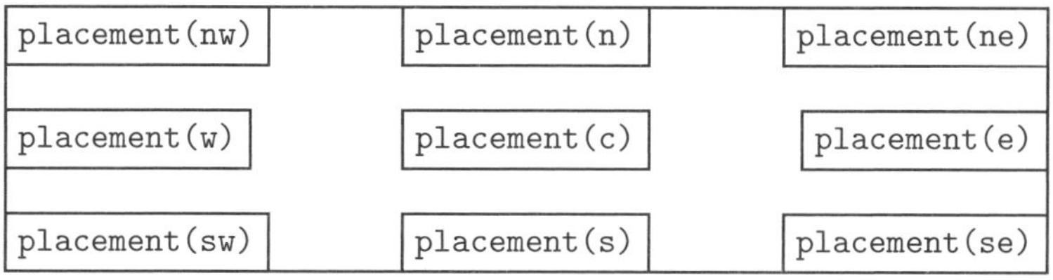

placement(nw)	placement(n)	placement(ne)
placement(w)	placement(c)	placement(e)
placement(sw)	placement(s)	placement(se)

Also see

[G-4] ***alignmentstyle*** — Choices for vertical alignment of text

[G-4] ***colorstyle*** — Choices for color

[G-4] ***compassdirstyle*** — Choices for location

[G-4] ***justificationstyle*** — Choices for how text is justified

[G-4] ***linepatternstyle*** — Choices for whether lines are solid, dashed, etc.

[G-4] ***linewidthstyle*** — Choices for thickness of lines

[G-4] ***marginstyle*** — Choices for size of margins

[G-4] ***relativesize*** — Choices for sizes of objects

[G-4] ***orientationstyle*** — Choices for orientation of textboxes

[G-4] ***text*** — Text in graphs

[G-4] ***textboxstyle*** — Choices for the overall look of text including border

[G-4] ***textsizestyle*** — Choices for the size of text

[G-3] ***title_options*** — Options for specifying titles

Title

[G-3] ***tif_options*** — Options for exporting to tagged image file format (TIFF)

Syntax

tif_options	Description
`width(#)`	width of graph in pixels
`height(#)`	height of graph in pixels

Description

The *tif_options* are used with `graph export` when creating TIFF graphs; see [G-2] **graph export**.

Options

`width(#)` specifies the width of the graph in pixels. `width()` must contain an integer between 8 and 16,000.

`height(#)` specifies the height of the graph in pixels. `height()` must contain an integer between 8 and 16,000.

Remarks

Remarks are presented under the following headings:

Using tif_options
Specifying the width or height

Using tif_options

You have drawn a graph and wish to create a TIFF file to include in a document. You wish, however, to set the width of the graph to 800 pixels and the height to 600 pixels:

```
. graph ...                                     (draw a graph)
. graph export myfile.tif, width(800) height(600)
```

Specifying the width or height

If the width is specified but not the height, Stata determines the appropriate height from the graph's aspect ratio. If the height is specified but not the width, Stata determines the appropriate width from the graph's aspect ratio. If neither the width nor the height is specified, Stata will export the graph on the basis of the current size of the Graph window.

Also see

[G-2] **graph export** — Export current graph

[G-2] **graph set** — Set graphics options

Title

[G-3] ***title_options*** — Options for specifying titles

Syntax

title_options	Description
title(*tinfo*)	overall title
subtitle(*tinfo*)	subtitle of title
note(*tinfo*)	note about graph
caption(*tinfo*)	explanation of graph
t1title(*tinfo*) t2title(*tinfo*)	rarely used
b1title(*tinfo*) b2title(*tinfo*)	rarely used
l1title(*tinfo*) l2title(*tinfo*)	vertical text
r1title(*tinfo*) r2title(*tinfo*)	vertical text

The above options are *merged-explicit*; see [G-4] **concept: repeated options**.

{ t | b | l | r }{ 1 | 2 }title() are allowed with graph twoway only.

where *tinfo* is

"*string*" ["*string*" [...]] [, *suboptions*]

suboptions	Description
prefix and suffix	add to title text
position(*clockposstyle*)	position of title—side
ring(*ringposstyle*)	position of title—distance
span	"centering" of title
textbox_options	rendition of title

See [G-3] ***textbox_options*** for a description of *textbox_options*.

Option position() is not allowed with { t | b | l | r }{ 1 | 2 }title().

Examples include

```
title("My graph")
note(`"includes both "high" and "low" priced items"')

title("First line" "Second line")
title("Third line", suffix)
title("Fourth line" "Fifth line", suffix)
```

The definition of *ringposstyle* and the default positioning of titles is

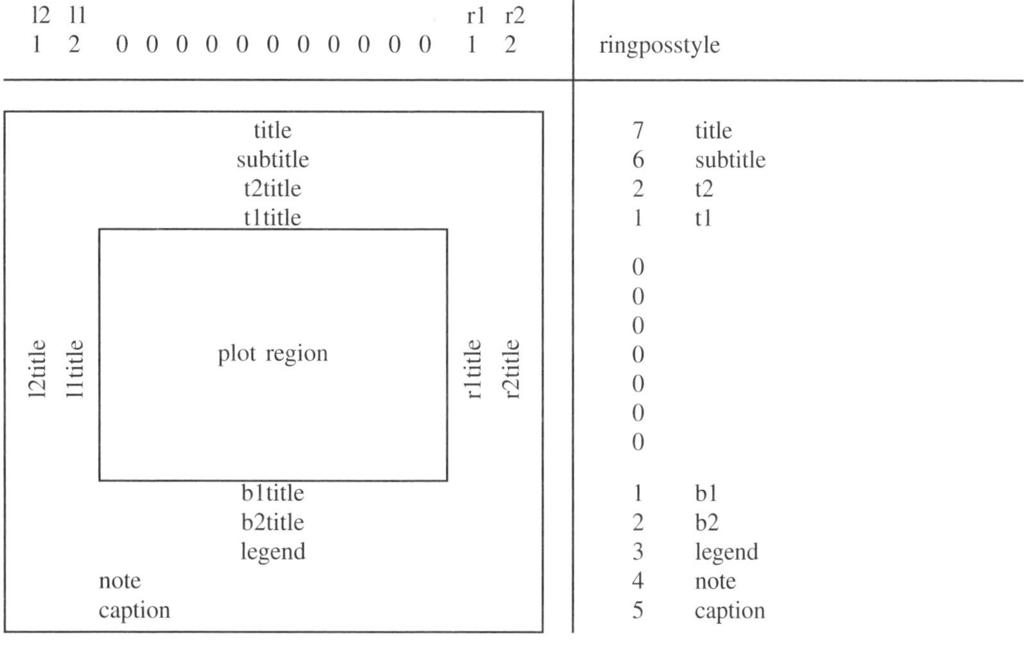

where titles are located is controlled by the scheme

Description

Titles are the adornment around a graph that explains the graph's purpose.

Options

`title(`*tinfo*`)` specifies the overall title of the graph. The title usually appears centered at the top of the graph. It is sometimes desirable to specify the `span` suboption when specifying the title, as in

```
. graph ..., ... title("Life expectancy", span)
```

See *Spanning* under *Remarks* below.

`subtitle(`*tinfo*`)` specifies the subtitle of the graph. The subtitle appears near the title (usually directly under it) and is presented in a slightly smaller font. `subtitle()` is used in conjunction with `title()`, and `subtitle()` is used by itself when the `title()` seems too big. For instance, you might type

```
. graph ..., ... title("Life expectancy") subtitle("1900-1999")
```

or

```
. graph ..., ... subtitle("Life expectancy" "1900-1999")
```

If `subtitle()` is used in conjunction with `title()` and you specify suboption `span` with `title()`, remember also to specify `span` with `subtitle()`.

`note(`*tinfo*`)` specifies notes to be displayed with the graph. Notes are usually displayed in a small font placed at the bottom-left corner of the graph. By default, the left edge of the note will align with the left edge of the plot region. Specify suboption `span` if you wish the note moved all the way left; see *Spanning* under *Remarks* below.

`caption(`*tinfo*`)` specifies an explanation to accompany the graph. Captions are usually displayed at the bottom of the graph, below the `note()`, in a font slightly larger than used for the `note()`. By default, the left edge of the caption will align with the left edge of the plot region. Specify suboption `span` if you wish the caption moved all the way left; see *Spanning* under *Remarks* below.

{`t`|`b`|`l`|`r`}{`1`|`2`}`title()` are rarely specified. It is generally better to specify the *axis_title_options* `ytitle()` or `xtitle()`; see [G-3] ***axis_title_options***. The {`t`|`b`|`l`|`r`}{`1`|`2`}`title()` options are included for backward compatibility with previous versions of Stata.

Suboptions

`prefix` and `suffix` specify that the specified text be added as separate lines either before or after any existing title of the specified type. See *Interpretation of repeated options* below.

`position(`*clockposstyle*`)` and `ring(`*ringposstyle*`)` override the default location of the title; see [G-4] ***clockposstyle*** and [G-4] ***ringposstyle***. `position()` specifies a direction *[sic]* according to the hours on the dial of a 12-hour clock, and `ring()` specifies how far from the plot region the title is to appear.

`ring(0)` is defined as inside the plot region and is for the special case when you are placing a title directly on top of the plot. `ring(`*k*`)`, $k>0$, specifies positions outside the plot region; the larger the `ring()` value, the farther away from the plot region. `ring()` values may be integer or noninteger and are treated ordinally.

`position(12)` puts the title directly above the plot region (assuming `ring()`>0), `position(3)` puts the title directly to the right of the plot region, and so on.

`span` specifies that the title be placed in an area spanning the entire width (or height) of the graph rather than an area spanning the plot region. See *Spanning* under *Remarks* below.

textbox_options are any of the options allowed with a textbox. Important options include

`justification(left` | `center` | `right)`: determines how the text is to be centered;

`orientation(horizontal` | `vertical)`: determines whether the text in the box reads from left to right or from bottom to top (there are other alternatives as well);

`color()`: determines the color of the text;

`box`: determines whether a box is drawn around the text;

`width(`*relativesize*`)`: overrides the calculated width of the text box and is used in cases when text flows outside the box or when there is too much space between the text and the right border of the box; see *Width and height* under [G-3] ***textbox_options***.

See [G-3] ***textbox_options*** for a description of each of the above options.

Remarks

Titles is the generic term we will use for titles, subtitles, keys, etc., and title options is the generic term we will use for `title()`, `subtitle()`, `note()`, `caption()`, and `{ t | b | l | r }{ 1 | 2 }title()`. Titles and title options all work the same way. In our examples, we will most often use the `title()` option, but we could equally well use any of the title options.

Remarks are presented under the following headings:

Multiple-line titles
Interpretation of repeated options
Positioning of titles
Alignment of titles
Spanning
Using the textbox options box and bexpand

Multiple-line titles

Titles can have multiple lines:

```
. graph ..., title("My title") ...
```

specifies a one-line title,

```
. graph ..., title("My title" "Second line") ...
```

specifies a two-line title, and

```
. graph ..., title("My title" "Second line" "Third line") ...
```

specifies a three-line title. You may have as many lines in your titles as you wish.

Interpretation of repeated options

Each of the title options can be specified more than once in the same command. For instance,

```
. graph ..., title("One") ... title("Two") ...
```

This does not produce a two-line title. Rather, when you specify multiple title options, the rightmost option is operative and the earlier options are ignored. The title in the above command will be "Two".

That is, the earlier options will be ignored unless you specify `prefix` or `suffix`. In

```
. graph ..., title("One") ... title("Two", suffix) ...
```

the title will consist of two lines, the first line being "One" and the second, "Two". In

```
. graph ..., title("One") ... title("Two", prefix) ...
```

the first line will be "Two" and the second line, "One".

Repeatedly specifying title options may seem silly, but it is easier to do than you might expect. Consider the command

```
. twoway (sc y1 x1, title("x1 vs. y1")) (sc y2 x2, title("x2 vs. y2"))
```

`title()` is an option of `twoway`, not `scatter`, and graphs have only one `title()` (although it might consist of multiple lines). Thus the above is probably not what the user intended. Had the user typed

```
. twoway (sc y1 x1) (sc y2 x2), title("x1 vs. y1") title("x2 vs. y2")
```

he would have seen his mistake. It is, however, okay to put `title()` options inside the `scatters`; `twoway` knows to pull them out. Nevertheless, only the rightmost one will be honored (because neither `prefix` nor `suffix` was specified), and thus the title of this graph will be "x2 vs. y2".

Multiple title options arise usefully when you are using a command that draws graphs that itself is written in terms of `graph`. For instance, the command `sts graph` (see [ST] **sts**) will graph the Kaplan–Meier survivor function. When you type

```
. sts graph
```

with the appropriate data in memory, a graph will appear, and that graph will have a `title()`. Yet, if you type

```
. sts graph, title("Survivor function for treatment 1")
```

your title will override `sts graph`'s default. Inside the code of `sts graph`, both `title()` options appear on the `graph` command. First appears the default, and second, appears the one that you specified. This programming detail is worth understanding because, as an implication, if you type

```
. sts graph, title("for treatment 1", suffix)
```

your title will be suffixed to the default. Most commands work this way, so if you use some command and it produces a title you do not like, specify `title()` (or `subtitle()`, ...) to override it, or specify `title(..., suffix)` (or `subtitle(..., suffix)`, ...) to add to it.

❑ Technical note

Title options before the rightmost ones are not completely ignored. Their options are merged and honored, so if a title is moved or the color changed early on, the title will continue to be moved or the color changed. You can always specify the options to change it back.

❑

Positioning of titles

Where titles appear is determined by the scheme that you choose; see [G-4] **schemes intro**. Options `position(`*clockposstyle*`)` and `ring(`*ringposstyle*`)` override that location and let you place the title where you want it.

`position()` specifies a direction *(sic)* according to the hours of a 12-hour clock and `ring()` specifies how far from the plot region the title is to appear.

Interpretation of clock `position()ring(`k`)`, $k > 0$ and `ring(0)`

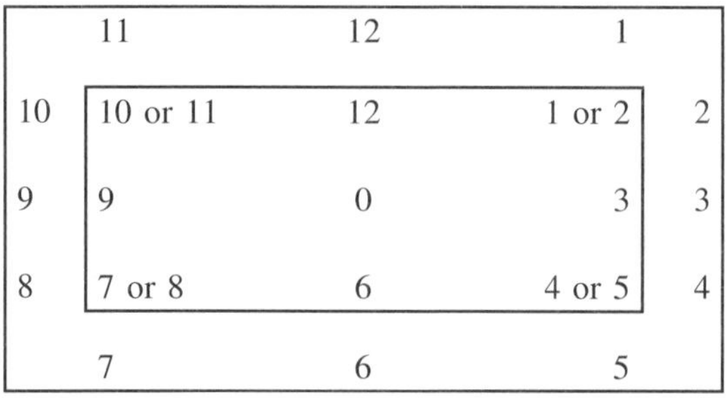

	11	12	1	
10	10 or 11	12	1 or 2	2
9	9	0	3	3
8	7 or 8	6	4 or 5	4
	7	6	5	

Interpretation of ring()

plot region	0	ring(0) = plot region
{t|b|l|r}1title()	1	
{t|b|l|r}2title()	2	ring(k), $k > 0$, is outside
legend()	3	the plot region
note()	4	
caption()	5	the larger the ring()
subtitle()	6	value, the farther
title()	7	away

position() has two interpretations, one for ring(0) and another for ring(k), $k > 0$. ring(0) is for the special case when you are placing a title directly on top of the plot. Put that case aside; titles usually appear outside the plot region.

A title directly above the plot region is at position(12), a title to the right at position(3), and so on. If you put your title at position(1), it will end up above and to the right of the plot region.

Now consider two titles—say title() and subtitle()—both located at position(12). Which is to appear first? That is determined by their respective ring() values. ring() specifies ordinally how far a title is from the plot region. The title with the larger ring() value is placed farther out. ring() values may be integer or noninteger.

For instance, legend() (see [G-3] ***legend_options***) is closer to the plot region than caption() because, by default, legend() has a ring() value of 4 and caption() a ring() value of 5. Because both appear at position(7), both appear below the plot region and because $4 < 5$, the legend() appears above the caption(). These defaults assume that you are using the default scheme.

If you wanted to put your legend below the caption, you could specify

```
. graph ..., legend(... ring(5.5)) caption("My caption")
```

or

```
. graph ..., legend(...) caption("My caption", ring(3.5))
```

The plot region itself is defined as ring(0), and if you specified that, the title would appear inside the plot region, right on top of what is plotted! You can specify where inside the plot region you want the title with position(), and the title will put itself on the corresponding edge of the plot region. In ring(0), the clock positions 1 and 2, 4 and 5, 7 and 8, and 10 and 11 are treated as being the same. Also, position(0) designates the center of the plot region.

Within the plot region—within ring(0)—given a position(), you can further shift the title up or down or left or right by specifying the title's margin() *textbox_option*. For instance, you might specify

```
. graph ..., caption(..., ring(0) pos(7)) ...
```

and then discover that the caption needed to be moved up and right a little and so change the caption() option to read

```
. graph ..., caption(..., ring(0) pos(7) margin(medium)) ...
```

See [G-3] ***textbox_options*** and [G-4] ***marginstyle*** for more information on the margin() option.

Alignment of titles

How should the text be placed in the textbox: left-justified, centered, or right-justified? The defaults that have been set vary according to title type:

title type	default justification
`title()`	centered
`subtitle()`	centered
`{t|b|l|r}{1|2}title()`	centered
`note()`	left-justified
`caption()`	left-justified

Actually, how a title is justified is, by default, determined by the scheme, and in the above, we assume that you are using a default scheme.

You can change the justification using the *textbox_option* `justification(left | center | right)`. For instance,

```
. graph ..., title("My title", justification(left)) ...
```

See [G-3] ***textbox_options***.

Spanning

Option `span` specifies that the title is to be placed in an area spanning the entire width (or height) of the graph rather than an area spanning the plot region. That is,

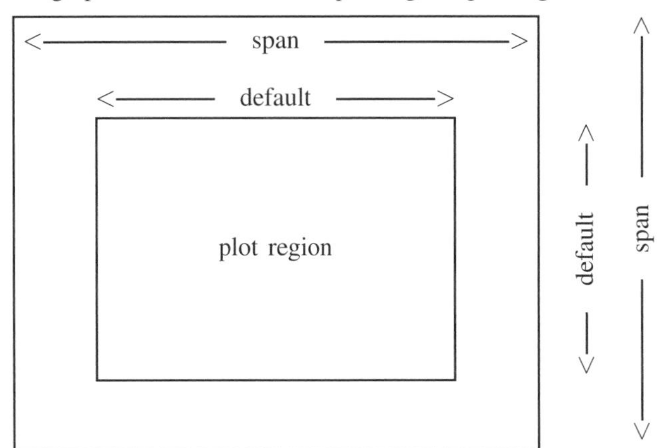

For instance, the `title()` is usually centered at the top of the graph. Is it to be centered above the plot region (the default) or between the borders of the entire available area (`title(..., span)` specified)? The `note()` is usually presented left-justified below the plot region. Is it left-justified to align with the border of the plot region (the default), or left-justified to the entire available area (`note(..., span)` specified)?

Do not confuse `span` with the *textbox* option `justification(left | center | right)` which places the text left-justified, centered, or right-justified in whatever area is being spanned; see *Alignment of titles* above.

Using the textbox options box and bexpand

The *textbox_options* box and bexpand—see [G-3] ***textbox_options***—can be put to effective use with titles. Look at three graphs:

```
. scatter mpg weight, title("Mileage and weight")
```

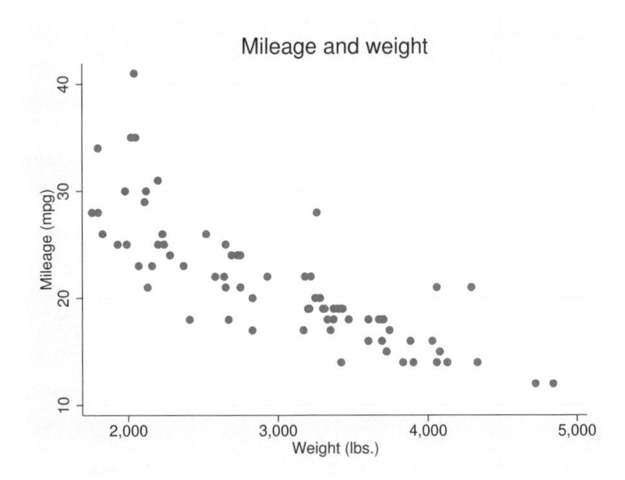


```
. scatter mpg weight, title("Mileage and weight", box)
```

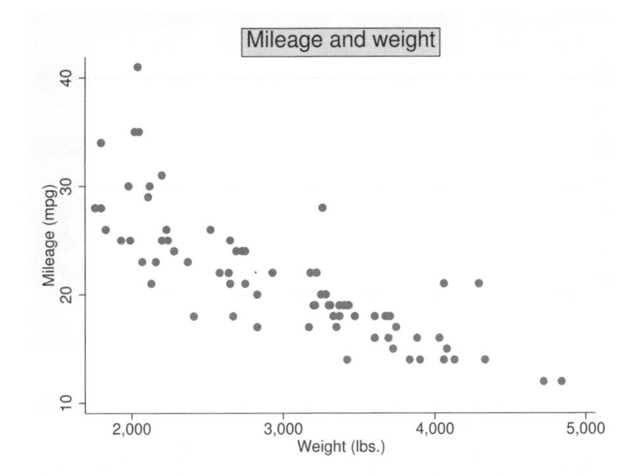

```
. scatter mpg weight, title("Mileage and weight", box bexpand)
```

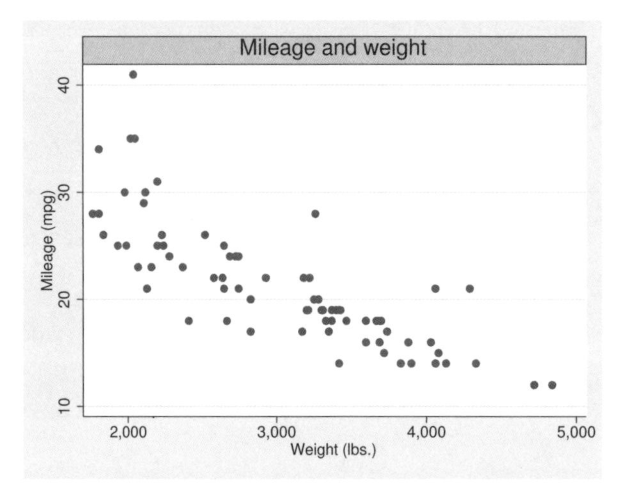

We want to direct your attention to the treatment of the title, which will be

Mileage and weight

Mileage and weight

Mileage and weight

Without options, the title appeared as is.

The textbox option `box` drew a box around the title.

The textbox options `bexpand` expanded the box to line up with the plot region and drew a box around the expanded title.

In both the second and third examples, in the graphs you will also note that the background of the textbox was shaded. That is because most schemes set the textbox option `bcolor()`, but `bcolor()` becomes effective only when the textbox is `box`ed.

Also see

[G-3] ***legend_options*** — Options for specifying legends

[G-4] ***text*** — Text in graphs

[G-3] ***textbox_options*** — Options for textboxes and concept definition

[G-4] **schemes intro** — Introduction to schemes

Title

[G-3] ***twoway_options*** — Options for twoway graphs

Syntax

The *twoway_options* allowed with all `twoway` graphs are

twoway_options	Description
added_line_options	draw lines at specified y or x values
added_text_options	display text at specified (y,x) value
axis_options	labels, ticks, grids, log scales
title_options	titles, subtitles, notes, captions
legend_options	legend explaining what means what
`scale(`*#*`)`	resize text, markers, line widths
region_options	outlining, shading, graph size
aspect_option	constrain aspect ratio of plot region
`scheme(`*schemename*`)`	overall look
`play(`*recordingname*`)`	play edits from *recordingname*
`by(`*varlist*`, ...)`	repeat for subgroups
`nodraw`	suppress display of graph
`name(`*name*`, ...)`	specify name for graph
`saving(`*filename*`, ...)`	save graph in file
advanced_options	difficult to explain

See [G-3] ***added_line_options***, [G-3] ***added_text_options***, [G-3] ***axis_options***, [G-3] ***title_options***, [G-3] ***legend_options***, [G-3] ***scale_option***, [G-3] ***region_options***, [G-3] ***aspect_option***, [G-3] ***scheme_option***, [G-3] ***by_option***, [G-3] ***nodraw_option***, [G-3] ***name_option***, [G-3] ***saving_option***, [G-3] ***advanced_options***.

Description

The above options are allowed with all *plottypes* (`scatter`, `line`, etc.) allowed by `graph twoway`; see [G-2] **graph twoway**.

Options

added_line_options specify that horizontal or vertical lines be drawn on the graph; see [G-3] ***added_line_options***. If your interest is in drawing grid lines through the plot region, see *axis_options* below.

added_text_options specifies text to be displayed on the graph (inside the plot region); see [G-3] ***added_text_options***.

axis_options specify how the axes are to look, including values to be labeled or ticked on the axes. These options also allow you to obtain logarithmic scales and grid lines. See [G-3] ***axis_options***.

title_options allow you to specify titles, subtitles, notes, and captions to be placed on the graph; see [G-3] ***title_options***.

legend_options specifies whether a legend is to appear and allows you to modify the legend's contents. See [G-3] ***legend_options***.

`scale(`*#*`)` specifies a multiplier that affects the size of all text, markers, and line widths in a graph. `scale(1)` is the default, and `scale(1.2)` would make all text, markers, and line widths 20% larger. See [G-3] ***scale_option***.

region_options allow outlining the plot region (such as placing or suppressing a border around the graph), specifying a background shading for the region, and controlling the graph size. See [G-3] ***region_options***.

aspect_option allows you to control the relationship between the height and width of a graph's plot region; see [G-3] ***aspect_option***.

`scheme(`*schemename*`)` specifies the overall look of the graph; see [G-3] ***scheme_option***.

`play(`*recordingname*`)` applies the edits from *recordingname* to the graph, where *recordingname* is the name under which edits previously made in the Graph Editor have been recorded and stored. See *Graph Recorder* in [G-1] **graph editor**.

`by(`*varlist*`, ...)` specifies that the plot be repeated for each set of values of *varlist*; see [G-3] ***by_option***.

`nodraw` causes the graph to be constructed but not displayed; see [G-3] ***nodraw_option***.

`name(`*name*[`, replace`]`)` specifies the name of the graph. `name(Graph, replace)` is the default. See [G-3] ***name_option***.

`saving(`*filename*[`, asis replace`]`)` specifies that the graph be saved as *filename*. If *filename* is specified without an extension, `.gph` is assumed. `asis` specifies that the graph be saved just as it is. `replace` specifies that, if the file already exists, it is okay to replace it. See [G-3] ***saving_option***.

advanced_options are not so much advanced as they are difficult to explain and are rarely used. They are also invaluable when you need them; see [G-3] ***advanced_options***.

Remarks

The above options may be used with any of the `twoway` plottypes—see [G-2] **graph twoway**—for instance,

```
. twoway scatter mpg weight, by(foreign)
. twoway line le year, xlabel(,grid) saving(myfile, replace)
```

The above options are options of `twoway`, meaning that they affect the entire twoway graph and not just one or the other of the plots on it. For instance, in

```
. twoway lfitci  mpg weight, stdf ||
         scatter mpg weight, ms(O) by(foreign, total row(1))
```

the `by()` option applies to the entire graph, and in theory you should type

```
. twoway lfitci  mpg weight, stdf  ||
         scatter mpg weight, ms(O) ||, by(foreign, total row(1))
```

or

```
. twoway (lfitci  mpg weight, stdf)
         (scatter mpg weight, ms(O)), by(foreign, total row(1))
```

to demonstrate your understanding of that fact. You need not do that, however, and in fact it does not matter to which plot you attach the *twoway_options*. You could even type

```
. twoway lfitci  mpg weight, stdf by(foreign, total row(1)) ||
        scatter mpg weight, ms(O)
```

and, when specifying multiple *twoway_options*, you could even attach some to one plot and the others to another:

```
. twoway lfitci  mpg weight, stdf by(foreign, total row(1)) ||
        scatter mpg weight, ms(O) saving(myfile)
```

Also see

[G-2] **graph twoway** — Twoway graphs

[G-3] ***axis_options*** — Options for specifying numeric axes

[G-3] ***title_options*** — Options for specifying titles

[G-3] ***legend_options*** — Options for specifying legends

[G-3] ***scale_option*** — Option for resizing text, markers, and line widths

[G-3] ***region_options*** — Options for shading and outlining regions and controlling graph size

[G-3] ***scheme_option*** — Option for specifying scheme

[G-3] ***by_option*** — Option for repeating graph command

[G-3] ***nodraw_option*** — Option for suppressing display of graph

[G-3] ***name_option*** — Option for naming graph in memory

[G-3] ***saving_option*** — Option for saving graph to disk

[G-3] ***advanced_options*** — Rarely specified options for use with graph twoway

[G-4] Styles, concepts, and schemes

Title

[G-4] *addedlinestyle* — Choices for overall look of added lines

Syntax

addedlinestyle	Description
`default`	determined by scheme
`extended`	extends through plot region margins
`unextended`	does not extend through margins

Other *addedlinestyles* may be available; type

```
. graph query addedlinestyle
```

to obtain the full list installed on your computer.

Description

Added lines are those added by the *added_line_options*. *addedlinestyle* specifies the overall look of those lines. See [G-3] ***added_line_options***.

Remarks

Remarks are presented under the following headings:

What is an added line?
What is an addedlinestyle?
You do not need to specify an addedlinestyle

What is an added line?

Added lines are lines added by the *added_line_options* that extend across the plot region and perhaps across the plot region's margins, too.

What is an addedlinestyle?

Added lines are defined by

1. whether the lines extend into the plot region's margin;
2. the style of the lines, which includes the lines' thickness, color, and whether solid, dashed, etc.; see [G-4] ***linestyle***.

The *addedlinestyle* specifies both these attributes.

You do not need to specify an addedlinestyle

The *addedlinestyle* is specified in the options

`yline(... , style(`*addedlinestyle*`) ...) xline(... , style(`*addedlinestyle*`) ...)`

Correspondingly, other `yline()` and `xline()` suboptions allow you to specify the individual attributes; see [G-3] ***added_line_options***.

You specify the *addedlinestyle* when a style exists that is exactly what you desire or when another style would allow you to specify fewer changes to obtain what you want.

Also see

[G-3] ***added_line_options*** — Options for adding lines to twoway graphs

Title

[G-4] *alignmentstyle* — Choices for vertical alignment of text

Syntax

alignmentstyle	Description
`baseline`	bottom of textbox = baseline of letters
`bottom`	bottom of textbox = bottom of letters
`middle`	middle of textbox = middle of letters
`top`	top of textbox = top of letters

Other *alignmentstyles* may be available; type

```
. graph query alignmentstyle
```

to obtain the full list installed on your computer.

Description

See [G-3] ***textbox_options*** for a description of textboxes. *alignmentstyle* specifies how the text is vertically aligned in a textbox. Think of the textbox as being horizontal, even if it is vertical when specifying this option.

alignmentstyle is specified inside options such as the `alignment()` suboption of `title()` (see [G-3] ***title_options***):

```
. graph ..., title("My title", alignment(alignmentstyle)) ...
```

Sometimes an *alignmentstylelist* is allowed. An *alignmentstylelist* is a sequence of *alignmentstyles* separated by spaces. Shorthands are allowed to make specifying the list easier; see [G-4] ***stylelists***.

Remarks

Think of the text as being horizontal, even if it is not, and think of the textbox as containing one line, such as

```
Hpqgxyz
```

`alignment()` specifies how the bottom of the textbox aligns with the bottom of the text.

`alignment(baseline)` specifies that the bottom of the textbox be the baseline of the letters in the box. That would result in something like

```
....Hpqgxyz....
```

where dots represent the bottom of the textbox. Periods in most fonts are located on the baseline of letters. Note how the letters `p`, `q`, `g`, and `y` extend below the baseline.

`alignment(bottom)` specifies that the bottom of the textbox be the bottom of the letters, which would be below the dots in the above example, lining up with the lowest part of the `p`, `q`, `g`, and `y`.

`alignment(middle)` specifies that the middle of the textbox line up with the middle of a capital `H`. This is useful when you want to align text with a line.

`alignment(top)` specifies that the top of the textbox line up with the top of a capital `H`.

Also see

[G-3] ***textbox_options*** — Options for textboxes and concept definition

[G-4] ***justificationstyle*** — Choices for how text is justified

Title

[G-4] *anglestyle* — Choices for the angle at which text is displayed

Syntax

anglestyle	Description
`horizontal`	horizontal; reads left to right
`vertical`	vertical; reads bottom to top
`rvertical`	vertical; reads top to bottom
`rhorizontal`	horizontal; upside down
`0`	0 degrees; same as `horizontal`
`45`	45 degrees
`90`	90 degrees; same as `vertical`
`180`	180 degrees; same as `rhorizontal`
`270` or `-90`	270 degrees; same as `rvertical`
#	*#* degrees; whatever you desire; *#* may be positive or negative

Note: under Unix, only angles `0`, `90`, `180`, and `270` display correctly on the screen. Angles are correctly rendered when printed.

Other *anglestyles* may be available; type

```
. graph query anglestyle
```

to obtain the full list installed on your computer. If other *anglestyles* do exist, they are merely words attached to numeric values.

Description

anglestyle specifies the angle at which text is to be displayed.

Remarks

anglestyle is specified inside options such as the marker-label option `mlabangle()` (see [G-3] ***marker_label_options***),

```
. graph ..., ... mlabel(...) mlabangle(anglestylelist) ...
```

or the axis-label suboption `angle()` (see [G-3] ***axis_label_options***):

```
. graph ..., ... ylabel(..., angle(anglestyle) ...) ...
```

For `mlabangle()`, an *anglestylelist* is allowed. An *anglestylelist* is a sequence of *anglestyles* separated by spaces. Shorthands are allowed to make specifying the list easier; see [G-4] ***stylelists***.

Also see

[G-3] ***marker_label_options*** — Options for specifying marker labels

Title

[G-4] ***areastyle*** — Choices for look of regions

Syntax

areastyle	Description
`background`	determined by scheme
`foreground`	determined by scheme
`outline`	`foreground` outline with no fill
`plotregion`	default for plot regions
`histogram`	default used for bars of histograms
`ci`	default used for confidence interval
`ci2`	default used for second confidence interval
`none`	no outline and no background color
`p1bar–p15bar`	used by first–fifteenth "bar" plot
`p1box–p15box`	used by first–fifteenth "box" plot
`p1pie–p15pie`	used by first–fifteenth "pie" plot
`p1area–p15area`	used by first–fifteenth "area" plot
`p1–p15`	used by first–fifteenth "other" plot

Other *areastyles* may be available; type

```
. graph query areastyle
```

to obtain the complete list of *areastyles* installed on your computer.

Description

The shape of the area is determined by context. The *areastyle* determines whether the area is to be outlined and filled and, if so, how and in what color.

Remarks

Remarks are presented under the following headings:

Overview of areastyles
Numbered styles
Using numbered styles
When to use areastyles

Overview of areastyles

areastyle is used to determine the look of

1. the entire region in which the graph appears
   (see option `style(`*areastyle*`)` in [G-3] ***region_options***)
2. bars
   (see option `bstyle(`*areastyle*`)` in [G-3] ***barlook_options***)
3. an area filled under a curve
   (see option `bstyle(`*areastyle*`)` in [G-3] ***barlook_options***)
4. most other enclosed areas, such as the boxes in box plots
   (see [G-2] **graph box**)

For an example of the use of the *areastyle* `none`, see *Suppressing the border around the plot region* in [G-3] ***region_options***.

Numbered styles

`p1bar–p15bar` are the default styles used for bar charts, including `twoway bar` charts and bar charts. `p1bar` is used for filling and outlining the first set of bars, `p2bar` for the second, and so on.

`p1box–p15box` are the default styles used for box charts. `p1box` is used for filling and outlining the first set of boxes, `p2box` for the second, and so on.

`p1pie–p15pie` are the default styles used for pie charts. `p1pie` is used for filling the first pie slice, `p2pie` for the second, and so on.

`p1area–p15area` are the default styles used for area charts, including `twoway area` charts and `twoway rarea` charts. `p1area` is used for filling and outlining the first filled area, `p2area` for the second, and so on.

`p1–p15` are the default area styles used for other plot types, including `twoway dropline` charts, `twoway spike` charts, `twoway rspike` charts, `twoway rcap` charts, `twoway rcapsym` charts, and `twoway rline` charts. `p1` is used for filling and outlining the first plot, `p2` for the second, and so on. For all the plots listed above, only lines are drawn, so the shade settings have no effect.

Using numbered styles

The look defined by a numbered style, such as `p1bar` or `p2area`, is determined by the scheme selected. By "look" we mean such things as color, width of lines, and patterns used.

Numbered styles provide default "looks" that can be controlled by a scheme. They can also be useful when you wish to make, say, the third element on a graph look like the first. You can, for example, specify that the third bar on a bar graph be drawn with the style of the first bar by specifying the option `barstyle(3, bstyle(p1bar))`.

When to use areastyles

You can often achieve an identical result by specifying an *areastyle* or using more specific options, such as `fcolor()` or `lwidth()`, that change the components of an *areastyle*—the fill color and outline attributes. You can even specify an *areastyle* as the base and then modify the attributes by using more specific options. It is often easiest to specify options that affect only the fill color or one outline characteristic rather than to specify an *areastyle*. If, however, you are trying to make many elements on a graph look the same, specifying the overall *areastyle* may be preferred.

Also see

[G-3] ***region_options*** — Options for shading and outlining regions and controlling graph size

[G-2] **graph bar** — Bar charts

[G-2] **graph pie** — Pie charts

[G-2] **graph twoway area** — Twoway line plot with area shading

[G-2] **graph twoway bar** — Twoway bar plots

[G-2] **graph twoway rarea** — Range plot with area shading

Title

[G-4] *axisstyle* — Choices for overall look of axes

Syntax

textstyle	Description
`horizontal_default`	default standard horizontal axis
`horizontal_notick`	default horizontal axis without ticks
`horizontal_nogrid`	default horizontal axis without gridlines
`horizontal_withgrid`	default horizontal axis with gridlines
`horizontal_noline`	default horizontal axis without an axis line
`horizontal_nolinetick`	default horizontal axis with neither an axis line nor ticks
`vertical_default`	default standard vertical axis
`vertical_notick`	default vertical axis without ticks
`vertical_nogrid`	default vertical axis without gridlines
`vertical_withgrid`	default vertical axis with gridlines
`vertical_noline`	default vertical axis without an axis line
`vertical_nolinetick`	default vertical axis with neither an axis line nor ticks

Other *axisstyles* may be available; type

```
. graph query axisstyle
```

to obtain the complete list of *axisstyles* installed on your computer.

Description

Axis styles are used only in scheme files (see `help scheme files`) and are not accessible from `graph` commands. You would rarely want to change axis styles.

axisstyle is a composite style that holds and sets all attributes of an axis, including the look of ticks and tick labels (see [G-4] ***ticksetstyle***) for that axis's major and minor labeled ticks and major and minor unlabeled ticks, the axis line style (see [G-4] ***linestyle***), rules for whether the axis extends through the plot region margin (both at the low and high end of the scale), whether grids are drawn for each of the labeled and unlabeled major and minor ticks, the gap between the tick labels and axis title, and any extra space beyond the axis title.

Remarks

When changing the look of an axis in a scheme file, you would rarely want to change the *axisstyle* entries. Instead, you should change the entries for the individual components making up the axis style.

Also see

[G-4] ***ticksetstyle*** — Choices for overall look of axis ticks

[G-4] ***tickstyle*** — Choices for the overall look of axis ticks and axis tick labels

[G-4] ***linestyle*** — Choices for overall look of lines

Title

[G-4] *bystyle* — Choices for look of by-graphs

Syntax

bystyle	Description
`default`	determined by scheme
`compact`	a more compact version of `default`
`stata7`	like that provided by Stata 7

Other *bystyles* may be available; type

```
. graph query bystyle
```

to obtain the complete list of *bystyles* installed on your computer.

Description

bystyles specify the overall look of by-graphs.

Remarks

Remarks are presented under the following headings:

What is a by-graph?
What is a bystyle?

What is a by-graph?

A by-graph is one graph (image, really) containing an array of separate graphs, each of the same type, and each reflecting a different subset of the data. For instance, a by-graph might contain graphs of miles per gallon versus weight, one for domestic cars and the other for foreign.

By-graphs are produced when you specify the `by()` option; see [G-3] ***by_option***.

What is a bystyle?

A *bystyle* determines the overall look of the combined graphs, including

1. whether the individual graphs have their own axes and labels or if instead the axes and labels are shared across graphs arrayed in the same row and/or in the same column;
2. whether the scales of axes are in common or allowed to be different for each graph; and
3. how close the graphs are placed to each other.

There are options that let you control each of the above attributes—see [G-3] ***by_option***—but the *bystyle* specifies the starting point.

You need not specify a *bystyle* just because there is something you want to change. You specify a *bystyle* when another style exists that is exactly what you desire or when another style would allow you to specify fewer changes to obtain what you want.

Also see

[G-3] ***by_option*** — Option for repeating graph command

Title

[G-4] *clockposstyle* — Choices for location: Direction from central point

Syntax

clockposstyle is

#	$0 \leq \# \leq 12$, # an integer

Description

clockposstyle specifies a location or a direction.

clockposstyle is specified inside options such as the `position()` *title_option* (see [G-3] ***title_options***) or the `mlabposition()` *marker_label_option* (see [G-3] ***marker_label_options***):

```
. graph ..., title(..., position(clockposstyle)) ...

. graph ..., mlabposition(clockposlist) ...
```

In cases where a *clockposlist* is allowed, you may specify a sequence of *clockposstyle* separated by spaces. Shorthands are allowed to make specifying the list easier; see [G-4] ***stylelists***.

Remarks

clockposstyle is used to specify a location around a central point:

```
  11  12  1
10          2
 9     0    3
8           4
   7   6   5
```

Sometimes the central position is a well-defined object (for example, for `mlabposition()`, the central point is the marker itself), and sometimes the central position is implied (for example, for `position()`, the central point is the center of the plot region).

clockposstyle `0` is always allowed: it refers to the center.

Also see

[G-3] ***marker_label_options*** — Options for specifying marker labels

[G-3] ***title_options*** — Options for specifying titles

[G-4] ***compassdirstyle*** — Choices for location

Title

[G-4] ***colorstyle*** — Choices for color

Syntax

colorstyle	Description
black	
gs0	gray scale: 0 = black
gs1	gray scale: very dark gray
gs2	
.	
.	
gs15	gray scale: very light gray
gs16	gray scale: 16 = white
white	
blue	
bluishgray	
brown	
cranberry	
cyan	
dimgray	between gs14 and gs15
dkgreen	dark green
dknavy	dark navy blue
dkorange	dark orange
eggshell	
emerald	
forest_green	
gold	
gray	equivalent to gs8
green	
khaki	
lavender	
lime	
ltblue	light blue
ltbluishgray	light blue-gray, used by scheme s2color
ltkhaki	light khaki
magenta	
maroon	
midblue	
midgreen	
mint	
navy	
olive	
olive_teal	

`orange`	
`orange_red`	
`pink`	
`purple`	
`red`	
`sand`	
`sandb`	bright sand
`sienna`	
`stone`	
`teal`	
`yellow`	
	colors used by *The Economist* magazine:
`ebg`	background color
`ebblue`	bright blue
`edkblue`	dark blue
`eltblue`	light blue
`eltgreen`	light green
`emidblue`	midblue
`erose`	rose
`none`	no color; invisible; draws nothing
`background` or `bg`	same color as background
`foreground` or `fg`	same color as foreground
# # #	RGB value; `white` = `"255 255 255"`
# # # #	CMYK value; `yellow` = `"0 0 255 0"`
hsv # # #	HSV value; `white` = `"hsv 255 255 255"`
color`*`*#*	color with adjusted intensity
`*`*#*	default color with adjusted intensity

When specifying RGB, CMYK, or HSV values, it is best to enclose the values in quotes; type `"128 128 128"` and not `128 128 128`.

For a color palette showing an individual color, type

`palette color` *colorstyle* [`, scheme(`*schemename*`)`]

and for a palette comparing two colors, type

`palette color` *colorstyle colorstyle* [`, scheme(`*schemename*`)`]

For instance, you might type

```
. palette color red green
```

See [G-2] **palette**.

Wherever a *colorstyle* appears, you may specify an RGB value by specifying three numbers in sequence. Each number should be between 0 and 255, and the triplet indicates the amount of red, green, and blue to be mixed. Each of the *colorstyles* in the table above is equivalent to an RGB value.

You can also specify a CMYK value wherever *colorstyle* appears, but the four numbers representing a CMYK value must be enclosed in quotes, for example, `"100 0 22 50"`.

You can also specify an HSV (hue, saturation, and value) color wherever *colorstyle* appears. HSV colors measure hue on a circular 360-degree scale with saturation and hue as proportions between 0 and 1. You must prefix HSV colors with `hsv` and enclose the full HSV specification in quotes, for example, `"hsv 180 .5 .5"`.

Other *colorstyles* may be available; type

```
. graph query colorstyle
```

to obtain the complete list of *colorstyles* installed on your computer.

Description

colorstyle specifies the color of a graphical component. You can specify *colorstyle* with many different `graph` options; all have the form

⟨*object*⟩`color(`*colorstyle*`)`

or

`color(`*colorstyle*`)`

For instance, option `mcolor()` specifies the color of markers, option `lcolor()` specifies the colors of connecting lines, and option `fcolor()` specifies the colors of the fill area. Option `color()` is equivalent to specifying all three of these options. Anywhere you see *colorstyle*, you can choose from the list above.

You will sometimes see that a *colorstylelist* is allowed, as in

```
. scatter ..., msymbol(colorstylelist) ...
```

A *colorstylelist* is a sequence of *colorstyles* separated by spaces. Shorthands are allowed to make specifying the list easier; see [G-4] ***stylelists***. When specifying RGB, CMYK, or HSV values in *colorstylelists*, remember to enclose the numbers in quotes.

Remarks

Remarks are presented under the following headings:

Colors are independent of the background color
White backgrounds and black backgrounds
RGB values
CMYK values
HSV values
Adjusting intensity

Colors are independent of the background color

Except for the colors `background` and `foreground`, colors do not change because of the background color. Colors `background` and `foreground` obviously do change, but otherwise, black means black, red means red, white means white, and so on. White on a black background has high visibility; white on a white background is invisible. Inversely, black on a white background has high visibility; black on a black background is invisible.

Color `foreground` always has high visibility.

Color `background` is the background color. If you draw something in this color, you will erase whatever is underneath it.

Color `none` is no color at all. If you draw something in this color, whatever you draw will be invisible. Being invisible, it will not hide whatever is underneath it.

White backgrounds and black backgrounds

The colors do not change because of the background color, but the colors that look best depend on the background color.

Graphs on the screen look best against a black background. With a black background, light colors stand out, and dark colors blend into the background.

Graphs on paper are usually presented against a white background. Dark colors stand out and light colors blend into the background.

Because most users need to make printed copies of their graphs, Stata's default is to present graphs on a white background, but you can change that; see [G-4] **schemes intro**.

If you want a dark background, it is better to choose a dark background rather than attempt to darken the background by using the *region_option* `graphregion(fcolor())` (see [G-3] ***region_options***); everything else about the graph's scheme will assume a background similar to how it was originally.

`graphregion(fcolor())` (and `graphregion(ifcolor())`, `plotregion(fcolor())`, and `plotregion(ifcolor())`) are best used for adding a little tint to the background.

RGB values

In addition to colors such as `red`, `green`, `blue`, or `cyan`, you can mix your own colors by specifying RGB values. An RGB value is a triplet of numbers, each of which specifies, on a scale of 0–255, the amount of red, green, and blue to be mixed. That is,

```
red     =  255    0    0
green   =    0  255    0
blue    =    0    0  255

cyan    =    0  255  255
magenta =  255    0  255
yellow  =  255  255    0
white   =  255  255  255
black   =    0    0    0
```

The overall scale of the triplet affects intensity; thus, changing 255 to 128 in all of the above would keep the colors the same but make them dimmer. (Color `128 128 128` is what most people call gray.)

CMYK values

In addition to mixing your own colors by using RGB values, you can mix your own colors by using CMYK values. If you have not heard of CMYK values or been asked to produce CMYK color separations, you can safely skip this section. CMYK is provided primarily to assist those doing color separations for mass printings. Although most inkjet printers use the more common RGB color values, printing presses almost always require CMYK values for color separation.

RGB values represent a mixing of red, green, and blue light, whereas CMYK values represent a mixing of pigments—cyan, magenta, yellow, and black. Thus, as the numbers get bigger, RGB colors go from dark to bright, whereas the CMYK colors go from light to dark.

CMYK values can be specified either as integers from 0 to 255, or as proportions of ink using real numbers from 0.0 to 1.0. If all four values are 1 or less, the numbers are taken to be proportions of ink. Thus, `127 0 127 0` and `0.5 0 0.5 0` specify almost equivalent colors.

Some examples of CMYK colors are

`red`	=	0	255	255	0	or, equivalently,	`0 1 1 0`
`green`	=	255	0	255	0	or, equivalently,	`1 0 1 0`
`blue`	=	255	255	0	0	or, equivalently,	`1 1 0 0`
`cyan`	=	255	0	0	0	or, equivalently,	`1 0 0 0`
`magenta`	=	0	255	0	0	or, equivalently,	`0 1 0 0`
`yellow`	=	0	0	255	0	or, equivalently,	`0 0 1 0`
`white`	=	0	0	0	0	or, equivalently,	`0 0 0 0`
`black`	=	0	0	0	255	or, equivalently,	`0 0 0 1`

For color representation, there is no reason for the K (black) component of the CMYK values, `255 255 255 0` and `0 0 0 255` both specify the color black. With pigments such as printer inks, however, using 100% of cyan, magenta, and yellow rarely produces a pure black. For that reason, CMYK values include a specific black component.

Internally, Stata stores all colors as RGB values, even when CMYK values are specified. This allows colors to be easily shown on most display devices. In fact, `graph export` will produce graph files using RGB values, even when CMYK values were specified as input. Only a few devices and graphics formats understand CMYK colors, with PostScript and EPS formats being two of the most important. To obtain CMYK colors in these formats, use the `cmyk(on)` option of the `graph export` command. You can also specify that all PostScript export files permanently use CMYK colors with the command `translator set Graph2ps cmyk on` or `translator set Graph2eps cmyk on` for EPS files.

Stata uses, for lack of a better term, normalized CMYK values. That simply means that at least one of the CMY values is normalized to 0 for all CMYK colors, with the K (black) value “absorbing” all parts of CM and Y where they are all positive. An example may help: `10 10 5 0` is taken to be the normalized CMYK value `5 5 0 5`. That is, all CMY colors were 5 or greater, so this component was moved to black ink, and 5 was subtracted from each of the CMY values. If you specify your CMYK colors in normalized form, these will be exactly the values output by `graph export`, and you should never be surprised by the resulting colors.

HSV values

You can also mix your own colors by specifying HSV values. These are also sometimes called HSL (hue, saturation, and luminance) or HSB (hue, saturation, and brightness). An HSV value is a triplet of numbers. The first number specifies the hue and is specified on a circular 360-degree scale. Any number can be specified for the hue, but numbers above 360 are taken as modulo 360. The second number specifies the saturation of the color as a proportion between 0 and 1, and the third number specifies the value (luminance/brightness) between 0 and 1. HSV colors must be prefaced with `hsv`.

Some examples of HSV colors are

```
red     =  hsv   0   1   1
green   =  hsv 120   1   1
blue    =  hsv 240   1   1

cyan    =  hsv 180   1   1
magenta =  hsv 300   1   1
yellow  =  hsv  60   1   1
white   =  hsv   0   0   1
black   =  hsv   0   0   0
```

Putting the primary colors in their HSV hue order,

```
red     =  hsv   0   1   1
yellow  =  hsv  60   1   1
green   =  hsv 120   1   1
cyan    =  hsv 180   1   1
blue    =  hsv 240   1   1
magenta =  hsv 300   1   1
```

With the exception of `black`, all the listed colors specify a saturation and value of `1`. This is because these are the primary colors in the RGB and CMYK spaces and therefore have full saturation and brightness. Reducing the saturation will reduce the amount of color. Reducing the brightness will make the color dimmer.

Adjusting intensity

To specify a color and modify its intensity (brightness), you might specify things such as

```
green*.8
red*1
purple*1.2
0 255 255*.8
```

Multiplying a color by 1 leaves the color unchanged. Multiplying by a number greater than 1 makes the color stand out from the background more; multiplying by a number less than 1 makes the color blend into the background more. For an example using the intensity adjustment, see *Typical use* in [G-2] **graph twoway kdensity**.

When modifying intensity, the syntax is

*color**#*

or the color may be omitted:

**#*

If the color is omitted, the intensity adjustment is applied to the default color, given the context. For instance, you specify `bcolor(*.7)` with `graph twoway bar`—or any other `graph twoway` command that fills an area—to use the default color at 70% intensity. Or you specify `bcolor(*2)` to use the default color at twice its usual intensity.

When you specify both the color and the adjustment, you must type the color first: `.8*green` will not be understood. Also, do not put a space between the *color* and the `*`, even when the *color* is an RGB or CMYK value.

color`*0` makes the color as dim as possible, but it is not equivalent to color `none`. *color*`*255` makes the color as bright as possible, although values much smaller than 255 usually achieve the same result.

Also see

[G-2] **palette** — Display palettes of available selections

[G-4] **schemes intro** — Introduction to schemes

Title

[G-4] ***compassdirstyle*** — Choices for location

Syntax

compassdirstyle	Synonyms First	Second	Third	Fourth
north	n	12		top
neast	ne	1	2	
east	e	3		right
seast	se	4	5	
south	s	6		bottom
swest	sw	7	8	
west	w	9		left
nwest	nw	10	11	
center	c	0		

Other *compassdirstyles* may be available; type

```
. graph query compassdirstyle
```

to obtain the complete list of *compassdirstyle*s installed on your computer.

Description

compassdirstyle specifies a direction.

compassdirstyle is specified inside options such as the `placement()` textbox suboption of `title()` (see [G-3] ***title_options*** and [G-3] ***textbox_options***):

```
. graph ..., title(..., placement(compassdirstyle)) ...
```

Sometimes you may see that a *compassdirstylelist* is allowed: a *compassdirstylelist* is a sequence of *compassdirstyles* separated by spaces. Shorthands are allowed to make specifying the list easier; see [G-4] ***stylelists***.

Remarks

Two methods are used for specifying directions—the compass and the clock. Some options use the compass and some use the clock. For instance, the textbox option `position()` uses the compass (see [G-3] ***textbox_options***), but the title option `position()` uses the clock (see [G-3] ***title_options***). The reason for the difference is that some options need only the eight directions specified by the compass, whereas others need more. In any case, synonyms are provided so that you can use the clock notation in all cases.

Also see

[G-3] ***textbox_options*** — Options for textboxes and concept definition

Title

[G-4] concept: gph files — Using gph files

Description

.gph files contain Stata graphs and, in fact, even include the original data from which the graph was drawn. Below we discuss how to replay graph files and to obtain the data inside them.

Remarks

Remarks are presented under the following headings:

Background
Gph files are machine/operating system independent
Gph files come in three forms
Advantages of live-format files
Advantages of as-is format files
Retrieving data from live-format files

Background

.gph files are created either by including the `saving()` option when you draw a graph,

```
. graph ..., ... saving(myfile)
```

or by using the `graph save` command afterward:

```
. graph ...
. graph save myfile
```

Either way, file `myfile.gph` is created; for details see [G-3] ***saving_option*** and [G-2] **graph save**.

At some later time, in the same session or in a different session, you can redisplay what is in the .gph file by typing

```
. graph use myfile
```

See [G-2] **graph use** for details.

Gph files are machine/operating system independent

The .gph files created by `saving()` and `graph save` are binary files written in a machine-and-operating-system independent format. You may send .gph files to other users, and they will be able to read them, even if you use, say, a Mac and your colleague uses a Windows or Unix computer.

Gph files come in three forms

There are three forms of `graph` files:

1. an old-format Stata 7 or earlier .gph file
2. a modern-format graph in as-is format
3. a modern-format graph in live format

You can find out which type a `.gph` file is by typing

```
. graph describe filename
```

See [G-2] **graph describe**.

Live-format files contain the data and other information necessary to recreate the graph. As-is format files contain a recording of the picture. When you save a graph, unless you specify the `asis` option, it is saved in live format.

Advantages of live-format files

A live-format file can be edited later and can be displayed using different schemes; see [G-4] **schemes intro**. Also, the data used to create the graph can be retrieved from the `.gph` file.

Advantages of as-is format files

As-is format files are generally smaller than live-format files.

As-is format files cannot be modified; the rendition is fixed and will appear on anyone else's computer just as it last appeared on yours.

Retrieving data from live-format files

First, verify that you have a live-format file by typing

```
. graph describe filename.gph
```

Then type

```
. discard
```

This will close any open graphs and eliminate anything stored having to do with previously existing graphs. Now display the graph of interest,

```
. graph use filename
```

and then type

```
. serset dir
```

From this point on, you are going to have to do a little detective work, but usually it is not much. Sersets are how `graph` stores the data corresponding to each plot within the graph. You can see [P] **serset**, but unless you are a programmer curious about how things work, that will not be necessary. We will show you below how to load each of the sersets (often there is just one) and to treat it from then on just as if it came from a `.dta` file.

Let us set up an example. Pretend that previously we have drawn the graph and saved it by typing

```
. use http://www.stata-press.com/data/r12/lifeexp
(Life expectancy 1998)
. scatter lexp gnppc, by(region)
```

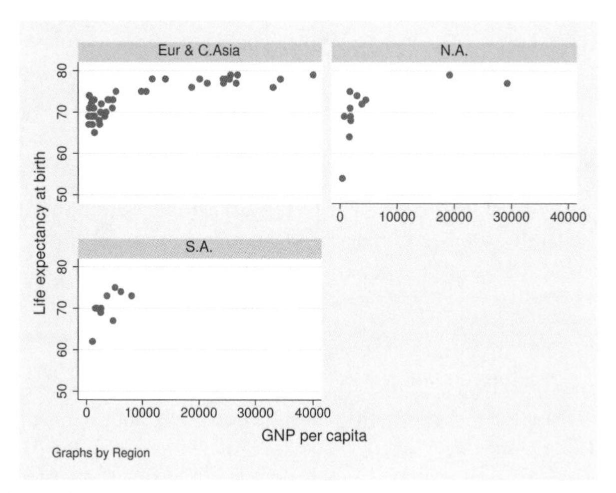


```
. graph save legraph
(file legraph.gph saved)
```

Following the instructions, we now type

```
. graph describe legraph.gph
```

legraph.gph stored on disk

```
        name:  legraph.gph
      format:  live
     created:  20 Jun 2011 13:04:30
      scheme:  s2gmanual
        size:  2.392 x 3.12
    dta file:  C:\Program Files\Stata12\ado\base\lifeexp.dta dated 26 Mar 2011 09:40
     command:  twoway scatter lexp gnppc, by(region)
. discard
. graph use legraph
. serset dir
  0.  44 observations on 2 variables
      lexp gnppc
  1.  14 observations on 2 variables
      lexp gnppc
  2.  10 observations on 2 variables
      lexp gnppc
```

We discover that our graph has three sersets. Looking at the graph, that should not surprise us. Although we might think of

```
. scatter lexp gnppc, by(region)
```

as being one plot, it is in fact three if we were to expand it:

```
. scatter lexp gnppc if region==1 ||
  scatter lexp gnppc if region==2 ||
  scatter lexp gnppc if region==3
```

The three sersets numbered 0, 1, and 2 correspond to three pieces of the graph. We can look at the individual sersets. To load a serset, you first set its number and then you type `serset use, clear`:

```
. serset 0
. serset use, clear
```

If we were now to type `describe`, we would discover that we have a 44-observation dataset containing two variables: `lexp` and `gnppc`. Here are a few of the data:

```
. list in 1/5
```

	lexp	gnppc
1.	72	810
2.	74	460
3.	79	26830
4.	71	480
5.	68	2180

These are the data that appeared in the first plot. We could similarly obtain the data for the second plot by typing

```
. serset 1
. serset use, clear
```

If we wanted to put these data back together into one file, we might type

```
. serset 0
. serset use, clear
. generate region=0
. save region0
. serset 1
. serset use, clear
. generate region=1
. save region1
. serset 2
. serset use, clear
. generate region=2
. save region2
. use region0
. append using region1
. append using region2
. erase region0.dta
. erase region1.dta
. erase region2.dta
```

In general, it will not be as much work to retrieve the data because in many graphs, you will find that there is only one serset. We chose a complicated `.gph` file for our demonstration.

Also see

[G-2] **graph display** — Display graph stored in memory

[G-2] **graph manipulation** — Graph manipulation commands

[G-2] **graph save** — Save graph to disk

[G-3] ***saving_option*** — Option for saving graph to disk

[P] **serset** — Create and manipulate sersets

Title

[G-4] concept: lines — Using lines

Syntax

The following affects how a line appears:

linestyle	overall style
linepatternstyle	whether solid, dashed, etc.
linewidthstyle	its thickness
colorstyle	its color

See [G-4] ***linestyle***, [G-4] ***linepatternstyle***, [G-4] ***linewidthstyle***, and [G-4] ***colorstyle***.

Description

Lines occur in many contexts—in borders, axes, the ticks on axes, the outline around symbols, the connecting of points in a plot, and more. *linestyle*, *linepatternstyle*, *linewidthstyle*, and *colorstyle* define the look of the line.

Remarks

Remarks are presented under the following headings:

linestyle
linepatternstyle
linewidthstyle
colorstyle

linestyle, *linepatternstyle*, *linewidthstyle*, and *colorstyle* are specified inside options that control how the line is to appear. Regardless of the object, these options usually have the same names:

`lstyle(`*linestyle*`)`
`lpattern(`*linepatternstyle*`)`
`lwidth(`*linewidthstyle*`)`
`lcolor(`*colorstyle*`)`

Though for a few objects, such as markers, the form of the names is

⟨*object*⟩`lstyle(`*linestyle*`)`
⟨*object*⟩`lpattern(`*linepatternstyle*`)`
⟨*object*⟩`lwidth(`*linewidthstyle*`)`
⟨*object*⟩`lcolor(`*linecolorstyle*`)`

For instance,

- The options to specify how the lines connecting points in a plot are to appear are specified by the options `lstyle()`, `lpattern()`, `lwidth()`, and `lcolor()`; see [G-3] ***connect_options***.
- The suboptions to specify how the border around a textbox, such as a title, are to appear are named `lstyle()`, `lpattern()`, `lwidth()`, and `lcolor()`; see [G-3] ***textbox_options***.
- The options to specify how the outline around markers is to appear are specified by the options `mlstyle()`, `mlpattern()`, `mlwidth()`, and `mlcolor()`; see [G-3] ***marker_options***.

Wherever these options arise, they always come in a group of four, and the four pieces have the same meaning.

linestyle

linestyle is specified inside the `lstyle()` option or sometimes inside the ⟨*object*⟩`lstyle()` option.

linestyle specifies the overall style of the line: its pattern (solid, dashed, etc.), thickness, and color.

You need not specify the `lstyle()` option just because there is something you want to change about the look of the line and, in fact, most of the time you do not. You specify `lstyle()` when another style exists that is exactly what you desire or when another style would allow you to specify fewer changes to obtain what you want.

See [G-4] ***linestyle*** for the list of what may be specified inside the `lstyle()` option.

linepatternstyle

linepatternstyle is specified inside the `lpattern()` or ⟨*object*⟩`lpattern()` option.

linepatternstyle specifies whether the line is solid, dashed, etc.

See [G-4] ***linepatternstyle*** for the list of what may be specified inside the `lpattern()` option.

linewidthstyle

linewidthstyle is specified inside the `lwidth()` or ⟨*object*⟩`lwidth()` option.

linewidthstyle specifies the thickness of the line.

See [G-4] ***linewidthstyle*** for the list of what may be specified inside the `lwidth()` option.

colorstyle

colorstyle is specified inside the `lcolor()` or ⟨*object*⟩`lcolor()` option.

colorstyle specifies the color of the line.

See [G-4] ***colorstyle*** for the list of what may be specified inside the `lcolor()` option.

Also see

[G-4] ***linestyle*** — Choices for overall look of lines

[G-4] ***linepatternstyle*** — Choices for whether lines are solid, dashed, etc.

[G-4] ***linewidthstyle*** — Choices for thickness of lines

[G-4] ***colorstyle*** — Choices for color

[G-3] ***connect_options*** — Options for connecting points with lines

Title

[G-4] concept: repeated options — Interpretation of repeated options

Syntax

Options allowed with `graph` are categorized as being

unique
rightmost
merged-implicit
merged-explicit

What this means is described below.

Remarks

It may surprise you to learn that most `graph` options can be repeated within the same `graph` command. For instance, you can type

```
. graph twoway scatter mpg weight, msymbol(Oh) msymbol(O)
```

and rather then getting an error, you will get back the same graph as if you omitted typing the `msymbol(Oh)` option. `msymbol()` is said to be a *rightmost* option.

`graph` allows that because so many other commands are implemented in terms of `graph`. Imagine that an ado-file that constructs the "`scatter mpg weight, msymbol(Oh)`" part, and you come along and use that ado-file, and you specify to it the option "`msymbol(O)`". The result is that the ado-file constructs

```
. graph twoway scatter mpg weight, msymbol(Oh) msymbol(O)
```

and, because `graph` is willing to ignore all but the rightmost specification of the `msymbol()` option, the command works and does what you expect.

Options in fact come in three forms, which are

1. *rightmost*: take the rightmost occurrence;
2. *merged*: merge the repeated instances together;
3. *unique*: the option may be specified only once; specifying it more than once is an error.

You will always find options categorized one of these three ways; typically that is done in the syntax diagram, but sometimes the categorization appears in the description of the option.

`msymbol()` is an example of a *rightmost* option. An example of a *unique* option is `saving()`; it may be specified only once.

Concerning *merged* options, they are broken into two subcategories:

2a. *merged-implicit*: always merge repeated instances together,

2b. *merged-explicit*: treat as *rightmost* unless an option within the option is specified, in which case it is merged.

`merged` can apply only to options that take arguments because otherwise there would be nothing to merge. Sometimes those options themselves take suboptions. For instance, the syntax of the `title()` option (the option that puts titles on the graph) is

`title("`*string*`"` [`"`*string*`"` [...]] [`,` *suboptions*]`)` `title()` has suboptions that specify how the title is to look and among them is, for instance, `color()`; see [G-3] ***title_options***. `title()` also has two other suboptions, `prefix` and `suffix`, that specify how repeated instances of the `title()` option are to be merged. For instance, specify

```
...title("My title") ...title("Second line", suffix)
```

and the result will be the same as if you specified

```
...title("My title" "Second line")
```

at the outset. Specify

```
...title("My title") ...title("New line", prefix)
```

and the result will be the same as if you specified

```
...title("New line" "My title")
```

at the outset. The `prefix` and `suffix` options specify exactly how repeated instances of the option are to be merged. If you do not specify one of those options,

```
...title("My title") ...title("New title")
```

the result will be as if you never specified the first option:

```
...title("New title")
```

`title()` is an example of a *merged-explicit* option. The suboption names for handling *merged-explicit* are not always `prefix` and `suffix`, but anytime an option is designated *merged-explicit*, it will be documented under the heading *Interpretation of repeated options* exactly what and how the merge options work.

❑ Technical note

Even when an option is *merged-explicit* and its merge suboptions are not specified, its other suboptions are merged. For instance, consider

```
...title("My title", color(red)) ...title("New title")
```

`title()` is *merged-explicit*, but because we did not specify one of its merge options, it is being treated as *rightmost*. Actually, it is almost being treated as rightmost because, rather than the `title()` being exactly what we typed, it will be

```
...title("New title", color(red))
```

This makes ado-files work as you would expect. Say that you run the ado-file `xyz.ado`, which constructs some graph and the command

```
graph ..., ...title("Std. title", color(red)) ...
```

You specify an option to `xyz.ado` to change the title:

```
. xyz ..., ...title("My title")
```

The overall result will be just as you expect: your title will be substituted, but the color of the title (and its size, position, etc.) will not change. If you wanted to change those things, you would have specified the appropriate suboptions in your `title()` option. ❑

Also see

[G-2] **graph** — The graph command

Title

[G-4] *connectstyle* — Choices for how points are connected

Syntax

connectstyle	Synonym	Description
`none`	`i`	do not connect
`direct`	`l`	connect with straight lines
`ascending`	`L`	`direct`, but only if $x_{j+1} > x_j$
`stairstep`	`J`	flat, then vertical
`stepstair`		vertical, then flat

Other *connectstyles* may be available; type

```
. graph query connectstyle
```

to obtain the full list installed on your computer.

Description

connectstyle specifies if and how points in a scatter are to be connected, for example, via straight lines or stairsteps.

connectstyle is specified inside the `connect()` option which is allowed, for instance, with `scatter`:

```
. scatter ..., connect(connectstylelist) ...
```

Here a *connectstylelist* is allowed. A *connectstylelist* is a sequence of *connectstyles* separated by spaces. Shorthands are allowed to make specifying the list easier; see [G-4] ***stylelists***.

Remarks

Points are connected in the order of the data, so be sure that data are in the desired order (which is usually ascending value of x) before specifying the `connect(`*connectstyle*`)` option. Commands that provide `connect()` also provide a `sort` option, which will sort by the x variable for you.

`connect(l)` is the most common choice.

`connect(J)` is an appropriate way to connect the points of empirical cumulative distribution functions (CDFs).

Also see

[G-3] ***connect_options*** — Options for connecting points with lines

Title

[G-4] *gridstyle* — Choices for overall look of grid lines

Syntax

gridstyle	Description
`default`	determined by scheme
`major`	determined by scheme; `default` or bolder
`minor`	determined by scheme; `default` or fainter
`dot`	dotted line

Other *gridstyles* may be available; type

```
. graph query gridstyle
```

to obtain the complete list of *gridstyles* installed on your computer.

Description

Grids are lines that extend from an axis across the plot region. *gridstyle* specifies the overall look of grids. See [G-3] ***axis_label_options***.

Remarks

Remarks are presented under the following headings:

What is a grid?
What is a gridstyle?
You do not need to specify a gridstyle
Turning off and on the grid

What is a grid?

Grids are lines that extend from an axis across the plot region.

What is a gridstyle?

Grids are defined by

1. whether the grid lines extend into the plot region's margin;
2. whether the grid lines close to the axes are to be drawn;
3. the line style of the grid, which includes the line's thickness, color, and whether they are solid, dashed, etc.; see [G-4] ***linestyle***.

The *gridstyle* specifies all three of these attributes.

You do not need to specify a gridstyle

The *gridstyle* is specified in the options named

{ y|x } { label|tick|mlabel|mtick } (...gstyle(*gridstyle*) ...)

Correspondingly, other { y|x }{ label|tick|mlabel|mtick }() suboptions allow you to specify the individual attributes; see [G-3] ***axis_label_options***.

You specify the *gridstyle* when a style exists that is exactly what you desire or when another style would allow you to specify fewer changes to obtain what you want.

Turning off and on the grid

Whether grid lines are included by default is a function of the scheme; see [G-4] **schemes intro**. Regardless of the default, whether grid lines are included is controlled not by the *gridstyle* but by the { y|x }{ label|tick|mlabel|mtick }() suboptions grid and nogrid.

Grid lines are nearly always associated with the ylabel() and/or xlabel() options. Specify { y|x }label(,grid) or { y|x }label(,nogrid). See [G-3] ***axis_label_options***.

Also see

[G-3] ***axis_label_options*** — Options for specifying axis labels

Title

[G-4] *intensitystyle* — Choices for the intensity of a color

Syntax

intensitystyle	Description
`inten0`	0% intensity, no color at all
`inten10`	
`inten20`	
...	
`inten90`	
`inten100`	100% intensity, full color
#	*#*% intensity, 0 to 100

Other *intensitystyles* may be available; type

```
. graph query intensitystyle
```

to obtain the complete list of *intensitystyles* installed on your computer. If other *intensitystyles* do exist, they are merely words attached to numeric values.

Description

intensitystyles specify the intensity of colors as a percentage from 0 to 100 and are used in *shadestyles*; see [G-4] ***shadestyle***.

Remarks

intensitystyle is used primarily in scheme files and is rarely specified interactively, though some options, such as the `intensity()` option, may accept the style names in addition to numeric values.

Also see

[G-4] ***shadestyle*** — Choices for overall look of filled areas

Title

[G-4] *justificationstyle* — Choices for how text is justified

Syntax

justificationstyle	Description
`left`	left-justified
`center`	centered
`right`	right-justified

Other *justificationstyles* may be available; type

```
. graph query justificationstyle
```

to obtain the complete list of *justificationstyles* installed on your computer.

Description

justificationstyle specifies how the text is "horizontally" aligned in the textbox. Choices include `left`, `right`, and `center`. Think of the textbox as being horizontal, even if it is vertical when specifying this option.

justificationstyle is specified in the `justification()` option nested within another option, such as `title()`:

```
. graph ..., title("Line 1" "Line 2", justification(justificationstyle)) ...
```

See [G-3] ***textbox_options*** for more information on textboxes.

Sometimes you will see that a *justificationstylelist* is allowed. A *justificationstylelist* is a sequence of *justificationstyles* separated by spaces. Shorthands are allowed to make specifying the list easier; see [G-4] ***stylelists***.

Remarks

justificationstyle typically affects the alignment of multiline text within a textbox and not the justification of the placement of the textbox itself; see *Justification* in [G-3] ***textbox_options***.

Also see

[G-3] ***textbox_options*** — Options for textboxes and concept definition

[G-4] ***alignmentstyle*** — Choices for vertical alignment of text

Title

[G-4] *legendstyle* — Choices for look of legends

Syntax

legendstyle	Description
`default`	determined by scheme

Other *legendstyles* may be available; type

```
. graph query legendstyle
```

to obtain the complete list of *legendstyles* installed on your computer.

Description

legendstyle specifies the overall style of legends and is specified in the `legend(style())` option:

```
. graph ..., legend( ... style(legendstyle) ...)
```

Remarks

Remarks are presented under the following headings:

What is a legend?
What is a legendstyle?
You do not need to specify a legendstyle

What is a legend?

A legend is a table that shows the symbols used in a graph along with text describing their meaning. Each symbol/text entry in a legend is called a key. See [G-3] ***legend_options*** for more information.

What is a legendstyle?

The look of a legend is defined by 14 attributes:

1. The number of columns or rows of the table
2. Whether, in a multicolumn table, the first, second, ..., keys appear across the rows or down the columns
3. Whether the symbol/text of a key appears horizontally adjacent or vertically stacked
4. The gap between lines of the legend
5. The gap between columns of the legend
6. How the symbol of a key is aligned and justified

7. The gap between the symbol and text of a key
8. The height to be allocated in the table for the symbol of the key
9. The width to be allocated in the table for the symbol of the key
10. The width to be allocated in the table for the text of the key
11. Whether the above-specified height and width are to be dynamically adjusted according to contents of the keys
12. The margin around the legend
13. The color, size, etc., of the text of a key (17 features)
14. The look of any titles, subtitles, notes, and captions placed around the table (23 characteristics each)

The *legendstyle* specifies all 14 of these attributes.

You do not need to specify a legendstyle

The *legendstyle* is specified in the option

`legend(style(`*legendstyle*`))`

Correspondingly, option `legend()` has other suboptions that will allow you to specify the 14 attributes individually; see [G-3] ***legend_options***.

Specify the *legendstyle* when a style exists that is exactly what you desire or when another style would allow you to specify fewer changes to obtain what you want.

Also see

[G-3] ***legend_options*** — Options for specifying legends

Title

[G-4] *linepatternstyle* — Choices for whether lines are solid, dashed, etc.

Syntax

linepatternstyle	Description
`solid`	solid line
`dash`	dashed line
`dot`	dotted line
`dash_dot`	
`shortdash`	
`shortdash_dot`	
`longdash`	
`longdash_dot`	
`blank`	invisible line
`"`*formula*`"`	e.g., `"-."` or `"--.."` etc.

A *formula* is composed of any combination of

`l`	solid line
`_`	(underscore) a long dash
`-`	(hyphen) a medium dash
`.`	short dash (almost a dot)
`#`	small amount of blank space

For a palette displaying each of the above named line styles, type

`palette linepalette` [`,` `scheme(`*schemename*`)`]

Other *linepatternstyles* may be available; type

```
. graph query linepatternstyle
```

to obtain the complete list of *linepatternstyles* installed on your computer.

Description

A line's look is determined by its pattern, thickness, and color; see [G-4] **concept: lines**. *linepatternstyle* specifies the pattern.

linepatternstyle is specified via options named

⟨*object*⟩⟨`l` or `li` or `line`⟩`pattern()`

or, just

⟨`l` or `li` or `line`⟩`pattern()`

For instance, for connecting lines (the lines used to connect points in a plot) used by `graph twoway function`, the option is named `lpattern()`:

```
. twoway function ..., lpattern(linepatternstyle) ...
```

Sometimes you will see that a *linepatternstylelist* is allowed:

```
. twoway line ..., lpattern(linepatternstylelist) ...
```

A *linepatternstylelist* is a sequence of *linepatterns* separated by spaces. Shorthands are allowed to make specifying the list easier; see [G-4] ***stylelists***.

Remarks

Although you may choose a prerecorded pattern (for example, `solid` or `dash`), you can build any pattern you wish by specifying a line-pattern formula. For example,

Formula	Description
`"l"`	solid line, same as `solid`
`"_"`	a long dash
`"_-"`	a long dash followed by a short dash
`"_--"`	a long dash followed by two short dashes
`"_--_#"`	a long dash, two short dashes, a long dash, and a bit of space
etc.	

When you specify a formula, you must enclose it in double quotes.

Also see

[G-4] **concept: lines** — Using lines

[G-4] ***linestyle*** — Choices for overall look of lines

[G-4] ***linewidthstyle*** — Choices for thickness of lines

[G-4] ***colorstyle*** — Choices for color

[G-4] ***connectstyle*** — Choices for how points are connected

Title

[G-4] ***linestyle*** — Choices for overall look of lines

Syntax

linestyle	Description
`foreground`	borders, axes, etc., in foreground color
`grid`	grid lines
`minor_grid`	a lesser grid line or same as `grid`
`major_grid`	a bolder grid line or same as `grid`
`refline`	reference lines
`yxline`	`yline()` or `xline()`
`none`	nonexistent line
`p1–p15`	used by first–fifteenth "line" plot
`p1bar–p15bar`	used by first–fifteenth "bar" plot
`p1box–p15box`	used by first–fifteenth "box" plot
`p1area–p15area`	used by first–fifteenth "area" plot
`p1solid–p15solid`	same as `p1–p15` but always solid
`p1mark–p15mark`	markers for first–fifteenth plot
`p1boxmark–p15boxmark`	markers for outside values of box plots
`p1dotmark–p15dotmark`	markers for dot plots
`p1other–p15other`	"other" lines, such as spikes and range plots

Other *linestyles* may be available; type

```
. graph query linestyle
```

to obtain the full list installed on your computer.

Description

linestyle sets the overall pattern, thickness, and color of a line; see [G-4] **concept: lines** for more information.

linestyle is specified via options named

⟨*object*⟩⟨`l` or `li` or `line`⟩`style()`

or

⟨`l` or `li` or `line`⟩`style()`

For instance, for connecting lines (the lines used to connect points in a plot) used by `graph twoway function`, the option is named `lstyle()`:

```
. twoway function ..., lstyle(linestyle) ...
```

Sometimes you will see that a *linestyle list* is allowed:

```
. twoway line ..., lstyle(linestylelist) ...
```

A *linestylelist* is a sequence of *linestyles* separated by spaces. Shorthands are allowed to make specifying the list easier; see [G-4] ***stylelists***.

Remarks

Remarks are presented under the following headings:

What is a line?
What is a linestyle?
You do not need to specify a linestyle
Specifying a linestyle can be convenient
What are numbered styles?
Suppressing lines

What is a line?

Nearly everything that appears on a graph is a line, the exceptions being markers, fill areas, bars, and the like, and even they are outlined or bordered by a line.

What is a linestyle?

Lines are defined by three attributes:

1. *linepattern*—whether it is solid, dashed, etc.;
   see [G-4] ***linepatternstyle***
2. *linewidth*—how thick the line is;
   see [G-4] ***linewidthstyle***
3. *linecolor*—the color of the line;
   see [G-4] ***colorstyle***

The *linestyle* specifies all three of these attributes.

You do not need to specify a linestyle

The *linestyle* is specified in options named

⟨*object*⟩⟨`l` or `li` or `line`⟩`style(`*linestyle*`)`

Correspondingly, three other options are available:

⟨*object*⟩⟨`l` or `li` or `line`⟩`pattern(`*linepatternstyle*`)`

⟨*object*⟩⟨`l` or `li` or `line`⟩`width(`*linewidthstyle*`)`

⟨*object*⟩⟨`l` or `li` or `line`⟩`color(`*colorstyle*`)`

Often the ⟨*object*⟩ prefix is not required.

You specify the *linestyle* when a style exists that is exactly what you want or when another style would allow you to specify fewer changes to obtain what you want.

Specifying a linestyle can be convenient

Consider the command

```
. line y1 y2 x
```

Assume that you wanted the line for y2 versus x to be the same as y1 versus x. You might set the pattern, width, and color of the line for y1 versus x and then set the pattern, width, and color of the line for y2 versus x to be the same. It would be easier, however, to type

```
. line y1 y2 x, lstyle(p1 p1)
```

lstyle() is the option that specifies the style of connected lines. When you do not specify the lstyle() option, results are the same as if you specified

lstyle(p1 p2 p3 p4 p5 p6 p7 p8 p9 p10 p11 p12 p13 p14 p15)

where the extra elements are ignored. In any case, p1 is one set of pattern, thickness, and color values; p2 is another set; and so on.

Say that you wanted y2 versus x to look like y1 versus x, except that you wanted the line to be green; you could type

```
. line y1 y2 x, lstyle(p1 p1) lcolor(. green)
```

There is nothing special about the *linestyles* p1, p2, ...; they merely specify sets of pattern, thickness, and color values, just like any other named *linestyle*. Type

```
. graph query linestyle
```

to find out what other line styles are available. You may find something pleasing, and if so, that is more easily specified than each of the individual options to modify the individual elements.

Also see *Appendix: Styles and composite styles* in [G-2] **graph twoway scatter** for more information.

What are numbered styles?

p1–p15 are the default styles for connecting lines in all twoway graphs, for example, twoway line, twoway connected, and twoway function. p1 is used for the first plot, p2 for the second, and so on. Some twoway graphs do not have connecting lines.

p1bar–p15bar are the default styles used for outlining the bars on bar charts; this includes twoway bar charts and bar charts. p1bar is used for the first set of bars, p2bar for the second, and so on.

p1box–p15box are the default styles used for outlining the boxes on box charts. p1box is used for the first set of boxes, p2box for the second, and so on.

p1area–p15area are the default styles used for outlining the areas on area charts; this includes twoway area charts and twoway rarea. p1area is used for the first filled area, p2area for the second, and so on.

p1solid–p15solid are the same as p1–p15, but the lines are always solid; they have the same color and same thickness as p1–p15.

p1mark–p15mark are the default styles for lines used to draw markers in all twoway graphs, for example, twoway scatter, twoway connected, and twoway rcapsym. p1mark is used for the first plot, p2mark for the second, and so on.

The *linepatternstyle* attribute is always ignored when drawing symbols.

p1boxmark–p15boxmark are the default styles for drawing the markers for the outside values on box charts. p1box is used for the first set of dots, p2box for the second, and so on.

`p1dotmark`–`p15dotmark` are the default styles for drawing the markers on dot charts. `p1dot` is used for the first set of dots, `p2dot` for the second, and so on.

`p1other`–`p15other` are the default styles used for "other" lines for some `twoway` plottypes, including the spikes for `twoway spike` and `twoway rspike` and the lines for `twoway dropline`, `twoway rcap`, and `twoway rcapsym`. `p1other` is used for the first set of lines, `p2other` for the second, and so on.

The "look" defined by a numbered style, such as `p1`, `p1mark`, `p1bar`, etc.—by "look" we mean width (see [G-4] ***linewidthstyle***), color (see [G-4] ***colorstyle***), and pattern (see [G-4] ***linepatternstyle***)—is determined by the scheme (see [G-4] **schemes intro**) selected.

Numbered styles provide default looks that can be controlled by a scheme. They can also be useful when you wish to make, say, the second "thing" on a graph look like the first. See *Specifying a linestyle can be convenient* above for an example.

Suppressing lines

Sometimes you want to suppress lines. For instance, you might want to remove the border around the plot region. There are two ways to do this: You can specify

⟨*object*⟩⟨`l` or `li` or `line`⟩`style(none)`

or

⟨*object*⟩⟨`l` or `li` or `line`⟩`color(`*color*`)`

The first usually works well; see *Suppressing the axes* in [G-3] ***axis_scale_options*** for an example.

For the outlines of solid objects, however, remember that lines have a thickness. Removing the outline by setting its line style to `none` sometimes makes the resulting object seem too small, especially when the object was small to begin with. In those cases, specify

⟨*object*⟩⟨`l` or `li` or `line`⟩`color(`*color*`)`

and set the outline color to be the same as the interior color.

Reference

Cox, N. J. 2009. Stata tip 82: Grounds for grids on graphs. *Stata Journal* 9: 648–651.

Also see

[G-4] **concept: lines** — Using lines

[G-4] ***linepatternstyle*** — Choices for whether lines are solid, dashed, etc.

[G-4] ***linewidthstyle*** — Choices for thickness of lines

[G-4] ***colorstyle*** — Choices for color

[G-4] ***connectstyle*** — Choices for how points are connected

Title

[G-4] *linewidthstyle* — Choices for thickness of lines

Syntax

linewidthstyle	Description
`none`	line has zero width; it vanishes
`vvthin`	thinnest
`vthin`	
`thin`	
`medthin`	
`medium`	
`medthick`	
`thick`	
`vthick`	
`vvthick`	
`vvvthick`	thickest
relativesize	any size you want

See [G-4] ***relativesize***.

Other *linewidthstyles* may be available; type

```
. graph query linewidthstyle
```

to obtain the full list installed on your computer.

Description

A line's look is determined by its pattern, thickness, and color; see [G-4] **concept: lines**. *linewidthstyle* specifies the line's thickness.

linewidthstyle is specified via options named

⟨*object*⟩⟨`l` or `li` or `line`⟩`width()`

or, just

⟨`l` or `li` or `line`⟩`width()`

For instance, for connecting lines (the lines used to connect points in a plot) used by `graph twoway function`, the option is named `lwidth()`:

```
. twoway function ..., lwidth(linewidthstyle) ...
```

Sometimes you will see that a *linewidthstylelist* is allowed:

```
. twoway line ..., lwidth(linewidthstylelist) ...
```

A *linewidthstylelist* is a sequence of *linewidths* separated by spaces. Shorthands are allowed to make specifying the list easier; see [G-4] ***stylelists***.

Remarks

If you specify the line width as `none`, the line will vanish.

Also see

[G-4] **concept: lines** — Using lines

[G-4] ***linepatternstyle*** — Choices for whether lines are solid, dashed, etc.

[G-4] ***colorstyle*** — Choices for color

[G-4] ***linestyle*** — Choices for overall look of lines

[G-4] ***connectstyle*** — Choices for how points are connected

Title

[G-4] ***marginstyle*** — Choices for size of margins

Syntax

marginstyle	Description	
`zero`	no margin	
`tiny`	tiny margin, all four sides	(smallest)
`vsmall`		
`small`		
`medsmall`		
`medium`		
`medlarge`		
`large`		
`vlarge`	very large margin, all four sides	(largest)
`bottom`	`medium` on the bottom	
`top`	`medium` on the top	
`top_bottom`	`medium` on bottom and top	
`left`	`medium` on the left	
`right`	`medium` on the right	
`sides`	`medium` on left and right	
# # # #	specified margins; left, right, bottom, top	
marginexp	specified margin or margins	

where *marginexp* is one or more elements of the form

{ `l` | `r` | `b` | `t` } [*space*] [`+` | `-` | `=`]#

such as

```
l=5
l=5 r=5
l+5
l+5 r=7.2  b-2 t+1
```

In both the *# # # #* syntax and the { `l` | `r` | `b` | `t` } [`+` | `-` | `=`]# syntax, # is interpreted as a percentage of the minimum of the width and height of the graph. Thus a distance of 5 is the same in both the vertical and horizontal directions.

When you apply margins to rotated textboxes, the terms *left*, *right*, *bottom*, and *top* refer to the box before rotation; see [G-3] ***textbox_options***.

Other *marginstyles* may be available; type

```
. graph query marginstyle
```

to obtain the complete list of *marginstyles* installed on your computer. If other *marginstyles* do exist, they are merely names associated with *# # # #* margins.

Description

marginstyle is used to specify margins (areas to be left unused).

Remarks

marginstyle is used, for instance, in the `margin()` suboption of `title()`:

```
. graph ..., title("My title", margin(marginstyle)) ...
```

marginstyle specifies the margin between the text and the borders of the textbox that will contain the text (which box will ultimately be placed on the graph). See [G-3] ***title_options*** and [G-3] ***textbox_options***.

As another example, *marginstyle* is allowed by the `margin()` suboption of `graphregion()`:

```
. graph ..., graphregion(margin(marginstyle)) ...
```

It allows you to put margins around the plot region within the graph. See *Controlling the aspect ratio* in [G-3] ***region_options*** for an example.

Also see

[G-3] ***textbox_options*** — Options for textboxes and concept definition

[G-3] ***region_options*** — Options for shading and outlining regions and controlling graph size

Title

[G-4] *markerlabelstyle* — Choices for overall look of marker labels

Syntax

markerlabelstyle	Description
`p1–p15`	used by first to fifteenth plot
`p1box–p15box`	used by first to fifteenth "box" plot

Other *markerlabelstyles* may be available; type

```
. graph query markerlabelstyle
```

to obtain the complete list of *markerlabelstyles* installed on your computer.

Description

markerlabelstyle defines the position, gap, angle, size, and color of the marker label. See [G-3] ***marker_label_options*** for more information.

markerlabelstyle is specified in the `mlabstyle()` option,

```
. graph ..., mlabstyle(markerlabelstyle) ...
```

Sometimes (for example, with `twoway scatter`), a *markerlabelstylelist* is allowed: a *markerlabelstylelist* is a sequence of *markerlabelstyles* separated by spaces. Shorthands are allowed to make specifying the list easier; see [G-4] ***stylelists***.

Remarks

Remarks are presented under the following headings:

What is a markerlabel?
What is a markerlabelstyle?
You do not need to specify a markerlabelstyle
Specifying a markerlabelstyle can be convenient
What are numbered styles?

What is a markerlabel?

A marker label is identifying text that appears next to (or in place of) a marker. Markers are the ink used to mark where points are on a plot.

What is a markerlabelstyle?

The look of marker labels is defined by four attributes:

1. the marker label's position—where the marker is located relative to the point; see [G-4] ***clockposstyle***
2. the gap between the marker label and the point; see [G-4] ***clockposstyle***
3. the angle at which the identifying text is presented; see [G-4] ***anglestyle***
4. the overall style of the text; see [G-4] ***textstyle***
   a. the size of the text; see [G-4] ***textsizestyle***
   b. the color of the text; see [G-4] ***colorstyle***

The *markerlabelstyle* specifies all four of these attributes.

You do not need to specify a markerlabelstyle

The *markerlabelstyle* is specified by the option

`mstyle(`*markerlabelstyle*`)`

Correspondingly, you will find other options available:

`mlabposition(`*clockposstyle*`)`
`mlabgap(`*relativesize*`)`
`mlabangle(`*anglestyle*`)`
`mlabtextstyle(`*textstyle*`)`
`mlabsize(`*textstyle*`)`
`mlabcolor(`*colorstyle*`)`

You specify the *markerlabelstyle* when a style exists that is exactly what you want or when another style would allow you to specify fewer changes to obtain what you want.

Specifying a markerlabelstyle can be convenient

Consider the command

```
. scatter y1 y2 x, mlabel(country country)
```

Assume that you want the marker labels for `y2` versus `x` to appear the same as for `y1` versus `x`. (An example of this can be found under *Eliminating overprinting and overruns* and under *Advanced use* in [G-3] ***marker_label_options***.) You might set all the attributes for the marker labels for `y1` versus `x` and then set all the attributes for `y2` versus `x` to be the same. It would be easier, however, to type

```
. scatter y1 y2 x, mlabel(country country) mlabstyle(p1 p1)
```

When you do not specify `mlabstyle()`, results are the same as if you specified

`mlabstyle(p1 p2 p3 p4 p5 p6 p7 p8 p9 p10 p11 p12 p13 p14 p15)`

where the extra elements are ignored. In any case, `p1` is one set of marker-label attributes, `p2` is another set, and so on.

Say that you wanted y2 versus x to look like y1 versus x, except that you wanted the line to be green; you could type

```
. scatter y1 y2 x, mlabel(country country) mlabstyle(p1 p1)
                    mlabcolor(. green)
```

There is nothing special about *markerlabelstyles* p1, p2, . . . ; they merely specify sets of marker-label attributes, just like any other named *markerlabelstyle*. Type

```
. graph query markerlabelstyle
```

to find out what other marker-label styles are available.

Also see *Appendix: Styles and composite styles* in [G-2] **graph twoway scatter** for more information.

What are numbered styles?

p1–p15 are the default styles for marker labels in twoway graphs that support marker labels, for example, twoway scatter, twoway dropline, and twoway connected. p1 is used for the first plot, p2 for the second, and so on.

p1box–p15box are the default styles used for markers showing the outside values on box charts. p1box is used for the outside values on the first set of boxes, p2box for the second set, and so on.

The "look" defined by a numbered style, such as p1 or p3box—by look we include such things as text color, text size, and position around marker—is determined by the scheme (see [G-4] **schemes intro**) selected.

Numbered styles provide default looks that can be controlled by a scheme. They can also be useful when you wish to make, say, the second set of labels on a graph look like the first. See *Specifying a markerlabelstyle can be convenient* above for an example.

Also see

[G-3] ***marker_label_options*** — Options for specifying marker labels

Title

[G-4] *markersizestyle* — Choices for the size of markers

Syntax

markersizestyle	Description
`vtiny`	the smallest
`tiny`	
`vsmall`	
`small`	
`medsmall`	
`medium`	
`medlarge`	
`large`	
`vlarge`	
`huge`	
`vhuge`	
`ehuge`	the largest
relativesize	any size you want, including size modification

See [G-4] ***relativesize***.

Other *markersizestyles* may be available; type

```
. graph query markersizestyle
```

to obtain the complete list of *markersizestyles* installed on your computer.

Description

Markers are the ink used to mark where points are on a plot; see [G-3] ***marker_options***. *markersizestyle* specifies the size of the markers.

Remarks

markersizestyle is specified inside the `msize()` option:

```
. graph ..., msize(markersizestyle) ...
```

Sometimes you will see that a *markersizestylelist* is allowed:

```
. scatter ..., msymbol(markersizestylelist) ...
```

A *markersizestylelist* is a sequence of *markersizestyles* separated by spaces. Shorthands are allowed to make specifying the list easier; see [G-4] ***stylelists***.

Also see

[G-3] ***marker_options*** — Options for specifying markers

[G-4] ***symbolstyle*** — Choices for the shape of markers

[G-4] ***colorstyle*** — Choices for color

[G-4] ***linewidthstyle*** — Choices for thickness of lines

[G-4] ***linepatternstyle*** — Choices for whether lines are solid, dashed, etc.

[G-4] ***linestyle*** — Choices for overall look of lines

[G-4] ***markerstyle*** — Choices for overall look of markers

Title

[G-4] *markerstyle* — Choices for overall look of markers

Syntax

markerstyle	Description
p1–p15	used by first to fifteenth "scatter" plot
p1box–p15box	used by first to fifteenth "box" plot
p1dot–p15dot	used by first to fifteenth "dot" plot

Other *markerstyles* may be available; type

```
. graph query markerstyle
```

to obtain the full list installed on your computer.

Description

Markers are the ink used to mark where points are on a plot. *markerstyle* defines the symbol, size, and color of a marker. See [G-3] ***marker_options*** for more information.

markerstyle is specified in the `mstyle()` option,

```
. graph ..., mstyle(markerstyle) ...
```

Sometimes you will see that a *markerstylelist* is allowed:

```
. twoway scatter ..., mstyle(markerstylelist) ...
```

A *markerstylelist* is a sequence of *markerstyles* separated by spaces. Shorthands are allowed to make specifying the list easier; see [G-4] ***stylelists***.

Remarks

Remarks are presented under the following headings:

What is a marker?
What is a markerstyle?
You do not have to specify a markerstyle
Specifying a markerstyle can be convenient
What are numbered styles?

What is a marker?

Markers are the ink used to mark where points are on a plot. Some people use the word *point* or *symbol*, but a point is where the marker is placed, and a symbol is merely one characteristic of a marker.

What is a markerstyle?

Markers are defined by five attributes:

1. *symbol*—the shape of the marker; see [G-4] ***symbolstyle***
2. *markersize*—the size of the marker; see [G-4] ***markersizestyle***
3. overall color of the marker; see [G-4] ***colorstyle***
4. interior (fill) color of the marker; see [G-4] ***colorstyle***
5. the line that outlines the shape of the marker:
   a. the overall style of the line; see [G-4] ***linestyle***
   b. the thickness of the line; see [G-4] ***linewidthstyle***
   c. the color of the line; see [G-4] ***colorstyle***

The *markerstyle* defines all five (seven) of these attributes.

You do not have to specify a markerstyle

The *markerstyle* is specified via the

`mstyle(`*markerstyle*`)`

option. Correspondingly, you will find seven other options available:

`msymbol(`*symbolstyle*`)`
`msize(`*markersizestyle*`)`
`mcolor(`*colorstyle*`)`
`mfcolor(`*colorstyle*`)`
`mlstyle(`*linestyle*`)`
`mlwidth(`*linewidthstyle*`)`
`mlcolor(`*colorstyle*`)`

You specify the *markerstyle* when a style exists that is exactly what you want or when another style would allow you to specify fewer changes to obtain what you want.

Specifying a markerstyle can be convenient

Consider the command

```
. scatter y1var y2var xvar
```

Say that you wanted the markers for `y2var` versus `xvar` to be the same as `y1var` versus `xvar`. You might set all the characteristics of the marker for `y1var` versus `xvar` and then set all the characteristics of the marker for `y2var` versus `xvar` to be the same. It would be easier, however, to type

```
. scatter y1var y2var xvar, mstyle(p1 p1)
```

`mstyle()` is the option that specifies the overall style of the marker. When you do not specify the `mstyle()` option, results are the same as if you specified

`mstyle(p1 p2 p3 p4 p5 p6 p7 p8 p9 p10 p11 p12 p13 p14 p15)` where the extra elements are ignored. In any case, p1 is one set of marker characteristics, p2 another, and so on.

Say that you wanted y2var versus xvar to look like y1var versus xvar, except that you wanted the symbols to be green; you could type

```
. scatter y1var y2var xvar, mstyle(p1 p1) mcolor(. green)
```

There is nothing special about the *markerstyles* p1, p2, ...; they merely specify sets of marker attributes just like any other named *markerstyle*. Type

```
. graph query markerstyle
```

to find out what other marker styles are available. You may find something pleasing, and if so, that is more easily specified than each of the individual options to modify the shape, color, size, ... elements.

What are numbered styles?

p1–p15 are the default styles for marker labels in twoway graphs that support marker labels, for example, twoway scatter, twoway dropline, and twoway connected. p1 is used for the first plot, p2 for the second, and so on.

p1box–p15box are the default styles used for markers showing the outside values on box charts. p1box is used for the outside values on the first set of boxes, p2box for the second set, and so on.

The "look" defined by a numbered style, such as p1 or p3dot—and by "look" we include such things as color, size, or symbol,—is determined by the scheme (see [G-4] **schemes intro**) selected.

Numbered styles provide default looks that can be controlled by a scheme. They can also be useful when you wish to make, say, the second set of markers on a graph look like the first. See *Specifying a markerstyle can be convenient* above for an example.

Also see

[G-3] ***marker_options*** — Options for specifying markers

[G-4] ***symbolstyle*** — Choices for the shape of markers

[G-4] ***colorstyle*** — Choices for color

[G-4] ***markersizestyle*** — Choices for the size of markers

[G-4] ***linewidthstyle*** — Choices for thickness of lines

[G-4] ***markerstyle*** — Choices for overall look of markers

[G-4] ***linestyle*** — Choices for overall look of lines

[G-4] ***stylelists*** — Lists of style elements and shorthands

Title

[G-4] ***orientationstyle*** — Choices for orientation of textboxes

Syntax

orientationstyle	Description
`horizontal`	text reads left to right
`vertical`	text reads bottom to top
`rhorizontal`	text reads left to right (upside down)
`rvertical`	text reads top to bottom

Other *orientationstyles* may be available; type

```
. graph query orientationstyle
```

to obtain the complete list of *orientationstyles* installed on your computer.

Description

A textbox contains one or more lines of text. *orientationstyle* specifies whether the textbox is horizontal or vertical.

orientationstyle is specified in the `orientation()` option nested within another option, such as `title()`:

```
. graph ..., title("My title", orientation(orientationstyle)) ...
```

See [G-3] ***textbox_options*** for more information on textboxes.

Remarks

orientationstyle specifies whether the text and box are oriented horizontally or vertically, vertically including text reading from bottom to top or from top to bottom.

Also see

[G-3] ***textbox_options*** — Options for textboxes and concept definition

Title

[G-4] *plotregionstyle* — Choices for overall look of plot regions

Syntax

plotregionstyle	Description
`twoway`	default for `graph twoway`
`transparent`	used for overlaid plot regions by `graph twoway`
`bargraph`	default for `graph bar`
`hbargraph`	default for `graph hbar`
`boxgraph`	default for `graph box`
`hboxgraph`	default for `graph hbox`
`dotgraph`	default for `graph dot`
`piegraph`	default for `graph pie`
`matrixgraph`	default for `graph matrix`
`matrix`	`graph matrix` interior region
`matrix_label`	`graph matrix` diagonal labels
`combinegraph`	default for `graph combine`
`combineregion`	`graph combine` interior region
`bygraph`	default for by graphs
`legend_key_region`	key and label region of legends

Other *plotregionstyle* may be available; type

```
. graph query plotregionstyle
```

to obtain the complete list of *plotregionstyles* installed on your computer.

Description

A *plotregionstyle* controls the overall look of a plot region.

Plot region styles are used only in scheme files (see `help scheme files`) and are not accessible from `graph` commands (see [G-2] **graph**). To learn about the `graph` options that affect plot styles, see [G-3] ***region_options***.

Remarks

The look of plot regions is defined by four sets of attributes:

1. *marginstyle*—the internal margin of the plot region; see [G-4] ***marginstyle***
2. overall *areastyle*—the look of the total area of the plot region; see [G-4] ***areastyle***
3. internal *areastyle*—the look of the area within the margin; see [G-4] ***areastyle***
4. positioning—horizontal and vertical positioning of the plot region if the space where the region is located is larger than the plot region itself

A *plotregionstyle* specifies all these attributes.

Also see

[G-4] ***marginstyle*** — Choices for size of margins

[G-4] ***areastyle*** — Choices for look of regions

Title

[G-4] *pstyle* — Choices for overall look of plot

Syntax

pstyle	Description
`ci`	first plot used as confidence interval
`ci2`	second plot used as confidence interval
`p1–p15`	used by first to fifteenth "other" plot
`p1line–p15line`	used by first to fifteenth "line" plot
`p1bar–p15bar`	used by first to fifteenth "bar" plot
`p1box–p15box`	used by first to fifteenth "box" plot
`p1dot–p15dot`	used by first to fifteenth "dot" plot
`p1pie–p15pie`	used by first to fifteenth "pie" plot
`p1area–p15area`	used by first to fifteenth "area" plot
`p1arrow–p15arrow`	used by first to fifteenth "arrow" plot

Other *pstyles* may be available; type

```
. graph query pstyle
```

to obtain the complete list of *pstyles* installed on your computer.

Description

A *pstyle*—always specified in option `pstyle(`*pstyle*`)`—specifies the overall style of a plot and is a composite of *markerstyle*; *markerlabelstyle*; *areastyle*; connected lines, *linestyle*, *connectstyle*; and the *connect_option* `cmissing()`. See [G-4] ***markerstyle***, [G-4] ***markerlabelstyle***, [G-4] ***areastyle***, [G-4] ***linestyle***, [G-4] ***connectstyle***, and [G-3] ***connect_options***.

Remarks

Remarks are presented under the following headings:

What is a plot?
What is a pstyle?
The pstyle( ) option
Specifying a pstyle
What are numbered styles?

What is a plot?

When you type

```
. scatter y x
```

y versus x is called a plot. When you type

```
. scatter y1 x || scatter y2 x
```

or

```
. scatter y1 y2 x
```

y1 versus x is the first plot, and y2 versus x is the second.

A plot is one presentation of a data on a graph.

What is a pstyle?

The overall look of a plot—the *pstyle*—is defined by the following attributes:

1. The look of markers, including their shape, color, size, etc.; see [G-4] ***markerstyle***
2. The look of marker labels, including the position, angle, size, color, etc.; see [G-4] ***marker-labelstyle***
3. The look of lines that are used to connect points, including their color, width, and style (solid, dashed, etc.); see [G-4] ***linestyle***
4. The way points are connected by lines (straight lines, stair step, etc.) if they are connected; see [G-4] ***connectstyle***
5. Whether missing values are ignored or cause lines to be broken when the points are connected
6. The way areas such as bars or beneath or between curves are filled, colored, or shaded, including whether and how they are outlined; see [G-4] ***areastyle***
7. The look of the "dots" in dot plots
8. The look of arrow heads

The *pstyle* specifies these seven attributes.

The pstyle( ) option

The *pstyle* is specified by the option

pstyle(*pstyle*)

Correspondingly, other options are available to control each of the attributes; see, for instance, [G-2] **graph twoway scatter**.

You specify the *pstyle* when a style exists that is exactly what you want or when another style would allow you to specify fewer changes to obtain what you want.

Specifying a pstyle

Consider the command

```
. scatter y1 y2 x, ...
```

and further, assume that many options are specified. Now imagine that you want to make the plot of y1 versus x look just like the plot of y2 versus x: you want the same marker symbols used, the same colors, the same style of connecting lines (if they are connecting), etc. Whatever attributes there are, you want them treated the same.

One solution would be to track down every little detail of how the things that are displayed appear and specify options to make sure that they are specified the same. It would be easier, however, to type

```
. scatter y1 y2 x, ... pstyle(p1 p1)
```

When you do not specify the `pstyle()` option, results are the same as if you specified

```
pstyle(p1 p2 p3 p4 p5 p6 p7 p8 p9 p10 p11 p12 p13 p14 p15)
```

where the extra elements are ignored. In any case, `p1` is one set of plot-appearance values, `p2` is another set, and so on. So when you type

```
. scatter y1 y2 x, ... pstyle(p1 p1)
```

all the appearance values used for `y2` versus `x` are the same as those used for `y1` versus `x`.

Say that you wanted `y2` versus `x` to look like `y1` versus `x`, except that you wanted the markers to be green; you could type

```
. scatter y1 y2 x, ... pstyle(p1 p1) mcolor(. green)
```

There is nothing special about the *pstyles* `p1`, `p2`, ...; they merely specify sets of plot-appearance values just like any other *pstyles*. Type

```
. graph query pstyle
```

to find out what other plot styles are available.

Also see *Appendix: Styles and composite styles* in [G-2] **graph twoway scatter** for more information.

What are numbered styles?

`p1`–`p15` are the default styles for all `twoway` graphs except `twoway line` charts, `twoway bar` charts, and `twoway area` charts. `p1` is used for the first plot, `p2` for the second, and so on.

`p1line`–`p15line` are the default styles used for line charts, including `twoway line` charts and `twoway rline`. `p1line` is used for the first line, `p2line` for the second, and so on.

`p1bar`–`p15bar` are the default styles used for bar charts, including `twoway bar` charts and bar charts. `p1bar` is used for the first set of bars, `p2bar` for the second, and so on.

`p1box`–`p15box` are the default styles used for box charts. `p1box` is used for the first set of boxes, `p2box` for the second, and so on.

`p1dot`–`p15dot` are the default styles used for dot charts. `p1dot` is used for the first set of dots, `p2dot` for the second, and so on.

`p1pie`–`p15pie` are the default styles used for pie charts. `p1pie` is used for the first pie slice, `p2pie` for the second, and so on.

`p1area`–`p15area` are the default styles used for area charts, including `twoway area` charts and `twoway rarea`. `p1area` is used for the first filled area, `p2area` for the second, and so on.

`p1arrow`–`p15arrow` are the default styles used for arrow plots, including `twoway pcarrow` plots and `twoway pcbarrow`. `p1arrow` is used for the first arrow plot, `p2arrow` for the second, and so on.

The "look" defined by a numbered style, such as `p1bar`, `p3`, or `p2area`, is determined by the scheme (see [G-4] **schemes intro**) selected. By "look" we mean such things as color, width of lines, or patterns used.

Numbered styles provide default looks that can be controlled by a scheme. They can also be useful when you wish to make, say, the second element on a graph look like the first. You can, for example, specify that markers for the second scatter on a scatterplot be drawn with the style of the first scatter by using the option `pstyle(p1 p1)`. See *Specifying a pstyle* above for a more detailed example.

Also see

[G-2] **graph twoway scatter** — Twoway scatterplots

[G-4] ***markerstyle*** — Choices for overall look of markers

[G-4] ***markerlabelstyle*** — Choices for overall look of marker labels

[G-4] ***areastyle*** — Choices for look of regions

[G-4] ***linestyle*** — Choices for overall look of lines

[G-4] ***connectstyle*** — Choices for how points are connected

[G-3] ***connect_options*** — Options for connecting points with lines

Title

[G-4] ***relativesize*** — Choices for sizes of objects

Syntax

relativesize	Description
#	specify size; size 100 = minimum of width and height of graph; # must be ≥ 0, depending on context
*#	specify size change via multiplication; `*1` means no change, `*2` twice as large, `*.5` half; # must be ≥ 0, depending on context

Negative sizes are allowed in certain contexts, such as for gaps; in other cases, such as the size of symbol, the size must be nonnegative, and negative sizes, if specified, are ignored.

Examples:

example	Description
`msize(2)`	make marker diameter 2% of g
`msize(1.5)`	make marker diameter 1.5% of g
`msize(.5)`	make marker diameter .5% of g
`msize(*2)`	make marker size twice as large as default
`msize(*1.5)`	make marker size 1.5 times as large as default
`msize(*.5)`	make marker size half as large as default
`xsca(titlegap(2))`	make gap 2% of g
`xsca(titlegap(.5))`	make gap .5% of g
`xsca(titlegap(-2))`	make gap −2% of g
`xsca(titlegap(-.5))`	make gap −.5% of g
`xsca(titlegap(*2))`	make gap twice as large as default
`xsca(titlegap(*.5))`	make gap half as large as default
`xsca(titlegap(*-2))`	make gap −2 times as large as default
`xsca(titlegap(*-.5))`	make gap −.5 times as large as default

where g = min(width of graph, height of graph)

Description

A *relativesize* specifies a size relative to the graph (or subgraph) being drawn. Thus as the size of the graph changes, so does the size of the object.

Remarks

relativesize is allowed, for instance, as a *textsizestyle* or a *markersizestyle*—see [G-4] ***textsizestyle*** and [G-4] ***markersizestyle***—and as the size of many other things, as well.

Relative sizes are not restricted to being integers; relative sizes of .5, 1.25, 15.1, etc., are allowed.

Also see

[G-4] ***markersizestyle*** — Choices for the size of markers

[G-4] ***textsizestyle*** — Choices for the size of text

Title

[G-4] *ringposstyle* — Choices for location: Distance from plot region

Syntax

ringposstyle is

#	$0 \leq \# \leq 100$, # real

Description

ringposstyle is specified inside options such as `ring()` and is typically used in conjunction with *clockposstyle* (see [G-4] ***clockposstyle***) to specify a position for titles, subtitles, etc.

Remarks

See *Positioning of titles* under *Remarks* of [G-3] ***title_options***.

Also see

[G-3] ***title_options*** — Options for specifying titles

[G-4] ***clockposstyle*** — Choices for location: Direction from central point

Title

[G-4] schemes intro — Introduction to schemes

Syntax

`set scheme` *schemename* [, `permanently`]

`graph` ... [, ... `scheme(`*schemename*`)` ...]

schemename	Foreground	Background	Description
s2color	color	white	**factory setting**
s2mono	monochrome	white	s2color in monochrome
s2manual	monochrome	white	used in the Stata manuals
s2gmanual	monochrome	white	used in this manual
s2gcolor	color	white	s2gmanual in color
s1rcolor	color	black	a plain look on black background
s1color	color	white	a plain look
s1mono	monochrome	white	a plain look in monochrome
s1manual	monochrome	white	a plain look, but smaller; used in some Stata manuals
economist	color	white	*The Economist* magazine
sj	monochrome	white	*Stata Journal*

See [G-4] **scheme s1**, [G-4] **scheme s2**, [G-4] **scheme economist**, and [G-4] **scheme sj**.

Other *schemenames* may be available; type

```
. graph query, schemes
```

to obtain the complete list of schemes installed on your computer.

Description

A scheme specifies the overall look of the graph.

`set scheme` sets the default scheme; see [G-2] **set scheme** for more information about this command.

Option `scheme()` specifies the graphics scheme to be used with this particular `graph` command without changing the default.

Remarks

Remarks are presented under the following headings:

The role of schemes
Finding out about other schemes
Setting your default scheme
The scheme is applied at display time
Background color
Foreground color
Obtaining new schemes

See `help scheme files` for a discussion of how to create your own schemes.

The role of schemes

When you type, for instance,

```
. scatter yvar xvar
```

results are the same as if you typed

```
. scatter yvar xvar, scheme(your_default_scheme)
```

If you have not used the `set scheme` command to change your default scheme, *your_default_scheme* is `s2color`.

The scheme specifies the overall look for the graph, and by that we mean just about everything you can imagine. It determines such things as whether y axes are on the left or the right, how many values are by default labeled on the axes, and the colors that are used. In fact, almost every statement made in other parts of this manual stating how something appears, or the relationship between how things appear, must not be taken too literally. How things appear is in fact controlled by the scheme:

- In [G-4] ***symbolstyle***, we state that markers—the ink that denotes the position of points on a plot—have a default size of `msize(medium)` and that small symbols have a size of `msize(small)`. That is generally true, but the size of the markers is in fact set by the scheme, and a scheme might specify different default sizes.
- In [G-3] ***axis_choice_options***, we state that when there is one "y axis", which appears on the left, and when there are two, the second appears on the right. What is in fact true is that where axes appear is controlled by the scheme and that most schemes work the way described. Another scheme named `economist`, however, displays things differently.
- In [G-3] ***title_options***, we state where the titles, subtitles, etc., appear, and we provide a diagram so that there can be no confusion. But where the titles, subtitles, etc., appear is in fact controlled by the scheme, and what we have described is what is true for the scheme named `s2color`.

The list goes on and on. If it has to do with the look of the result, it is controlled by the scheme.

To understand just how much difference the scheme can make, you should type

```
. scatter yvar xvar, scheme(economist)
```

`scheme(economist)` specifies a look similar to that used by *The Economist* magazine (http://www.economist.com), whose graphs we believe to be worthy of emulation. By comparison with the `s2color` scheme, the `economist` scheme moves y axes to the right, makes titles left justified, defaults grid lines to be on, sets a background color, and moves the note to the top right and expects it to be a number.

Finding out about other schemes

A list of schemes is provided in the syntax diagram above, but do not rely on the list being up to date. Instead, type

```
. graph query, schemes
```

to obtain the complete list of schemes installed on your computer.

Try drawing a few graphs with each:

```
. graph ..., ... scheme(schemename)
```

Setting your default scheme

If you want to set your default scheme to, say, `economist`, type

```
. set scheme economist
```

The `economist` scheme will now be your default scheme for the rest of this session, but the next time you use Stata, you will be back to using your old default scheme. If you type

```
. set scheme economist, permanently
```

`economist` will become your default scheme both the rest of this session and in future sessions.

If you want to change your scheme back to `s2color`—the default scheme in Stata as originally shipped—type

```
. set scheme s2, permanently
```

See [G-2] **set scheme**.

The scheme is applied at display time

Say that you type

```
. graph mpg weight, saving(mygraph)
```

to create and save the file `mygraph.gph` (see [G-3] ***saving_option***). If later you redisplay the graph by typing

```
. graph use mygraph
```

the graph will reappear as you originally drew it. It will be displayed using the same scheme with which it was originally drawn, regardless of your current `set scheme` setting. If you type

```
. graph use mygraph, scheme(economist)
```

the graph will be displayed using the `economist` scheme. It will be the same graph but will look different. You can change the scheme with which a graph is drawn beforehand, on the original `graph` command, or later.

Background color

In the table at the beginning of the entry, we categorized the background color as being white or black, although actually what we mean is light or dark because some of the schemes set background tinting. We mean that "white" background schemes are suitable for printing. Printers (both the mechanical ones and the human ones) prefer that you avoid dark backgrounds because of the large amounts of ink required and the corresponding problems with bleed-through. On the other hand, dark backgrounds look good on monitors.

In any case, you may change the background color of a scheme by using the *region_options* `graphregion(fcolor())`, `graphregion(ifcolor())`, `plotregion(fcolor())`, and `plotregion(ifcolor())`; see [G-3] ***region_options***. When overriding the background color, choose light colors for schemes that naturally have white backgrounds and dark colors for regions that naturally have black backgrounds.

Schemes that naturally have a black background are by default printed in monochrome. See [G-2] **set printcolor** if you wish to override this.

If you are producing graphs for printing on white paper, we suggest that you choose a scheme with a naturally white background.

Foreground color

In the table at the beginning of this entry, we categorized the foreground as being color or monochrome. This refers to whether lines, markers, fills, etc., are presented by default in color or monochrome. Regardless of the scheme you choose, you can specify options such as `mcolor()` and `lcolor()`, to control the color for each item on the graph.

Just because we categorized the foreground as monochrome, this does not mean you cannot specify colors in the options.

Obtaining new schemes

Your copy of Stata may already have schemes other than those documented in this manual. To find out, type

```
. graph query, schemes
```

Also, new schemes are added and existing schemes updated along with all the rest of Stata, so if you are connected to the Internet, type

```
. update query
```

and follow any instructions given; see [R] **update**.

Finally, other users may have created schemes that could be of interest to you. To search the Internet, type

```
. findit scheme
```

From there, you will be able to click to install any schemes that interest you; see `findit` in [R] **search**.

Once a scheme is installed, which can be determined by verifying that it appears in the list shown by

```
. graph query, schemes
```

you can use it with the `scheme()` option

```
. graph ..., ... scheme(newscheme)
```

or you can set it as your default, temporarily

```
. set scheme newscheme
```

or permanently

```
. set scheme newscheme, permanently
```

Also see

[G-3] ***scheme_option*** — Option for specifying scheme

[G-2] **set scheme** — Set default scheme

[G-4] **scheme economist** — Scheme description: economist

[G-4] **scheme s1** — Scheme description: s1 family

[G-4] **scheme s2** — Scheme description: s2 family

[G-4] **scheme sj** — Scheme description: sj

Title

[G-4] scheme economist — Scheme description: economist

Syntax

schemename	Foreground	Background	Description
`economist`	color	white	*The Economist* magazine

For instance, you might type

```
. graph ..., ... scheme(economist)
. set scheme economist [, permanently ]
```

See [G-3] ***scheme_option*** and [G-2] **set scheme**.

Description

Schemes determine the overall look of a graph; see [G-4] **schemes intro**.

Scheme `economist` specifies a look similar to that used by *The Economist* magazine.

Remarks

The Economist magazine (http://www.economist.com) uses a unique and clean graphics style that is both worthy of emulation and different enough from the usual to provide an excellent example of just how much difference the scheme can make.

Among other things, *The Economist* puts the y axis on the right rather than on the left of scatterplots.

Also see

[G-4] **schemes intro** — Introduction to schemes

[G-3] ***scheme_option*** — Option for specifying scheme

[G-2] **set scheme** — Set default scheme

Title

[G-4] scheme s1 — Scheme description: s1 family

Syntax

s1 family	Foreground	Background	Description
s1rcolor	color	black	color on black
s1color	color	white	color on white
s1mono	monochrome	white	gray on white
s1manual	monochrome	white	s1mono, but smaller; used in some Stata manuals

For instance, you might type

```
. graph ..., ... scheme(s1color)
. set scheme s1rcolor [, permanently ]
```

See [G-3] ***scheme_option*** and [G-2] **set scheme**.

Description

Schemes determine the overall look of a graph; see [G-4] **schemes intro**.

The s1 family of schemes is similar to the s2 family—see [G-4] **scheme s2**—except that s1 uses a plain background, meaning that no tint is applied to any part of the background.

Remarks

s1 is a conservative family of schemes that some people prefer to s2.

Of special interest is s1rcolor, which displays graphs on a black background. Because of pixel bleeding, monitors have higher resolution when backgrounds are black rather than white. Also, many users experience less eye strain viewing graphs on a monitor when the background is black. Scheme s1rcolor looks good when printed, but other schemes look better.

Schemes s1color and s1mono are derived from s1rcolor. Either of these schemes will deliver a better printed result. The important difference between s1color and s1mono is that s1color uses solid lines of different colors to connect points, whereas s1mono varies the line-pattern style.

Scheme s1manual is used in some of the Stata manuals, although it is not used in this one. s1manual is the same as s1mono but presents graphs at a smaller overall size.

Also see

[G-4] **schemes intro** — Introduction to schemes

[G-3] ***scheme_option*** — Option for specifying scheme

Title

[G-4] scheme s2 — Scheme description: s2 family

Syntax

schemename	Foreground	Background	Description
s2color	color	white	factory setting
s2mono	monochrome	white	s2color in monochrome
s2manual	monochrome	white	used in the Stata manuals
s2gmanual	monochrome	white	used in this manual
s2gcolor	color	white	s2gmanual in color

For instance, you might type

```
. graph ..., ... scheme(s1mono)
. set scheme s2mono [, permanently ]
```

See [G-3] ***scheme_option*** and [G-2] **set scheme**.

Description

Schemes determine the overall look of a graph; see [G-4] **schemes intro**.

The s2 family of schemes is Stata's default scheme.

Remarks

s2 is the family of schemes that we like for displaying data. It provides a light background tint to give the graph better definition and make it visually more appealing. On the other hand, if you feel the tinting distracts from the graph, see [G-4] **scheme s1**; the s1 family is nearly identical to s2 but does away with the extra tinting.

In particular, we recommend that you consider scheme s1rcolor; see [G-4] **scheme s1**. s1rcolor uses a black background, and for working at the monitor, it is difficult to find a better choice.

In any case, scheme s2color is Stata's default scheme. It looks good on the screen, good when printed on a color printer, and more than adequate when printed on a monochrome printer.

Scheme s2mono has been optimized for printing on monochrome printers. Also, rather than using the same symbol over and over and varying the color, s2mono will vary the symbol's shape, and in connecting points, s2mono varies the line pattern (s2color varies the color).

Scheme s2gmanual is the scheme used in printing this manual. It is similar to s2manual except that connecting lines are solid and gray scales rather than patterned and black.

Scheme s2gmanual is the scheme used in printing the Stata Graphics manual. It is similar to s2manual except that connecting lines are solid and gray scales rather than patterned and black.

Scheme s2gcolor is the same scheme as s2gmanual except that color is used.

❑ Technical note

The colors used in the s2color scheme were changed slightly after Stata 8 to improve printing on color inkjet printers and printing presses—the amount of cyan in the some colors was reduced to prevent an unintended casting toward purple. You probably will not notice the difference, but if you want the original colors, they are available in the scheme s2color8.

❑

Also see

[G-4] **schemes intro** — Introduction to schemes

[G-3] ***scheme_option*** — Option for specifying scheme

Title

[G-4] scheme sj — Scheme description: sj

Syntax

schemename	Foreground	Background	Description
sj	monochrome	white	*Stata Journal*

For instance, you might type

```
. graph ..., ... scheme(sj)
. set scheme sj [, permanently ]
```

See [G-3] ***scheme_option*** and [G-2] **set scheme**.

Description

Schemes determine the overall look of a graph; see [G-4] **schemes intro**.

Scheme `sj` is the official scheme of the *Stata Journal*; see [R] **sj**.

Remarks

When submitting articles to the *Stata Journal*, graphs should be drawn using the scheme `sj`.

Before drawing graphs for inclusion with submissions, make sure that scheme `sj` is up to date. Schemes are updated along with all the rest of Stata, so you just need to type

```
. update query
```

and follow any instructions given; see [R] **update**.

Also visit the *Stata Journal* website for any special instructions. Point your browser to http://www.stata-journal.com.

Also see

[G-4] **schemes intro** — Introduction to schemes

[G-3] ***scheme_option*** — Option for specifying scheme

[G-2] **set scheme** — Set default scheme

[R] **sj** — Stata Journal and STB installation instructions

Title

[G-4] *shadestyle* — Choices for overall look of filled areas

Syntax

shadestyle	Description
`foreground`	areas in the default foreground color
`plotregion`	plot region area
`legend`	legend area
`none`	nonexistent area
`ci`	areas representing confidence intervals
`histogram`	histogram bars
`sunflowerlb`	light sunflowers
`sunflowerdb`	dark sunflowers
`p1bar–p15bar`	used by first to fifteenth "bar" plot
`p1box–p15box`	used by first to fifteenth "box" plot
`p1area–p15area`	used by first to fifteenth "area" plot
`p1pie–p15pie`	used for first to fifteenth pie slices

Other *shadestyle* may be available; type

```
. graph query shadestyle
```

to obtain the complete list of *shadestyles* installed on your computer.

Description

shadestyle sets the color and intensity of the color for a filled area.

Shadestyles are used only in scheme files (see `help scheme files`) and are not accessible from `graph` commands (see [G-2] **graph**).

Remarks

Remarks are presented under the following headings:

What is a shadestyle?
What are numbered styles?

What is a shadestyle?

Shaded areas are defined by two attributes:

1. *colorstyle*—the color of the shaded area; see [G-4] ***colorstyle***
2. *intensity*—the intensity of the color; see [G-4] ***intensitystyle***

The *shadestyle* specifies both of these attributes.

The intensity attribute is not truly necessary because any intensity could be reached by changing the RGB values of a color; see [G-4] ***colorstyle***. An intensity, however, can be used to affect the intensity of many different colors in some scheme files.

What are numbered styles?

`p1bar`–`p15bar` are the default styles used for filling the bars on bar charts, including `twoway bar` charts and bar charts. `p1bar` is used for the first set of bars, `p2bar` for the second, and so on.

`p1box`–`p15box` are the default styles used for filling the boxes on box charts. `p1box` is used for the first set of boxes, `p2box` for the second, and so on.

`p1area`–`p15area` are the default styles used for filling the areas on area charts, including `twoway area` charts and `twoway rarea`. `p1area` is used for the first filled area, `p2area` for the second, and so on.

`p1pie`–`p15pie` are the default styles used for filling pie slices, including pie charts. `p1pie` is used by the first slice, `p2pie` for the second, and so on.

The look defined by a numbered style, such as `p1bar`, `p1box`, or `p1area`, is determined by the scheme (see schemes intro) selected. By "look" we mean *colorstyle* and intensity (see [G-4] ***colorstyle*** and [G-4] ***intensitystyle***).

Also see

[G-4] ***colorstyle*** — Choices for color

[G-4] ***intensitystyle*** — Choices for the intensity of a color

Title

[G-4] *stylelists* — Lists of style elements and shorthands

Syntax

A *stylelist* is a generic list of style elements and shorthands; specific examples of *stylelists* include *symbolstylelist*, *colorstylelist*, etc.

A *stylelist* is

el [*el* [...]]

where each *el* may be

el	Description
as_defined_by_style	what *symbolstyle*, *colorstyle*, ... allows
"*as defined by style*"	must quote *el*s containing spaces
`"*as "defined" by style*"'	compound quote *el*s containing quotes
.	specifies the "default"
=	repeat previous *el*
..	repeat previous *el* until end
...	same as ..

If the list ends prematurely, it is as if the list were padded out with . (meaning the default for the remaining elements).

If the list has more elements than required, extra elements are ignored.

= in the first element is taken to mean . (period).

If the list allows numbers including missing values, if missing value is not the default, and if you want to specify missing value for an element, you must enclose the period in quotes: ".".

Examples:

```
. ..., ... msymbol(O d p o) ...
. ..., ... msymbol(O . p) ...
. ..., ... mcolor(blue . green green) ...
. ..., ... mcolor(blue . green =) ...
. ..., ... mcolor(blue blue blue blue) ...
. ..., ... mcolor(blue = = =) ...
. ..., ... mcolor(blue ...) ...
```

Description

Sometimes an option takes not a *colorstyle* but a *colorstylelist*, or not a *symbolstyle* but a *symbolstylelist*. *colorstyle* and *symbolstyle* are just two examples; there are many styles. Whether an option allows a list is documented in its syntax diagram. For instance, you might see graph matrix ... [, ... mcolor(*colorstyle*) ...]

in one place and

graph twoway scatter ... [, ... mcolor(*colorstylelist*) ...]

in another. In either case, to learn about *colorstyles*, you would see [G-4] ***colorstyle***. Here we have discussed how you would generalize a *colorstyle* into a *colorstylelist* or a *symbolstyle* into a *symbolstylelist*, etc.

Also see

[G-2] **graph twoway** — Twoway graphs

Title

[G-4] *symbolstyle* — Choices for the shape of markers

Syntax

symbolstyle	Synonym (if any)	Description
circle	O	solid
diamond	D	solid
triangle	T	solid
square	S	solid
plus	+	
X	X	
smcircle	o	solid
smdiamond	d	solid
smsquare	s	solid
smtriangle	t	solid
smplus		
smx	x	
circle_hollow	Oh	hollow
diamond_hollow	Dh	hollow
triangle_hollow	Th	hollow
square_hollow	Sh	hollow
smcircle_hollow	oh	hollow
smdiamond_hollow	dh	hollow
smtriangle_hollow	th	hollow
smsquare_hollow	sh	hollow
point	p	a small dot
none	i	a symbol that is invisible

For a symbol palette displaying each of the above symbols, type

palette symbolpalette [, scheme(*schemename*)]

Other *symbolstyles* may be available; type

```
. graph query symbolstyle
```

to obtain the complete list of *symbolstyles* installed on your computer.

Description

Markers are the ink used to mark where points are on a plot; see [G-3] ***marker_options***. *symbolstyle* specifies the shape of the marker.

You specify the *symbolstyle* inside the `msymbol()` option allowed with many of the `graph` commands:

```
. graph twoway ..., msymbol(symbolstyle) ...
```

Sometimes you will see that a *symbolstylelist* is allowed:

```
. scatter ..., msymbol(symbolstylelist) ...
```

A *symbolstylelist* is a sequence of *symbolstyles* separated by spaces. Shorthands are allowed to make specifying the list easier; see [G-4] ***stylelists***.

Remarks

Remarks are presented under the following headings:

Typical use
Filled and hollow symbols
Size of symbols

Typical use

`msymbol(`*symbolstyle*`)` is one of the more commonly specified options. For instance, you may not be satisfied with the default rendition of

```
. scatter mpg weight if foreign ||
  scatter mpg weight if !foreign
```

and prefer

```
. scatter mpg weight if foreign, msymbol(oh) ||
  scatter mpg weight if !foreign, msymbol(x)
```

When you are graphing multiple y variables in the same plot, you can specify a list of *symbolstyles* inside the `msymbol()` option:

```
. scatter mpg1 mpg2 weight, msymbol(oh x)
```

The result is the same as typing

```
. scatter mpg1 weight, msymbol(oh) ||
  scatter mpg2 weight, msymbol(x)
```

Also, in the above, we specified the symbol-style synonyms. Whether you type

```
. scatter mpg1 weight, msymbol(oh) ||
  scatter mpg2 weight, msymbol(x)
```

or

```
. scatter mpg1 weight, msymbol(smcircle_hollow) ||
  scatter mpg2 weight, msymbol(smx)
```

makes no difference.

Filled and hollow symbols

The *symbolstyle* specifies the *shape* of the symbol, and in that sense, one of the styles `circle` and `hcircle`—and `diamond` and `hdiamond`, etc.—is unnecessary in that each is a different rendition of the same shape. The option `mfcolor(`*colorstyle*`)` (see [G-3] ***marker_options***) specifies how the inside of the symbol is to be filled. `hcircle()`, `hdiamond`, etc., are included for convenience and are equivalent to specifying

`msymbol(Oh)`: `msymbol(O) mfcolor(none)`

`msymbol(dh)`: `msymbol(d) mfcolor(none)`

etc.

Using `mfcolor()` to fill the inside of a symbol with different colors sometimes creates what are effectively new symbols. For instance, if you take `msymbol(O)` and fill its interior with a lighter shade of the same color used to outline the shape, you obtain a pleasing result. For instance, you might try

`msymbol(O) mlcolor(yellow) mfcolor(.5*yellow)`

or

`msymbol(O) mlcolor(gs5) mfcolor(gs12)`

as in

```
. scatter mpg weight, msymbol(O) mlcolor(gs5) mfcolor(gs14)
```

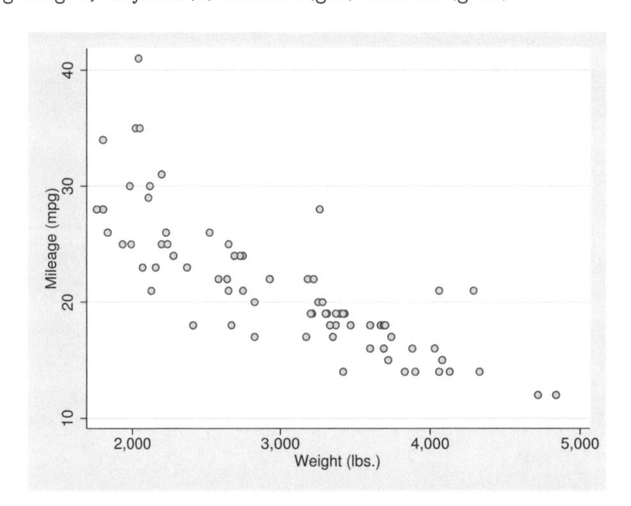

Size of symbols

Just as `msymbol(O)` and `msymbol(Oh)` differ only in `mfcolor()`, `msymbol(O)` and `msymbol(o)`—symbols `circle` and `smcircle`—differ only in `msize()`. In particular,

`msymbol(O)`: `msymbol(O) msize(medium)`

`msymbol(o)`: `msymbol(O) msize(small)`

and the same is true for all the other large and small symbol pairs.

`msize()` is interpreted as being relative to the size of the graph region (see [G-3] ***region_options***), so the same symbol size will in fact be a little different in

```
. scatter mpg weight
```

and

```
. scatter mpg weight, by(foreign total)
```

Also see

[G-3] ***marker_options*** — Options for specifying markers

[G-4] ***markersizestyle*** — Choices for the size of markers

[G-4] ***colorstyle*** — Choices for color

[G-4] ***linepatternstyle*** — Choices for whether lines are solid, dashed, etc.

[G-4] ***linewidthstyle*** — Choices for thickness of lines

[G-4] ***linestyle*** — Choices for overall look of lines

[G-4] ***markerstyle*** — Choices for overall look of markers

Title

[G-4] *text* — Text in graphs

Description

All text elements in Stata graphs support the use of certain SMCL markup directives, or tags, to affect how they appear on the screen. SMCL, which stands for Stata Markup and Control Language and is pronounced "smickle", is Stata's output language and is discussed in detail in [P] **smcl**.

All text output in Stata, including text in graphs, can be modified with SMCL.

For example, you can italicize a word in a graph title:

```
. scatter mpg weight, title("This is {it:italics} in a graph title")
```

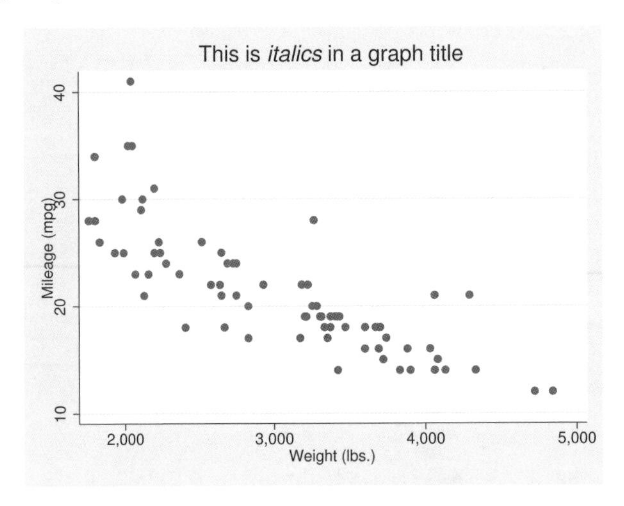

This entry documents the features of SMCL that are unique to graphs. We recommend that you have a basic understanding of SMCL before reading this entry; see [P] **smcl**.

Remarks

Remarks are presented under the following headings:

Overview

Assuming you read [P] **smcl** before reading this entry, you know about the four syntaxes that SMCL tags follow. As a refresher, the syntaxes are

Syntax 1: `{xyz}`

Syntax 2: `{xyz:`*text*`}`

Syntax 3: `{xyz` *args*`}`

Syntax 4: `{xyz` *args*`:`*text*`}`

Syntax 1 means "do whatever it is that `{xyz}` does". Syntax 2 means "do whatever it is that `{xyz}` does, do it on the text *text*, and then stop doing it". Syntax 3 means "do whatever it is that `{xyz}` does, as modified by *args*". Finally, syntax 4 means "do whatever it is that `{xyz}` does, as modified by *args*, do it on the text *text*, and then stop doing it".

Most SMCL tags useful in graph text follow syntax 1 and syntax 2, and one (`{fontface}`) follows syntax 3 and syntax 4.

Bold and italics

Changing text in graphs to **bold** or *italics* is done in exactly the same way as in the Results window. Simply use the SMCL `{bf}` and `{it}` tags:

```
. scatter mpg weight,
> caption("{bf:Source}: {it:Consumer Reports}, used with permission")
```

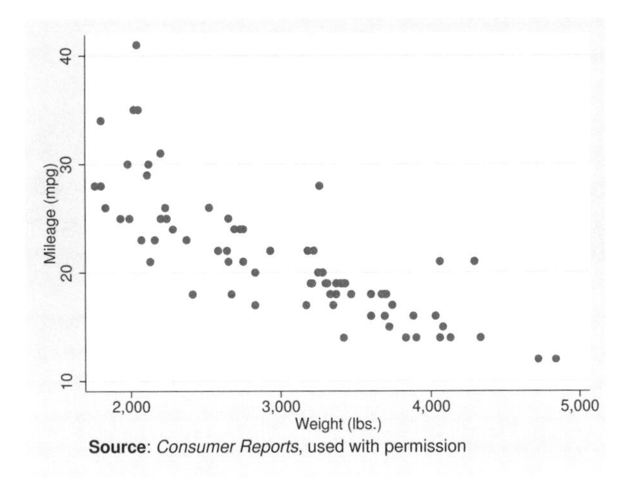

`{bf}` and `{it}` follow syntaxes 1 and 2.

Superscripts and subscripts

You can include superscripts and subscripts in text in graphs. This may surprise you, because it is not possible to do so with text in the Results window. Because graphs are not constrained to use fixed-width fonts and fixed-height lines like output in the Results window, it is possible to allow more features for text in graphs.

It is simple to use the {superscript} and {subscript} tags to cause a piece of text to be displayed as a superscript or a subscript. Here we will plot a function and will change the title of the graph to something appropriate:

```
. twoway function y = 2*exp(-2*x), range(0 2)
> title("{&function}(x)=2e{superscript:-2x}")
```

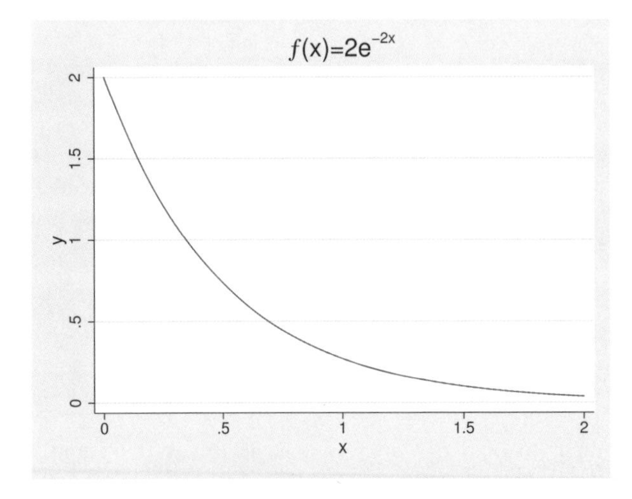

{superscript} and {subscript} follow syntaxes 1 and 2. {sup} and {sub} may be used as shorthand for {superscript} and {subscript}.

The example above also demonstrates the use of a symbol, {&function}; symbols will be discussed in more detail below.

Fonts, standard

Stata provides four standard font faces for graphs to allow text to be displayed in a sans-serif font (the default), a serif font, a monospace (fixed-width) font, or a symbol font. These fonts have been chosen to work across operating systems and in graphs exported to PostScript and Encapsulated PostScript files.

The SMCL tags used to mark text to be displayed in any of these fonts and the fonts that are used on each type of system are shown below:

SMCL	{stSans}	{stSerif}	{stMono}	{stSymbol}
Windows	Arial	Times New Roman	Courier New	Symbol
Mac	Helvetica	Times	Courier	Symbol
Unix	Sans	Serif	Monospace	Sans
PS/EPS	Helvetica	Times	Courier	Symbol

Note: We recommend that you leave in place the mapping from these four SMCL tags to the fonts we have selected for each operating system. However, you may override the default fonts if you wish. See [G-2] **graph set** for details.

Changing fonts within text on a graph is easy:

```
. scatter mpg weight, title("Here are {stSerif:serif},
> {stSans:sans serif}, and {stMono:monospace}")
```

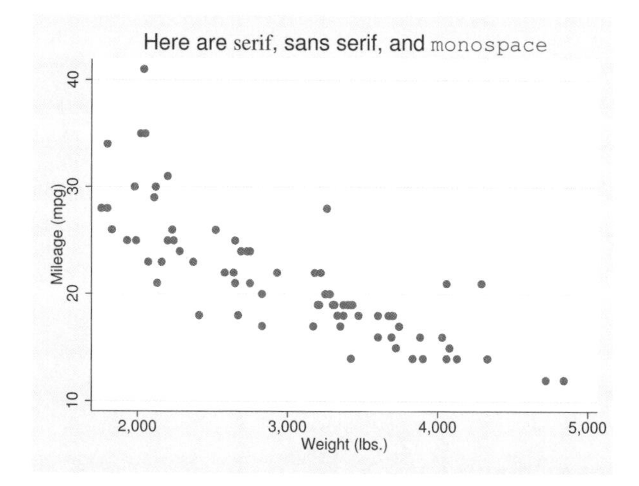

`{stSans}`, `{stSerif}`, `{stMono}`, and `{stSymbol}` follow syntaxes 1 and 2.

The `{stSymbol}` tag lets you display hundreds of different symbols, such as Greek letters and math symbols. There are so many possibilities that symbols have their own shorthand notation to help you type them and have their own section describing how to use them. See *Greek letters and other symbols* below.

Fonts, advanced

In addition to the four standard fonts, you may display text in a graph using any font available on your operating system by using the `{fontface}` tag. If the font face you wish to specify contains spaces in its name, be sure to enclose it in double quotes within the `{fontface}` tag. For example, to display text using a font on your system named "Century Schoolbook", you would type

```
. scatter mpg weight,
> title(`"Text in {fontface "Century Schoolbook":a different font}"')
```

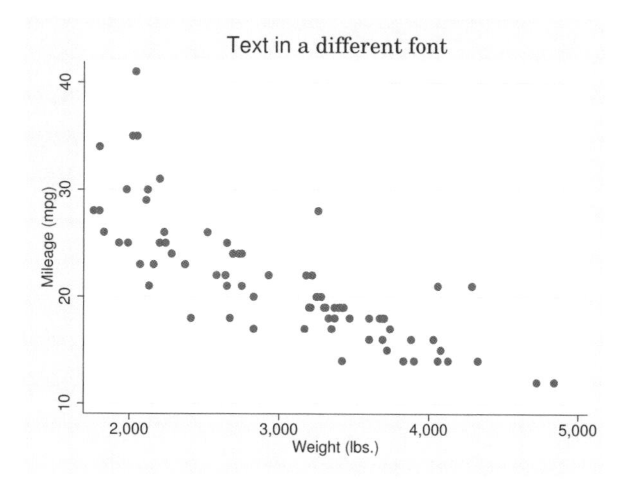

If the font face you specify does not exist on your system, the operating system will substitute another font.

`{fontface}` follows syntaxes 3 and 4.

The four standard fonts may also be specified using the `{fontface}` tag. For example, you can specify the default serif font with `{fontface "stSerif"}`; in fact, `{stSerif}` is shorthand for exactly that.

If you choose to change fonts in graphs by using the `{fontface}` tag, keep in mind that if you share your Stata `.gph` files with other Stata users, they must have the exact same fonts on their system for the graphs to display properly. Also, if you need to export your graphs to PostScript or Encapsulated PostScript files, Stata will have to try to convert your operating system's fonts to PostScript fonts and embed them in the exported file. It is not always possible to properly convert and embed all fonts, which is why we recommend using one of the four standard fonts provided by Stata.

In Stata for Unix, if you use fonts other than the four standard fonts and you wish to export your graphs to PostScript or Encapsulated PostScript files, you may need to specify the directory where your system fonts are located; see [G-3] ***ps_options***.

Greek letters and other symbols

Stata provides support for many symbols in text in graphs, including both capital and lowercase forms of the Greek alphabet and many math symbols.

You may already be familiar with the `{char}` tag—synonym `{c}`—which follows syntax 3 and allows you to output any ASCII character. If not, see *Displaying characters using ASCII code* in [P] **smcl**. All the features of `{char}`, except for the line-drawing characters, may be used in graph text.

Graph text supports even more symbols than `{char}`. For the symbols Stata supports, we have chosen to define SMCL tags with names that parallel HTML character entity references. HTML character entity references have wide usage and, for the most part, have very intuitive names for whatever symbol you wish to display.

In HTML, character entity references are all of the form "`&`*name*`;`", where *name* is supposed to be an intuitive name for the given character entity. In SMCL, the tag for a given character entity is "`{&`*name*`}`".

For example, in HTML, the character reference for a capital Greek Sigma is `Σ`. In SMCL, the tag for a capital Greek Sigma is `{&Sigma}`.

In some cases, the HTML character reference for a particular symbol has a name that is not so intuitive. For example, HTML uses `ƒ` for the "function" symbol (f). SMCL provides `{&fnof}` to match the HTML character reference, as well as the more intuitive `{&function}`.

All SMCL symbol tags follow syntax 1.

See *Full list of SMCL tags useful in graph text* for a complete list of symbols supported by SMCL in graphs.

As an example, we will graph a function and give it an appropriate title:

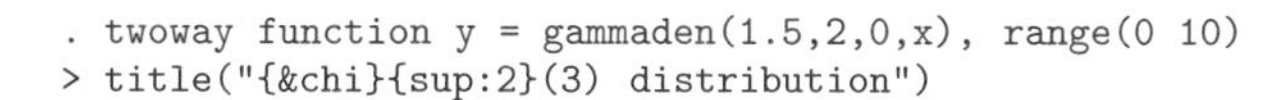

```
. twoway function y = gammaden(1.5,2,0,x), range(0 10)
> title("{&chi}{sup:2}(3) distribution")
```

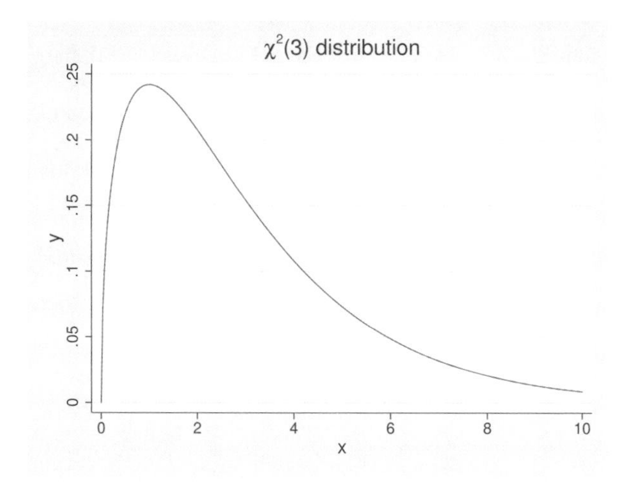

Greek letters and other math symbols are displayed using the `{stSymbol}` font. For example, `{&Alpha}` is equivalent to `{stSymbol:A}`.

Full list of SMCL tags useful in graph text

The SMCL tags that are useful in graph text are the following:

SMCL tag	Description
`{bf}`	Make text bold
`{it}`	Make text italic
`{superscript}`	Display text as a superscript
`{sup}`	Synonym for `{superscript}`
`{subscript}`	Display text as a subscript
`{sub}`	Synonym for `{subscript}`
`{stSans}`	Display text with the default sans serif font
`{stSerif}`	Display text with the default serif font
`{stMono}`	Display text with the default monospace (fixed-width) font
`{stSymbol}`	Display text with the default symbol font
`{fontface "`*fontname*`"}`	Display text with the specified *fontname*
`{char `*code*`}`	Display ASCII character
`{&`*symbolname*`}`	Display a Greek letter, math symbol, or other symbol

The Greek letters supported by SMCL in graph text are the following:

SMCL tag	Symbol	Description
{&Alpha}	A	Capital Greek letter Alpha
{&Beta}	B	Capital Greek letter Beta
{&Gamma}	Γ	Capital Greek letter Gamma
{&Delta}	Δ	Capital Greek letter Delta
{&Epsilon}	E	Capital Greek letter Epsilon
{&Zeta}	Z	Capital Greek letter Zeta
{&Eta}	H	Capital Greek letter Eta
{&Theta}	Θ	Capital Greek letter Theta
{&Iota}	I	Capital Greek letter Iota
{&Kappa}	K	Capital Greek letter Kappa
{&Lambda}	Λ	Capital Greek letter Lambda
{&Mu}	M	Capital Greek letter Mu
{&Nu}	N	Capital Greek letter Nu
{&Xi}	Ξ	Capital Greek letter Xi
{&Omicron}	O	Capital Greek letter Omicron
{&Pi}	Π	Capital Greek letter Pi
{&Rho}	R	Capital Greek letter Rho
{&Sigma}	Σ	Capital Greek letter Sigma
{&Tau}	T	Capital Greek letter Tau
{&Upsilon}	Υ	Capital Greek letter Upsilon
{&Phi}	Φ	Capital Greek letter Phi
{&Chi}	X	Capital Greek letter Chi
{&Psi}	Ψ	Capital Greek letter Psi
{&Omega}	Ω	Capital Greek letter Omega
{&alpha}	α	Lowercase Greek letter alpha
{&beta}	β	Lowercase Greek letter beta
{&gamma}	γ	Lowercase Greek letter gamma
{&delta}	δ	Lowercase Greek letter delta
{&epsilon}	ϵ	Lowercase Greek letter epsilon
{&zeta}	ζ	Lowercase Greek letter zeta
{&eta}	η	Lowercase Greek letter eta
{&theta}	θ	Lowercase Greek letter theta
{&thetasym}	ϑ	Greek theta symbol
{&iota}	ι	Lowercase Greek letter iota
{&kappa}	κ	Lowercase Greek letter kappa
{&lambda}	λ	Lowercase Greek letter lambda
{&mu}	μ	Lowercase Greek letter mu
{&nu}	ν	Lowercase Greek letter nu
{&xi}	ξ	Lowercase Greek letter xi
{&omicron}	o	Lowercase Greek letter omicron
{&pi}	π	Lowercase Greek letter pi
{&piv}	ϖ	Greek pi symbol
{&rho}	ρ	Lowercase Greek letter rho
{&sigma}	σ	Lowercase Greek letter sigma
{&sigmaf}	ς	Greek 'final' sigma symbol
{&tau}	τ	Lowercase Greek letter tau

SMCL tag	Symbol	Description
`{&upsilon}`	υ	Lowercase Greek letter upsilon
`{&upsih}`	Υ	Greek upsilon with a hook symbol
`{&phi}`	ϕ	Lowercase Greek letter phi
`{&chi}`	χ	Lowercase Greek letter chi
`{&psi}`	ψ	Lowercase Greek letter psi
`{&omega}`	ω	Lowercase Greek letter omega

Math symbols supported by SMCL in graph text are the following:

SMCL tag	Symbol	Description
`{&weierp}`	$\wp$	Weierstrass p, power set
`{&image}`	$\Im$	Imaginary part
`{&imaginary}`		Synonym for `{&image}`
`{&real}`	$\Re$	Real part
`{&alefsym}`	$\aleph$	Alef, first transfinite cardinal
`{&}`	&	Ampersand
`{<}`	$<$	Less than
`{>}`	$>$	Greater than
`{&le}`	$\leq$	Less than or equal to
`{&ge}`	$\geq$	Greater than or equal to
`{&ne}`	$\neq$	Not equal to
`{&fnof}`	f	Function
`{&function}`		Synonym for `{&fnof}`
`{&forall}`	$\forall$	For all
`{&part}`	∂	Partial differential
`{&exist}`	$\exists$	There exists
`{&empty}`	$\emptyset$	Empty set, null set, diameter
`{&nabla}`	∇	Nabla, backward difference
`{&isin}`	$\in$	Element of
`{&element}`		Synonym for `{&isin}`
`{¬in}`	$\notin$	Not an element of
`{&prod}`	$\prod$	N-ary product, product sign
`{&sum}`	$\sum$	N-ary summation
`{&minus}`	$-$	Minus sign
`{±}`	$\pm$	Plus-or-minus sign
`{&plusminus}`		Synonym for `{±}`
`{&lowast}`	$\ast$	Asterisk operator
`{&radic}`	$\surd$	Radical sign, square root
`{&sqrt}`		Synonym for `{&radic}`
`{&prop}`	$\propto$	Proportional to
`{&infin}`	∞	Infinity
`{&infinity}`		Synonym for `{&infin}`
`{&ang}`	$\angle$	Angle
`{&angle}`		Synonym for `{&ang}`
`{&and}`	$\wedge$	Logical and, wedge
`{&or}`	$\vee$	Logical or, vee

SMCL tag	Symbol	Description
{&cap}	∩	Intersection, cap
{&intersect}		Synonym for {&cap}
{&cup}	∪	Union, cup
{&union}		Synonym for {&cup}
{&int}	∫	Integral
{&integral}		Synonym for {&int}
{&there4}	∴	Therefore
{&therefore}		Synonym for {&there4}
{&sim}	∼	Tilde operator, similar to
{&cong}	≅	Approximately equal to
{&asymp}	≍	Almost equal to, asymptotic to
{&equiv}	≡	Identical to
{&sub}	⊂	Subset of
{&subset}		Synonym for {&sub}
{&sup}	⊃	Superset of
{&superset}		Synonym for {&sup}
{&nsub}	⊄	Not a subset of
{&nsubset}		Synonym for {&nsub}
{&sube}	⊆	Subset of or equal to
{&subsete}		Synonym for {&sube}
{&supe}	⊇	Superset of or equal to
{&supersete}		Synonym for {&supe}
{&oplus}	⊕	Circled plus, direct sum
{&otimes}	⊗	Circled times, vector product
{&perp}	⊥	Perpendicular, orthogonal to, uptack
{&orthog}		Synonym for {&perp}
{&sdot}	·	Dot operator
{&dot}		Synonym for {&sdot}
{&prime}	′	Prime, minutes, feet
{&Prime}	″	Double prime, seconds, inches
{&frasl}	⁄	Fraction slash
{&larr}	←	Leftward arrow
{&uarr}	↑	Upward arrow
{&rarr}	→	Rightward arrow
{&darr}	↓	Downward arrow
{&harr}	↔	Left–right arrow
{&crarr}	↵	Downward arrow with corner leftward, carriage return
{&lArr}	⇐	Leftward double arrow, is implied by
{&uArr}	⇑	Upward double arrow
{&rArr}	⇒	Rightward double arrow, implies
{&dArr}	⇓	Downward double arrow
{&hArr}	⇔	Left–right double arrow

Other symbols supported by SMCL in graph text are the following:

SMCL tag	Symbol	Description
`{&trade}`	™	Trademark
`{&trademark}`		Synonym for `{&trade}`
`{®}`	®	Registered trademark
`{©}`	©	Copyright
`{©right}`		Synonym for `{©}`
`{&bull}`	•	Bullet
`{&bullet}`		Synonym for `{&bull}`
`{&hellip}`	...	Horizontal ellipsis
`{&ellipsis}`		Synonym for `{&hellip}`
`{&loz}`	◊	Lozenge, diamond
`{&lozenge}`		Synonym for `{&loz}`
`{&diamond}`		Synonym for `{&loz}`
`{&spades}`	♠	Spades card suit
`{&clubs}`	♣	Clubs card suit
`{&hearts}`	♡	Hearts card suit
`{&diams}`	♢	Diamonds card suit
`{&diamonds}`		Synonym for `{&diams}`

Also see

[G-2] **graph set** — Set graphics options

[G-3] ***ps_options*** — Options for exporting or printing to PostScript

[G-3] ***eps_options*** — Options for exporting to Encapsulated PostScript

[P] **smcl** — Stata Markup and Control Language

Title

[G-4] *textboxstyle* — Choices for the overall look of text including border

Syntax

textboxstyle	Description
`heading`	large text suitable for headings
`subheading`	medium text suitable for subheadings
`body`	medium text
`smbody`	small text

Other *textboxstyles* may be available; type

```
. graph query textboxstyle
```

to obtain the complete list of *textboxstyles* installed on your computer.

Description

A textbox contains one or more lines of text. *textboxstyle* specifies the overall style of the textbox.

textboxstyle is specified in the `style()` option nested within another option, such as `title()`:

```
. graph ..., title("My title", style(textboxstyle)) ...
```

See [G-3] ***textbox_options*** for more information on textboxes.

Sometimes you will see that a *textboxstylelist* is allowed. A *textboxstylelist* is a sequence of *textboxstyles* separated by spaces. Shorthands are allowed to make specifying the list easier; see [G-4] ***stylelists***.

Remarks

Remarks are presented under the following headings:

What is a textbox?
What is a textboxstyle?
You do not need to specify a textboxstyle

What is a textbox?

A textbox is one or more lines of text that may or may not have a border around it.

What is a textboxstyle?

Textboxes are defined by 11 attributes:

1. Whether the textbox is vertical or horizontal; see [G-4] ***orientationstyle***
2. The size of the text; see [G-4] ***textsizestyle***
3. The color of the text; see [G-4] ***colorstyle***

4. Whether the text is left-justified, centered, or right-justified; see [G-4] ***justificationstyle***
5. How the text aligns with the baseline; see [G-4] ***alignmentstyle***
6. The margin from the text to the border; see [G-4] ***marginstyle***
7. The gap between lines; see [G-4] ***relativesize***
8. Whether a border is drawn around the box, and if so
    a. The color of the background; see [G-4] ***colorstyle***
    b. The overall style of the line used to draw the border, which includes its color, width, and whether solid or dashed, etc.; see [G-4] ***linestyle***
9. The margin from the border outward; see [G-4] ***marginstyle***
10. Whether the box is to expand to fill the box in which it is placed
11. Whether the box is to be shifted when placed on the graph; see [G-4] ***compassdirstyle***

The *textboxstyle* specifies all 11 of these attributes.

You do not need to specify a textboxstyle

The *textboxstyle* is specified in option

`tstyle(`*textboxstyle*`)`

Correspondingly, you will find other options are available for setting each attribute above; see [G-3] ***textbox_options***.

You specify the *textboxstyle* when a style exists that is exactly what you desire or when another style would allow you to specify fewer changes to obtain what you want.

Also see

[G-4] ***text*** — Text in graphs

[G-3] ***textbox_options*** — Options for textboxes and concept definition

[G-4] ***textstyle*** — Choices for the overall look of text

Title

[G-4] *textsizestyle* — Choices for the size of text

Syntax

textsizestyle	Description
`zero`	no size whatsoever, vanishingly small
`minuscule`	smallest
`quarter_tiny`	
`third_tiny`	
`half_tiny`	
`tiny`	
`vsmall`	
`small`	
`medsmall`	
`medium`	
`medlarge`	
`large`	
`vlarge`	
`huge`	
`vhuge`	largest
`tenth`	one-tenth the size of the graph
`quarter`	one-fourth the size of the graph
`third`	one-third the size of the graph
`half`	one-half the size of the graph
`full`	text the size of the graph
relativesize	any size you want

See [G-4] ***relativesize***.

Other *textsizestyles* may be available; type

```
. graph query textsizestyle
```

to obtain the complete list of *textsizestyles* installed on your computer.

Description

textsizestyle specifies the size of the text.

textsizestyle is specified inside options such as the `size()` suboption of `title()` (see [G-3] ***title_options***):

```
. graph ..., title("My title", size(textsizestyle)) ...
```

Also see [G-3] ***textbox_options*** for information on other characteristics of text.

Also see

[G-3] ***marker_label_options*** — Options for specifying marker labels

[G-4] ***text*** — Text in graphs

[G-3] ***textbox_options*** — Options for textboxes and concept definition

Title

[G-4] ***textstyle*** — Choices for the overall look of text

Syntax

textstyle	Description
`heading`	large text suitable for headings; default used by `title()`
`subheading`	medium text suitable for subheadings; default used by `subtitle()`
`body`	medium-sized text; default used by `caption()`
`small_body`	small text; default used by `note()`
`axis_title`	default for axis titles
`label`	text suitable for labeling
`key_label`	default used to label keys in legends
`small_label`	default used to label points
`tick_label`	default used to label major ticks
`minor_ticklabel`	default used to label minor ticks

Other *textstyles* may be available; type

```
. graph query textboxstyle                    (sic)
```

to obtain the complete list of all *textstyles* installed on your computer. The *textstyle* list is the same as the *textboxstyle* list.

Description

textstyle specifies the overall look of single lines of text. *textstyle* is specified in options such as the marker-label option `mltextstyle()` (see [G-3] ***marker_label_options***):

```
. twoway scatter ..., mlabel(...) mltextstyle(textstylelist) ...
```

In the example above, a *textstylelist* is allowed. A *textstylelist* is a sequence of *textstyles* separated by spaces. Shorthands are allowed to make specifying the list easier; see [G-4] ***stylelists***.

A *textstyle* is in fact a *textboxstyle*, but only a subset of the attributes of the textbox matter; see [G-4] ***textboxstyle***.

Remarks

Remarks are presented under the following headings:

What is text?
What is a textstyle?
You do not need to specify a textstyle
Relationship between textstyles and textboxstyles

What is text?

Text is one line of text.

What is a textstyle?

How text appears is defined by five attributes:

1. Whether the text is vertical or horizontal; see [G-4] ***orientationstyle***
2. The size of the text; see [G-4] ***textsizestyle***
3. The color of the text; see [G-4] ***colorstyle***
4. Whether the text is left-justified, centered, or right-justified; see [G-4] ***justificationstyle***
5. How the text aligns with the baseline; see [G-4] ***alignmentstyle***

The *textstyle* specifies these five attributes.

You do not need to specify a textstyle

The *textstyle* is specified in options such as

`mltextstyle(`*textstyle*`)`

Correspondingly, you will find other options are available for setting each attribute above; see [G-3] ***marker_label_options***.

You specify the *textstyle* when a style exists that is exactly what you desire or when another style would allow you to specify fewer changes to obtain what you want.

Relationship between textstyles and textboxstyles

textstyles are in fact a subset of the attributes of *textboxstyles*; see [G-4] ***textboxstyle***. A textbox allows multiple lines, has an optional border around it, has a background color, and more. By comparison, text is just a line of text, and *textstyle* is the overall style of that single line.

Most textual graphical elements are textboxes, but there are a few simple graphical elements that are merely text, such as the marker labels mentioned above. The `mltextstyle(`*textstyle*`)` option really should be documented as `mltextstyle(`*textboxstyle*`)` because it is in fact a *textboxstyle* that `mltextstyle()` accepts. When `mltextstyle()` processes the *textboxstyle*, however, it looks only at the five attributes listed above and ignores the other attributes *textboxstyle* defines.

Also see

[G-3] ***marker_label_options*** — Options for specifying marker labels

[G-4] ***text*** — Text in graphs

[G-4] ***textboxstyle*** — Choices for the overall look of text including border

Title

[G-4] *ticksetstyle* — Choices for overall look of axis ticks

Syntax

ticksetstyle	Description
`major_horiz_default`	default major tickset for horizontal axes, including both ticks and labels but no grid
`major_horiz_withgrid`	major tickset for horizontal axes, including a grid
`major_horiz_nolabel`	major tickset for horizontal axes, including ticks but not labels
`major_horiz_notick`	major tickset for horizontal axes, including labels but not ticks
`major_vert_default`	default major tickset for vertical axes, including both ticks and labels but no grid
`major_vert_withgrid`	major tickset for vertical axes, including a grid
`major_vert_nolabel`	major tickset for vertical axes, including ticks but not labels
`major_vert_notick`	major tickset for vertical axes, including labels but not ticks
`minor_horiz_default`	default minor tickset for horizontal axes, including both ticks and labels but no grid
`minor_horiz_nolabel`	minor tickset for horizontal axes, including ticks but not labels
`minor_horiz_notick`	minor tickset for horizontal axes, including labels but not ticks
`minor_vert_default`	vertical axes default, having both ticks and labels but no grid
`minor_vert_nolabel`	minor tickset for vertical axes, including ticks but not labels
`minor_vert_notick`	minor tickset for vertical axes, including labels but not ticks

Other *ticksetstyles* may be available; type

```
. graph query ticksetstyle
```

to obtain the complete list of *ticksetstyles* installed on your computer.

Description

Tickset styles are used only in scheme files (see `help scheme files`) and are not accessible from `graph` commands; see [G-2] **graph**.

ticksetstyle is a composite style that holds and sets all attributes of a set of ticks on an axis, including the look of ticks and tick labels ([G-4] ***tickstyle***), the default number of ticks, the angle of the ticks, whether the labels for the ticks alternate their distance from the axis and the size of that alternate distance, the *gridstyle* (see [G-4] ***gridstyle***) if a grid is associated with the tickset, and whether ticks are labeled.

Also see

[G-4] ***tickstyle*** — Choices for the overall look of axis ticks and axis tick labels

[G-4] ***gridstyle*** — Choices for overall look of grid lines

Title

[G-4] *tickstyle* — Choices for the overall look of axis ticks and axis tick labels

Syntax

tickstyle	Description
`major`	major tick and major tick label
`major_nolabel`	major tick with no tick label
`major_notick`	major tick label with no tick
`minor`	minor tick and minor tick label
`minor_nolabel`	minor tick with no tick label
`minor_notick`	minor tick label with no tick
`none`	no tick, no tick label

Other *tickstyles* may be available; type

```
. graph query tickstyle
```

to obtain the complete list of *tickstyles* installed on your computer.

Description

Ticks are the marks that appear on axes. *tickstyle* specifies the overall look of ticks. See [G-3] ***axis_label_options***.

Remarks

Remarks are presented under the following headings:

What is a tick? What is a tick label?

A tick is the small line that extends or crosses an axis and next to which, sometimes, numbers are placed.

A tick label is the text (typically a number) that optionally appears beside the tick.

What is a tickstyle?

tickstyle is really misnamed; it ought to be called a *tick_and_tick_label_style* in that it controls both the look of ticks and their labels.

Ticks are defined by three attributes:

1. The length of the tick; see [G-4] ***relativesize***
2. Whether the tick extends out, extends in, or crosses the axis
3. The line style of the tick, including its thickness, color, and whether it is to be solid, dashed, etc.; see [G-4] ***linestyle***

Labels are defined by two attributes:

1. The size of the text
2. The color of the text

Ticks and tick labels share one more attribute:

1. The gap between the tick and the tick label

The *tickstyle* specifies all six of these attributes.

You do not need to specify a tickstyle

The *tickstyle* is specified in the options named

{ `y` | `x` }{ `label` | `tick` | `mlabel` | `mtick` } `(tstyle(`*tickstyle*`))`

Correspondingly, there are other { `y` | `x` }{ `label` | `tick` | `mlabel` | `mtick` } `()` suboptions that allow you to specify the individual attributes; see [G-3] ***axis_label_options***.

You specify the *tickstyle* when a style exists that is exactly what you desire or when another style would allow you to specify fewer changes to obtain what you want.

Suppressing ticks and/or tick labels

To suppress the ticks that usually appear, specify one of these styles

tickstyle	Description
`major_nolabel`	major tick with no tick label
`major_notick`	major tick label with no tick
`minor_nolabel`	minor tick with no tick label
`minor_notick`	minor tick label with no tick
`none`	no tick, no tick label

For instance, you might type

```
. scatter ..., ylabel(,tstyle(major_notick))
```

Suppressing the ticks can be useful when you are creating special effects. For instance, consider a case where you wish to add grid lines to a graph at $y =$ 10, 20, 30, and 40, but you do not want ticks or labels at those values. Moreover, you do not want even to interfere with the ordinary ticking or labeling of the graph. The solution is

```
. scatter ..., ymtick(10(10)40, grid tstyle(none))
```

We "borrowed" the `ymtick()` option and changed it so that it did not output ticks. We could just as well have borrowed the `ytick()` option. See [G-3] ***axis_label_options***.

Also see

[G-3] ***axis_label_options*** — Options for specifying axis labels

Subject and author index

This is the subject and author index for the *Graphics Reference Manual*. Readers interested in topics other than graphics should see the combined subject index (and the combined author index) in the *Quick Reference and Index*.

Semicolons set off the most important entries from the rest. Sometimes no entry will be set off with semicolons, meaning that all entries are equally important.

A

B

C

D

E

H

I

J

K

L

M

N

O

P

Q

R

S

Y

Z